The Berkeley Lectures on Energy – Vol. 1

Introduction to
Carbon Capture
and **Sequestration**

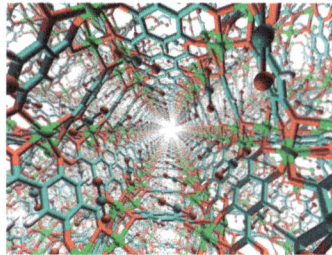

The Berkeley Lectures on Energy

ISSN: 2054-4189

Vol. 1: Introduction to Carbon Capture and Sequestration
by Berend Smit, Jeffrey A. Reimer, Curtis M. Oldenburg and Ian C. Bourg

The Berkeley Lectures on Energy – Vol. 1

Introduction to
Carbon Capture
and Sequestration

Berend Smit
Jeffrey A. Reimer
University of California, Berkeley, USA

Curtis M. Oldenburg
Ian C. Bourg
Lawrence Berkeley National Laboratory, USA

Imperial College Press

ICP

Published by

Imperial College Press
57 Shelton Street
Covent Garden
London WC2H 9HE

Distributed by

World Scientific Publishing Co. Pte. Ltd.
5 Toh Tuck Link, Singapore 596224
USA office: 27 Warren Street, Suite 401-402, Hackensack, NJ 07601
UK office: 57 Shelton Street, Covent Garden, London WC2H 9HE

Library of Congress Cataloging-in-Publication Data
Smit, Berend, 1962–   author.
    Introduction to carbon capture and sequestration / by Berend Smit (University of California,
Berkeley, USA), Jeffrey A. Reimer (University of California, Berkeley, USA), Curtis M. Oldenburg
(Lawrence Berkeley National Laboratory, USA) & Ian C. Bourg (Lawrence Berkeley National
Laboratory, USA).
        pages cm. -- (The Berkeley lectures on energy ; vol. 1)
    Includes bibliographical references and index.
    ISBN 978-1-78326-327-1 (hardcover : alk. paper) --
    ISBN 978-1-78326-328-8 (softcover : alk. paper)
    1. Carbon sequestration. 2. Separation (Technology) I. Reimer, Jeffrey A., author.
II. Oldenburg, Curtis M., author. III. Bourg, Ian C., author. IV. Title.
    TP156.S45S62 2014
    660'.2842--dc23
                                                                2013046563

British Library Cataloguing-in-Publication Data
A catalogue record for this book is available from the British Library.

In-house Editor: Lee Xin Ying

Typeset by Stallion Press
Email: enquiries@stallionpress.com

Contents

Preface

Sustainable energy generation is one of the biggest challenges of our generation. All long-term solutions rely on direct or indirect conversion of solar energy. However, these solutions appear to be years from implementation. In the coming decades then, while the relative importance of fossil fuels will decrease, absolute use of fossil fuels will not. Whether we like it or not, we simply won't be able to transition overnight to a world that is carbon neutral for its energy. Although the best way of sequestering carbon is to leave fossil fuels in the ground, this is not realistic, and sober consideration of the problem demands an alternative involving drastically lower carbon emissions from fossil fuels during a relatively short transition period to carbon-free energy. However, there is in our view a substantial gap between our hope for a fast transition to sustainable energy and the energy scenario that is in fact most likely to emerge. As researchers, we simply have very little control on how much coal the world may decide to burn. Carbon Capture and Sequestration (CCS), employed on a global scale, can sustain a transition period in the world's energy use and help mitigate alarmingly high CO_2 levels in the atmosphere. Some may argue that this technology will provide the excuse to prolong the use of fossil fuels. Others, including us, argue that we need all the help we can get in reducing CO_2 emissions, and that by implementing CCS we can exploit an already available fossil fuel infrastructure. The latter viewpoint may be less idealistic, but given the enormous scale of the current fossil fuel infrastructure, it is a more pragmatic approach.

UC Berkeley and Lawrence Berkeley National Laboratory have large research programs addressing the world's energy future. The Berkeley Energy Lectures are aimed at introducing these research programs to our undergraduate and graduate students in the sciences and engineering. In the present text, the topics include discussion of our current understanding of CO_2 in and around the planet, the geological storage of CO_2, and the science and technology of capturing CO_2.

If you glance through the book, you will quickly see that the lectures are very different from a typical science or engineering course. The book covers relevant topics in the geosciences, climate science, chemical engineering, materials science, and chemistry. The text covers topics as diverse as how to estimate the number of gigatonnes of carbon dioxide that can be stored in a geological formation; how to use a molecular model of pore architecture to optimize the performance of a membrane; what insights can be gained from the planetary science question known as the faint young star paradox; and how to consider the importance of heat integration in amine scrubbing.

Is it important that students know about all these ostensibly disparate topics? For us, this is completely the wrong question. The real question is "Why would someone interested in science and energy not want to know all this?" When our graduate students talk about their research and people realize it is about CCS, the questions they are likely to get are along the lines of "Is it really safe to sequester CO_2?"; "Is CO_2 really causing global warming?"; and "How much will CCS influence the cost of electricity?" One answer to these questions is, "I have no idea. I am working on quantum chemical calculations to determine the binding energy of CO_2 in molecular organic frameworks." This answer is not acceptable for us. Our students have to realize that research on energy has the potential to impact people's daily lives and any researcher in this field must be aware of this. It is therefore incumbent on our researchers to know the state of the art in the entire field of CCS, even if their research is as fundamental as developing a novel experimental NMR method to unravel the dynamics of CO_2 hopping from one site to another in a capture material! Even more importantly, our researchers need to appreciate that these are very difficult problems. We may not even know what the solutions are, but what will be clear from this book is that any solution requires a concerted effort of researchers in many different disciplines. For such a collaboration to work, one does need to know the basics of each relevant discipline and what drives research progress in each field. If a chemist develops a beautiful new material, she has to realize that her material will be used to separate many megatonnes of CO_2. The chapters on separations will help her get some insight into what engineers consider important in materials for separations. For her to value her new material, she also needs to see how and why CO_2 molecules ultimately become incorporated into limestone.

Finally, we should warn the readers: this text is not a typical survey that glances over topics. Admittedly, some of the topics will be easy to understand, but not always. If we get really excited about a topic, we discuss the results of the most recent scientific publications in the field. In some cases we could not resist introducing some new ideas that have not yet been published! We strongly feel that the combination of a molecular understanding translated by chemists into molecules and fully integrated with clever engineering and the geological sciences, together with a sense of the scale of the energy landscape, is what will define the next generation of scientists. And we hope you will be one of them.

The Campanile Bell Tower of UC Berkeley.

Acknowledgments

First of all we would like to acknowledge our colleagues who contributed guest lectures during the course, as much of the material you will find in this book is based on these lectures: Gary Rochelle (UT Austin), Joan Brennecke (Notre Dame), Hongcai Zhou (Texas A&M), Bill Collins, Christer Jansson, Don DePaolo, Jonathan Ajo-Franklin, Sergi Molins, Jiamin Wan, Alejandro Fernandez-Martinez, Tim Kneafsey, Seiji Nakagawa (all from Lawrence Berkeley National Laboratory), Ronny Pini (Stanford), Dave Luebke (National Energy Technology Laboratory), Abhoyjit Bhown (Electric Power Research Institute), Richard Baker (Membrane Technology and Research), Kurt Zenz House (C12 Energy), Jeffrey Long, and Ron Cohen (UC Berkeley).

We are very grateful for the editorial support from Aster Tang and Teresa Chin, and Richard Martin for preparing several of the figures. Special thanks go to Marjorie Went for her critical reading and editing of the entire book, as well as for many useful suggestions.

This book is based on the research of the authors and for this Berend Smit and Jeffrey Reimer would like to acknowledge support of the Center for Gas Separations Relevant to Clean Energy Technologies (http://www.cchem.berkeley.edu/co2efrc/), an Energy Frontier Research Center funded by the US Department of Energy, Office of Science, Office of Basic Energy Sciences under Award Number DE-SC0001015. Curtis M. Oldenburg receives the majority of his research support from the Assistant Secretary for Fossil Energy, Office of Sequestration, Hydrogen, and Clean Coal Fuels, through the National Energy Technology Laboratory,

US Department of Energy. Ian C. Bourg would like to acknowledge support from the Center for Nanoscale Control of Geologic CO_2 (http://esd. lbl.gov/research/facilities/ncgc/), an Energy Frontier Research Center funded by the US Department of Energy, Office of Science, Office of Basic Energy Sciences under Award Number DE-AC02-05CH11231.

About the Authors

Berend Smit (Berend-Smit@Berkeley.edu) studied Chemical Engineering and Physics at the Technical University in Delft (The Netherlands). His PhD is in chemistry at Utrecht University. He worked at Shell Research in Amsterdam before he became Professor of Computational Chemistry at the University of Amsterdam. He was elected director of the Centre Européen de Calcul Atomique Moléculaire (CECAM) in Lyon. At present, he works in the Department of Chemical and Biomolecular Engineering and Department of Chemistry at UC Berkeley, where he is directing an Energy Frontier Research Center for the US Department of Energy (DOE) focused on Carbon Capture. Berend's research interests are in the development and applications of molecular simulation techniques.

More details on his research can be found on his homepage: http://www.cchem.berkeley.edu/molsim/personal_pages/berend/

Jeffrey Reimer (reimer@berkeley.edu) received his degrees in chemistry from UC Santa Barbara and Caltech. After a postdoc at IBM Yorktown Heights, he joined the faculty at UC Berkeley where he is the C. Judson King Endowed Professor of Chemical Engineering. Professor Reimer is a Fellow of the American Association for the Advancement of Science and the American Physical Society and has won every teaching award given on the UC Berkeley campus. His research is focused on NMR spectroscopy, with particular attention to its design for, and application to, problems in materials physics and chemistry.

For details, visit his homepage: http://india.cchem.berkeley.edu/~reimer/

Curtis M. Oldenburg (cmoldenburg@lbl.gov) received his degrees in geology from UC Berkeley and UC Santa Babara with emphasis on igneous petrology and modeling magma dynamics. He came to the Lawrence Berkeley National Laboratory (LBNL) in 1990 as a post-doc and worked on model development and applications for strongly coupled flow problems in geothermal energy and vadose zone hydrogeology. When geologic carbon sequestration research began in the late 1990s, he applied his experience with coupled flow problems to specialize in modeling and simulation of CO_2 injection, trapping, leakage, and related risk assessment. Curt is the Head of the Geologic Carbon Sequestration Program at LBNL, and Editor in Chief of the Wiley journal, *Greenhouse Gases: Science and Technology*.

For more information, visit his webpage: http://esd.lbl.gov/about/staff/curtisoldenburg/

Ian C. Bourg (icbourg@lbl.gov) received his bachelor's degree in Industrial Process Engineering from the National Institute of Applied Sciences in Toulouse (France) and his doctorate in Civil and Environmental Engineering from UC Berkeley. In 2009, he joined the Lawrence Berkeley National Laboratory. Since 2011, he has served on the executive committee of the Center for Nanoscale Control of Geologic CO_2, a DOE-supported Energy Frontiers Research Center.

The goal of Dr. Bourg's research is to develop a fundamental understanding of the properties of water at interfaces by using atomistic simulations and continuum-scale models. At the present time, his group is investigating the nanoscience of geologic carbon sequestration, the aquatic geochemistry of nanoporous media, and the molecular scale origin of kinetic isotope effects.

More details on his research can be found on his homepage: http://esd.lbl.gov/about/staff/ianbourg/

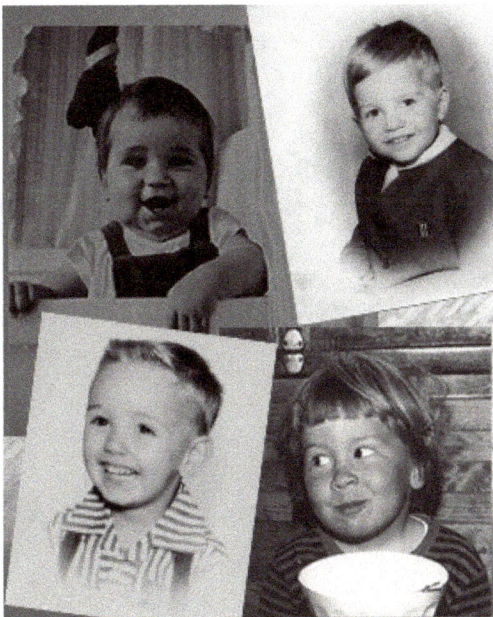

The authors blissfully unaware of the amount of CO_2 in the air.

Chapter 1

Energy and Electricity

In this chapter we will be counting carbon atoms. How many carbon atoms do we emit each year? How many of those can we capture? The questions are simple but the answers are enormous. How enormous? That is the topic of this chapter.

Section 1

Introduction

The idea of carbon capture and storage is simple. As we continue to rely on fossil fuels for our energy, we mitigate the emission of CO_2 by capturing it at the point of combustion and subsequently storing it in geological formations (see **Figure 1.1.1**). The CO_2 is separated from the flue gas emitted by a power plant ("capture") and subsequently compressed and transported through pipelines. At a nearby site, the CO_2 is injected through a deep well into a geological formation ("sequestration" or "storage").

In this text our main focus is on the science related to carbon capture and sequestration (CCS):

- The most expensive part of CCS is carbon capture. The separation of CO_2 from flue gas and its subsequent compression requires equipment to be added to an existing power plant, or design changes when building a new power plant. This new equipment, whether added on or introduced in construction, requires capital investments and operating costs that will significantly increase the price of the energy produced. The goals of current CCS research are to improve the separation process and to develop novel materials that can be used in the capture process as efficiently as possible.
- Geological storage involves injecting large amounts of CO_2 into geological formations. Here science must focus on ensuring that this process is done in such a way that the CO_2 is permanently stored with minimal impact on the environment. Given the inaccessibility and heterogeneity inherent in the deep subsurface, there will always be uncertainty about CO_2 trapping and brine displacement. To address this, research on geological CO_2 storage is focused on understanding the relevant physical and chemical processes over a wide range of length and time scales to ensure storage permanence and efficiency while making sure environmental impacts are acceptable.

The research and technology associated with CCS is directly related to the science of global warming, energy economics and policy, and

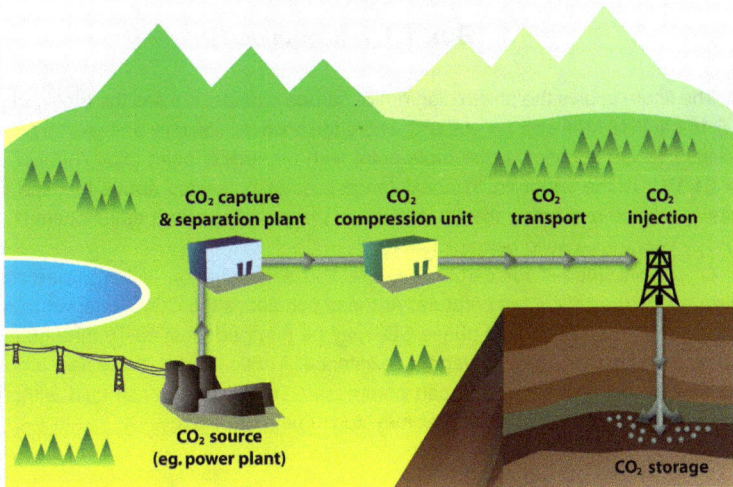

Figure 1.1.1 Schematic drawing of the carbon capture and storage process
Figure by Hyun Jung Kim, based on the IPCC Report [1.1].

Box 1.1.1 Carbon dioxide

The protagonist in this text is CO_2, so we need some basic data about this molecule. CO_2 is colorless and, at low concentrations, odorless. The molecular mass of CO_2 is 44 g/mol. At standard temperature and pressure, the density of carbon dioxide is around 1.8 kg/m³, about 1.5 times that of air. Because it is heavier than air, it has a propensity to sink, at least in the absence of external convective mixing (e.g., breezes and winds). This makes CO_2 a safety hazard for asphyxiation.

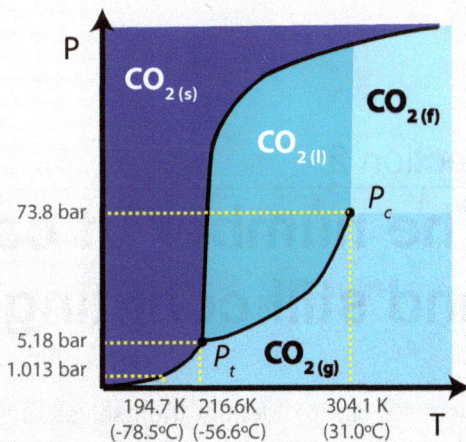

(Continued)

Box 1.1.1 (Continued)

The figure shows the phase diagram of carbon dioxide. We see the triple point at $p = 5.18$ bar (518 kPa) and $T = -56.6°C$ where the solid (s), liquid (l), and gas (g) phases coexist. The line connecting the triple point with the critical point ($p_c = 73.8$ bar and $T_c = 31.1°C$) is the vapor-liquid coexistence line. At pressures above this line CO_2 exists as a liquid and below the line it exists in the gas phase. At 1 atmosphere (1.013 bar) CO_2 becomes a solid below 194.7 K.

Of special interest for carbon capture and sequestration is the **supercritical region** above the critical temperature. At these conditions the CO_2 is referred to as in the fluid (f) phase. In this fluid phase CO_2 may be pumped most easily through pipelines because there is no liquid-vapor coexistence. A critical temperature near ambient conditions also implies that one can compress CO_2 gas to a dense fluid without it going through a phase transition that may consume more energy.

many other factors that are not addressed in traditional science or engineering courses. These topics are a context for CCS, and understanding them is essential to appreciating the technical directions and choices that are made in CCS research.

As CO_2 plays a central role in the text, we have summarized some of the properties of carbon dioxide in **Box 1.1.1**.

Section 2

The number of carbon atoms and still counting...

Chemists like to identify themselves with molecules. If anything, the holy grail of a chemist is a perfect intuition on how molecules would react in different environments. In the later chapters we will see some beautiful examples of this intuition put into action to make new materials to capture CO_2. Chemists have the entire periodic table at their disposal to make the most innovative materials. Certainly, we would like the world's most brilliant chemists to use their impressive tools to make a novel material such that every atom is placed at exactly the right position to efficiently capture CO_2 from flue gas. At first sight this challenge looks very similar to finding the perfect drug, in which each atom is put at exactly the right position to intercept a virus or bacterium. We will argue that carbon capture research, however, is burdened by an additional constraint: the enormous scale of energy production.

This may sound somewhat abstract, so let us give an example. Suppose your start-up company has come up with the perfect material to capture CO_2. Of course, you keep the exact formula of your material secret, but you have many CEOs of the world's leading electricity companies lined up to build a carbon capture unit next to their power plants. In fact, your material is so successful that 80% of all power plants in the world would like to implement CCS based on your material. This looks like the dream of any start-up until you realize how much of your material needs to be synthesized. The scale of the energy landscape is gigantic, so if your magic capture material contains an element that is not sufficiently abundant, then no matter how superior your carbon capture chemistry might be, your company will fail because the solution you deliver is *not sustainable.*

The chemical sciences in the 21st century invoke modern sensibilities for the design of materials, with particular attention to the sources of raw materials, their abundance, the impact on the environment to secure them, and the role raw materials sourcing plays in political, social, and economic structures worldwide. This is particularly striking for the case of carbon capture, where proposed innovations are intimately coupled with energy production. The implications of these innovations are gigantic: gigantic in volume, gigantic in environmental impact, gigantic in costs, in fact gigantic in any aspect one can imagine.

To get some intuition about the scale on which we use energy, let us ask some very simple questions:

1. How much CO_2 do we produce at the moment?
2. How much CO_2 will be produced 1, 5, or 30 years from now?

3. How much CO_2 comes from one power plant?
4. How much CO_2 can we capture?

Current CO_2 production

The most important source of atmospheric CO_2 is the burning of fossil fuels as part of our energy consumption. In the literature one can find many statistics on energy consumption. **Figure 1.2.1** is one example in which the total world energy consumption is divided according to its different sources [1.2]. This graph illustrates the importance of fossil fuels: over 80% of the energy sources we use emit CO_2. In addition, we see that the total energy consumption has doubled during the last 30 years.

The total energy production is given in **Mtoe**, which stands for 10^6 tonnes of equivalent oil, i.e., the produced energy is equivalent to the energy released by burning 10^6 tonnes (10^9 kg) of oil (in SI units 1 toe = 42 GJ). The energy content of different fuels is different and the conversion to Mtoes allows us to compare them. **Table 1.2.1** shows how many tonnes of CO_2 are produced by the combustion of three types of fossil fuels due to their different stoichiometries. These differences can be understood if we look at the basic combustion equation:

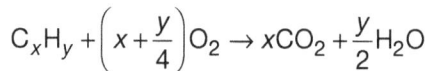

$$C_xH_y + \left(x+\frac{y}{4}\right)O_2 \rightarrow xCO_2 + \frac{y}{2}H_2O$$

Natural gas has a much higher hydrogen content compared to coal, and therefore produces less CO_2 per tonne of fuel.

Figure 1.2.1 shows that the 2011 global energy consumption is more than 10,000 Mtoe. If all that energy were from oil, we would annually produce approximately $30,000 \times 10^6$ tonnes of CO_2. Often, we express this quantity in terms of gigatonnes of carbon (Gt C), which is 8.2 Gt C. The real CO_2 production is higher, as the different fossil fuels do not produce identical amounts of CO_2 per unit energy. If we correct for these differences, we obtain the result that our current global energy consumption produces over 31,000 M tonne (31,000,000,000,000 kg) of CO_2 each year.

The production of cement also produces significant amounts of CO_2 that is emitted to the atmosphere. The historical CO_2 emission data from different sources are given in **Figure 1.2.2** [1.3].

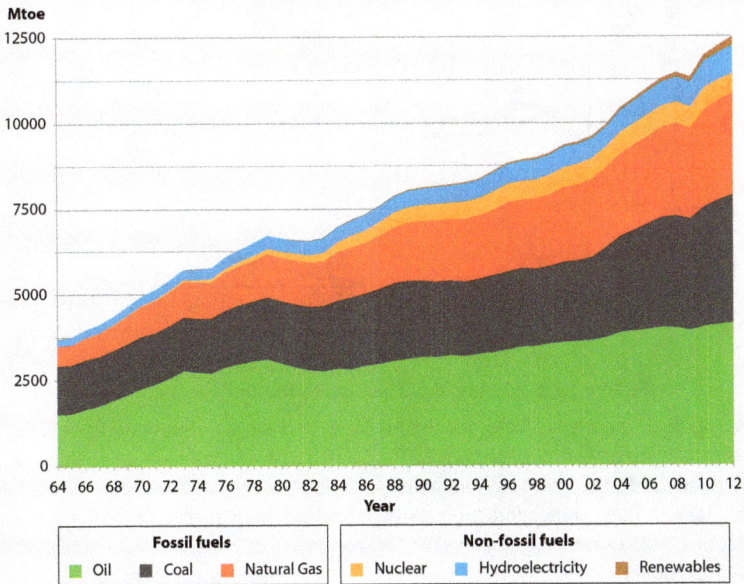

Figure 1.2.1 World energy consumption

World primary energy consumption grew by 2.5% in 2011, less than half the growth rate experienced in 2010 but close to the historical average. Oil remains the world's leading fuel, accounting for 33.1% of global energy consumption, which is the lowest share on record. Coal's market share of 30.3% was the highest since 1969. Oil, coal and natural gas are fossil fuels while the others are non-fossil fuels. *Figure based on data from BP Statistical Review of World Energy* [1.2].

Table 1.2.1 CO_2 production in tonne per tonne fuel

Source	CO_2
oil	3.0
gas	2.7
coal	4.4

Important questions remain. Where does all the CO_2 go after it has been emitted? Is the total amount of CO_2 we have emitted large or small compared to the CO_2 that is already present in the atmosphere? We will come back to these questions in Chapter 3, where we discuss the carbon cycle.

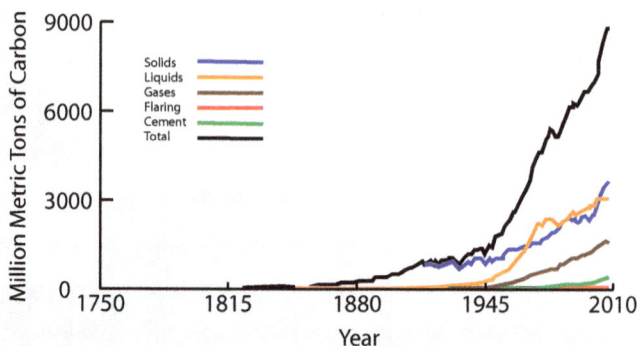

Figure 1.2.2 Historical CO_2 emission from different sources

Globally, liquid and solid fuels accounted for 76.3% of the emissions from fossil-fuel burning and cement production in 2007. Combustion of gas fuels (e.g., natural gas) accounted for 18.5% (1551 million metric tonnes of carbon) of the total emissions from fossil fuels in 2007, reflecting a gradually increasing global utilization of natural gas. Emissions from cement production (377 million metric tonnes of carbon in 2007) have more than doubled since the mid 1970s and now represent 4.5% of global CO_2 releases from fossil-fuel burning and cement production. Gas flaring, which accounted for roughly 2% of global emissions during the 1970s, now accounts for less than 1% of global CO_2 release. The units are in mass of carbon instead of mass of CO_2. To convert these estimates to units of CO_2, multiply by 3.667. *Figure adapted from the Carbon Dioxide Information Analysis Center* [1.3].

Question 1.2.1 Renewables and Energy

What is the percentage of renewable energy in our current energy consumption? How is this percentage divided between the different sources of renewable energies? What was this percentage in 1800?

Future CO_2 production

In this chapter we focus on the production of CO_2. In the next two chapters we discuss the relation between CO_2 levels in the atmosphere and the average temperature. In our discussion about future CO_2 production, we borrow from these chapters the concern that rising CO_2 levels cause climate change.

Let us look again at the CCS process. One may wonder whether it is really a good idea to dig up coal, burn it, capture the resulting CO_2 and subsequently store the CO_2 in the ground. All scientific challenges for CCS would be solved if we were to simply leave the coal in the ground; indeed, coal is one of nature's own efficient ways of capturing and storing carbon for millions of years. Perhaps one could capture CO_2 from the air and use sunlight to convert it back to fuel. Alternatively, the world could stop using fossil fuels altogether as an energy source. Implementing these scenarios, however, requires the development of alternative energy sources or significant new technologies.

In this context Pacala and Socolow have developed a very useful concept for thinking about reducing CO_2 emissions [1.4]. Their starting point is the business as usual scenario; if we do not take any additional action, the increase in our energy consumption will result in a doubling of the 2006 CO_2 emissions by 2056 (see **Figure 1.2.3**). Suppose we set as a target that over the next 50 years the increase in our energy consumption should not increase our CO_2 emissions above 2006 levels. This target implies that we need to have the technologies in place to avoid the emission of 7 billion tonnes of carbon a year by 2056, compared to the business as usual scenario. Pacala and Socolow argue that we need a realistic scenario to achieve this. For example, in theory one can achieve this aim by imposing a policy that all energy above 2006 levels can only be generated using nuclear energy. Such a scenario, however, would require the construction of new nuclear power plants at an unrealistic pace — one every other month or so for the coming 50 years (see also Section 1.5). Pacala and Socolow analyze many different scenarios and argue that it might be possible to achieve a lower target, on the order of 1 billion tonnes of avoided carbon per year by 2056, with a single existing technology. Then, Pacala and Socolow identify 15 technologies, each of which has the potential to achieve 1 billion tonnes of avoided carbon per year by 2056 (see **Figure 1.2.4**). By combining any 7 of these "wedges" one can achieve the overall reduction target while maintaining energy growth and avoiding economic disruption. It has now been 10 years since Pacala and Socolow published their article; **Box 1.2.1** gives an update by Davis *et al.* [1.5].

From our perspective it is nearly impossible to envision a scenario that limits total carbon emission in the atmosphere without relying on CCS (see **Figure 1.2.5**).

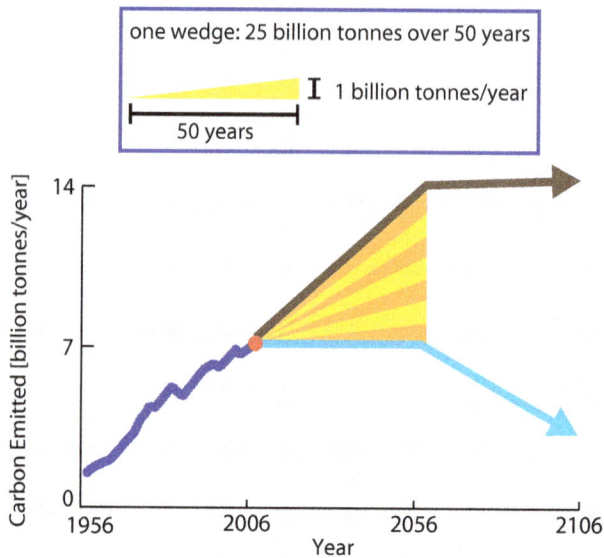

one wedge: 25 billion tonnes over 50 years

1 billion tonnes/year

50 years

Figure 1.2.3 Reducing CO_2 emissions

Past (dark blue line) and future worldwide carbon emissions. The brown line gives the business as usual scenario in which we assume that our future energy needs are met using the same resources we use today. The concept of wedges is based on the idea that we can achieve a total emission reduction of 7 billion tonnes of carbon per year by implementing seven technologies. In the figure, each wedge represents the ramping up of a CO_2 abatement technology, over a period of fifty years, to a total of 1 billion tonnes of CO_2 avoided per year. Overall, the seven wedges represent a total emission reduction of 175 Gt C between 2006 and 2056. *Figure adapted from Pacala and Socolow* [1.4].

Energy and population

Arguably the single most important issue underlying the rise in energy consumption is the growth of the population. It is no surprise that at present there exist large disparities in annual energy consumption per capita around the globe. **Figure 1.2.6** illustrates these large differences in energy consumption, and hence carbon emission per capita, between the different countries [1.6]. What future emissions might look like can be extrapolated from **Movie 1.2.1**, which shows how emission correlates with per capita income [1.7].

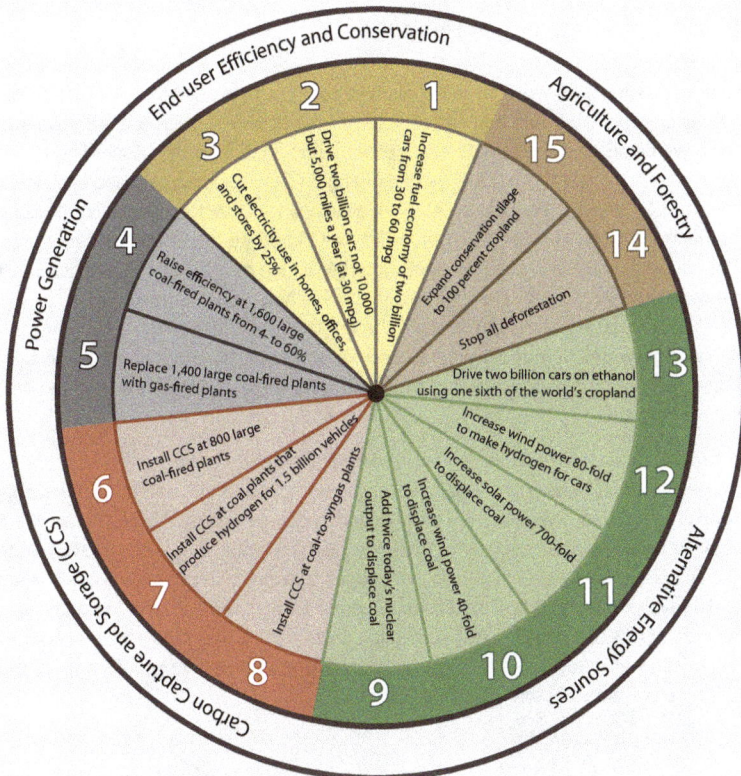

Figure 1.2.4 Options to reduce carbon emissions

Fifteen possible wedges based on existing technologies that each avoid 25 billion tonnes of carbon production over a period of 50 years. It is important to note that achieving each wedge is not easy; it is in fact at the edge of what is feasible with current technology. Success requires global implementation. *Figure adapted from Pacala and Socolow [1.4].*

Geographic, cultural, and global political issues loom as we consider limiting carbon emissions. Because the bulk of past carbon emissions was confined to a few regions (e.g., the USA and Europe), it is very difficult to argue against the logic that the responsibility for the reductions should be shouldered by the historically highest emitters. However, if this logic means that other countries agree to take action only if their own emissions per capita exceed those of the USA, then a global solution will remain elusive.

Box 1.2.1 The Wedges Revisited

It has been almost 10 years since Pacala and Socolow published their wedges article in *Science*. This work has had significant impact because it offered a framework in which we could address climate change using known technology. According to the "wedges model," we should have been ramping up CO_2 reduction and decreasing our emissions by 0.1 Gt C per year by 2015. However, since 2004 the growth in emissions has neither stopped nor slowed, but in fact has increased so much that Davis *et al.* [1.5] posed the question as to whether the original idea of 7 wedges would still result in a maximum increase of less than 2°C in the average global temperature.

Davis *et al.* argue that to achieve the original goal we would need 21 instead of 7 wedges, and further we would need an additional 10 wedges to completely phase out CO_2 emissions. The figure shows how these wedges are grouped into different classes:

- 12 hidden wedges — in the figure we see a difference between the scenario with frozen technology and the business as usual scenario equivalent to 12 wedges, reflecting that innovation continues to decarbonize our energy. Davis *et al.* argue that making these wedges explicit reduces the danger of double counting innovations.
- 9 stabilization wedges — these are the wedges needed to stabilize the emissions at the 2010 level. They include 2 additional wedges to account for new insights into the amount of CO_2 that needs to be avoided to ensure that we stay below the 500 ppm level.
- 10 phase-out wedges — to replace the entire energy infrastructure and land-use practice by methods that do not emit CO_2.

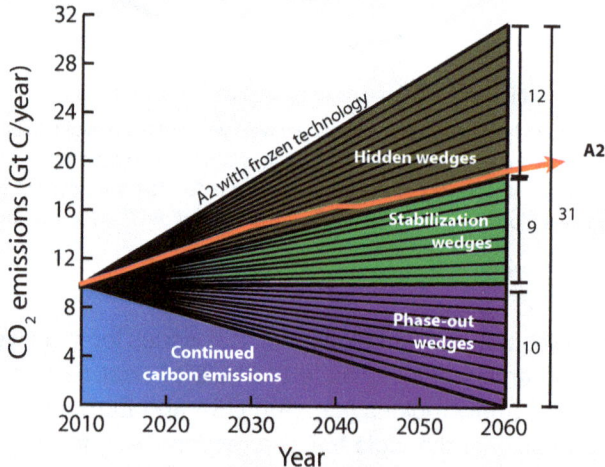

Future CO_2 emission scenarios (SRES, see **Box 2.5.1** in Chapter 2). Wedges expand linearly from 0 to 1 Gt C/year from 2010 to 2060. The total avoided emissions per wedge is 25 Gt C, such that altogether the hidden, stabilization and phase-out wedges represent 775 Gt C of cumulative emissions. *Figure modified from* [1.5].

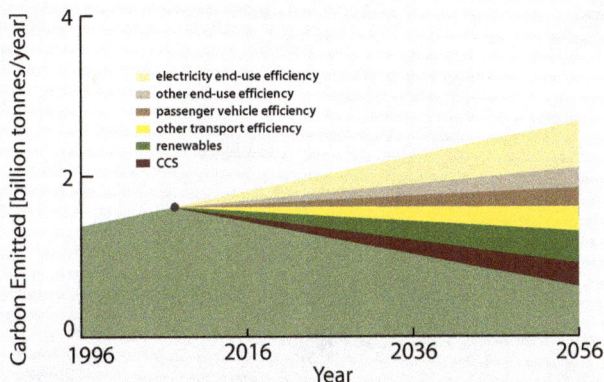

Figure 1.2.5 Possible scenarios to reduce carbon emissions

A possible selection of wedges to reduce the carbon emissions in the USA. *Figure adapted from the illustration in* [1.4].

Figure 1.2.6 Global annual energy consumption per capita by country

A Btu is a British thermal unit — a measurement of thermal energy. A Btu is roughly 1,055 J. Fuels are often measured in Btu to show how much potential they have to heat water into steam or to provide energy in other ways, for example to engines. *Figure from BURN, data from the International Energy Agency* [1.6].

Chakravarty *et al.* [1.8] argue that carbon emissions are a global problem where each person has an individual responsibility to mitigate CO_2 emissions. These authors have therefore proposed a plan for global emissions reduction that is based upon individuals, as opposed to nations. **Figure 1.2.7**

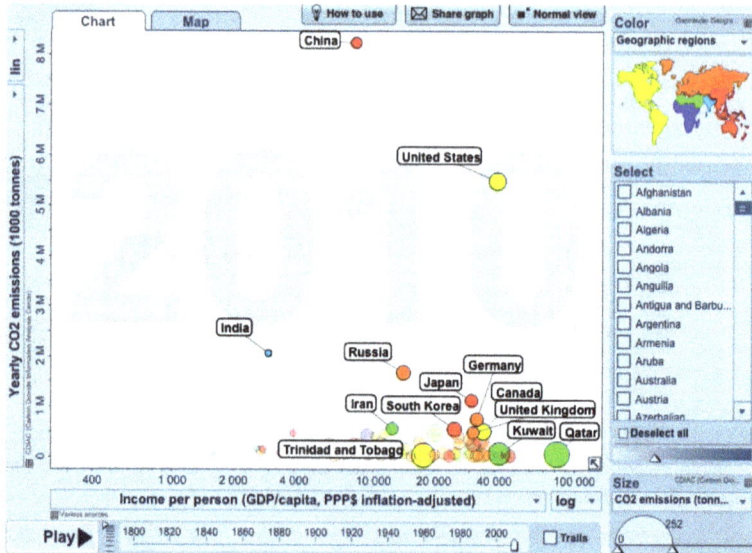

Movie 1.2.1 Historical carbon emissions for different countries

In 1820, at the dawn of the industrial revolution, the UK emitted the most CO_2 — both per person and in total. The USA overtook the UK by 1900, and recently in 2006 China became the biggest emitter of CO_2 in the world. The movie shows the total CO_2 emission as a function of income per person. The color of the circle represents the continent and its size indicates CO_2 emission per person. *Source: Free material from* www.gapminder. org, *data from the Carbon Dioxide Information Analysis Center* [1.7]. *This movie can be viewed at*: http://www.worldscientific.com/worldscibooks/10.1142/p911#t=suppl

illustrates the authors' logic. The graph gives annual individual emissions (calculated for the year 2030) on the ordinate, and the population contributing to that level of emissions on the abscissa. The blue-shaded region, for example, shows that 1.1 billion people will be responsible for emissions in excess of 10.8 tonnes per year of CO_2. A plan to reduce the total carbon emissions by 13 Gt CO_2 to 30 Gt CO_2 corresponds to a cap on emissions for those 1.1 billion people to 10.8 t CO_2 per year. Affording the poorest of the world an additional 1 t CO_2 per year would require the highest emitters to decrease their contributions from 10.8 to 9.6 t CO_2 per year. A policy based on individual carbon emissions is deemed "fair and pragmatic." If we think about emission reductions in these terms, each country has an incentive to limit the number of people in its population that emit above the world target.

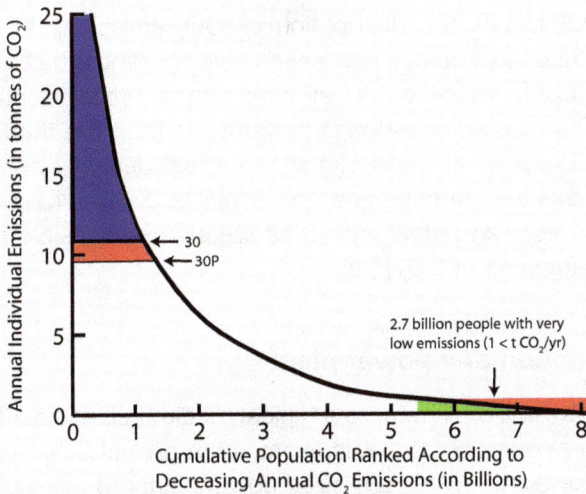

Figure 1.2.7 Individual carbon emissions

Prediction of the individual carbon emissions in 2030. The figure shows the total number of people that have an annual emission greater than the point on the curve. The global reduction targets can be achieved if emission by individuals in the blue area are capped. This blue area corresponds to a global annual emissions reduction of 13 Gt CO_2 from the 30 Gt CO_2 target. If one slightly increases this level to the 30P arrow, this would allow the poorest in the world to create a minimum individual emission of 1 t CO_2. This would raise the emissions of 2.7 billion people who emit less than 1 t CO_2 (green area at the right). *Figure adapted from Chakravarty et al. [1.8].*

How much CO_2 can we capture?

Until now, we have assumed that all CO_2 emissions are equivalent. From an environmental point of view there is, of course, no difference whether the source is a car or a power plant. From a carbon capture point of view, however, there is a large difference in the costs associated with capturing CO_2. As we will see in the carbon capture chapters, the technology of capturing CO_2 requires large equipment, which makes it very difficult to install a carbon capture facility on a car or an airplane. For example, if we install such a facility on an airplane, the collected CO_2 after a flight would weigh three times the weight of the kerosene used to power the aircraft in the first place! This illustrates one of the practical difficulties with carbon capture for mobile sources. Carbon capture is much more feasible at stationary sources, most of which are electric power plants. **Figure 1.2.8** shows the distribution of CO_2 emissions from different sources. Stationary sources are responsible for almost half of the global emissions. Furthermore, out of all stationary sources, electrical power generated by coal combustion is responsible for more than half of the emissions. Because of these statistics a great deal of attention will be focused in this text on coal-fired power plants. **Figure 1.2.8** shows that the cement, iron, and steel industries are also responsible for emitting significant amounts of CO_2 [1.9].

CO_2 production of a power plant

In this text, we will often refer to a typical medium-sized coal-fired power plant; the first question to ask is, what is the annual CO_2 emission from such a power plant? A medium-sized power plant produces 500 MW and emits about 400 m^3/s of flue gas containing about 12% of CO_2 (volume). This gives us about 2.6 million tonnes of CO_2 per year. If we assume a lifetime of 50 years, we have to sequester 130 million tonnes of CO_2 coming from 6.3×10^{11} m^3 of flue gas. In **Table 1.2.2**, we relate this number to the top 10 USA power plants [1.10]. These numbers are staggering.

It is interesting to look at the emissions from our existing infrastructure. To do this let us assume that we can ban new CO_2-emitting

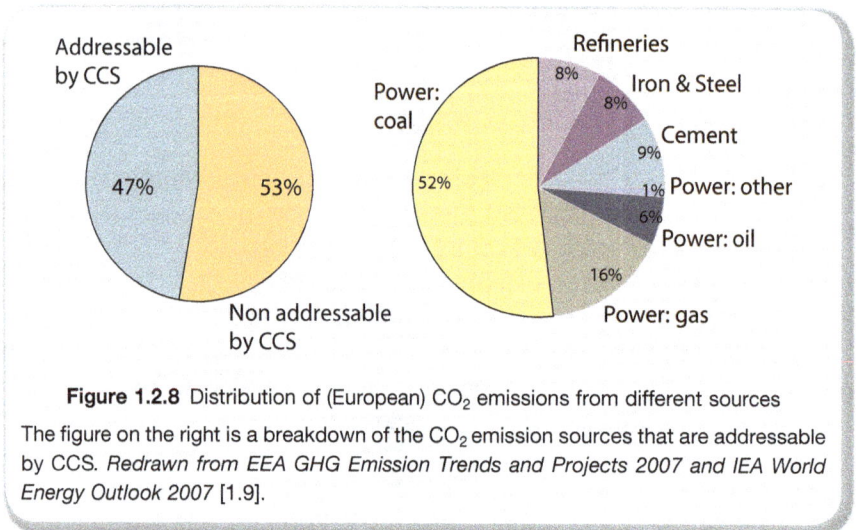

Figure 1.2.8 Distribution of (European) CO_2 emissions from different sources

The figure on the right is a breakdown of the CO_2 emission sources that are addressable by CCS. *Redrawn from EEA GHG Emission Trends and Projects 2007 and IEA World Energy Outlook 2007* [1.9].

Table 1.2.2 The top 10 USA CO_2 emitters in 2011

Rank	Plant	Company	Parent	Location	CO_2, millions metric tonnes	Capacity, Gigawatt electrical
1	Scherer	Georgia Power	Southern Company	Juliette, GA	21.90	3.56
2	James H Miller Jr	Alabama Power	Southern Company	Quinton, AL	21.89	2.64
3	Martin Lake	Luminant	Energy Future Holding	Tatum, TX	18.35	2.25
4	Labadie	Union Electric Company	Ameren	Labadie, MO	18.09	2.38
5	W A Parish	Texas Genco II	NRG Energy	Thompsons, TX	17.60	2.70
6	Gen J M Gavin	Ohio Power	American Electric Power	Cheshire, OH	17.52	2.60
7	Navajo Generating Station	Salt River Project	Multiple Parties	Page, AZ	16.80	2.25
8	Bruce Mansfield	First Energy	First Energy	Shippingport, PA	16.19	2.74
9	Monroe	Detroit Edison	DTE Energy	Monroe, MI	15.81	3.29
10	Gibson	Duke Energy Indiana	Duke Energy	Owensville, IN	15.70	3.15

Data from Environment Protection Agency [1.10].

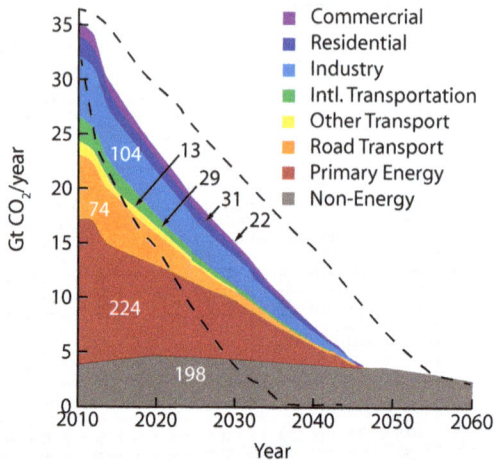

Figure 1.2.9 Future emissions from existing infrastructure by sector

Scenario of expected CO_2 emissions from existing energy and transportation infrastructure by industry sector. Projections are for a scenario where existing infrastructure is used to the end of its normal lifetime, and then replaced by non-CO_2 emitting infrastructure. The dashed lines indicate total emissions from upper- and lower-bounding scenarios (282 and 710 Gt CO_2, respectively). The numbers are cumulative emissions. *Figure adapted from Davis et al.* [1.11].

infrastructure (electricity and transportation). But this ban does not include the existing infrastructure; we suppose that the current infrastructure will only be replaced at the end of its normal lifetime by non-CO_2-emitting infrastructure. Davis *et al.* [1.11] calculated that under these assumptions the expected temperature increase of the planet would be 1.1–1.4 °C and the atmosphere would exhibit a concentration of CO_2 below 430 ppm. This remaining increase in CO_2 levels is due to the cumulative emission of 496 Gt of CO_2 resulting from the burning of fossil fuels for the current energy infrastructure that is still in use between 2010 and 2060 (see **Figure 1.2.9**).

Recall that in 2010 annual emission was about 35 Gt of CO_2. Even if we were to ban new infrastructure, we would continue to emit carbon at the annual rate of 35 Gt of CO_2 per year! Our current infrastructure is committed to emit CO_2 until the end of its natural lifetime. For example, the average lifetime of a coal-fired power plant is 38.6 years. A coal-fired power plant that was built in 2007 is committed to emit CO_2 for another

35.3 years. A similar persistence of emitting infrastructure holds for transportation and other industries.

Figure 1.2.10 shows the emissions per country due to committed infrastructure. If we compare the USA and China we see that both countries emitted about the same amount in 2010. However, the committed emissions of China (182 Gt) are more than two times larger than those of the USA (74 Gt). The reason is that the infrastructure of China is more recent.

In this context, we can understand why the most threatening CO_2 emissions are from fossil energy-based infrastructure that is being built now. It will commit the largest amount of future CO_2 emissions, as most of the new projects are being built with the business as usual philosophy.

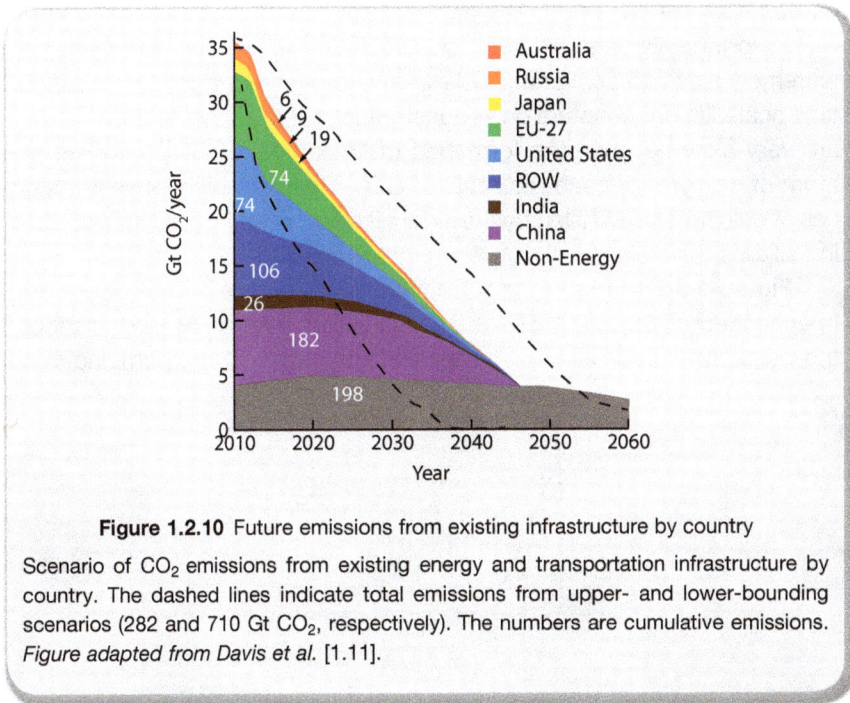

Figure 1.2.10 Future emissions from existing infrastructure by country

Scenario of CO_2 emissions from existing energy and transportation infrastructure by country. The dashed lines indicate total emissions from upper- and lower-bounding scenarios (282 and 710 Gt CO_2, respectively). The numbers are cumulative emissions. *Figure adapted from Davis et al. [1.11].*

Section 3

Making dreamium™

Storing CO_2 in geological formations almost looks like a waste of resources. Why is conversion of CO_2 into a useful product not the main topic of this book?

To answer this question we first look at the thermodynamics of carbon, shown in **Figure 1.3.1**. We see that by burning carbon we decrease its energy. However, CO_2 is not the lowest energy thermodynamic state; the lowest energy state is the mineral carbonate (e.g., limestone). Indeed, as this diagram implies, most of the carbon on earth is in the form of **limestone**. The famous white cliffs of Dover actually contain much more carbon than we can find in the atmosphere!

Unfortunately the kinetics of converting CO_2 into limestone are extremely slow, so converting CO_2 to limestone is not possible on a short time scale. (In our chapters on sequestration, we will show that the kinetics are very slow because the formation of limestone ($CaCO_3$) requires one atom of calcium for each molecule of CO_2. The availability of vast quantities of calcium is controlled by the slow dissolution rate of calcium-containing silicate minerals.)

Figure 1.3.2 illustrates the current uses of CO_2. The bulk of CO_2 is used in enhanced oil recovery, where CO_2 is injected into an oil field to reduce the viscosity of crude oil and thus facilitate its transport and pumping to the

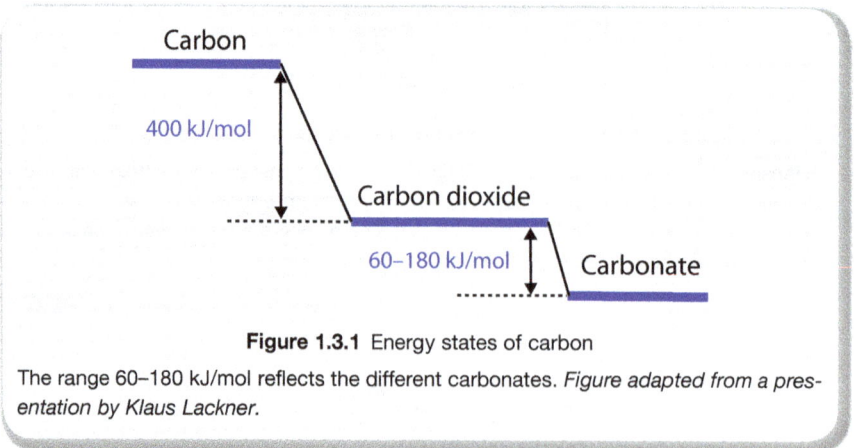

Figure 1.3.1 Energy states of carbon

The range 60–180 kJ/mol reflects the different carbonates. *Figure adapted from a presentation by Klaus Lackner.*

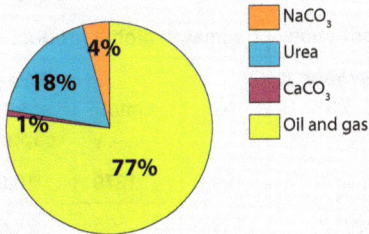
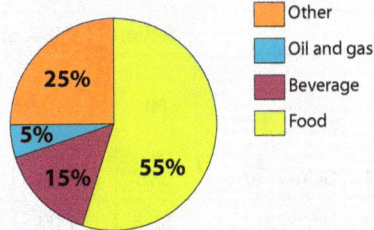

Figure 1.3.2 Annual consumption of CO_2 in the USA

The annual use of CO_2 is 100 Mt, which can be supplied by 3–4 coal-fired power plants. *Figure based on data from SRI Consulting, MIT, UT Austin.*

surface. The total yearly use of CO_2 for this purpose is about 100 Mt. This number is equivalent to the emission from four large power plants. Of course, it would be beneficial to use CO_2 from flue gasses instead of natural reservoirs for enhanced oil recovery, but it is clear that our power plants jointly produce much, much more CO_2 than we can use for oil recovery.

Let us assume as an alternative that we have some magic chemistry to take a molecule ZZ and connect it to a CO_2 molecule. We call our product, $ZZCO_2$, **"dreamium"** in anticipation of the many beautiful properties our new material will have. At present we have to trust that we will discover the chemistry to make dreamium. But let us assume that this chemistry exists and that this chemistry is surprisingly simple; we can use any method for the production of ZZ, and dreamium will follow. Suppose furthermore that we can make ZZ from any commodity chemical that is produced in the world. Bhown and Freeman looked at the consequences [1.12]. In **Table 1.3.1** we list the top 50 chemicals produced in the world in 2009. As the chemistry of making dreamium is so simple, we can use all of them. The interesting observation is that by **using all of the top 50 chemicals being produced in the world, we cannot even capture 10% of the global CO_2 emissions**. Let us think about these numbers. Our production of dreamium[TM] barely makes a dent in our CO_2 emissions, yet we now have a product, dreamium, that we need to market on a scale that is larger than the top 50 chemicals combined! In addition, by making dreamium we have depleted all supplies of other chemicals in the world.

This example illustrates the sheer size of CO_2 emissions. Converting CO_2 into something useful on a scale that is relevant to climate change

Table 1.3.1 Comparison of CO_2 production and the production of chemicals

		Estimated USA production			Estimated global production		
		Mt	Gmol	GWe-yr at 90% capture	Mt	Gmol	GWe-yr at 90% capture
1	Sulfuric acid	38.7	394	2.1	199.9	1879	10.0
2	Nitrogen	32.5	1159	6.2	139.6	4595	24.5
3	Ethylene	25.0	781	4.2	112.6	3243	17.3
4	Oxygen	23.3	829	4.4	100.0	3287	17.5
5	Lime	19.4	347	1.8	283.0	4653	24.8
6	Polyethylene	17.0	530	2.8	60.0	1729	9.2
7	Propylene	15.3	354	1.9	53.0	1134	6.0
8	Ammonia	13.9	818	4.4	153.9	8332	44.3
9	Chlorine	12.0	169	0.9	61.2	795	4.2
10	Phosphoric acid	11.4	116	0.6	22.0	207	1.1
...	...						
50	Nylon	1.9	8	0.0	2.3	8	0.0
	Total	419	8,681	46	2,412	48,385	257
	2009 coal-fired generation GWe-yr			200			>1000
	Approximate CO_2 emissions	6,000	136,000		31,000	750,000	

Approximate production of world's top 50 chemicals in 2009 compared with CO_2 emissions. The column GWe-yr at 90% capture indicates how much total energy can be generated per year by capturing 90% of the CO_2 and converting it all to Dreamium using the particular chemical. *Data from Bhown and Freeman* [1.12].

mitigation will saturate any market and deplete any supply. We will simply produce too much! Actually, there are only two processes that operate on the same scale as CO_2: water treatment and energy.

As an alternative, one could envision converting the CO_2 back into fuel. However, **Figure 1.3.1** shows that converting CO_2 into carbon costs energy; if we have this energy, why not convert it directly to electricity instead of using it to upgrade CO_2 into a fuel to be burned for making electricity? There is considerable scientific interest in "CO_2 to fuels" research, particularly if one imagines capturing existing atmospheric CO_2 and converting it back to fuel. Again, we will say more about this topic in our chapter on future sequestration directions.

Section 4

Making electricity

In the previous section we stated that electric power generation is responsible for the bulk of stationary CO_2 emissions. **Figure 1.4.1** illustrates the relative importance of the various energy sources for making electricity. This is a complex figure that reveals a great deal of information. For example, it reveals the paucity of renewable energy sources for electricity production. We can hope that in the future this share will increase significantly, but as the starting point is so low, ramping up the efforts such that its share is comparable to, say, natural gas, will be a major challenge. What is important in the present context is the prevalence of coal (and natural gas) in our current electricity production.

So let us now focus on coal and natural gas for the production of electricity. In Chapter 4 we will demonstrate that it is much more expensive to capture CO_2 at very low concentrations, and therefore it is more cost-effective to capture carbon from flue gas streams with higher CO_2 concentrations. **Table 1.4.1** compares the CO_2 concentrations in the flue gasses from coal and natural gas power plants to the CO_2 concentration in air. The comparison of coal and natural gas is interesting. In natural gas the hydrogen-to-carbon ratio is much higher (natural gas is largely CH_4), hence the burning of natural gas generates relatively more water than CO_2. This suggests that one can decrease the CO_2 emissions by a factor of two simply by changing from coal-fired power plants to natural gas-fired power plants. Indeed, making this change is one of the wedges in the scheme by Pacala and Socolow [1.4]. Currently (as of 2012) this is

Table 1.4.1 CO_2 concentration in flue gasses and air

Gas	CO_2 Concentrations
Air	380–580 ppm
Natural Gas	5–8%
Coal	10–15%

happening throughout the USA, owing to the glut of natural gas obtained from hydraulic fracturing and drilling technologies.

One expects, therefore, that phasing out coal-fired power plants and replacing them with natural gas-fired power plants will be an important contribution to the reduction of **greenhouse gasses**.

Such schemes confront an economic reality. The average lifetime of a coal-fired power plant can be as much as 50 years and the investment outlay is on the order of $2 billion for a new 600 MW coal-fired power plant [1.13]. Electricity companies will be very reluctant to write off such an investment, so we most likely have to accept that it is only economically feasible to replace the very old and inefficient coal-fired power plants with natural gas plants (see Section 1.2).

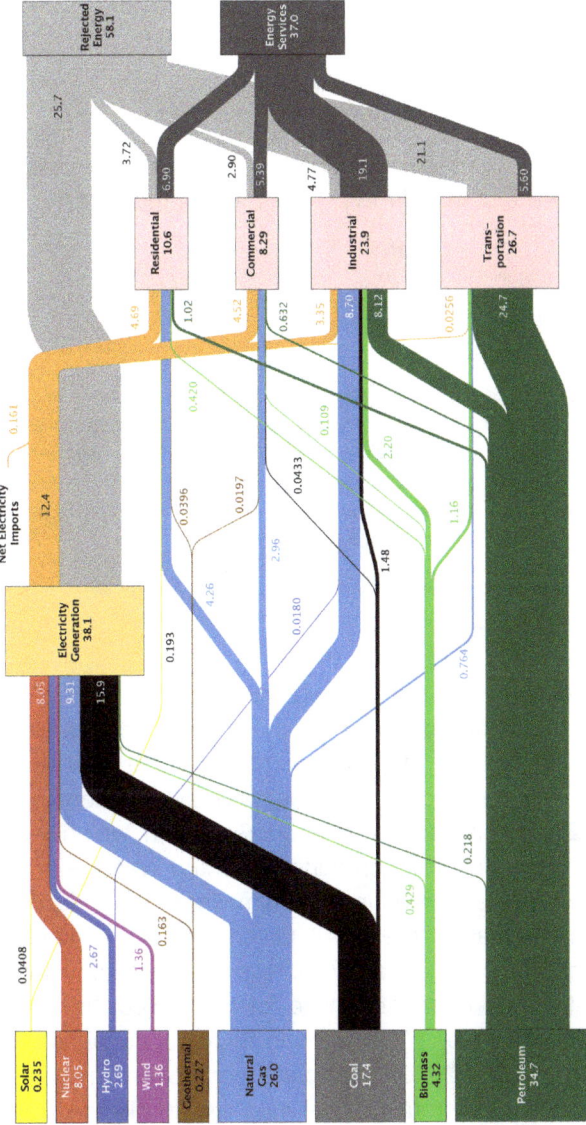

Figure 1.4.1 Energy flow, CO_2 emissions, and reserves

Data is based on DOE/EIA-0384(2011), October 2012. Distributed electricity represents only retail electricity sales and does not include self-generation. EIA reports flows for non-thermal resources (i.e., hydro, wind and solar) in BTU-equivalent values by assuming a typical fossil fuel plant "heat rate." The efficiency of electricity production is calculated as the total retail electricity delivered divided by the primary energy input into electricity generation. End use efficiency is estimated at 80% for the residential, commercial and industrial sectors, and as 25% for the transportation sector. Totals may not equal sum of components due to independent rounding. LLNL-MI-410527. *Source: LLNL, 2012.*

Section 5

Case studies: on ideas and reality

In this section we present some calculations carried out by the students at Berkeley which illustrate some of the points made in this chapter.

The carbon bicycle

Question: The Berkeley Solve Energy@home project has proposed that every person in the world be issued a stationary bicycle which serves as an electric generator. Every person is required to run the cycle to generate electricity for one hour each day. For this service, everyone will be paid half of the minimum wage. The money to fund this program would come from taxing oil. How much energy will we generate? And how much do we need to increase the price of oil to pay for this project?

Solution: Forrest Abouelnasr, Josh Howe, Vicky Jun, and Karthish Manthiram.

It would be very difficult to construct an exact calculation for this scenario because many of the numbers we need are not readily available. However, one does not need very accurate numbers to see whether this is a goofy idea or not. We need to make some basic assumptions.

The current (2012) world population is about 6.9 billion, which we will use as our potential work force. Estimating minimum wage is a little trickier due to the vast discrepancies in minimum wage from country to country (and from state to state in the USA!). Given that first-world countries like the USA use disproportionately more power than developing countries, we will set the compensation rate for our project to half the 2012 US federal minimum wage, $7.25 an hour. So in our scenario every person in the world is paid $3.625 for one hour's worth of energy generation per day. This amounts to a total cost of roughly $25 billion per day.

The next step is to calculate how much energy is produced by our cycling effort. The power-generating concept behind a stationary bike is a simple one: One pedals, turning the bike's rear wheel, which in turn spins a generator (essentially a spinning magnet within a coil of wire). As the magnet spins within this coil, electricity flows through the coil. This electricity can then be used immediately or stored in a battery. The amount of electricity produced depends on the resistance of the bicycle, the speed at which the cyclist can pedal, and the efficiency of the generator.

For our calculation, we will assume that each person generates a 125 W output with a 60% generator efficiency.

These estimates yield 270 kJ per person per 1 hour work day, or for the entire world population, 1.86×10^{15} J per day. Compared to the estimated 1.76×10^{17} J/day of global electricity consumption (17.8 trillion kWh per year), this project will produce roughly 1% of the required world electricity demand. (Or, if this experiment were arranged such that the entire world population were to generate electrical power for only the USA (which in 2009 consumed $3.7 Å \times 10^{16}$ J of electricity every day), it would produce only 5% of the USA's electricity usage.)

We can also translate this energy to reduced carbon emissions. This is a much more difficult calculation in practice as we also need to take into account the amount of CO_2 that is emitted in manufacturing our bicycles. Forgetting to factor in the contribution of manufacturing to carbon emissions is a classic blunder made by students, policy makers and scientists alike. Nevertheless, we forge ahead and ignore this contribution. In the USA, it is known that the population emits about 0.67 kg of CO_2 for each kWh of electricity generated. Using this information, we can easily quantify the carbon emissions avoided due to the cycling program:

$$1.86 \times 10^{15} \text{ J per day} \times \frac{1 \text{ kWh}}{3600000 \text{ J}} \times \frac{0.67 \text{ kg } CO_2}{1 \text{ kWh}}$$

$$= 3.46 \times 10^8 \text{ kg } CO_2 \text{ per day}$$

This means that the cycling program will save 346 Gg (346 million kg) of CO_2 emissions per day, assuming USA efficiencies apply worldwide. Compared to the 91.78 Tg (91.78 billion kg) of total carbon dioxide emitted per day, this program reduces carbon emissions by less than 0.5%. This is yet another reminder of the enormous scale of carbon emissions that we have to address.

Now let's look at sources of funding for our cycling project. What about paying for it by taxing oil? Worldwide oil consumption is about 96 million barrels of oil per day. We can calculate the necessary tax to support the program by spreading its cost of $25 billion per day over the total number of barrels of oil. This would equate to a tax of $260 per barrel, or approximately a 228% markup in oil prices (using $114 as the approximate April 2012 cost of a barrel of oil). This funding plan is not viable.

Conclusion: as to our original proposal, it seems the suggestion to mandate that each person in the world accept half the global minimum wage in order to power an electrical generator for one hour each day will not make a significant dent in global carbon emissions (cutting them by only 0.5%), and will not produce a significant amount of electricity (just 1% of the world's total demand).

On the plus side, the cycling program would have the benefit of educating people about the direct connection between energy consumption, carbon emissions, and energy production.

Kyoto Protocol progress

Question: How many nuclear power plants would need to be built per month worldwide to meet the Kyoto protocol by nuclear energy only?

Solution: Eun Hee Lim, Raven Julia McGuane, Joachim Seel, and Annie Teng.

The Kyoto Protocol is an international agreement linked to the United Nations Framework Convention on Climate Change (UNFCCC). The agreement did not specify absolute global climate goals, but rather relative performance goals for a number of developed "Annex I" countries. Developing countries, in contrast, were not bound by any concrete emission reduction goals. A list of selected Annex I countries and their CO_2 emissions targets is shown in **Table 1.5.1** [1.14].

In general, the Kyoto Protocol dictates that by the end of 2012, the average global CO_2 emissions for Annex I countries must decrease by 5.2% from their 1990 levels. However, some Annex I countries (notably the United States) did not ratify the Kyoto Protocol, effectively relieving them from fulfilling their original pledges.

Politics aside, for our calculations we will use the original goal of a 5.2% emissions reduction and calculate the number of power plants needed to fulfill that protocol. We will also make a few other broad assumptions, namely: (1) that creation of nuclear power plants and their

Table 1.5.1 Kyoto Protocol emissions targets for selected countries (data from UNFCCC [1.14])

Country/Region	Kyoto Target until 2012
EU-15*, Bulgaria, Czech Republic, Estonia, Latvia, Liechtenstein, Lithuania, Monaco, Romania, Slovakia, Slovenia, Switzerland	−8%
USA**	−7%
Canada, Hungary, Japan, Poland	−6%
Croatia	−5%
New Zealand, Russian Federation, Ukraine	0
Norway	+1%
Australia	+8%
Iceland	+10%

*The 15 States that were EU members in 1997 when the Kyoto Protocol was adopted took on the 8% target that will be distributed among themselves, taking advantage of a scheme under the Protocol known as a "bubble," whereby countries have different individual targets that when combined achieve an overall target for that group of countries. The EU has already reached agreement on how its targets will be distributed.
**The USA has indicated its intention not to ratify the Kyoto Protocol.

operation is carbon neutral; (2) that the Kyoto Protocol will be fully implemented at the end of 2012; and (3) that we must first consider the progress that Annex I states have made toward their Kyoto Protocol goals (using the most recent data at the time of this publication).

Out of the 36 ratifying countries, only seven have not been able to satisfy their country-specific requirement in the stationary source subsection by 2009 (the last year with available data reports to the IPCC). This takes into account the varying targets for each country, ranging from −8% (all of the EU) to +10% in Iceland. These seven countries include Australia (failure by 44%), Canada (19%), Japan (7%), Liechtenstein (10%), New Zealand (14%), Norway (40%) and Switzerland (2%). In order to fulfill their Kyoto Protocol goals, Annex I states would have to reduce an additional 205 Tg CO_2 between 2009 and the end of 2012.

The carbon savings achieved by building nuclear power plants depend on the power plants which they will replace. When we consider the solution of building more nuclear power plants, we must keep in mind that at the present time they can only replace stationary point sources, meaning coal- or natural gas-fired power plants, to alleviate the carbon burden. (Electrification of transportation, especially vehicles on the road, is still limited.) For this analysis, we compare two scenarios: one where the nuclear plants replace coal-fired power plants with average life-cycle emissions of 866 g CO_2 per kWh produced, and one where the nuclear plants replace combined natural gas cycle power plants with average life-cycle emissions of 439 g CO_2 per kWh produced. Both emission intensities represent the most efficient power plants and thus an upper limit to the number of required power plants (from an economic point of view it would be more sensible to replace older and less efficient power plants with higher emission factors).

Assuming that the average capacity of a new nuclear reactor would be 1 GW and each reactor would substitute coal/gas electricity, we find that between 9 (coal scenario) and 18 (gas scenario) nuclear power plants would need to have been built between 2009 and 2012. Assuming these plants had been built consecutively with equal temporal spacing, a new nuclear power plant would need to have been built every 4 or 2 months, respectively.

Searching for Bigfoot (calculating and comparing carbon footprints)

Question: *Calculate your own carbon footprint using an online calculator* (http://www.carbonfootprint.com) *and identify the two largest contributing factors for your lifestyle. Compare this to the carbon footprint of an average citizen in the USA, in India, and in China.*

Solution: *Angus Ming Yiu Chan, Sirine Constance Fakra, Jeffrey Kaut Krajewski, Yuguang She, Zhou Lin, Sophia Louise Shevick, Mohammad Haider Agha Hasan, Kristopher Enslow, and Anna Claire Harley-Trochimczy.*

A carbon footprint is the contribution to greenhouse gas emissions that a person, organization, country, or region emits over a specified length of time. The contributions to the carbon footprint can be broken down into two categories: primary and secondary. *Primary* emissions are those which result directly from one's actions, including the use of electricity and burning fossil fuels for transportation. *Secondary* emissions are those which result indirectly from one's activities, including the carbon emissions associated with the foods one eats and the products one purchases.

In comparing carbon emissions around the globe, Luxembourg had the highest per capita emissions at 33.8 tonnes CO_2. The United States was second with 28.6 tonnes CO_2 per person. In contrast, developing countries like China and India have per capita emissions of 3.1 tonnes and 1.8 tonnes CO_2, respectively.

California's Climate Plan aims to reduce CO_2 emissions to 1990 levels by 2020, which would mean reducing CO_2 emissions to 10 tonnes per person. To see how feasible this reduction would be, we calculated the personal carbon footprint of three UC Berkeley students. We used the calculator provided by Carbon Footprint Ltd. The average CO_2 emissions in our group is 11.92 tonnes CO_2 per person per year. This includes contributions from transportation, housing, and lifestyle choices (see **Table 1.5.2**). The secondary lifestyle factors include food and consumer choices, such as how often one recycles or how often one eats seasonal foods.

Table 1.5.2 Personal carbon footprints for three UC Berkeley students

Category	Student A	Student B	Student C	Average (metric tonnes CO_2)
House	0.63	0.33	0.6	0.52
Flights	2.28	3.55	1.64	2.49
Car	5.7	3.42	2.19	3.77
Bus & Rail	0.06	0.04	0.1	0.07
Secondary Lifestyle	5.17	5.9	4.15	5.07
Total	**13.84**	**13.24**	**8.68**	**11.92**

Source: Carbon Footprint Ltd, http://www.carbonfootprint.com

Due to Berkeley's fortuitous location in a fairly temperate zone, many people do not use air conditioning, and use limited electricity during the summer. While some heating may be necessary in the winter, the temperature almost never drops below freezing, making "house" a limited (~5%) component of a typical Berkeley student's carbon footprint. By far the largest contributors to the carbon footprint of the average Berkeley student seem to be the secondary lifestyle choices (37%), followed by car travel (27%). This result may be atypical to the rest of the United States due not only to the location, but to the nature of the student lifestyle. Students are more likely to walk or use a bike as their primary mode of transportation and are likely to share rooms or apartments with several roommates, which cuts the total carbon footprint per person.

It may be worth noting a few anomalies in the individual footprints. Student A just recently moved from Los Angeles, a city much more centered on driving, which explains the higher contributing percentage of car travel to her total. Student B had taken several flights in the last year for both conferences and vacations, increasing his flight total. Student C does not own a car and tends to make more environmentally conscious consumer choices, which ultimately leads to her lower overall footprint than those of the other two students.

In **Table 1.5.3** we calculate a fourth student's, Student D's, carbon footprint before and after his acquisition of a car. After acquiring a car, his carbon footprint increased by more than 50%. In **Table 1.5.4** we compare his results to the average carbon footprints per capita in the USA, China, and India. While his total levels are much lower than the USA

Table 1.5.3 Calculation of Student D's carbon footprint with and without a car

Category	Metric tonnes of CO_2 per year trial w/o car	Metric tonnes of CO_2 per year trial w/ car
House	0.11	0.11
Flights	1.44	1.44
Car	0.15	3.51
Bus/Rail	0.14	0.14
Secondary Sources	3.99	3.99
Total	**5.82**	**9.18**

Table 1.5.4 Comparison of students' footprints with various countries' per capita footprints

Comparison Groups	Metric tonnes of CO_2 per year per capita
Student D w/o car	5.82
Student D w/ car	9.18
USA Average*	28.6
China Average*	3.1
India Average*	1.8
World Target to Combat Climate Change	2

Source: Hertwich et al. [1.15].

average, they are still higher than the world target to combat climate change [1.15].

A concrete solution to carbon capture

Question: How much CO_2 can we capture if we use it as a source of concrete?

Solution: Samuel Taft Schloemer, Lingchen Fan, and Joseph Jung-Wen Chen.

Figure 1.3.1 shows that carbonate is the most stable form of carbon. Since carbonates are the major component of concrete, we will explore the use of CO_2 from flue gas emissions as a source of concrete.

Concrete is a composite material made from cement, sand, aggregate, and water. The cement is the binding agent, which reacts with the water to form calcium silicates that glue together the sand and aggregate particles. The aggregate, often composed of crushed stone, provides the compressive strength of the material (see **Figure 1.5.1**). Cement production is a major source of CO_2 emissions worldwide, accounting for around 829 million metric tonnes of CO_2, or 3.4% of all emissions, in the year 2000. However, concrete could be a net CO_2 sink if the aggregate were made of a substance that consumed CO_2. For example, a common aggregate is calcium carbonate ($CaCO_3$), also known as

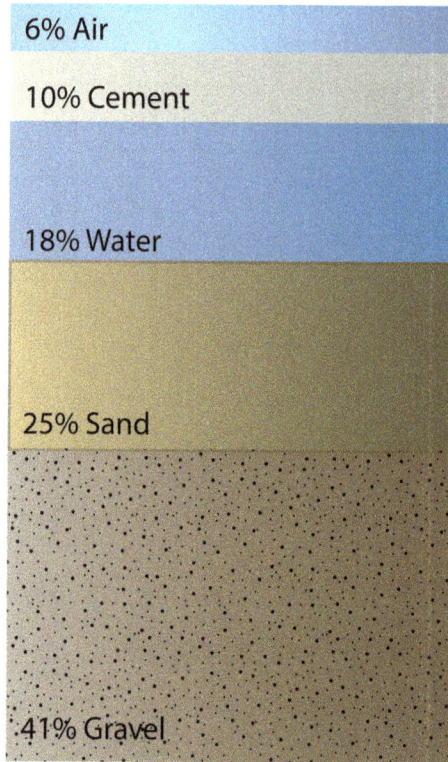

Figure 1.5.1 Typical composition of hydraulic cement concrete

limestone. To get a measure of the feasibility of producing concrete from CO_2, we will estimate the amount of CO_2 we could sequester if all the aggregate were made from limestone ($CaCO_3$) formed from CO_2.

Carbon dioxide can be sequestered in concrete via the formation of $CaCO_3$. This is accomplished in the presence of water, and various calcium phases such as $Ca(OH)_2$ and CSH (Calcium Silicate Hydrate), which are present in concrete. First, CO_2 reacts with water to form carbonic acid:

$$CO_2 + H_2O \rightarrow H_2CO_3$$

The carbonic acid will then react with the calcium phases to form calcium carbonate and water:

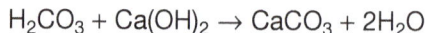

$$H_2CO_3 + Ca(OH)_2 \rightarrow CaCO_3 + 2H_2O$$

Similarly, CSH present in concrete can liberate CaO from $Ca(OH)_2$ once it has converted, again forming carbonate:

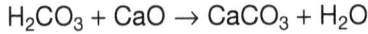

$$H_2CO_3 + CaO \rightarrow CaCO_3 + H_2O$$

A simplified way to write this calcination process is to combine the reaction of CaO and CO_2 to give:

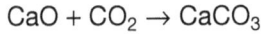

$$CaO + CO_2 \rightarrow CaCO_3$$

Let's imagine that we have an unlimited source of CaO. Each mole of CaO can absorb one mole of CO_2 according to the reaction above. Current global production of cement is about 3.3 billion metric tonnes per year. Given that typical concrete is composed of four parts aggregate to one part cement by weight, the global production of concrete consumes approximately 13.2 billion metric tonnes of aggregate per year. Assuming this is composed completely of $CaCO_3$, this is the equivalent of 5.8 Pg per year. We conclude that sequestering CO_2 emissions into the aggregate in concrete would absorb about 19% of our annual CO_2 emissions.

It turns out that concrete can sequester even more CO_2. The cement that is used as the binder in concrete is comprised mainly of CaO. The cement that has been placed into roads, buildings, dams, etc., will undergo chemical reactions as the concrete ages, and CO_2 from the atmosphere will diffuse into the concrete and react with CaO to form $CaCO_3$. This is a natural part of the concrete curing process and acts to increase the strength of the material. CaO accounts for about 63% of the cement by weight. Assuming this all eventually reacts with CO_2 in the atmosphere, cement curing and aging could account for an additional 1.6 Pg per year of CO_2 sequestration, or a total of 24% of global CO_2 emissions.

In summary, these calculations indicate that current global concrete production could absorb only 19% to 25% of net CO_2 emissions. These calculations are predicated on the availability of an unlimited supply of CaO. Given that the CaO for cement must be produced by driving the CO_2 off $CaCO_3$ with large quantities of heat, usually supplied by fossil fuel combustion, this process would be infeasible even if the scale of concrete production was greater than CO_2 emissions.

Paving the way for change

Question: If we convert all of the CO_2 emitted by a car over its lifetime into limestone and use this limestone to pave someone's yard, what is the surface area we can cover? How does this differ depending on whether the car is a Prius or Hummer?

Solution: Miguel Angel Garcia Jr., Eunice JiYoung An, Joshua Deitch, and Anton Mlinar.

Typically, cars run on internal combustion engines that rely on the combustion of fossil fuels to turn chemical energy into mechanical energy. As a by-product, carbon dioxide is produced and released as exhaust into the atmosphere. But there are alternatives to this process. The two methods which have garnered the most attention are the hybrid electric and electric vehicles. By relying either partially (hybrid) or almost wholly (EV) on electric power, these vehicles reduce the combustion of fossil fuels for power, thus reducing carbon emissions. Due to their high gas mileage and decreased emissions, many people are hailing hybrids (like the Prius) and new mid-sized electric vehicles as viable weapons to battle climate change.

There's no doubt that hybrid and electric vehicles represent some reduction in carbon emissions compared to gas-guzzling alternatives like the Hummer, but to what degree? Even if the electricity supplied to a EV is carbon free (or carbon neutral), there is still a significant energy cost in the production of these hybrid and electric vehicles, and thus the carbon savings gained from replacing internal combustion engines could be offset by emissions related to production. Also, there are many concerns over the batteries used by these cars. Not only is battery manufacturing energy-intensive, but considerable pollution results from the mining of essential rare earth metals that it involves. To determine the true environmental impact of alternative engines for transportation, we compared emissions from production and use between the Hummer and the Prius. To make it more interesting, we represented these values in terms of surface area of limestone that would be created by each, assuming we could capture all of the carbon emissions in this format.

On average, the lifetime mileage of a car is about 185,000 miles, and the average miles per gallon (mpg) for a car ranges from 14 to 29 mpg. We chose to compare a 2011 Prius to a 2010 H3 4WD Hummer. We used the combined city/highway EPA miles per gallon listing for both the Hummer (14 mpg) and the Prius (50 mpg). Assuming these conditions, a Prius will consume about 3,700 gallons and a Hummer 13,214 gallons of gasoline over its lifetime. We can calculate the number of moles of CO_2 produced per gallon of gasoline starting with the balanced equation for octane combustion (octane is a reasonable proxy for pure gasoline):

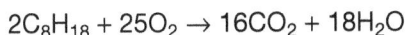

$$2C_8H_{18} + 25O_2 \rightarrow 16CO_2 + 18H_2O$$

A gallon of 87-octane gas contains approximately 2.8 kg of C_8H_{18} per gallon (density about $0.75g/cm^3$). That gives us:

$$\frac{2800 \text{ g } C_8H_{18}}{1 \text{ gallon}} \times \frac{1 \text{ mole } C_8H_{18}}{114.2 \text{ g } C_8H_{18}} \times \frac{16 \text{ moles } CO_2}{2 \text{ moles } C_8H_{18}}$$
$$= 196 \text{ moles } CO_2 \text{ per gallon}$$

We calculate that 1 gallon produces 196 moles of CO_2, or about 8,628 grams of CO_2. The EPA cites this number as slightly higher, around 8,800 grams (200 moles) of CO_2 per gallon of gasoline, though they may be assuming a higher octane gasoline. So, if anything, we are calculating a lower limit of this system.

Now on to our novel capture solution. We know from our calcination equation that:

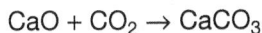

$$CaO + CO_2 \rightarrow CaCO_3$$

This means 1 gallon of gas producing 196 moles of CO_2 would yield 196 moles (about 19.6 kg) of $CaCO_3$ via calcination. Next, we apply this metric to the lifetime amount of gasoline used by our two varieties of cars.

In all categories, we see that the Hummer uses more than 3.5 times the resources compared to the Prius. To quantify this in terms of surface area of limestone, we need to assume a few building parameters. Most stone used for paving has a density between 3,000 and 3,300 kg/m^3, but limestone's density ranges from 1,741 to 2,800 kg/m^3. For better paving conditions, we will use a high density of 2,800 kg/m^3 for our limestone project. The thickness of the stone will vary from project to project, but we will assume an average thickness of 4 inches (0.1016 m) for our purposes.

Using these measures for the Hummer, we get:

$$259{,}000 \text{ kg CaCO}_3 \times \frac{1 \text{ m}^3}{2{,}800 \text{ kg CaCO}_3} \times \frac{1 \text{ m}^2 \text{ pavement}}{0.1016 \text{ m}^3 \text{ CaCO}_3}$$

$$= 910 \text{ m}^2 \text{ pavement}$$

and for the Prius:

$$72{,}500 \text{ kg CaCO}_3 \times \frac{1 \text{ m}^3}{2{,}800 \text{ kg CaCO}_3} \times \frac{1 \text{ m}^2 \text{ pavement}}{0.1016 \text{ m}^3 \text{ CaCO}_3}$$

$$= 255 \text{ m}^2 \text{ pavement}$$

In other words, it would take approximately 6.5 Hummers or 24 Priuses to generate the carbon emissions necessary to pave an entire football field (6,050 m^2) with limestone.

The professor, the plane, and the Prius

Question: Your Berkeley professors are likely to drive a Prius, but they are also very proud of their United 1k Frequent Flyer status, which is a perceived signature of their international importance. Does the carbon savings accrued by driving a Prius compensate their flying habits?

Solution: Angus Ming Yiu Chan, Sirine Constance Fakra, Jeffrey Kaut Krajewski, Yuguang She, and Zhou Lin.

The relative ease and automation of transportation has made it easy to forget the toll it takes on the environment. Whether it is a drive down the street or a flight across an ocean, all modern transportation has a carbon footprint. While it is possible to reduce carbon emissions through various lifestyle changes, the relative tradeoffs between modes of transportation make it difficult to judge the scale and net effectiveness of these savings. For example, a dramatic environmental change at home could be wholly negated by a single vacation flight.

Let's look more closely at the case of our Prius-driving, world-traveling professor. His United Airlines 1k Frequent Flyer Status requires a minimum of 100,000 km of yearly air travel, yet his Prius saves gas by getting 50 mpg versus around 16 mpg with an SUV. According to the Federal Highway Administration, American males of roughly professorial age (35–54) drove an average of 15,859 miles per year in 2011. Females in the same age bracket, on the other hand, drove an average of only 11,464 miles per year. We will use an average of these two estimates for our current calculation, and assume our professor drives about 13,661 miles per year.

According to the EPA, one gallon of gasoline produces around 8.92×10^{-3} metric tonnes of CO_2. We can use this information to construct a simple calculation of CO_2 savings based on our professor's choice of vehicle:

$$\text{SUV: } 13,661 \text{ miles} \times \frac{1 \text{ gallon gas}}{16 \text{ miles}} \times \frac{8.92 \times 10^{-3} \text{ metric tons of } CO_2}{1 \text{ gallon gas}}$$
$$= 7.61 \text{ metric tons of } CO_2$$

$$\text{Prius: } 13,661 \text{ miles} \times \frac{1 \text{ gallon gas}}{50 \text{ miles}} \times \frac{8.92 \times 10^{-3} \text{ metric tons of } CO_2}{1 \text{ gallon gas}}$$
$$= 2.43 \text{ metric tons of } CO_2$$

By driving a hybrid vehicle, our professor reduces his car's carbon emissions by a factor of 3 (see **Table 1.5.5**), saving approximately 5.18 tonnes of CO_2 per year. Yet how do his CO_2 savings on the road compare with his emissions in the air? To find out how much carbon is released on his behalf, we imagined the professor must make frequent trips between San Francisco and Amsterdam (6,230 miles), with a layover in

Table 1.5.5 Gas efficiency and CO_2 footprint by car type

Type of Car	Miles per gallon	Lifetime gas consumed (gal)	Lifetime CO_2 produced (g)	Limestone required for calcination (kg)
Hummer	14	13,214	114 million	259,000
Prius	50	3,700	31 million	72,500

New York. In order to maintain his 1k Flyer Status, he must make this trip roughly 16 times (8 round trips) per year.

According to the International Civil Aviation Organization (ICAO) carbon emissions calculator, a passenger making a round trip from San Francisco (SFO) to Amsterdam (AMS) in Economy class is responsible for a carbon footprint of approximately 1.472 metric tonnes of CO_2. Per year, this equates to 11.78 metric tonnes of CO_2 from flights for our professor, twice the amount of CO_2 saved annually by switching from an SUV to a Prius. Furthermore, due to other pollutants and the altitude at which they are released, the environmental impact of an aircraft is larger than just the net carbon dioxide released. According to the IPCC, we should apply an empirical multiplier of 2.7, called the radiative forcing index (RFI), to find the actual net greenhouse effect of flying. This brings our professor's flight-based carbon footprint up to a staggering 32 effective metric tonnes of CO_2 annually, dwarfing the 5 tonnes saved by switching to a Prius.

When it comes to reducing carbon emissions, it is very important to understand the scope of the savings. In the example of the traveling professor, a seemingly large cutback in one area may actually be insignificant compared to another. In order to combat global warming, it is important to know how to significantly reduce emissions by targeting the greatest contributors. There has been much debate over which is better for the environment: car or air travel. These factors are variable depending on the flying and driving habits, aircraft, and vehicle type for the traveler in question. In the end, both are significant contributors to our carbon footprints and need to be reexamined as a whole.

Section 6

Review

1.1. Reading self-test

1. At 33°C and 80 bar carbon dioxide is
   - a. Liquid
   - b. Solid
   - c. Gas
   - d. Supercritical

2. What is the lowest energy state of carbon?
   - a. Carbon
   - b. Methane
   - c. Coal
   - d. CO_2
   - e. Carbonate

3. 1 Mtoe is equivalent to
   - a. 3.968×10^{13} BTU
   - b. 4.187×10^{16} J
   - c. 11.63 billion kW h
   - d. All of the above
   - e. None of the above

4. What is not a proposed wedge?
   - a. Replace 1,400 large coal-fired plants with gas-fired plants
   - b. Install CCS at 80 large coal-fired power plants
   - c. Increase solar power 700-fold to displace coal
   - d. Cut electricity use in homes, offices and stores by 25%

5. How many people emit less than 1 t CO_2/year?
   - a. 1.7 billion
   - b. 2.7 billion
   - c. 3.7 billion
   - d. 4.7 billion

6. What is the total use of CO_2 in the USA?
   a. 10 Mt
   b. 100 Mt
   c. 1000 Mt
   d. 1,000 Mt

7. Out of the top ten largest CO_2 emitters in the USA, the ratio between the one that is emitting the most and the least per GWe is
   a. 3
   b. 2
   c. 1.5
   d. 0.5

8. What was the percentage of solar energy used in electricity production in 2011?
   a. 0–0.01%
   b. 0.01–0.05%
   c. 0.05–0.1%
   d. >0.1%

9. The ratio of the carbon footprint of an average USA student and a student from India is
   a. 1
   b. 1.5
   c. 15
   d. 50

1.2. CO_2 emissions

1. Which country was the largest emitter of CO_2 in the year 2011?
   a. USA
   b. China
   c. India
   d. Qatar

2. Which country was the second largest emitter of CO_2 in the year 2011?
   a. USA
   b. China
   c. India
   d. Qatar

3. Which country has the highest CO_2 emission per capita?
   a. USA
   b. China
   c. India
   d. Qatar

1.3. Global energy consumption

1. What is the world's leading fuel used to generate electricity?
   a. Coal
   b. Oil
   c. Natural gas
   d. Hydroelectricity

2. The following graph shows the world consumption of primary energy resources in million tonnes oil equivalent. Give the fuels corresponding to the letters A–F.

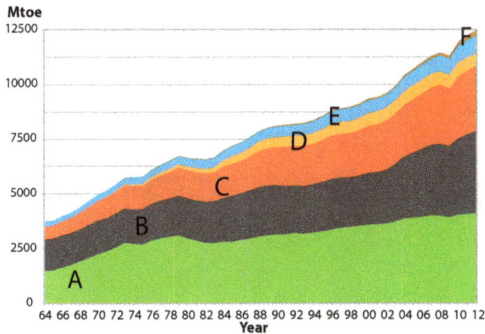

1.4. Capturing CO_2

1. According to the International Energy Agency's World Energy Outlook in 2007, what percentage of CO_2 emissions can be addressed by CCS?
   a. 37%
   b. 47%
   c. 57%
   d. 67%

2. The following graph shows the different sources of CO$_2$ emissions that are addressable by CCS. Give the fuel source corresponding to the letters A–G.

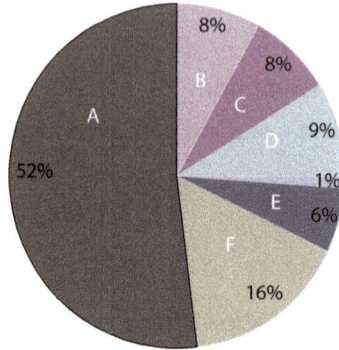

Section 7
References

1. IPCC, 2005. IPCC Special Report on Carbon Dioxide Capture and Storage. Prepared by Working Group III of the Intergovernmental Panel on Climate Change [Metz, B., O. Davidson, H.C. de Coninck, M. Loos, and L.A. Meyer (eds.)]. Cambridge University Press: Cambridge, United Kingdom and New York. http://www.ipcc.ch/pdf/special-reports/srccs/srccs_wholereport.pdf
2. BP Statistical Review of World Energy, June 2012. http://www.bp.com/assets/bp_internet/globalbp/globalbp_uk_english/reports_and_publications/statistical_energy_review_2011/STAGING/local_assets/pdf/statistical_review_of_world_energy_full_report_2012.pdf
3. Boden, T.A., G. Marland, and R.J. Andres, 2010. Global, Regional, and National Fossil-Fuel CO_2 Emissions. Carbon Dioxide Information Analysis Center, Oak Ridge National Laboratory, US Department of Energy. http://dx.doi.org/10.3334/CDIAC/00001_V2010
4. Pacala, S. and R. Socolow, 2006. A Plan to Keep Carbon in Check. Scientific American, 50–57. http://cmi.princeton.edu/resources/pdfs/carbon_plan.pdf
5. Davis, S.J., L. Cao, K. Caldeira, and M.I. Hoffert, 2013. Rethinking wedges. Environ. Research Letters, **8** (1), 011001. http://dx.doi.org/10.1088/1748-9326/8/1/011001
6. BURN (an energy journal). Data from the International Energy Agency.
7. Movie generated from gapminder.org, data from the Carbon Dioxide Information Analysis Center. http://www.gapminder.org/videos/gapminder-videos/gapcast-10-energy/
8. Chakravarty, S., A. Chikkatur, H. de Coninck, S. Pacala, R. Socolow, and M. Tavoni, 2009. Sharing global CO_2 emission reductions among one billion high emitters. P. Natl. Acad. Sci. USA, **106** (29), 11884. http://dx.doi.org/10.1073/Pnas.0905232106
9. Oliver, J.G.J, G. Janssens-Maenhout and J.A.H.W. Peters, 2012. Trends in Global CO_2 Emissions; 2012 Report, The Hague: PBL Netherlands Environmental Assessment Agency; Ispra: Joint Research Centre. http://edgar.jrc.ec.europa.eu/CO2REPORT2012.pdf
10. Environment Protection Agency. http://ghgdata.epa.gov/ghgp/main.do
11. Davis, S.J., K. Caldeira, and H.D. Matthews, 2010. Future CO_2 emissions and climate change from existing energy infrastructure. Science, **329** (5997), 1330. http://dx.doi.org/10.1126/Science.1188566

12. Bhown, A.S. and B.C. Freeman, 2011. Analysis and status of post-combustion carbon dioxide capture technologies. Environ. Sci. Technol., **45** (20), 8624. http://dx.doi.org/10.1021/es104291d
13. Schlissel, D., A. Smith, and R. Wilson, 2008. Coal-Fired Power Plant Construction Costs. Synapse Energy Economics Inc: Cambridge, MA. http://www.synapse-energy.com/Downloads/SynapsePaper.2008-07.0.Coal-Plant-Construction-Costs.A0021.pdf
14. United Nations Framework Convention on Climate Change: Kyoto Protocol Emission Targets. http://unfccc.int/kyoto_protocol/items/3145.php
15. Hertwich, E.G. and G.P. Peters, 2009. Carbon footprint of nations: A global, trade-linked analysis. Environ. Sci. Technol., **43** (16), 6414. http://dx.doi.org/10.1021/Es803496a

Chapter 2

The Atmosphere and Climate Modeling

There is little scientific overlap between carbon capture and sequestration and the science of the atmosphere and climate modeling. Yet, the justification for carbon capture and sequestration relies on our understanding of the role of CO_2 in the atmosphere and our predictions on how the increase in CO_2 levels will affect our future climate.

Section 1

Introduction

In the next chapters, we focus on the climate and the carbon cycle. These topics are not directly related to the science and technology of carbon capture and sequestration (CCS). Yet, it is exactly these two fields that make the scientific case for why we have to be concerned about future CO_2 levels. These topics are much discussed in the media, and it is our experience that discussions about CCS inevitably invite questions about climate change. We therefore include this topic in our CSS book. Lectures given by two of our colleagues, Professors Ron Cohen and William Collins, on the atmosphere and on climate modeling, respectively, along with Intergovernmental Panel on Climate Change (IPCC) reports, provide the basis for the material presented in this chapter.

Section 2

The earth with and without an atmosphere

We can do a very simple calculation to illustrate the role of the atmosphere in controlling the temperature of the surface of the earth. The earth receives its energy from the sun. The amount of energy is given by **Stefan-Boltzmann's law**, which relates the energy of a radiating body J to its surface temperature:

$$J = \sigma T_{Sun}^4$$

where σ is the Stefan-Boltzmann constant. Assuming that the earth is in equilibrium, all the energy the earth is receiving is ultimately emitted by radiation (i.e., earthshine):

$$J^{out} = \sigma T_{Earth}^4$$

We can now construct an energy balance (see **Figure 2.2.1**), where we have to take into account that only a very small fraction of the total energy emitted by the sun hits the surface of the earth. The amount of energy reaching the top of the earth's atmosphere on a square meter facing the sun each second during the daytime is about 1,370 J. If we average this out over the entire planet (day and night), we obtain 25% of this, or 341 W/m^2. According to Stefan-Boltzmann's law, this energy corresponds to a planet with a temperature of approximately 279K (6°C). The actual average temperature is about 14°C. We will show that our simple calculation gave a lower temperature because we have ignored the effect of the atmosphere.

Figure 2.2.2 shows the fate of the energy that hits the upper part of the atmosphere [2.1]. Not all solar energy reaches the surface of the earth, as part of this energy gets reflected by the clouds and aerosols

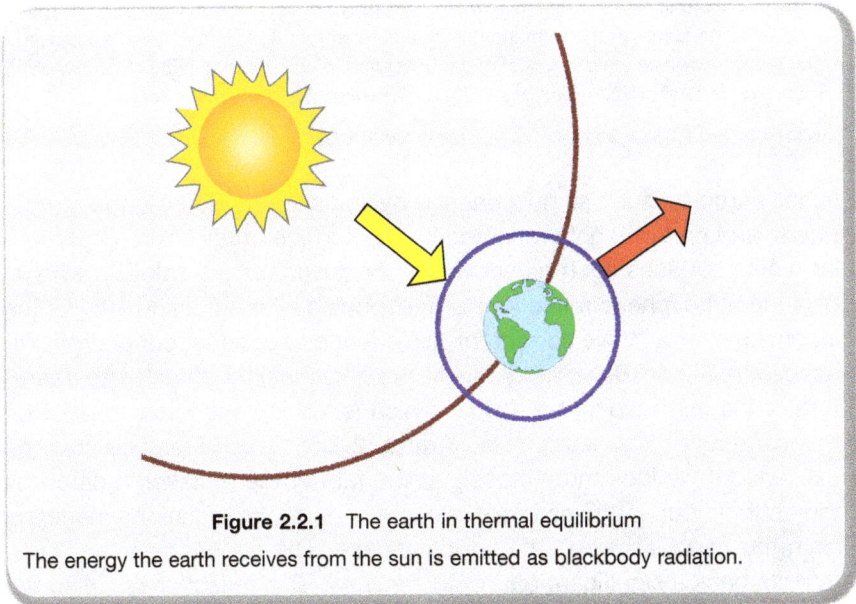

Figure 2.2.1 The earth in thermal equilibrium

The energy the earth receives from the sun is emitted as blackbody radiation.

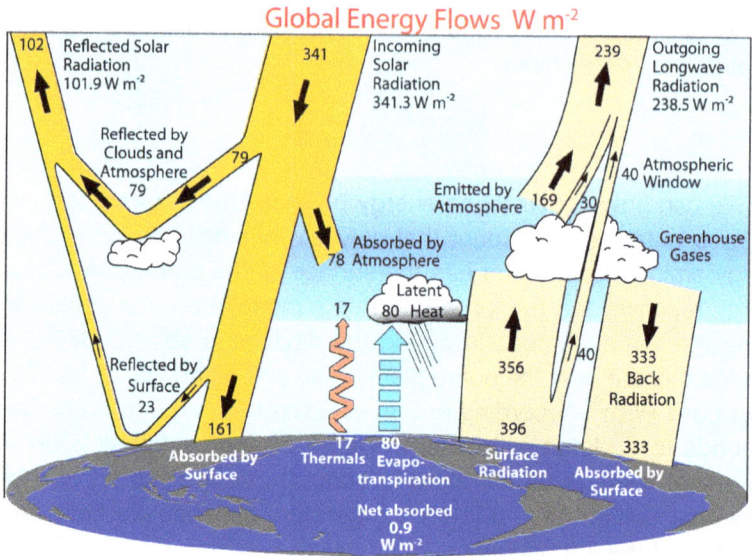

Figure 2.2.2 The earth's energy balance

The numbers give the annual energy flow per square meter, averaged over the entire earth for the period from March 2000 to May 2004. We see that ~30% of the 341 W/m² of incoming radiation is reflected and is not absorbed by the earth's surface. The broad arrows indicate schematically the flow of different types of energy in proportion to their relative magnitudes. In the formation of clouds, energy in the form of heat is released and this is known as the latent heat of condensation. *Figure reproduced with permission from Trenberth et al. [2.1]. Copyright (2009), American Meteorological Society.*

(small particles, such as dust caused by volcano eruptions) in the atmosphere, and part of it gets reflected by the surface of the earth (in particular, white surfaces such as snow, ice, or deserts). This "**albedo effect**" plus the atmospheric reflections are responsible for reflecting 30% of the incoming energy. If we correct for this effect and conduct our energy balance with Stefan-Boltzmann's Law, we find that the effective temperature of the earth would be −19°C, which is too low for water to exist on the surface of the earth (see **Figure 2.2.3**). The difference can be explained if we look more closely at the fate of the radiation emitted by the earth. In our −19°C estimate, we assumed all the radiation emanating from the earth is being emitted into space. The atmosphere, however, reflects back a significant part of this energy. This reflection is called the

(a)	(b)

Figure 2.2.3   With and without an atmosphere

(a) A picture of the moon which is considered to not have an atmosphere. Without an atmosphere, the earth would have a temperature of 107°C during the day and at night −153°C, giving an average temperature of −23°C. *Image of the moon from NASA*, https://solarsystem.nasa.gov/planets/profile.cfm?Object=Moon

(b) Because of the greenhouse gasses, the average temperature of the earth is 14°C. *Image of the earth from NASA*, http://blogs.smithsonianmag.com/science/files/2010/04/modis_wonderglobe_lrg.jpg

Question 2.2.1 What about fossil fuels?

In our calculations of the temperature of the earth we ignored the fact that we are burning fossil fuels. If we assume all fossil fuel energy eventually is converted into heat, how does this energy compare with the energy we receive from the sun?

natural greenhouse effect. If we take this effect into account in calculating the earthshine, we can explain why the actual temperature of the surface of the earth is much higher than −19°C. Given that the average temperature of the earth is 14°C, Stefan-Boltzmann's law tells us that the earth is emitting about 390 W/m². As we recall from **Figure 2.2.2**, the surface of the earth is receiving 161 W/m² from the sun. As the net energy balance for the earth is zero, the difference is the greenhouse effect. This effect is huge. It is interesting to compare this effect to the heating caused by the burning of fossil fuels (see **Question 2.2.1**).

Figure 2.2.2 also illustrates that the energy balance in the atmosphere is quite complex, as different mechanisms contribute to the temperature of the surface of the earth. That these effects are often counterintuitive can be illustrated with the following example. Suppose we were to eliminate all clouds in the atmosphere; would this increase or decrease the temperature of the earth? Clouds have two effects: they reflect the radiation from the sun, and they contribute to the greenhouse effect. The latter effect is something we have experienced; nights are less cool if the night sky is cloudy. Despite these local effects, the net balance is that the cooling effect of clouds is larger than their heating effect.

Section 3

The atmosphere

Let's consider the atmosphere in more detail. **Figure 2.3.1** gives an illustration of the different layers of the atmosphere. From the pressure and density curves we see that a substantive gas layer around the earth extends into the stratosphere. It is this gas layer that is responsible for the greenhouse effect.

Absorption of radiation by the atmosphere

An interesting question is hidden in the results of the last section. We have seen that because of the greenhouse effect the radiation from the earth gets reflected back to the earth, but the radiation from the sun does not get reflected back in the same proportion. How is this possible? How does the atmosphere differentiate between the radiation from the sun and that from the earth?

To answer this question we need to know the difference between the radiation from the sun and the radiation emitted by the earth. The sun has a higher temperature and therefore emits radiation at a shorter wavelength compared to radiation emitted by the earth (according to the blackbody radiation law). The radiation from the sun is in the ultra-violet

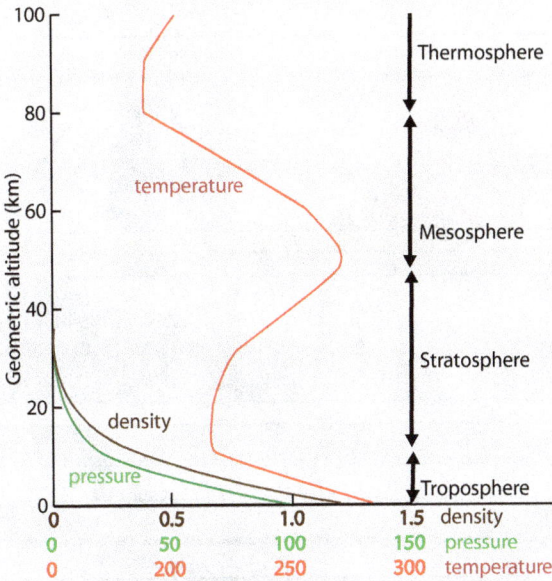

Figure 2.3.1 Properties of the atmosphere

The pressure (in kPa), density (in kg/dm^3), and temperature (in K) as a function of the altitude (in km).

(UV) and visible light part of the spectrum, while the earth emits in the infrared region. If we look at the sun, we of course see the difference between a cloudy and a clear day; in addition, the probability of sunburn is much lower when the sky is overcast. The clouds reflect most of the radiation from the sun, but without clouds the atmosphere is transparent to UV and visible light. This is very different for the infrared radiation emitted by the earth; at these wavelengths the atmosphere is absorbing. Thus the greenhouse effect is the differential absorption in the atmosphere between solar radiation and the earth's blackbody emission. In **Figure 2.3.2** the emission spectra of the sun and earth are compared with the absorption spectrum of the atmosphere.

We can also investigate which gasses contribute most to the greenhouse effect (**Table 2.3.1**). These gasses are, in decreasing order of their contribution: water, carbon dioxide, methane, and ozone. The concentrations of these and other gasses in the atmosphere are shown in

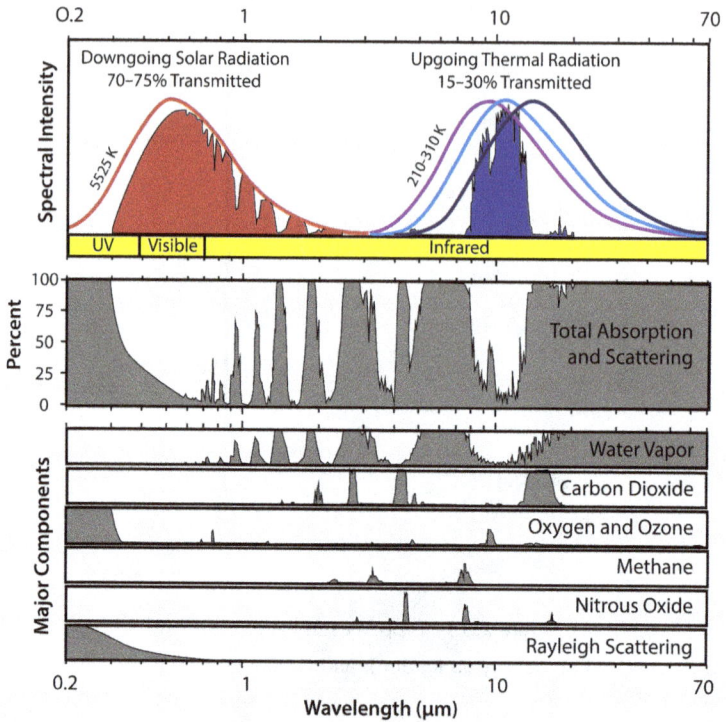

Figure 2.3.2 Radiation and the atmosphere

The top graph shows the solar radiation from an object at 5500K (red), and the earth's radiation at 288K (blue). Because the surface of the earth has slightly different temperatures at different locations, the blue curve is jagged. We see that the sun radiates in the ultra-violet to visible wavelengths, while the earth emits in the infrared. The middle graph shows the absorption bands in the earth's atmosphere; if the absorption is high at a certain wavelength, the atmosphere won't transmit at that wavelength. The bottom graphs show the contribution of individual gas absorption spectra to the total absorption spectrum. *Image based on the original created by Robert A. Rohde.*

Table 2.3.2. We see that water is the most important greenhouse gas. In addition, both CO_2 and CH4, despite their low concentrations, contribute significantly to the greenhouse effect.

There is one additional point to make in comparing the effects of the different gasses on the overall greenhouse effect. For this we need **Beer's law**, which states that there is a logarithmic dependence

Table 2.3.1 Contribution of different gasses to the greenhouse effect and their lifetime in the atmosphere

Gas	Contribution	Lifetime
Water vapor	36–72%	100 years
Carbon dioxide	9–26%	100–1,000 years
Methane	4–9%	10 years
Ozone	3–7%	hours to months

Table 2.3.2 Composition of the atmosphere

Gas	Volume (ppm)	Volume (%)
Nitrogen	781,000	78
Oxygen	209,500	21
Argon	9,340	0.934
Carbon dioxide	394	0.039
Neon	18	0.0018
Helium	5.24	0.0005
Methane	1.79	0.00018
Krypton	1.14	0.00011
Hydrogen	0.55	0.000055
Nitrous oxide	0.33	0.000033
Carbon monoxide	0.1	0.00001
Xenon	0.09	0.000009
Ozone	0–0.07	0–0.000007

These numbers are for a dry atmosphere; the water content of the atmosphere is on average 0.4% (1–4% at the surface).

between the intensity I of light through a substance and the product of the density of the gas, ρ, the absorption coefficient of the substance, σ, and the distance the light travels through the material (i.e., the path length), l:

$$\frac{I}{I_0} = e^{-\rho\sigma l}$$

If at a particular wavelength the atmosphere is already fully or mostly absorbing, then adding more of a substance with a similar absorption

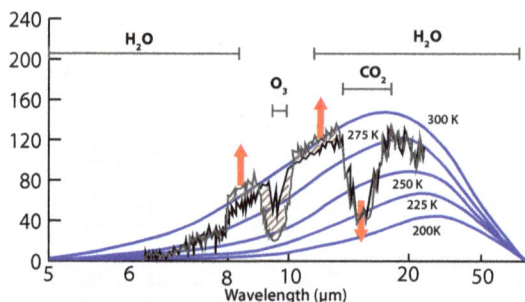

Figure 2.3.3 Absorption spectrum of the atmosphere

Details of the absorption spectrum of the atmosphere and the contributions of the various chemicals. The blue curves are the emission spectra at the indicated temperatures and black and grey curves the absorption spectra of the atmosphere.

coefficient will have a smaller effect. So chemicals that absorb at the same wavelength as water can be very strong greenhouse gasses, but their effect will be modest because water already is prevalent in the atmosphere. If, however, we add a gas that is strongly absorbing at a wavelength where the atmosphere is transparent, it will have a very strong effect. This is exactly the case with CO_2 and CH_4, which makes them strong greenhouse gasses.

The other important factor is the length of time that gasses reside in the atmosphere. A chemical can be a very strong greenhouse gas, but if this chemical is only stable for a few days, its effect will be much smaller compared to that of chemicals that persist for many years. **Table 2.3.1** shows some of these lifetimes; we see that CO_2 will be present in the atmosphere for a very long time. In the next chapter, we will explain why this time scale is so long.

Let us now summarize this section by making the previous discussion more quantitative. In **Figure 2.3.3** we look at the outgoing radiation from the earth (the earthshine), as if we are sitting at the outer edge of the atmosphere and looking down at the earth. We have identified the main components that are responsible for the absorption at a given wavelength and we have also shown the wavelengths at which the earth is transmitting. Let us now increase the CO_2 level in the atmosphere. We see that at 15 μm less radiation will pass, which will increase the temperature of the surface of the earth, and the corresponding blackbody radiation spectrum will effectively shift to correspond to a higher temperature. The net result is that more energy will be emitted at shorter wavelengths and less at the longer wavelengths. These shifts would theoretically continue until the total emitted energy equals the energy received from the sun.

Water as a greenhouse gas

Water vapor is the most abundant and important greenhouse gas in the atmosphere. Why are we not more worried about the water content of the atmosphere? As most of the surface of the earth is ocean, human activity has very little, if any, direct influence on the amount of water vapor in the atmosphere. That said, there is, however, an important amplification effect. For example, if the temperature of the surface of the earth decreases, the vapor pressure of water in the atmosphere decreases, its absorption of the earth's blackbody radiation decreases, the greenhouse effect decreases, and the earth gets cooler. Similarly, if the surface temperature of the earth increases, the increased water vapor concentration will amplify the greenhouse effect. In Chapter 3 we will discuss in detail how the earth naturally regulates the surface temperature by controlling CO_2 levels, but in processes with time scales on the order of thousands of years.

Other sources contributing to greenhouse effects

In the previous section, we focused on greenhouse gasses as the main cause of the increasing temperature of the surface of the earth. We have already seen that CO_2 is not the only greenhouse gas that plays a role; other gasses also contribute.

In addition, there are other factors that are not directly related to greenhouse gasses that also have an influence on the energy balance of the planet. For example, decreasing the amount of snow on the ground decreases the amount of energy reflected from the earth. Energy not reflected is absorbed, and thus decreasing ground snow has a net effect similar to a greenhouse gas. The concept of "**radiative forcing**" is introduced to quantitate these different effects. For example, how might one compare the effects of different forms of land use on CO_2 levels? Radiative forcing is a measure of how the energy balance of the earth-atmosphere system is influenced when factors that affect climate are altered. It is given

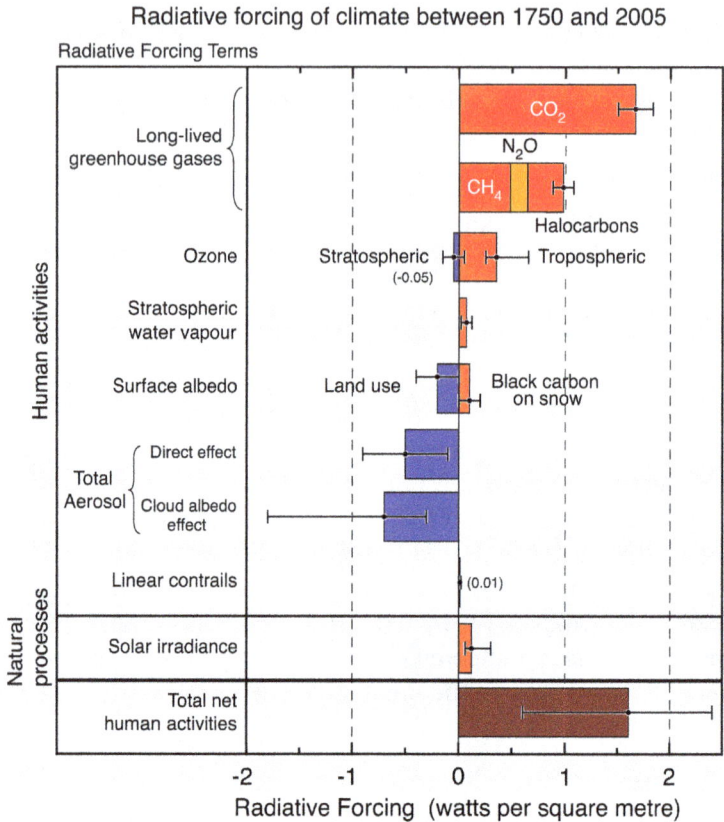

Figure 2.3.4   Radiative forcing

Summary of the principal components of the radiative forcing of climate change. All these radiative forcings are associated with human activities or natural processes. The values represent the forcings in 2005 relative to the start of the industrial era (about 1750). Human activities cause significant changes in long-lived gasses, ozone, water vapor, surface albedo, aerosols, and contrails. The only increase in natural forcing of any significance between 1750 and 2005 occurred in solar irradiance. Positive forcings lead to warming of the climate and negative forcings lead to a cooling. The thin black line attached to each colored bar represents the range of uncertainty for the respective value. *Figure from IPCC, reproduced with permission* [2.2].

as the measured rate of energy change per unit area (in W/m^2) of the globe. A positive value increases the temperature of the earth. **Figure 2.3.4** compares the radiative forcings of various human activities [2.2]. As we can see, not all human activities increase global temperature. For example, changes in land use have resulted in more energy being reflected from the surface of the earth.

Section 4

The climate

Is the climate changing? Is the climate changing because of human activity? These questions have been addressed by many scientists and have been documented in many reports. The answers are "yes" and "yes." Here we discuss some of the relevant data, mainly to illustrate that the notion of climate change is not based on a single observation, but rather on many independent measurements that point consistently in the same direction.

Climate and temperature

Before we do this, we need to define climate. Climate is the long-term average of the weather (including temperature, cloud cover, rainfall, drought, etc.). Weather would suggest, for example, that on some days it is cooler in Berkeley in California than it is in Anchorage in Alaska. Climate tells us that Anchorage is colder than Berkeley. Public opinion tends to be easily swayed by the weather, making discussion of climate change problematic for scientists and policy makers. Because the weather fluctuates significantly, it is impossible to associate a single weather event with climate change. A good illustration of this is given in **Figure 2.4.1** from the National Oceanic and Atmospheric Administration (NOAA) [2.3]. The year 2012 was the eighth warmest on the record, yet **Figure 2.4.1** shows that in many places on the planet the average temperature in 2012 was actually below the 30-year average. **Figure 2.4.2** shows selected significant climate anomalies and events in October 2012 [2.3].

The most famous curve indicating climate change is the "**hockey stick curve**" from the 2001 IPCC report [2.4], which is shown in **Figure 2.4.3.** This figure shows the deviation from the 1961–1990 average temperature on the Northern hemisphere. The importance of the

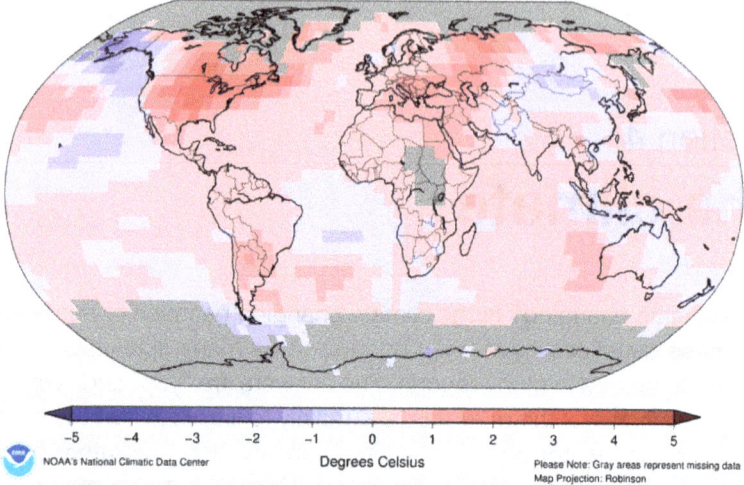

Land & Ocean Temperature Anomalies Jan–Nov 2012
(with respect to a 1981–2010 base period)
Data Source: GHCN-M version 3.2.0 & ERSST version 3b

NOAA's National Climatic Data Center   Degrees Celsius   Please Note: Gray areas represent missing data
Map Projection: Robinson

Figure 2.4.1 Temperature anomalies for the period Jan–Nov 2012

Deviations of average global temperature across land and ocean surfaces for January–November compared to a 30-year average (1981–2010). This period was the eighth warmest on record, at 0.59°C (1.06°F) above the 20th century average. 2012 surpassed 2011 as the warmest La Niña year since at least 1950, according to NOAA's Climate Prediction Center. *Figure from National Climatic Data Center* [2.3].

hockey stick is the suggestion that our current temperatures are outside the "normal" range. The fluctuations are large, in particular as the older temperature data are not as reliable as the data we can obtain now, creating significant uncertainty in the data. This uncertainty raises the question of whether the observed heating of the planet is real. And, if it is real, what is the cause? The scientific reaction to uncertainty is, of course, to ask if we can find other data that would support or contradict this hypothesis. At this point, it is important to mention that there has historically been a large scientific debate about whether climate change is real or not. The outcome of this debate has been a large number of different experiments all aimed at testing the hypothesis of global warming. As a result of these tests, the scientific consensus is now that climate change is real and that it is caused by human activity. Let's look at some of these tests and data and how they have borne out these hypotheses.

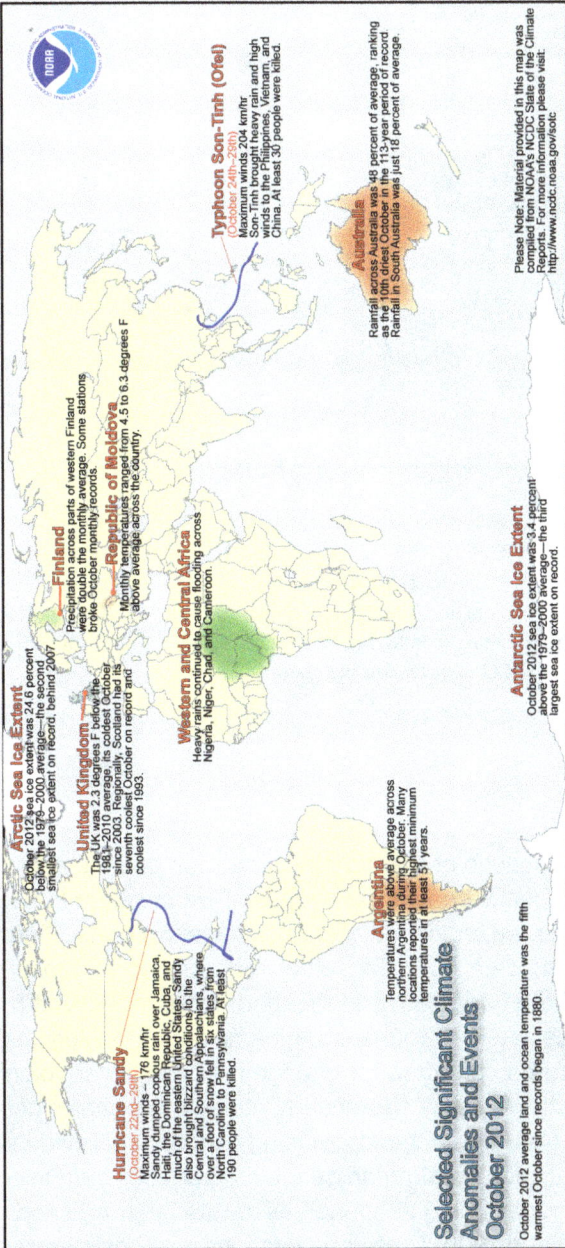

Figure 2.4.2 Selected significant climate anomalies and events in October 2012

Western parts of Finland experienced precipitation totals that were double the October monthly average. Some stations broke their all-time highest monthly precipitation records for October. Hurricane Sandy dumped copious rain over Jamaica, Haiti, the Dominican Republic, Cuba, and much of the eastern United States. Sandy also brought blizzard conditions to the central and southern Appalachians, shattering the all-time USA October monthly and single storm snowfall records. October was dry across Australia, with the continent experiencing rainfall that was 48% of average for the month. This was the 10th driest October since precipitation records were begun in 1900. South Australia experienced the fifth driest month on record, reporting just 18% of average rainfall. *Figure from National Climatic Data Center [2.3].*

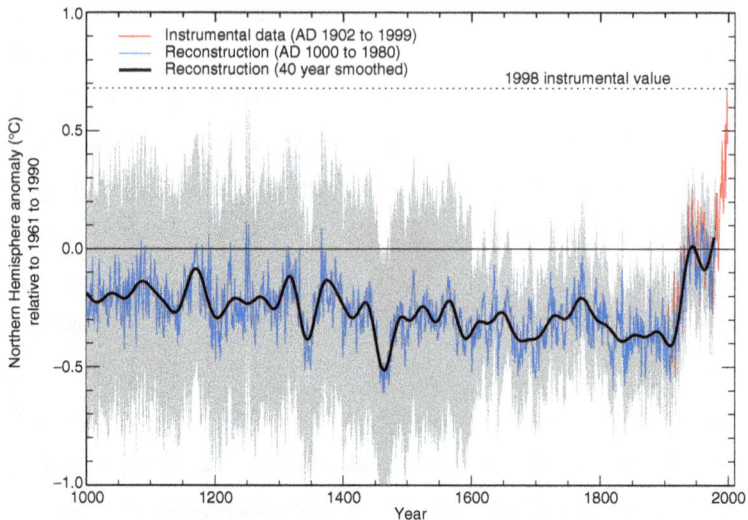

Figure 2.4.3   Average temperature on the Northern Hemisphere

Millennial Northern Hemisphere (NH) temperature reconstruction (blue — tree rings, corals, ice cores, and historical records) and instrumental data (red) from AD 1000 to 1999. A smoother version of the NH series (black) and two standard error limits (shaded in gray) are shown. *Figure from IPCC, reproduced with permission* [2.4].

Oceans

If the earth is heating, because of the thermal expansion and the melting of ice, the level of the oceans should rise. Hence, if we measure the sea level as a function of time we should be able to observe this trend. **Figure 2.4.4** shows two sets of data [2.2]. The older data are from measurements of the height of the tide relative to the land. More recently, it has become possible to use satellites to measure the height of the sea level. These tide data need to be corrected for change in ocean water volume and land motion, which increases the uncertainties in the earlier data. The conclusion of these studies is that convincing experimental evidence supports the claim that the global mean sea level gradually rose in the 20th century and is currently rising at an increased rate, after a period of little change between 0 and 1900. These observations are consistent with our expectations on the basis of the global temperatures.

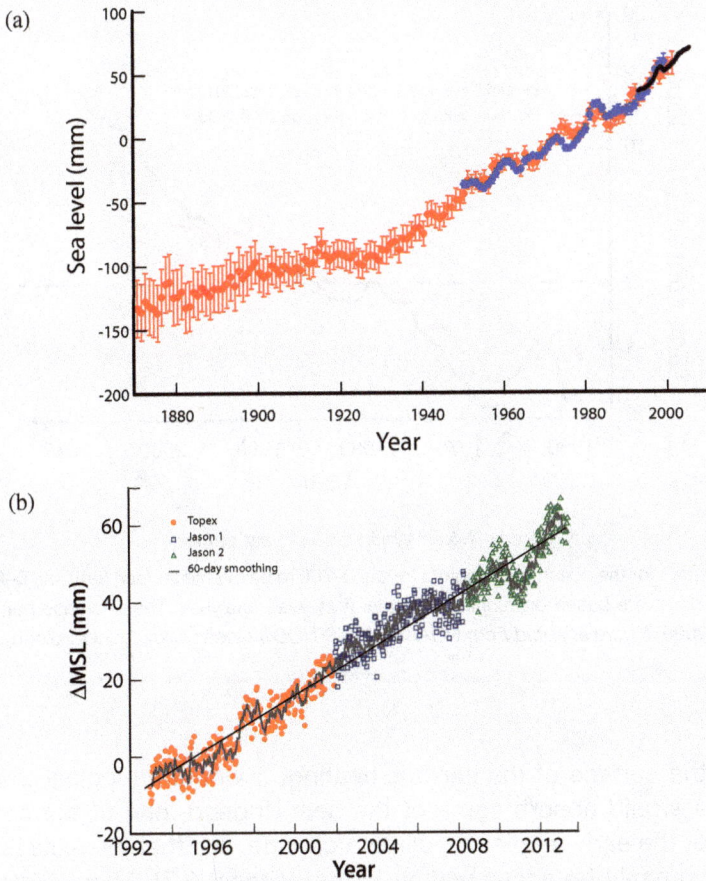

Figure 2.4.4 Global mean sea levels

(a) Annual deviations from averages of the global mean sea level (mm). The red curve shows reconstructed sea level fields since 1870; the blue curve shows coastal tide gauge measurements since 1950, and the black curve is based on satellite altimetry. *Figure from IPCC, reproduced with permission* [2.2].

(b) Variations in global mean sea level (differences from the mean, 1993 to mid-2001) computed from satellite altimetry from August 1992 to August 2012, averaged over 65°S to 65°N. Symbols are ten-day estimates from three different satellites (green, blue and orange). The Jason-2 altimeter mission was launched in June 2008, extending the record of precise altimetry initiated by the TOPEX/Poseidon (T/P) mission (August 1992) and later extended by the Jason-1 mission (December 2001). *Figure from CU Sea Level Research Group, University of Colorado* [2.5].

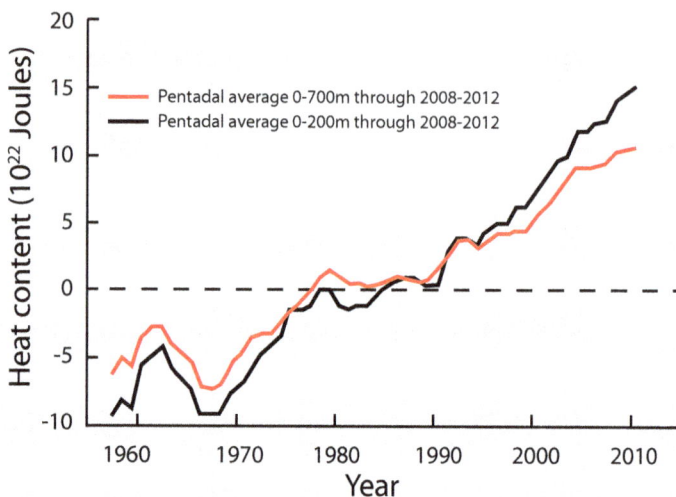

Figure 2.4.5 World Ocean heat content

Time series for the ocean heat content for the 0–700 m below the surface (red) and 0–2,000 m (black) layers based on running pentadal (five-year) analyses. The reference period is 1955–2006. *Figure adapted from NOAA/NESDIS/NODC Ocean Climate Laboratory.*

If the surface of the earth is heating, it would be logical that the oceans would absorb some of the heat. Indeed, one of the conclusions of the early studies on climate change was that the excess heat must primarily be accumulating in the oceans [2.7]. This initiated an experimental effort to measure the global sea heat content. The scientific challenge is to find the historical temperatures of the subsurface of the oceans [2.8]. **Figure 2.4.5** shows some recent experimental data [2.6].

Figure 2.4.5 shows that the heat content of the ocean is steadily increasing. Consistent with global atmospheric warming, most of the heat is stored in the upper 700 m and slowly mixes to greater depths. If we assume a linear increase with a rate of 0.43×10^{22} J per year in the 1955–2010 period, the total increase in the heat content is 24×10^{22} J and the mean increase in temperature is 0.09°C. **Figure 2.4.6** shows that all oceans contributed to this heating.

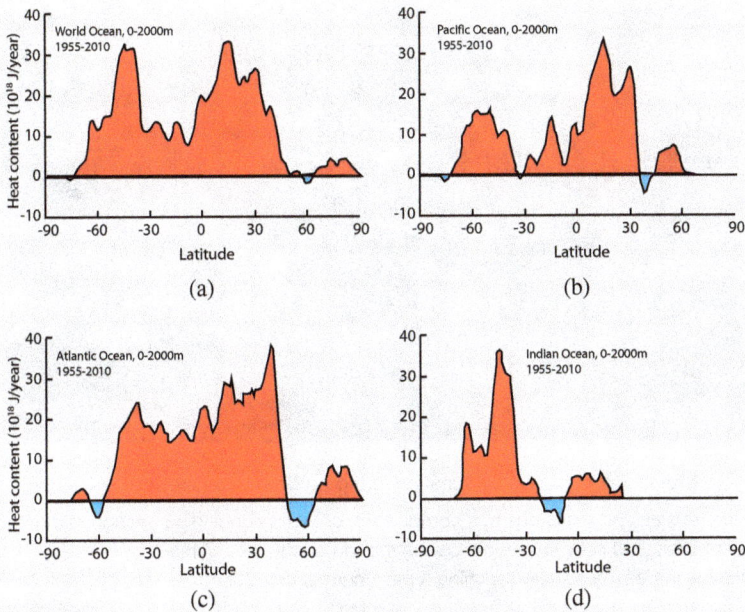

Figure 2.4.6 Heat content of the different oceans

Linear trend (1955–1959) to (2006–2010) in zonally integrated ocean heat content for the World Ocean and individual ocean basins as a function of latitude for the 0–2,000 m layer. Red indicates a positive trend and blue a negative trend. *Figure reproduced with permission from Levitus et al.* [2.6].

Another effect of increasing temperature is the melting of ice. **Movie 2.4.1** shows that ice caps go through freeze and melt cycles every year [2.9]. However, a warming climate would mean that the ice retreats further each summer. By comparing historical measurements of the extent of the ice caps, we can obtain an independent confirmation of climate change. **Figure 2.4.7** shows the amount of ice in the Arctic Sea. Indeed, the sea ice minimum summertime extent was reached in September 2012 [2.10]. **Figure 2.4.8** shows a representative selection of glacier length records from different parts of the world. The figure shows a worldwide retreat of ice. **Figure 2.4.9** summarizes the experimental data on the amount of ice at both poles. Observations consistently show a global-scale decline of snow and ice over many years.

Movie 2.4.1 Seasonal variation in the Arctic sea ice area 1978–2006

Sea ice freezes and melts due to a combination of factors, including the age of the ice, air temperatures, and solar insolation. During the winter, the area of the Arctic Ocean covered by sea ice increases, usually reaching a maximum extent during the month of March. The area covered by sea ice then decreases, reaching its minimum extent in September most years. The core of the ice cap is the perennial ice, which survives the summer because it is sufficiently thick. However, because it has been thinning year after year, the perennial ice has now become vulnerable to melt. The disappearing older ice gets replaced in winter with thinner seasonal ice that usually melts completely in the summer. *Movie from Arctic Climate Research Center at UIUC, with permission [2.9]. This movie can be viewed at:* http://www.worldscientific.com/worldscibooks/10.1142/p911#t=suppl

Extreme weather

For many years the conventional wisdom was that a single hot summer does not provide evidence for climate change; the weather fluctuates and occasional very cold winters or hot summers are exactly the type of fluctuations one could expect. However, one of the expected outcomes of climate change is that the probability of extreme weather is going to increase. **Figure 2.4.10** shows some of the recent climate events in the USA. Individually, each of them could be seen as bad luck of the climate, or better, weather dice. In a recent study Hansen and co-workers quantified the probability of bad luck [2.11]. As a reference, these authors used the period 1951–1980. During these three decades the average temperature was relatively constant. The authors used the data that were collected of the temperatures all over the earth during this period to calibrate

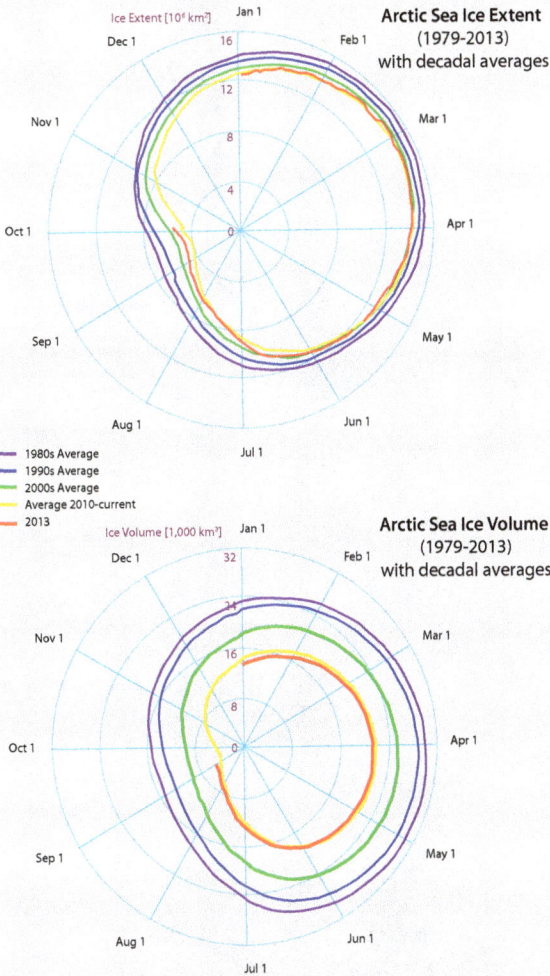

Figure 2.4.7 The Arctic Sea's ice extent and volume, 1979–2013

The top figure gives the sea ice extent and the bottom figure its volume. The sea ice minimum summertime extent, which is normally reached in September, has been decreasing over the last three decades as Arctic ocean and air temperatures have increased. The frozen cap of the Arctic Ocean appears to have reached its annual summertime minimum extent and broken a new record low on September 16, 2013. 2013's minimum extent is approximately half the size of the average extent from 1979 to 2000. 2013's minimum extent also marks the first time Arctic sea ice has dipped below 4 million km^2. *Figures redrawn from Jim Pettit* [2.10].

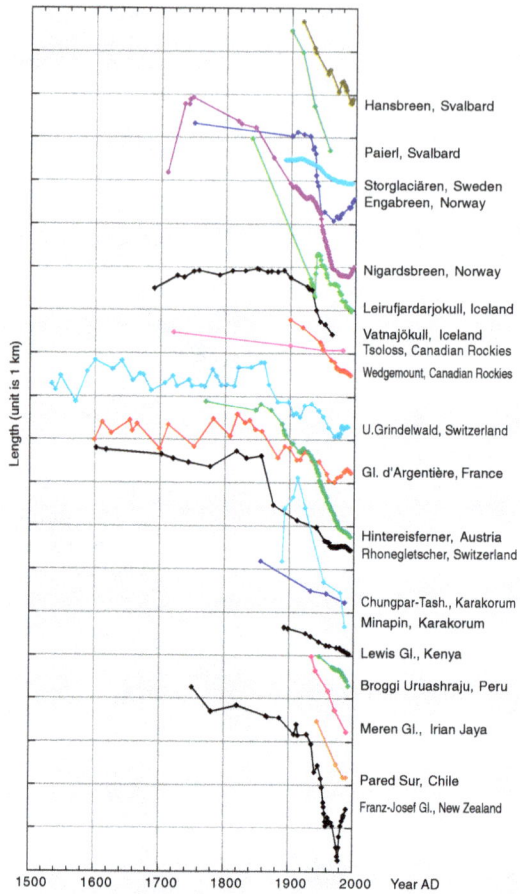

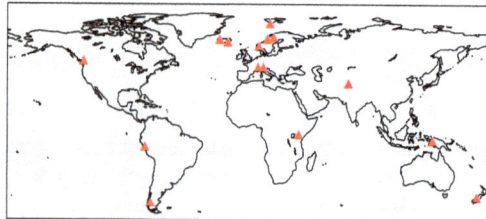

Figure 2.4.8 Length of glaciers

A graphical depiction of the length of twenty different glaciers from different parts of the world. Curves are offset vertically for clarity. The geographical distribution of the data is shown in the bottom figure. *Figure from IPCC, reproduced with permission* [2.4].

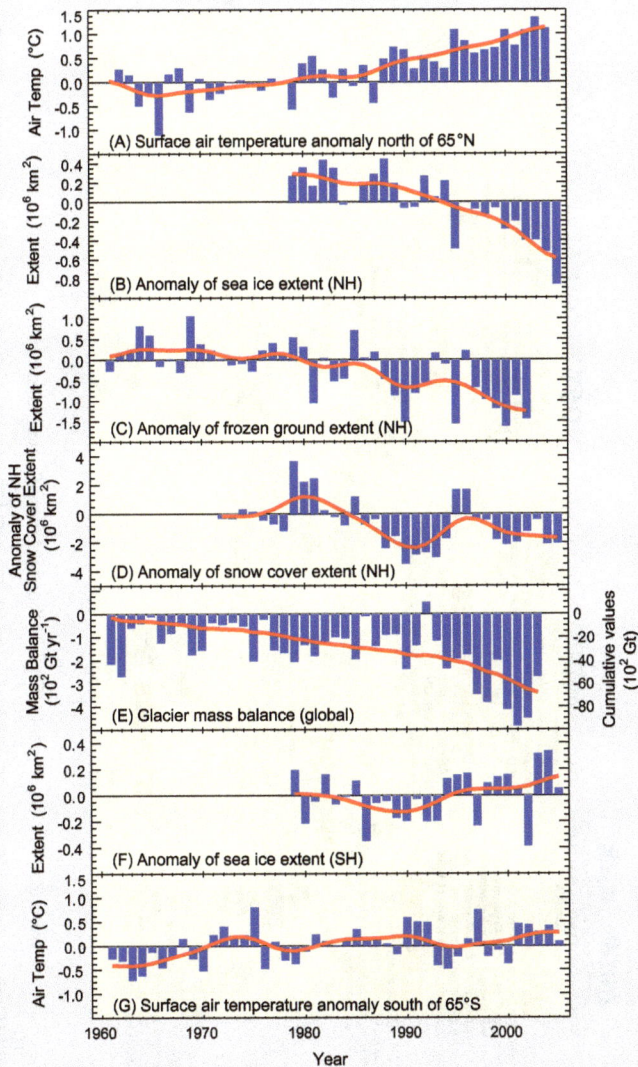

Figure 2.4.9 Ice extent on the North and South Poles

Deviations from the long-term averages of polar surface air temperature (A, G), arctic and antarctic sea ice extent (B, F), Northern Hemisphere (NH) frozen ground extent (C), NH snow cover extent (D) and global glacier mass balance (E). The solid red line in E denotes the cumulative global glacier mass balance; in the other panels it shows decadal variations. *Figure from IPCC, reproduced with permission* [2.2].

Preliminary Significant U.S. Weather and Climate Events for 2012

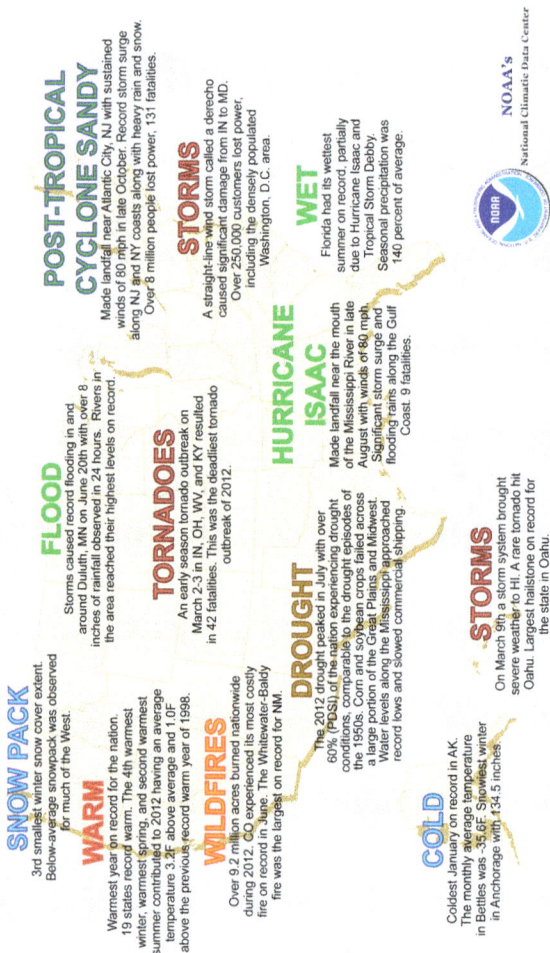

SNOW PACK
3rd smallest winter snow cover extent. Below-average snowpack was observed for much of the West.

WARM
Warmest year on record for the nation. 19 states record warm. The 4th warmest winter, warmest spring, and second warmest summer contributed to 2012 having an average temperature 3.2F above average and 1.0F above the previous record warm year of 1998.

WILDFIRES
Over 9.2 million acres burned nationwide during 2012. CO experienced its most costly fire on record in June. The Whitewater-Baldy fire was the largest on record for NM.

COLD
Coldest January on record in AK. The monthly average temperature in Bettles was -35.6F. Snowless winter in Anchorage with 134.5 inches.

FLOOD
Storms caused record flooding in and around Duluth, MN on June 20th with over 8 inches of rainfall observed in 24 hours. Rivers in the area reached their highest levels on record.

TORNADOES
An early season tornado outbreak on March 2-3 in IN, OH, WV, and KY resulted in 42 fatalities. This was the deadliest tornado outbreak of 2012.

DROUGHT
The 2012 drought peaked in July with over 60% (PDSI) of the nation experiencing drought conditions, comparable to the drought episodes of the 1950s. Corn and soybean crops failed across a large portion of the Great Plains and Midwest. Water levels along the Mississippi approached record lows and slowed commercial shipping.

STORMS
On March 9th a storm system brought severe weather to HI. A rare tornado hit Oahu. Largest hailstone on record for the state in Oahu.

POST-TROPICAL CYCLONE SANDY
Made landfall near Atlantic City, NJ with sustained winds of 80 mph in late October. Record storm surge along NJ and NY coasts along with heavy rain and snow. Over 8 million people lost power. 131 fatalities.

STORMS
A straight-line wind storm called a derecho caused significant damage from IN to MD. Over 250,000 customers lost power, including the densely populated Washington, D.C. area.

HURRICANE ISAAC
Made landfall near the mouth of the Mississippi River in late August with winds of 80 mph. Significant storm surge and flooding rains along the Gulf Coast. 9 fatalities.

WET
Florida had its wettest summer on record, partially due to Hurricane Isaac and Tropical Storm Debby. Seasonal precipitation was 140 percent of average.

NOAA's National Climatic Data Center

Figure 2.4.10   Significant USA weather and climate events for 2012

2012 marked the warmest year on record for the contiguous United States with the year consisting of a record warm spring, second warmest summer, fourth warmest winter and a warmer-than-average autumn. The average temperature for 2012 was 55.3F, 3.2F above the 20th century average, and 1.0F above the average for 1998, the previous warmest year. *Figure from NOAA's National Climatic Data Center* [2.12].

the climate dice; i.e., they assumed that the deviations of the temperature have a Gaussian distribution. The width of this distribution represents the likelihood that at a given place on earth the average temperature during a period of the year deviates from the average. On the basis of the 1951–1980 statistics one can estimate the fraction of the earth's surface area that should experience extreme weather at any given time. For example, for 2006–2011, the climate dice predicts that an average temperature that deviates more than three standard deviations from the 30-year average should affect 0.1–0.4% of the surface of the earth. The data from 2006–2011 in **Figure 2.4.11** show that 4–13% of the planet had such extreme weather. This is an order of magnitude more than one could expect from a normal climate dice. Hence, while an individual heat wave can be ignored as a natural fluctuation, the total number of anomalies in the climate cannot.

Carbon dioxide

A key question is, of course, whether these observations of global warming are related to CO_2 levels in the atmosphere. Data on CO_2 levels can also be obtained from ice core data. Ice cores represent the accumulation of snow for thousands of years (see **Figure 2.4.12**) [2.2]. With each snowfall, small gas bubbles get trapped in these cores. Analyzing the composition of the gas in the bubbles as a function of the depth of the core sample gives us detailed information about the concentration of various gasses in the air at the time these bubbles were formed. In these experiments one can also measure the $\delta^{18}O$ in shells, which is a proxy for the average temperature when these shells were formed (see **Box 2.4.1**). The data in **Figure 2.4.13** show that over the last 600,000 years the temperatures and CO_2 levels have fluctuated within a very narrow range [2.2]. One can clearly see the interglacial warm periods. In these periods CO_2 levels are higher and volumes of ice are smaller. The current greenhouse gas levels, indicated by stars, are significantly higher than the levels seen in the last 600,000 years. In **Figure 2.4.14** the concentration of greenhouse gasses in the atmosphere over the last 2,000 years is shown [2.2]. This figure emphasizes the anomaly of the current level of greenhouse gasses in the atmosphere. Current concentrations of atmospheric CO_2 and CH_4 far exceed pre-industrial values found from polar ice core records of atmospheric composition dating as far back as 650,000 years.

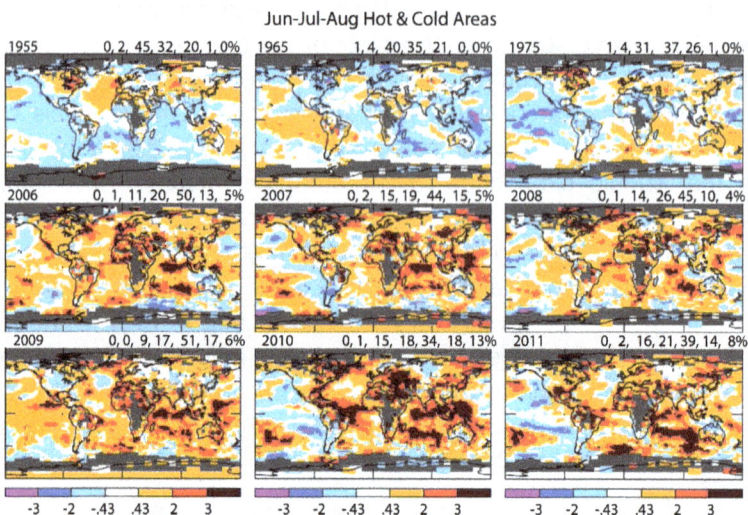

Jun-Jul-Aug Hot & Cold Areas

1955 — 0, 2, 45, 32, 20, 1, 0%	1965 — 1, 4, 40, 35, 21, 0, 0%	1975 — 1, 4, 31, 37, 26, 1, 0%
2006 — 0, 1, 11, 20, 50, 13, 5%	2007 — 0, 2, 15, 19, 44, 15, 5%	2008 — 0, 1, 14, 26, 45, 10, 4%
2009 — 0, 0, 9, 17, 51, 17, 6%	2010 — 0, 1, 15, 18, 34, 18, 13%	2011 — 0, 2, 16, 21, 39, 14, 8%

-3   -2   -.43   .43   2   3

Figure 2.4.11   Surface temperature anomalies

June–July–August surface temperature anomalies over Northern Hemisphere land in 1955, 1965, 1975, and 2006–2011 relative to a 1951–1980 base period in units of standard deviation. Numbers above each map show the percent of surface area covered by each category in the color bar. Brown and purple represent extreme warm and cold weather. *Figure from Hansen et al., with permission* [2.11].

Figure 2.4.12   Example of an ice core

Image by Guillaume Dargaud, with permission.

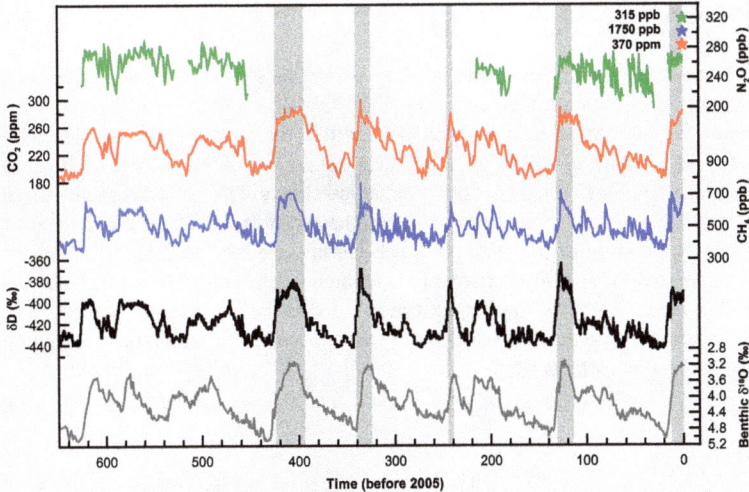

Figure 2.4.13 Ice core measurement

Variations (in thousand years) of deuterium (δD; black), a proxy for local temperature; $\delta^{18}O$ marine records (dark grey), a proxy for global ice volume fluctuations; and the atmospheric concentrations of the greenhouse gasses CO_2 (red), CH_4 (blue), and nitrous oxide (N_2O; green). Data derived from air trapped within ice cores from Antarctica and from recent atmospheric measurements. The shading indicates the last interglacial warm periods. Downward trends in the benthic $\delta^{18}O$ curve reflect increasing ice volumes on land. The stars and labels indicate atmospheric concentrations in 2000. *Figure from IPCC, reproduced with permission* [2.2].

Is this excess of CO_2 caused by human activity? This question has also been studied extensively, and several methods demonstrate that the answer is "yes." For example, in the atmosphere the ratio of heavy to light carbon isotopes (Carbon-13 versus Carbon-12) is different from that of fossil fuels. Why? Fossil fuels come from plant matter, and the chemistry of photosynthesis in plants results in a lower efficiency of ^{13}C incorporation into the plant compared to ^{12}C. Thus our fossil fuels are slightly depleted of ^{13}C. When we combust fossil fuels, then, we should reduce the $^{13}C:^{12}C$ ratio in the atmosphere. Indeed, experiments have shown that the decline of atmospheric ^{13}C levels is perfectly correlated with the increase in CO_2 levels (see **Figure 2.4.15**).

Another example of how we know that the current increase in atmospheric CO_2 is from anthropogenic activity comes from analyses of the

Box 2.4.1 $\delta^{18}O$ and δD

$\delta^{18}O$ analysis is a technique used in geochemistry as a proxy for paleotemperatures. This technique involves measuring the $^{18}O{:}^{16}O$ ratio found, for example, in corals, foraminifera and ice cores. It relies on the fact that water with ^{18}O has slightly different properties than water with ^{16}O. A practical consequence is that the $\delta^{18}O$ value of seawater correlates with the extent of polar ice sheets. Another consequence is that the foraminifera shells of calcium carbonate have different $^{18}O{:}^{16}O$ ratios depending on the temperature of the water in which shells were formed. Similar observations have been made for H (hydrogen) and D (deuterium), which is the basis determining the D:H ratio. Scientists throughout the world employ enormously sensitive instruments and considerable perspicacity in using isotopes to understand temperature changes over geological ages.

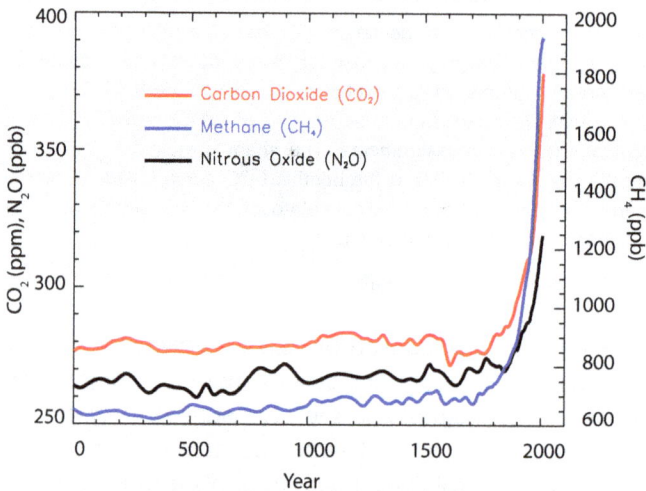

Figure 2.4.14 Concentration of the most important greenhouse gasses

Atmospheric concentrations of the important greenhouse gasses over the last 2,000 years, in units of parts per million (ppm) or parts per billion (ppb), indicate the number of molecules of the greenhouse gas per million or billion air molecules, respectively, in an atmospheric sample. *Figure from IPCC, reproduced with permission* [2.2].

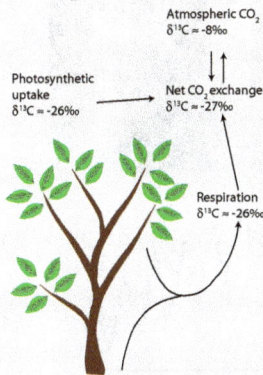

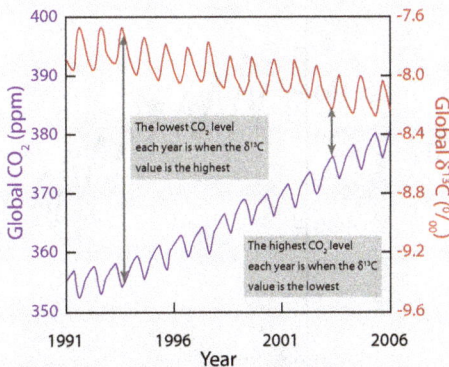

(a)   (b)

Figure 2.4.15   Carbon-13 versus Carbon-12

(a) Plant photosynthesis discriminates against ^{13}C. Plant carbon tends to have less ^{13}C than the CO_2 from which it is formed — fossil fuels are ^{13}C depleted!

(b) Experimental data of CO_2 levels correlated with the Carbon-13 ratio of the CO_2. Because of the combustion of Carbon-12 richer material, this ratio is decreasing. *Figure redrawn from NOAA Earth System Research Laboratory.*

oxygen content of the atmosphere. If we burn fossil fuels, stoichiometry tells us that for each molecule of CO_2 produced, we consume a molecule of O_2:

$$CH_n + \left(1 + \frac{n}{4}\right)O_2 \rightarrow CO_2 + \frac{n}{2}H_2O$$

If CO_2 is increasing, we should therefore expect the amount of oxygen in the atmosphere to decline in exactly the same proportion as CO_2 produced. The experimental data in **Figure 2.4.16** indeed confirm this.

When the dust has settled

We hope to have convinced you that the climate is changing. This conclusion started with a few scientists who proposed a very bold idea. This

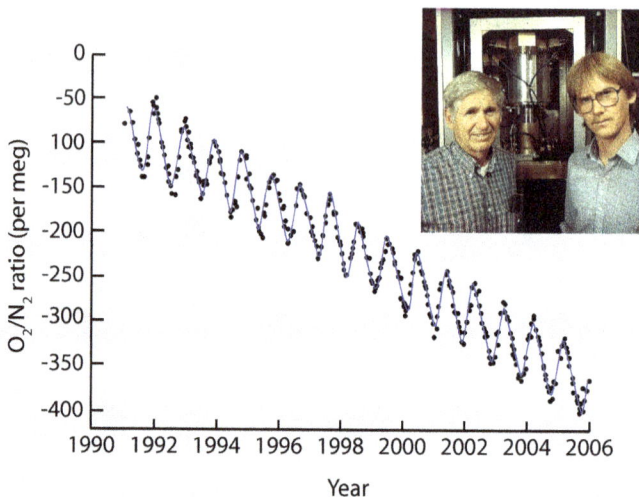

Figure 2.4.16   The second Keeling Curve: Oxygen

(inset) Charles Keeling (left) and son Ralph in 1983. A kitchen table conversation led to the younger's Keeling's interest in atmospheric oxygen measurements. *Figures from the Scripps Institution of Oceanography*.

idea was at first received very skeptically by the scientific community. The beauty of science is that this skepticism was translated into new experiments that could prove or disprove whether these bold ideas were correct. These new experiments have shown that global temperatures are increasing, the oceans are rising, glaciers are retracting, and the world's weather has become more extreme. And little by little the skepticism has been replaced by a large body of evidence that all points to the same conclusion. Whether we like it or not climate change is real, and humans are the cause of it.

Section 5

Climate models

The physics of the climate

The fundamental principles employed in modeling the earth's climate are well known. Climate models are based on the three fundamental principles that follow from Newton's laws: conservation of mass, conservation of energy, and conservation of momentum. If we consider the flow of material in and out a control volume, we can write a "mass balance," which tells us that we will accumulate mass if more material comes in than goes out. Similar balances can be written for energy and momentum. The resulting equations are commonly referred to as the **Navier-Stokes** equations. Solving the Navier-Stokes equations analytically is only possible for a very limited set of simple problems. Numerical solutions to these equations are necessary for "real-world" problems. Obtaining these numerical solutions is so sophisticated that it is a special field of science — "Computational Fluid Dynamics." This field of research has benefitted enormously from the extraordinary advances in computer power and software in the past 20 years.

The number of applications of Computational Fluid Dynamics is large. For example, if we would like to mathematically model the differences between toothpaste and water flowing through a pipe, we need to solve the Navier-Stokes equations parameterized with the properties of water and toothpaste. This field has had some remarkable successes. For example, virtually all modern airplanes are completely designed on a computer. The predictions from solutions to the Navier-Stokes equations accurately provide the details of air flow around the wing of a commercial airliner! Such details afford the design of new aircraft that are even more safe than their predecessors. These same equations are used to forecast the weather and to predict the climate.

At this point we would like to emphasize that, although the underlying physics of the flow around an airplane and the flow of the atmosphere is the same, the differences in size and complexity create enormous scientific challenges. We would like to highlight these challenges with a short historical perspective of the different climate models.

Climate models

Figure 2.5.1 illustrates the basic climate model; a spherical grid is constructed in the computer that mimics the atmosphere. The number of grid points we can use to describe our system depends on the size of the computer that we can use. In **Figure 2.5.2** the size of the grid for the different IPCC studies is shown. If we compare the first studies in 1990 with the more recent ones in 2007, the differences are striking in the amount of detail one can include [2.2]. For example, in the FAR (1990) model it

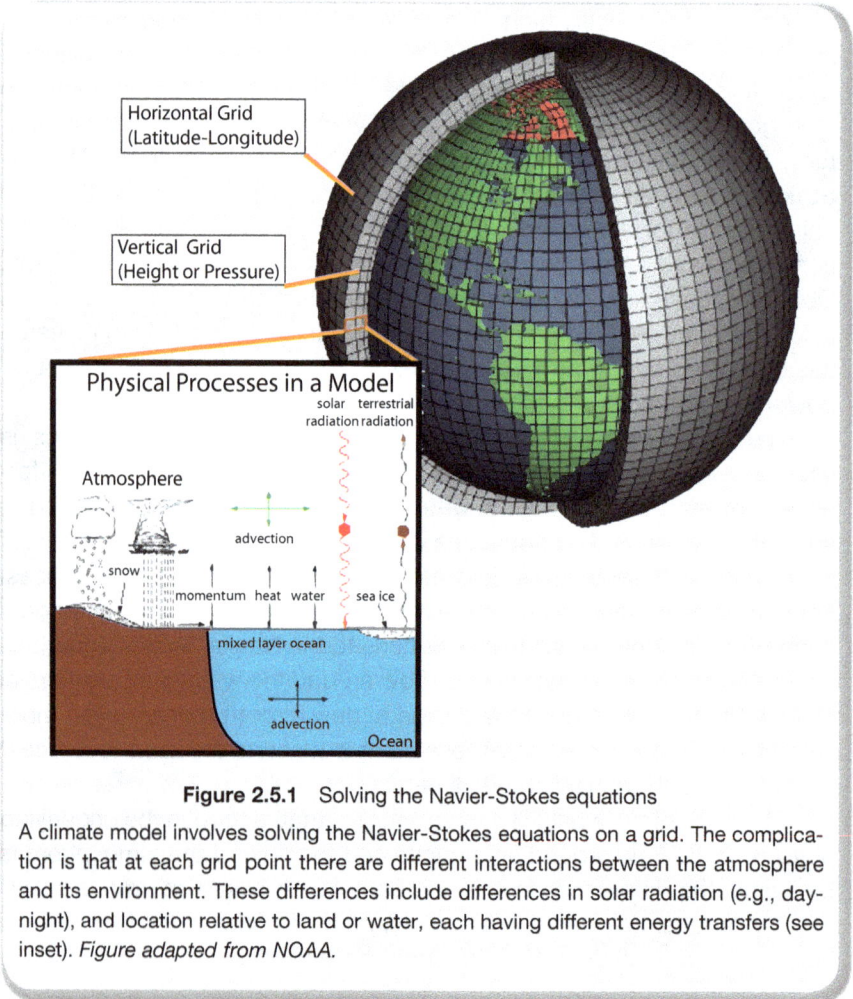

Figure 2.5.1 Solving the Navier-Stokes equations

A climate model involves solving the Navier-Stokes equations on a grid. The complication is that at each grid point there are different interactions between the atmosphere and its environment. These differences include differences in solar radiation (e.g., day-night), and location relative to land or water, each having different energy transfers (see inset). *Figure adapted from NOAA.*

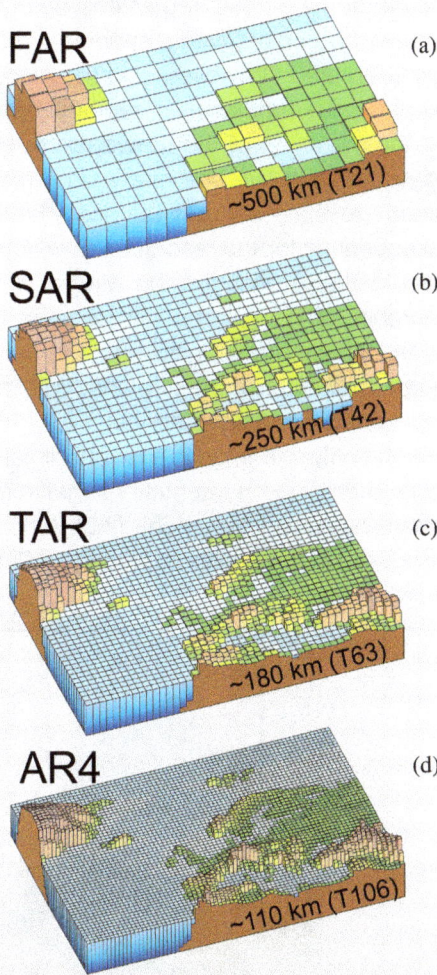

Figure 2.5.2 Geographic resolution used in the climate models

Geographic resolution used in the climate models for the different IPCC reports: FAR (IPCC, 1990), SAR (IPCC, 1996), TAR (IPCC, 2001a), and AR4 (2007). The figures show the resolution of northern Europe. The vertical resolution in both atmosphere and ocean models has increased comparably with the horizontal resolution, beginning typically with a single-layer slab ocean and ten atmospheric layers in the FAR and progressing to about thirty levels in both atmosphere and ocean. *Figure from IPCC, reproduced with permission* [2.2].

would be very difficult to recognize a particular country in Europe. The newest models for the coming IPCC report improve the resolution by a factor of 5, which gives grid points with a spacing of 50 km. With such a resolution, the prediction of air flow over, say, the Alps, improves significantly and this allows for the prediction of regional climate change.

If we compare the modeling of the flow patterns in the atmosphere with those around an airplane, we see many similarities. As mentioned before, the equations are the same but the scale is very different. This has important consequences for the resolution at which we can solve the equations. If we would like to describe the wing of an airplane, we can apply the same number of grid points that we use to describe all of Northern Europe. The spacing is not 50 km, but so small that it can describe every square millimeter of the airplane. With such a fine resolution the predictions can replace experiments.

Resolution is not the only challenge in a global climate model. One also has to take into account the various interactions that influence the climate. **Figure 2.5.3** summarizes these interactions [2.2]. Most of these interactions have been discussed in the previous section. In **Figure 2.5.4**, we see

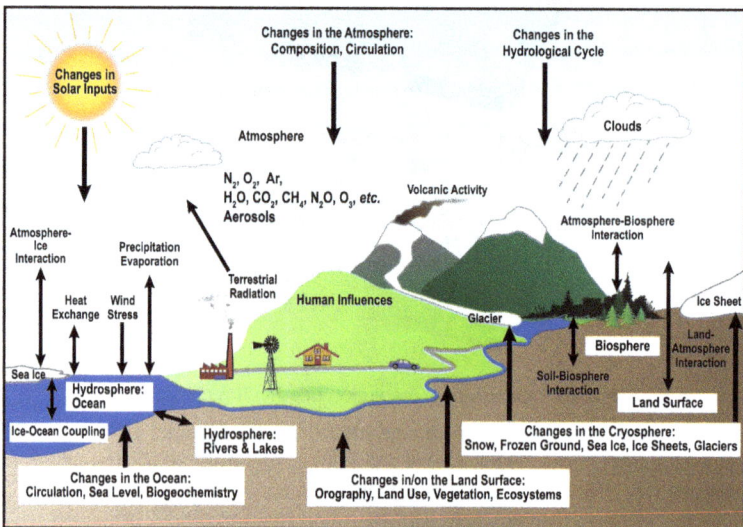

Figure 2.5.3   The climate system

Schematic view of the components of the climate system, their processes and interactions. *Figure from IPCC, reproduced with permission [2.2].*

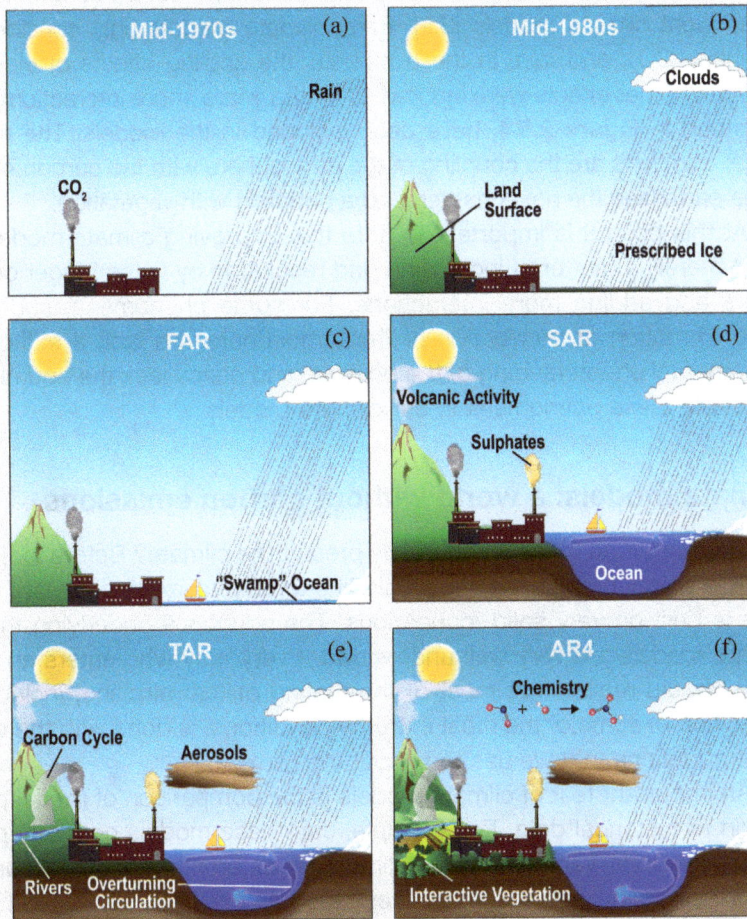

Figure 2.5.4 Complexity of the climate models

The complexity of climate models has increased over the last few decades. The additional physics incorporated in the models is shown pictorially by the different features of the modeled world: FAR (IPCC, 1990), SAR (IPCC, 1996), TAR (IPCC, 2001a), and AR4 (2007). *Figures from IPCC, reproduced with permission* [2.2].

how an increasing number of these interactions have been included in the subsequent climate models [2.2]. Early climate models only considered sunshine, CO_2, and rain. In these models, the cooling effects of clouds, and many other effects were ignored. Over the years, more interactions, as described in **Figure 2.5.4**, have been included in the models. The most recent additions are the coupling of the atmosphere with the carbon cycle of the earth (see the next chapter) and a coupling with vegetation.

At this point it is important to note that improving climate modeling is not merely a matter of increasing grid resolution by using bigger computers and adding more interactions. For some phenomena, such as cloud formation or convection of the atmosphere, we lack a sufficient fundamental understanding of the physics, and hence lack the equations to include these phenomena in our climate models.

Climate models: a world without carbon emissions

Can these climate models accurately predict the climate? Before looking at some actual predictions, we would like to emphasize that climate science is built on very solid foundations. The equations underlying these climate models are very well understood. Everybody who enters an airplane should realize that most of the testing of that airplane, including imposition of some of the most extreme conditions, is done with the very same equations that we use to predict the climate.

An important test of climate models is the comparison of model prediction to historical data. For example, can these models correctly predict the seasonal variations of the climates in different parts of the world? A more stringent test is whether these models can correctly predict the response of the climate to major perturbations. For example, major volcanic eruptions such as that of Mt. Pinatubo in 1991 increase the amount of dust in the atmosphere, which in turn leads to a faster increase in nighttime compared to daytime temperatures, a larger degree of warming in the Arctic, and small, short-term global cooling with subsequent recovery.

The results of such a comparison between climate data and models are illustrated in **Figure 2.5.5** [2.2]. The typical protocol in this field of research is for multiple computer models hosted by different groups to consider the same climate perturbation. Each of these models incorporates different computer code implementations and different assumptions.

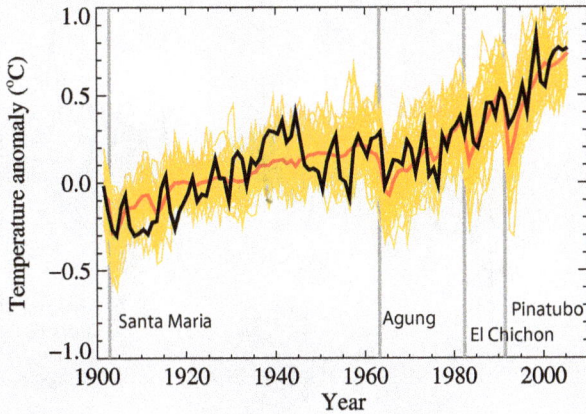

Figure 2.5.5 Temperature predictions of different climate models

Comparison between global mean surface temperature anomalies (°C) from observations and climate model simulations. The black line represents experimental data. The thin yellow lines are the results of 58 simulations produced by 14 models with both anthropogenic and natural forcings. The red line is the average of these 58 simulations. The thin vertical lines indicate major volcanic events. *Figure from IPCC, reproduced with permission* [2.2].

As a result, their predictions are slightly different. These differences are a very useful measure of the intrinsic uncertainties in the computer modeling of climate. **Figure 2.5.5** shows that the average of all models gives a surprisingly accurate prediction of the long-term increase of the average temperature as well as the drops in temperatures induced by the volcanic eruptions [2.2]. That these models give such an accurate prediction of the climate over a period of 100 years is impressive, and adds confidence to predictions based upon these models.

One of the most powerful applications of these climate models is comparison of "parallel earths." For example, we can rerun all the climate models that are used to generate **Figure 2.5.5**, but now with an earth in which no fossil fuels are used. The models depicted in **Figure 2.5.5** include the increase in CO_2 levels due to burning of fossil fuels. **Figure 2.5.6** shows the same predictions, *but now in a world in which there are no anthropogenic emissions of greenhouse gasses* [2.2]. These calculations show that, indeed, the recent temperature increase can only be explained if we include anthropogenic emissions. These calculations demonstrate that large scale CO_2 emissions influence the climate.

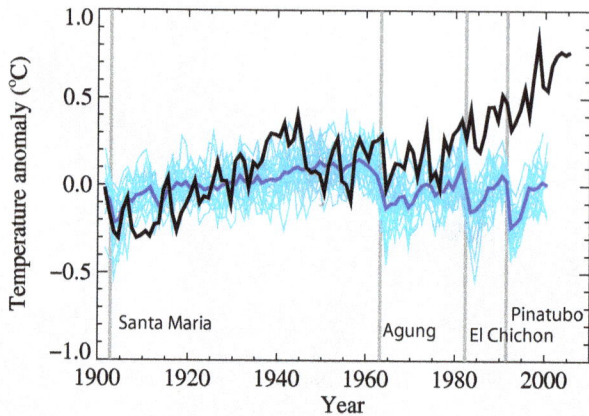

Figure 2.5.6 Temperatures on an earth without anthropogenic greenhouse gas emissions

Comparison between global mean surface temperature anomalies (°C) from observations and climate model simulations without anthropogenic CO_2 emissions. The black line represents experimental data. The thin blue lines are the results of 19 simulations produced by 5 models with only natural forcings. The thick blue line is the average of these 19 simulations. The thin vertical lines indicate volcanic events. *Figure from IPCC, reproduced with permission* [2.2].

Future emissions

Future climate predictions are predicated on models for how humans behave. In **Box 2.5.1** various scenarios of human behavior are discussed [2.13]. **Figure 2.5.7** shows the predictions of the average surface temperature for each of the scenarios described in **Box 2.5.1** [2.2]. All scenarios give a significant increase in global surface temperature. The orange line describes a scenario in which the concentration of greenhouse gasses is frozen at the 2000 level. Even in this case, the next two decades will still show a warming trend. The reason for the delay between the changes in the greenhouse gas emissions (e.g., ceasing all CO_2 emissions) and changes in climate is the "lag time" owing to the response of the oceans to changes in CO_2 levels. Similar predictions have been made for the amount of sea ice (**Figure 2.5.8**) or sea levels (**Figure 2.5.9**) [2.2]. The other important effect of climate change is the occurrence of more extreme weather patterns. In **Figure 2.5.10**, the

Box 2.5.1 Emission scenarios

To predict what the climate will look like in, say, 100 years, we have to make assumptions about future CO_2 emissions. These emissions are closely related to future energy use and possible changes in the use of fossil fuels. Future energy use (and sources of energy) is difficult to predict, so climate scientists adopt various scenarios to make climate predictions. The scenarios typically include a "business as usual scenario" where we assume that we continue to use energy exactly the same way as in the past. In a similar way, we can assume a scenario in which emissions are capped at some level that might be the result of an international treaty. The IPCC has developed a set of scenarios in their Special Report on Emissions Scenarios. These scenarios are referred to as the "SRES." The IPCC uses four different story lines of how the world might evolve:

A1 (an integrated world) is a scenario with rapid economic growth and a global population that reaches 9 billion in 2050, then gradually declines. In this world there is a quick spread of new and efficient technologies, and income and way of life converge between the different regions. Extensive social and cultural interactions are manifest worldwide. Within A1 there are three subset scenarios, each differing in their use of fossil fuels:

A1FI: emphasis on fossil fuels.
A1B: a balanced emphasis on all energy sources.
A1T: emphasis on non-fossil energy sources.

A2 (a heterogeneous world) is a scenario where nations are self-reliant and operate independently. The population continues to increase and economic developments are regional. Technological changes are slow and fragmented.

B1 (an integrated ecologically friendly world) is a scenario where the world sees the same rapid economic growth as in A1: the population will increase to 9 billion, and as in A1, it will decrease afterwards. Technological advances reduce the use of materials and result in the introduction of clean and resource-efficient technologies. The world adopts global solutions to ensure economic, social and environmental stability. This scenario does not include economic incentives to mitigate climate change.

B2 (a more divided but ecologically friendly world) is a scenario wherein the world shows intermediate economic growth and seeks local solutions to economic, social, and environmental problems. The population is continuously increasing, but at a slower rate than in A2.

The figure shows the global greenhouse gas emissions predicted in each of these scenarios.

(Continued)

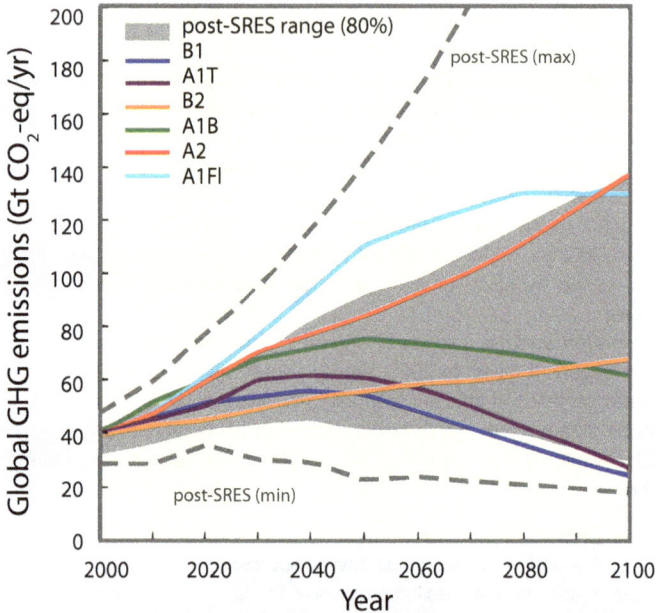

Box 2.5.1 (*Continued*)

Figure from IPCC, reproduced with permission [2.13].

predicted probabilities of excessive rains or droughts are shown for various parts of the world.

We see that climate models show that the effects of increasing CO_2 are not negligible. Many studies discuss at length the effects of these changes on vegetation and the habitats of animals. Rising sea levels will disproportionately affect humans living near oceans.

Can biology run fast enough?

One of the consequences of climate change is that plants and animals will see a changing environment. One can argue that this is nothing new, as it is exactly this adaptation to a changing environment that created current ecosystems. Indeed, we can examine historical records to see

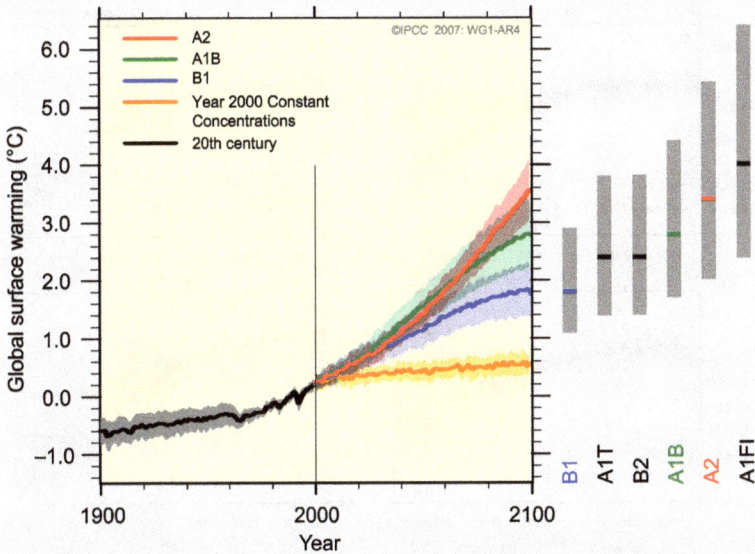

Figure 2.5.7 Predictions of global surface temperatures

Predictions of the change in the global surface temperature. The solid lines are multi-model global averages of surface warming for scenarios A2, A1B, and B1 (see **Box 2.5.1**). The shading denotes the uncertainty in the model predictions. The orange line represents a scenario in which the CO_2 concentration is held constant at its year 2000 value. The grey bars on the right indicate the best estimate (solid line within each bar) and the likely range assessed for the six scenarios. *Figure from IPCC, reproduced with permission* [2.2].

how, for example, vegetation changed during the glacial and interglacial periods.

It is interesting, however, to consider climate predictions for the increase of the average temperature in the coming years and how this might affects our ecosystems. We assume that a given biological species needs a specific temperature range to survive; and thus, when the temperature rises, they will need to move to higher latitudes or higher altitudes. One can see that the rate of climate change determines how fast these species will need to move in order to survive. How far they will need to move depends further upon the local geography: on mountain slopes a species has to move far less to reduce the temperature compared to on a flat area. If we combine a modest emission scenario (A1B, see **Box 2.5.1**)

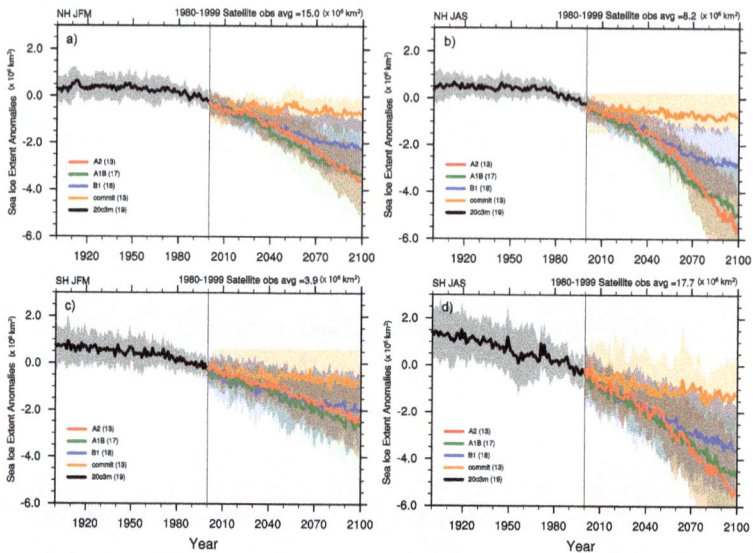

Figure 2.5.8 Predictions of the amount of sea ice

(a): The amount of sea ice in the Northern Hemisphere, January to March (JFM).

(b): The amount of sea ice in the Northern Hemisphere, July to September (JAS).

(c): The amount of sea ice in the Southern Hemisphere, January to March (JFM).

(d): The amount of sea ice in the Southern Hemisphere, July to September (JAS).

These are predictions for the scenarios of **Box 2.5.1** using different climate models. Sea ice extent is defined as the total area where sea ice concentration exceeds 15%. Anomalies are relative to the period 1980 to 2000. The number of models is given in the legend and is different for each scenario. *Figures from IPCC, reproduced with permission* [2.2].

with the local geology, we can translate the annual increase in temperature into a local velocity with which a species has to migrate to stay in a constant temperature region [2.14]. This "Velocity of Climate Change" is shown in **Figure 2.5.11**.

To put these numbers in historical perspective it is important to realize that the fastest migration that has ever been observed is 1 km/year (tree migration during the Holocene epoch). A more common

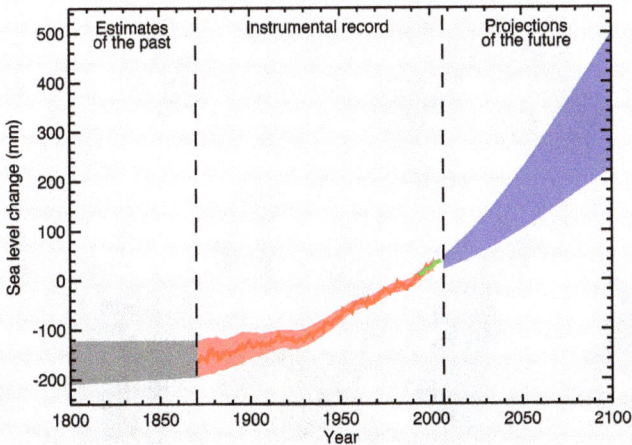

Figure 2.5.9 Sea levels: past and future

Deviation of the mean sea level from the 1980–1990 level. For the period before 1870, global measurements of sea levels are not available. The shading shows the uncertainty in the measurements or predictions. The blue shading represents the range of model projections for the A1B scenario for the 21st century. *Figure from IPCC, reproduced with permission* [2.2].

velocity is 0.1 km/year. **Figure 2.5.11** shows that 28% of the surface of the earth will experience climate changes that would require species to migrate faster than an (optimistic) maximum rate of 1 km/year. Hence, climate change produces unprecedented challenges for the ecosystems on a very large fraction of the earth.

The problem of Truth

Predictions of climate change rely on the results of very large simulations. On the basis of these predictions, we make far-reaching policy decisions, investing large sums of money in climate-related projects, and forcing all of society to make big changes in the way we use and produce energy. To warrant all the hassle, our predictions had better be right. So can we trust our models? Or, to phrase the question more dramatically: Are our climate models telling us the Truth, the whole Truth, and nothing but the Truth?

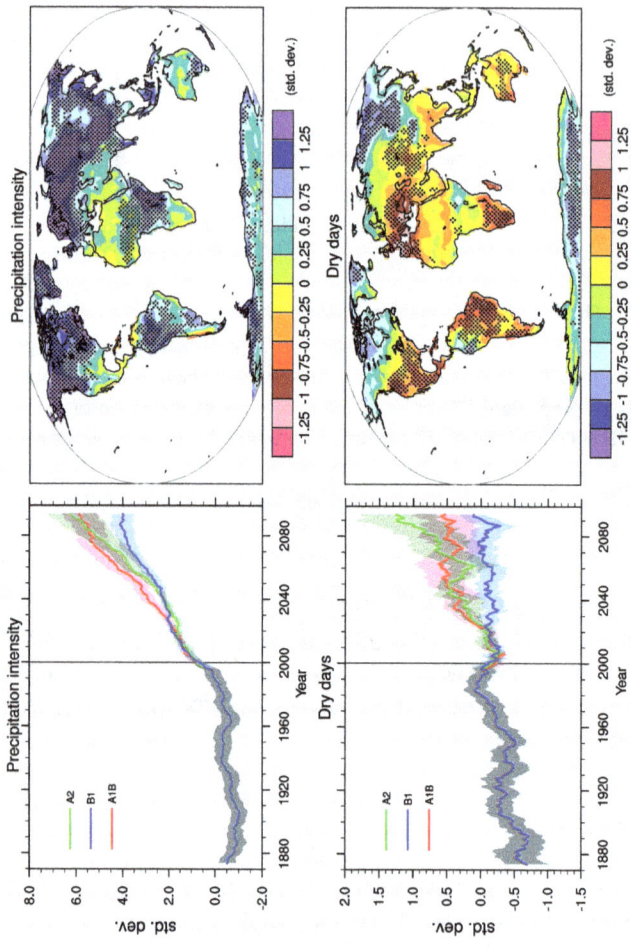

Figure 2.5.10 Changes in extreme weather

Changes in extreme weather (excessive rains or droughts) as predicted from 9 different climate models for the SRES scenarios: (a) Globally averaged changes in precipitation intensity (defined as the annual total precipitation divided by the number of wet days). (b) Changes in spatial patterns of simulated precipitation intensity between two 20-year means for the A1B scenario. (c) Globally averaged changes in dry days (defined as the annual maximum number of consecutive dry days). (d) Changes in spatial patterns of simulated dry days between two 20-year means for the A1B scenario. *Figures from IPCC, reproduced with permission* [2.2].

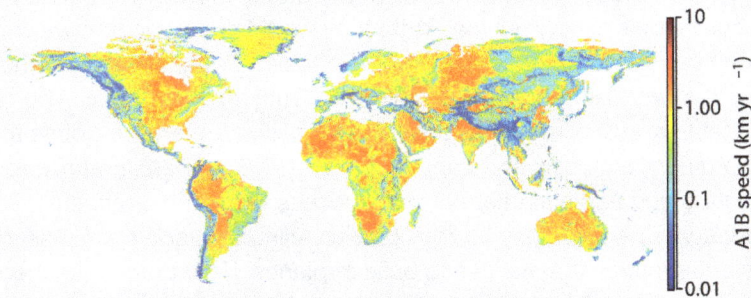

Figure 2.5.11 The velocity of climate change

The figure shows the average speed at which an ecosystem needs to move in km per year to ensure a constant temperature given the climate change predictions following the A1B scenario. *Figure by Loarie et al.* [2.14], *reproduced with permission from Macmillan Publishers.*

There are two types of answers to this question. Philosophers will tell you that the idea of a True theory is problematic to start with. Scientists will point to the various sorts of uncertainty in their models [2.15]. Both answers can be read as a version of *"We're not sure."* This might sound like grist to the mill of the climate skeptics, but it is not. To see why not, it is important to understand why *"we're not sure"* is not the same as *"our models may well be wrong, and until we're absolutely certain there's no reason to act on them."*

For philosophers, the main reason to doubt the idea of a single, True theory is this: given a set of data, there are always many different theories that are mutually incompatible, but all equally consistent with the data. If the data exhaust the available evidence, we have no evidential reason for believing one theory over another: our theory is underdetermined by the data. To complicate matters, if we find that our theory conflicts with experiment, there is no single right thing to do to make our theory more True. Because all the parts are interconnected, the experiment does not simply prove that one particular hypothesis is wrong; and because of underdetermination, there are no hypotheses that we must hold true come what may. From this point of view, making a climate model is not like building a house on the solid foundations of the laws of physics, but rather like a ship that we have to rebuild at open sea, replacing the parts with new scientific insights, and keeping it afloat as best as we can [2.16].

While philosophers would thus argue that it is preposterous to claim that our models simply tell the Truth, they would *not* argue that we should therefore dismiss science as our most reliable source of information about the world, the next best thing to Truth that we have. The main argument for this is known as the "no miracles argument." It points to the extraordinary success of science: our best theories are pretty good at predicting and explaining stuff. That success would be a miracle if those theories were completely wrong. Indeed, climate models are based on the same theories that we use to build airplanes. If we trust these theories enough to get in a plane and fly over the Atlantic, we might give some credence to their predictions on climate, too. Against that background, we might say that, if we are going to base crucial policy decisions on the predictions made by our best theories, our problem is not whether our models are absolutely True in some abstract philosophical sense: our problem is to be clear why the methods we use to accept claims are good enough to warrant the uses to which we put them [2.17].

Now the scientific sense of *"We're not sure"* becomes important. In the case of climate models, most of the underlying physics and chemistry is well understood, and we can trust that the algorithms, given the input parameters, generate the correct solutions.

However, these algorithms do require basic data on the interactions of molecules in the atmosphere, the flow of air over mountains, the effect of clouds, etc. In practice we will never have experimental data from all temperatures and conditions that occur to put into climate simulations. So these models rely on all kinds of clever schemes to estimate these input parameters. In addition, we cannot use an infinitely accurate grid (see **Figure 2.5.2**). The use of finite grid points results in a loss of information. To strengthen the no-miracles argument above: given all these assumptions and simplifications, it is impressive to see the accuracy with which climate models could predict the path of superstorm Sandy in October 2012. Is Sandy enough evidence that our current climate models give a sufficiently realistic picture of the climate that we can ask people to invest billions of dollars in CO_2 emission reductions?

Figure 2.5.12 gives an intriguing result: the uncertainty in our climate change predictions over the years has not decreased! We have tried to impress you with the fact that increased computer power has increased our resolution and complexity of climate models. The reason that this did not result in less uncertainty in the predictions is that each additional

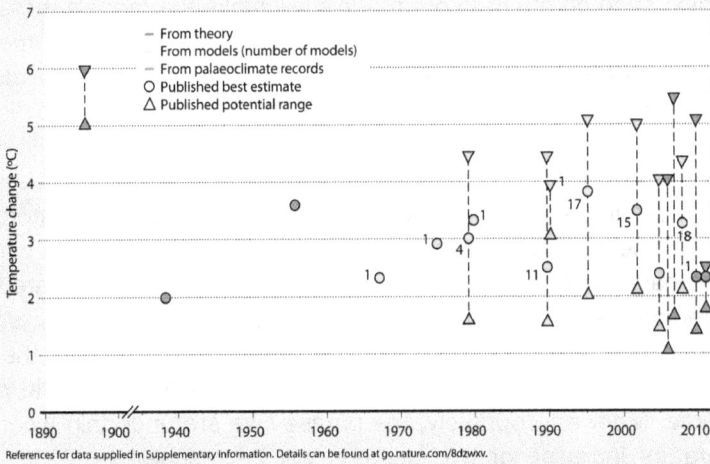

PREDICTION STABILITY

Estimates of climate sensitivity — the rise in global temperature caused by a doubling of atmospheric carbon dioxide levels — have remained fairly steady for decades.

Legend:
— From theory
From models (number of models)
— From palaeoclimate records
O Published best estimate
△ Published potential range

References for data supplied in Supplementary information. Details can be found at go.nature.com/8dzwxv.

Figure 2.5.12 Evolution of the accuracy of climate predictions

Evolution of the estimates of the change in average temperature of the earth if the CO_2 concentration is doubled (see Maslin and Austin [2.18] for details). Note the 1886 prediction derives from Svante Arrhenius' landmark 1886 work [2.19]. The remarkable feature of this figure is not the constancy of the predicted effect of CO_2 doubling. Rather, it is that despite an impressive increase in the complexity and details of the climate models, the uncertainty in the predictions has not decreased. *Figure from Maslin et al. [2.18], with permission from Macmillan Publishers.*

element to these models also requires more input parameters, resulting in new uncertainties. For example, if we describe the earth with a 500 km grid size, the interaction of the atmosphere with the surface of the earth or ocean will be lumped in effective parameters. In practice one can develop many different ways in which one lumps these parameters, and experimental data are used to calibrate these lumping schemes. If we now reduce the grid size to 50 km, we have a resolution in which we distinguish mountains, lakes, forests etc. and each has very different interactions with the atmosphere; instead of one lumped parameter, we have to incorporate all these interactions. These new insights often reveal the hidden uncertainties in the previous models. Similarly, a new generation of models might include, say, a feedback mechanism from the permafrost in Arctic regions, yet we have very little understanding of how much methane will be released from the permafrost if the temperature increases. However, the potential effect on the climate can be enormous, so adding such an effect to our models will make the models more realistic but also much more uncertain. To paraphrase the former US Secretary of Defense, Donald Rumsfeld, we could say that one aspect of climate research consists of turning "unknown unknowns" into "known unknowns."

From a public perception point of view this is a difficult message; our models have improved enormously and we have significantly more certainty that the outcomes of model predictions are accurate. Yet we continue to have uncertainties about the numerical values associated with our predictions. Communicating these uncertainties requires special attention [2.18]. The conclusion that "we are uncertain whether the temperature in 2040 will increase by two degrees" is not helpful in public discourse. More appropriately, the phrase "we are uncertain whether a temperature increase of two degrees will be reached by 2030 or by 2050," expresses confidence in the outcome, and appropriate uncertainty in the numerical estimates.

Section 6

Review

2.1. Reading self-test

1. Without an atmosphere the average temperature of the earth would be
   a. 42.8F
   b. 57.2F
   c. −2.2F
   d. 32F

2. The percentage of the radiation by the sun that is reflected by the clouds is
   a. 9%
   b. 23%
   c. 31%
   d. 41%

3. What is the lifetime of CO_2 in the atmosphere?
   a. 10 hours
   b. 1 year
   c. 10 years
   d. 500 years

4. Which sentence about radiative forcing is NOT correct?
   a. Soot on snow increases the global temperature
   b. Coal-fired power plants emit particulates in the atmosphere that enhance global warming
   c. Solar irradiance adds 0.1 W/m^2 to global warming
   d. Ozone can have a positive or negative effect on the radiative forcing

5. The statistics of the climate in the years 1951–1980 show that at any given moment in time the fraction of the total surface of the earth that has extreme weather is
   a. < 0.01%
   b. 0.1–0.5%
   c. 0.5–1%
   d. 1–5%
   e. 5–13%

6. The statistics of the climate in the years 2006–2011 show that at any given moment in time the fraction of the total surface of the earth that has extreme weather is
   a. < 0.01%
   b. 0.1–0.5%
   c. 0.5–1%
   d. 1–5%
   e. 4–13%

7. Advances of computer power have increased the resolution of a grid point in climate models
   a. from 500 km in 1990 to 180 km in 2001
   b. from 500 km in 1990 to 50 km in 2012
   c. from 5,000 km in 1990 to 100 km in 2001
   d. from 250 km in 1990 to 50 km in 2012

8. The complexity of a typical climate model in 2001 includes
   a. Sun, rain, CO_2
   b. Sun, clouds, land, ice, rain, CO_2
   c. Sun, clouds, land, ice, rain, ocean, aerosols, carbon cycle, CO_2
   d. Sun, clouds, land, ice, rain, ocean, chemistry of the atmosphere, carbon cycle, CO_2

9. What is NOT part of the A2 emission scenario?
   a. Nations are self-reliant and operate independently
   b. The population continues to increase and economic developments are regional
   c. Technological changes are fast
   d. GHG emissions in 2060 will reach 100 Gt of CO_2 a year

10. In the A2 emission scenario, the global surface temperature is predicted to increase in 2050 by
    a. 1°C
    b. 1.5°C
    c. 2°C
    d. 2.5°C

Section 7
References

1. Trenberth, K.E., J.T. Fasullo, J.T. Kiehl, 2009. Earth's annual global mean energy budget. Bull. Amer. Meteor. Soc., **90** (3), 311–324. http://dx.doi.org/10.1175/2008BAMS2634.1
2. Climate Change, 2007. The Physical Science Basis. Working Group I Contribution to the Fourth Assessment Report of the Intergovernmental Panel on Climate Change, Cambridge, UK, and NY: Cambridge University Press. (In this book, we have used the following figures: Figure 1.2; Figure 1.4; FAQ 1.2 Figure 1; FAQ 2.1 Figure 1; FAQ 2.1 Figure 2; FAQ 4.1 Figure 1; Figure 5.13; FAQ 5.1 Figure 1; Figure 6.3; Figure 9.5; Figure 10.13; Figure 10.18; Figure SPM.5.)
3. NOAA National Climatic Data Center, State of the Climate, December 2012. Global Analysis for November 2012, published online, retrieved on January 11, 2013 from http://www.ncdc.noaa.gov/sotc/global/
4. Climate Change, 2001. The Scientific Basis. Working Group I Contribution to the Third Assessment Report of the Intergovernmental Panel on Climate Change, Figure 2.18; Figure 2.20. Cambridge, UK, and NY: Cambridge University Press.
5. Nerem, R.S., D. Chambers, C. Choe, and G.T. Mitchum, 2010. Estimating mean sea level change from the TOPEX and Jason Altimeter missions. Marine Geodesy, **33** (1), 435.
6. Levitus, S., J.I. Antonov, T.P. Boyer, *et al.* 2012. World ocean heat content and thermosteric sea level change (0–2000 m), 1955–2010. Geophys. Res. Lett., **39** (10), 603. http://dx.doi.org/10.1029/2012gl051106
7. Hansen, J., M. Sato, R. Ruedy *et. al.*, 1997. Forcings and chaos in interannual to decadal climate change. J. Geophys. Res-Atmos., **102** (D22), 25679. http://dx.doi.org/10.1029/97jd01495
8. Levitus, S., J.I. Antonov, T.P. Boyer, and C. Stephens, 2000. Warming of the world ocean. Science, **287** (5461), 2225. http://dx.doi.org/10.1126/Science.287.5461.2225
9. Walsh, J.E., and W.L. Chapman, Arctic Climate Research Center at UIUC, Arctic Climate Research Center, UIUC, Sea ice animation. http://arctic.atmos.uiuc.edu/cryosphere/all.final.1978-2006.mov
10. Figures by Jim Petit; the sea ice extent data are from the Japan Aerospace Exploration Agency, and the sea ice volume data are from the Polar Science Center at the University of Washington. https://sites.google.com/site/pettitclimategraphs/sea-ice-volume

11. Hansen, J., M. Sato, and R. Ruedy, 2012. Perception of climate change. P. Natl. Acad. Sci. USA, **109** (37), E2415. http://dx.doi.org/10.1073/Pnas.1205276109

12. NOAA National Climatic Data Center, December 2012. State of the Climate: National Overview for Annual 2012, published online, retrieved on June 11, 2013 from http://www.ncdc.noaa.gov/sotc/national/2012/13

13. Climate Change, 2007. Synthesis Report. Contribution of Working Groups I, II and III to the Fourth Assessment Report of the Intergovernmental Panel on Climate Change, Figure 3.1. IPCC: Geneva, Switzerland.

14. Loarie, S.R., P.B. Duffy, H. Hamilton, G.P. Asner, C.B. Field, and D.D. Ackerly, 2009. The velocity of climate change. Nature, **462** (7276), 1052. http://dx.doi.org/10.1038/Nature08649

15. Oreskes, N., K. Shrader-Frechette, and K. Belitz, 1994. Verification, validation, and confirmation of numerical-models in the earth sciences. Science, **263** (5147), 641. http://dx.doi.org/10.1126/Science.263.5147.641

16. Neurath, O., 1932. Protokollsätze. Erkenntnis, **3**, 204.

17. Cartwright, N.D., 2010. Foreword. In *Fictions and Models: New Essays*, edited by J. Woods. Philosophia: Munich.

18. Maslin, M., and P. Austin, 2012. Climate models at their limit? Nature, **486** (7402), 183.

19. Arrhenius, S., 1896. On the influence of carbonic acid in the air upon the temperature of the ground. Phys. Mag. S., **41** (5), 237.

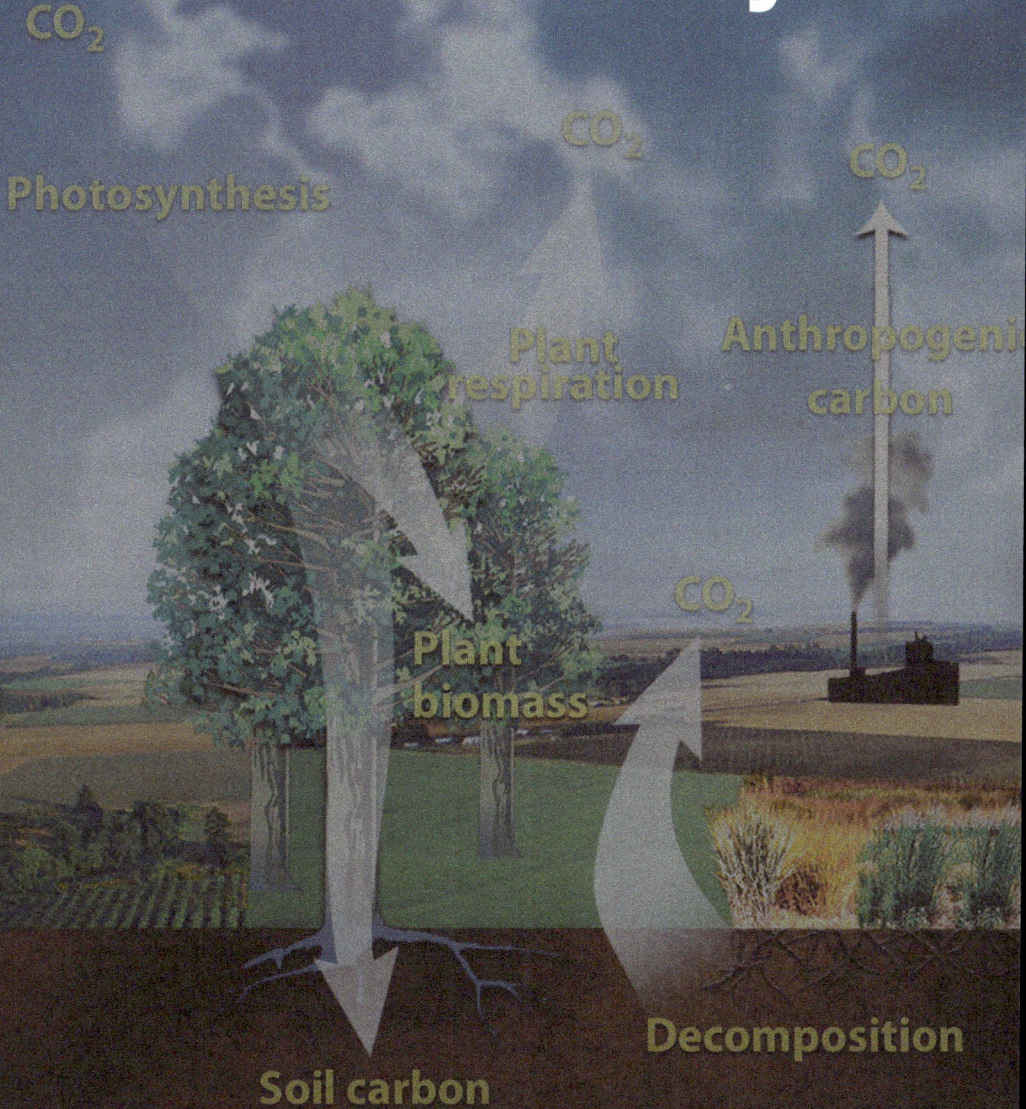

Chapter 3

The Carbon Cycle

CO_2

Photosynthesis

CO_2

CO_2

Plant
respiration

Anthropogeni
carbon

CO_2

Plant
biomass

Decomposition

Soil carbon

CO_2 in the atmosphere plays an important role in regulating the temperature of the earth. Many different mechanisms can influence the CO_2 concentration in the atmosphere. These mechanisms are part of what is commonly referred to as the carbon cycle.

Section 1

Introduction

In Chapter 2 we have seen that the CO_2 in the atmosphere plays an important role in regulating the temperature of the earth. It is therefore important to understand the biological, geological, and chemical mechanisms that control CO_2 levels in the atmosphere. In this chapter, we discuss these mechanisms.

Figure 3.1.1 illustrates the various mechanisms by which carbon is exchanged between different regions of the earth [3.1]. Photosynthesis takes CO_2 from the atmosphere and converts the carbon into biomass. The inverse process, the decomposition of biomass stored in the soils, converts the carbon of the biomass into CO_2 which is subsequently released into the atmosphere. If we look carefully at the figure, we see many different mechanisms that transfer carbon from one chemical state to another or transfer carbon from one "reservoir" to another. If we would like to understand how CO_2 concentrations in the atmosphere are regulated, we have to elucidate the interactions between the different carbon reservoirs. For example, an increased atmospheric CO_2 concentration will stimulate biosynthesis and hence increase the amount of biomass. Similarly, an increase in atmospheric CO_2 will increase the uptake of CO_2 in the oceans. Quantification of this carbon cycle should answer questions such as: how much of the anthropogenic CO_2 that is added to the atmosphere will end up in biomass or in the oceans?

To answer this and other questions we will look at the earth from a systems point of view. The idea of system analysis is to describe a cell, organism, or a chemical factory as a system of mathematical objects that interact with each other. This sounds abstract, but we will see that this approach gives us interesting insights. In the case of the carbon cycle, the first step is to quantify the different reservoirs of carbon.

In **Table 3.1.1** the amount of carbon in the various reservoirs is summarized [3.2] and in **Figure 3.1.2** these numbers are presented in a system representation [3.3]. The difference between **Table 3.1.1** and **Figure 3.1.1** is that in the latter, we did not consider the reservoir of carbon that exists more than 20 km below the surface of the earth. If we look at the numbers, we can already draw some important conclusions. It is apparent that most of the carbon on earth is in the form of carbonate

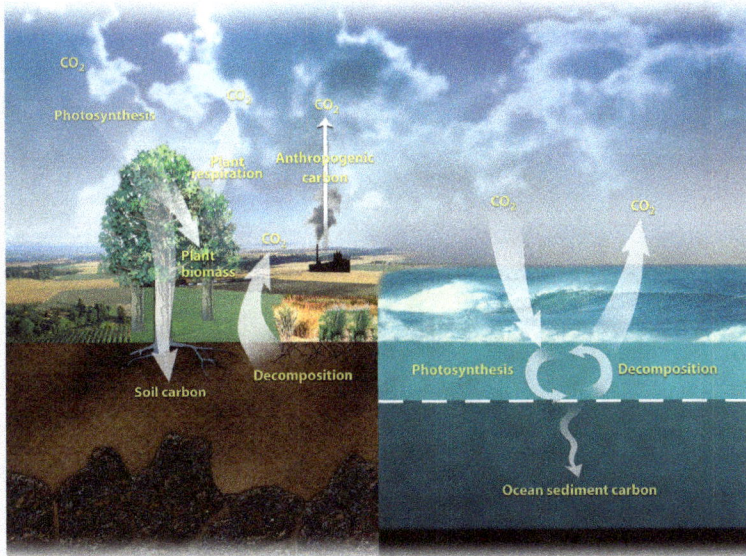

Figure 3.1.1 Part of the carbon cycle

A simplified representation of the contemporary global carbon cycle. *Figure from the US Department of Energy's Office of Science Genomic Science Program* [3.1].

minerals. These are partly found in the oceans, but the vast majority are in sediments and crust.

Our next step is to consider the various mechanisms that allow carbon to exchange between these different reservoirs. We first apply our chemical intuition. Chemists understand that in order to keep the concentration of a chemical nearly constant, they use a buffer. The most important property of buffers is that their concentration is much larger than that of the chemicals they seek to control. If we would like to buffer the CO_2 concentration in the atmosphere, we therefore need a large reservoir of carbon. **Figure 3.1.2** shows that terrestrial carbon and carbon in the surface of the ocean exist in amounts similar to the total amount of carbon in the atmosphere. And so these reservoirs do not act as a buffer. The carbon reservoirs in the deep ocean and in sediments and crust are, however, sufficiently large to act as buffers.

The next step in our system analysis is to describe the interactions between the different reservoirs. This will be the topic of the next section.

Table 3.1.1 Carbon pools in the major reservoirs

Pools	Quantity (Gt)
Atmosphere	720
Oceans	38,400
Total inorganic	37,400
Surface layer	670
Deep layer	36,730
Total organic	1,000
Lithosphere	
Sedimentary carbonates	>60,000,000
Kerogens	15,000,000
Terrestrial biosphere (total)	2,000
Living biomass	600–1,000
Dead biomass	1,200
Aquatic biosphere	1–2
Fossil fuels	4,130
Coal	3,510
Oil	230
Gas	140
Other (peat)	250

Data from Falkowski et al. [3.2].

However, before doing this we would like to jump to the main conclusion of this analysis, shown in **Figure 3.1.3** which summarizes our knowledge on the carbon cycle [3.2]. It is well known that CO_2 levels have shown very large fluctuations in the history of the earth. **Figure 3.1.3** depicts these fluctuations as a "variance spectrum" for CO_2. On the ordinate is the amplitude of the CO_2 fluctuation (in ppm of CO_2) and on the abscissa is the period of time associated with the fluctuation. The longest period of fluctuation, on the order of the age of the earth (broad "peak" near the origin of the graph) is associated with the beginning of the earth when the CO_2 levels were very large, about 6 orders of magnitude larger than the present values. This broad peak decays such that the current CO_2 levels are on the order of hundreds of ppm. In the sections that follow, we will see what geological forces are at play that cause this broad peak.

It is clear from this figure that, in addition to this very broad peak, there are well defined periods (e.g., daily, yearly, 10^5 years) on which the

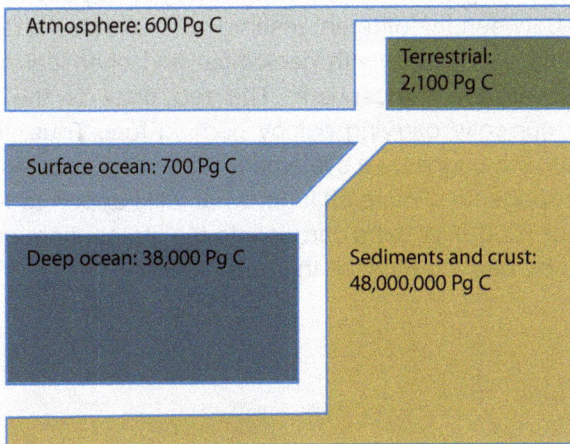

Figure 3.1.2 Carbon reservoirs

System representation of the carbon reservoirs of the earth. Pg C is petagrams (10^{15} grams) of carbon. *Figure based on data from Sigman and Boyle [3.3].*

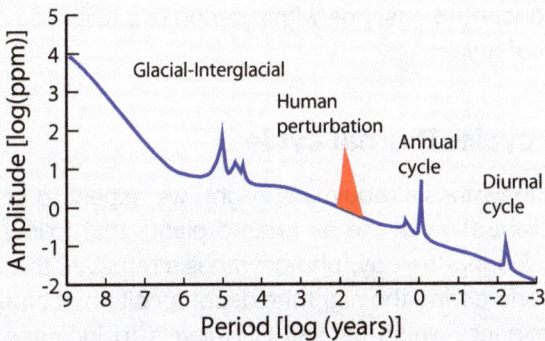

Figure 3.1.3 Fluctuation of the CO_2 concentrations in the atmosphere

Schematic variance spectrum for CO_2 over the course of the earth's history. The figure illustrates the size and time scale of the fluctuation of the CO_2 concentrations observed in the atmosphere. *Figure adapted from Falkowski et al. [3.2].*

CO_2 levels show significant fluctuations that are directly related to the interactions between the different reservoirs. These peaks represent the time constants associated with geological and chemical mechanisms affecting CO_2 in the atmosphere. The red peak is the experiment humans are currently carrying out by burning fuel. Thus, the systems analysis approach suggests a systematic way of approaching the question of anthropogenic climate change: what will happen to the CO_2 levels if we inject 1000 Gt of extra carbon into the atmosphere over a period of 200 years by burning most of the fossil fuels?

Section 2
The biological carbon cycle

Our elementary biology course informs us that green plants use photosynthesis to convert CO_2 into biomass as part of the biological carbon cycle. The natural rhythm of plant activity results in two cycles in atmospheric CO_2 concentrations, one with a period of a single day, and another with a period of a year.

Biological cycle: Diurnal cycle

Because photosynthesis requires sunlight, we expect to see changes in the CO_2 concentration of the air around plants that follow the day and night rhythm. During the day, photosynthesis removes the CO_2 from the atmosphere, and during the night the decomposition of biomass (e.g., by bacteria) continues, which causes CO_2 levels to increase. Indeed, the peak on the shortest time scale in **Figure 3.1.3** corresponds exactly to this process [3.4]. Compared to the other fluctuations, the amplitude of this cycle is relatively small. Experiments that support this observation are carried out in forests using large towers on which atmospheric CO_2 concentrations are measured at different heights (see **Figure 3.2.1**). An example of the results of these experiments is depicted in **Figure 3.2.2**.

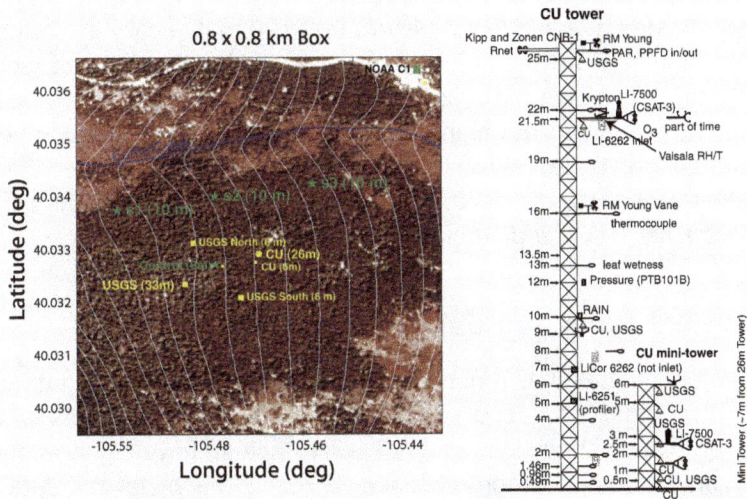

Figure 3.2.1 Biological CO_2 levels

Measurements at the Niwot Ridge AmeriFlux site on the east slope of the Rocky Mountains in Colorado (left). CU is the location of the tower illustrated on the right side, which allows for the measurement of CO_2 levels at different heights above the ground. *Figure by Sun et al. [3.4], reprinted with permission from Elsevier.*

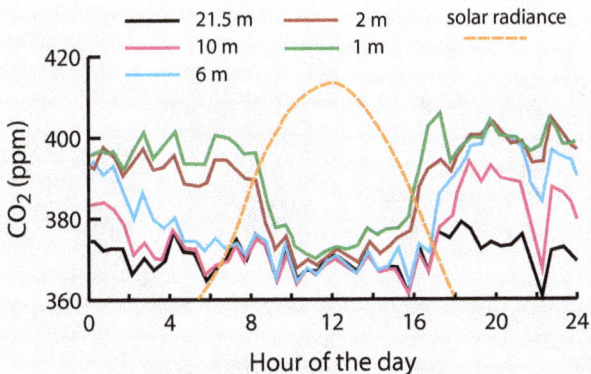

Figure 3.2.2 Daytime CO_2 levels at different heights

CO_2 levels at different heights above the ground and solar radiance as a function of the time of the day. *Figure by Sun et al. [3.4], reprinted with permission from Elsevier.*

The data give the CO_2 levels and the solar radiation as a function of the time of the day. We see that, at levels up to 20 m (typically the height of tree and forest growth) CO_2 levels decrease during the day and recover during the night. We also see that these diurnal variations become smaller as we get to the higher levels, with CO_2 concentrations nearly constant at the highest levels (21.5 m and above). This illustrates the fact that the effect of the diurnal cycle on the overall CO_2 levels in the atmosphere is small (see the diurnal peak in **Figure 3.1.3**).

Biological cycle: Annual cycle

The next fluctuation has a period of a year. The experimental data upon which this "peak" is based is demonstrated by measurements of CO_2 levels on Mauna Loa in Hawaii. We are fortunate to have highly accurate CO_2 concentrations recorded since 1970. These historical data are shown in **Figure 3.2.3** [3.5]. From this data, a steady increase of overall CO_2 levels in the atmosphere is immediately apparent. In addition to this general increase we observe a modulation with a period of a year. The data for the year 2012, shown in **Figure 3.2.3,** clearly demonstrate that during the summer CO_2 levels are lower than during winter. This is consistent with the consumption of CO_2 by growing biomass in the summer [3.5]. The size of the fluctuations is of the order of 10 ppm, or 3% of the total amount of carbon in the atmosphere.

Question 3.2.1 Hawaii is in the middle?

Those who have visited Hawaii know it is located very close to the equator. As winter and summer are inverted in the Northern and Southern hemispheres, one would expect to measure CO_2 levels that are the average of summer and winter in Hawaii. Why do the graphs in **Figure 3.2.3** show otherwise?

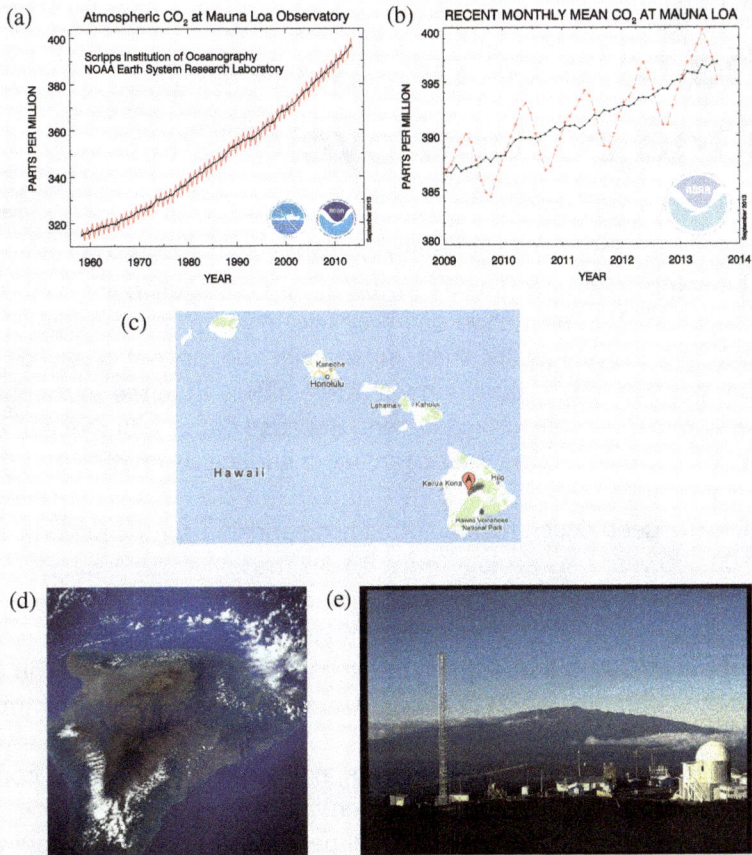

Figure 3.2.3 Atmospheric CO_2 Levels

(a) Historical CO_2 levels in the atmosphere as measured at the Mauna Loa Observatory. *Figure from Dr. Pieter Tans, NOAA/ESRL and Dr. Ralph Keeling, Scripps Institution of Oceanography* [3.5].

(b) The data show the modulation of the CO_2 levels over a year. During the summer the CO_2 levels are higher compared to the winter. *Figure from Dr. Pieter Tans, NOAA/ESRL and Dr. Ralph Keeling, Scripps Institution of Oceanography* [3.5].

(c) Location of the Mauna Loa observatory in Hawaii. *Map data from Google.*

(d) Satellite picture of the island of Hawaii. *Picture from Earth Sciences and Image Analysis, NASA-Johnson Space Center.*

(e) Mauna Loa Observatory. *Photo by NOAA/ESRL/GMD.*

Section 3

The role of oceans in the carbon cycle

The previous section described the interactions of atmospheric CO_2 with the biosphere. Carbon dioxide in the air is also exchanging with CO_2 in the ocean. This exchange can only take place at the surface of the ocean. The surface water then mixes with the deeper layers of the ocean, but this mixing occurs much more slowly than the exchange of atmospheric CO_2 with surface waters. Because of this difference in time scale, we have separated the ocean into a surface reservoir and a deep ocean reservoir.

It is the deep ocean reservoir that is responsible for the periodicity at the 100,000 year time scale (**Figure 3.1.3**). These fluctuations have been measured from air bubbles in ice cores from Antarctica (**Figures 3.3.1**) [3.6]. The ice core data show oscillations in the CO_2 levels on a time scale of ~100,000 years. We see that these fluctuations change the CO_2 levels by ~100 ppm, which is about 30% of the total CO_2 concentration in the air.

Such large changes in CO_2 levels in the atmosphere have consequences for the average temperature on the earth, and this periodicity is associated with glacial and interglacial periods on earth. The underlying mechanism for this periodic behavior has remained an important outstanding scientific question. The period that we find in the ice core data is statistically correlated to the solar cycles, and for many years the so-called Milankovitch hypothesis has been put forward as an explanation for the cyclic character of the glacial and interglacial periods and the 100,000 year period of CO_2 oscillations [3.7]. This theory posits that the annual cycle of the earth around the sun is not a perfect circle, but rather has a much more complicated pattern with characteristic periods of roughly 100,000, 41,000, and 23,000 years. Because the earth is warmer or cooler depending on how far it is from the sun, the oscillations in earth's orbit lead to a warmer and colder earth, hence growing or shrinking glaciers. This warming and cooling would also affect atmospheric CO_2 concentrations, as we discuss below. The main criticism of this

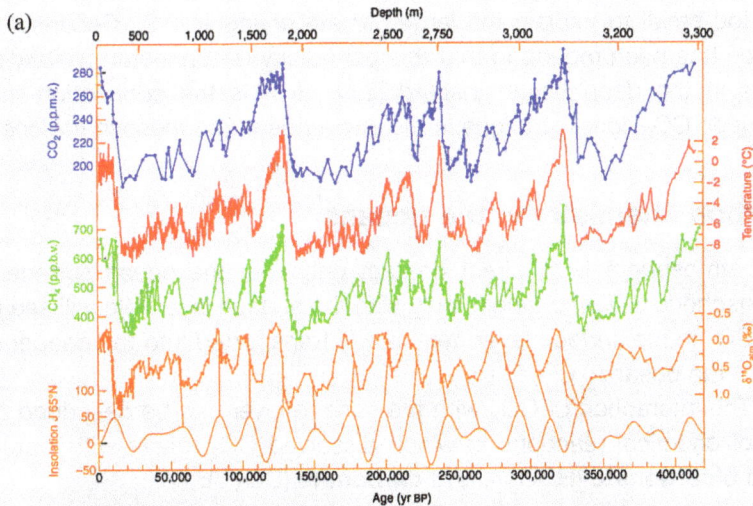

Figure 3.3.1 Vostok ice cores

(a) Ice core data spanning 420,000 years from the Vostok, Antarctica research station. Current time is at the left. From bottom to top: Solar variation at 65°N due to the Milankovitch cycles, ^{18}O isotope of oxygen, levels of methane (CH_4), relative temperature, and levels of carbon dioxide (CO_2). *Figure by Petit et al. [3.6], reproduced with permission from Macmillan Publishers.*

(b) Location of Lake Vostok in Antarctica. *Composite satellite photo from NASA.*

(c) French, Russian, and American scientists in the Vostok team photo with unprocessed ice cores. Cores coming out of the barrel are generally 4–6 m long and are cut into 1 m sections. The pictured ice columns are unprocessed cores. White containers in the background are used for transporting 1 m sections. *Photo by NASA taken at Vostok.*

theory is that the energy differences associated with the changes in orbit are too small to explain the large climate changes [3.2]. Recently, evidence has been mounting that this periodicity is connected to flow patterns in the deep ocean instead [3.8]. How is the connection made between CO_2 concentrations in the atmosphere and those in the ocean?

Carbon chemistry in the oceans

The atmosphere is in direct contact only with the ocean surface, so atmospheric CO_2 exchanges only with the surface layer. We will see that the CO_2 in the surface layer gets slowly transported into the deeper layers of the ocean.

The interaction of CO_2 with the surface layer can be described as a set of chemical reactions in which dissolved CO_2 reacts with water to form **bicarbonate** (HCO_3^-), and **carbonate** (CO_3^{2-}):

$$CO_2(g) \rightleftarrows CO_2(aq)$$

$$CO_2(aq) + H_2O \rightleftarrows HCO_3^- + H^+$$

$$HCO_3^- + H^+ \rightleftarrows CO_3^{2-} + 2H^+$$

The dissolved $CO_2(aq)$, bicarbonate, and carbonate anions are collectively called Dissolved Inorganic Carbon (DIC). We see from these equations that the CO_2 levels in the atmosphere are coupled to the acidity of the oceans: higher alkalinity (higher pH, more basic) causes the chemical equilibria in the bottom two equations to shift to the right, thus yielding higher carbonate (CO_3^{2-}) in the ocean. Because $CO_2(aq)$ is consumed by these shifts, the first equilibrium equation tells us that higher alkalinity increases the ability of surface oceans to take up atmospheric CO_2. A more acidic ocean will have the opposite effect: a decrease in the ability of surface waters to take up CO_2. The solubility of CO_2 in water also depends on temperature: the lower the temperature, the higher the solubility. At a typical surface sea water pH of 8.2, the concentrations of the various components $[CO_2]$, $[HCO_3^-]$, and $[CO_3^{2-}]$ are 0.5%, 89% and 10.5%, respectively [3.9].

The chemical equilibria affecting carbon in the ocean are further complicated by the many minerals containing carbon that can dissolve in water. For example, **calcite** ($CaCO_3$) is very abundant and can dissociate according to:

$$CaCO_3 \rightleftarrows Ca^{2+} + CO_3^{2-}$$

The presence of carbonate in this reaction means that it is coupled to the CO_2 equations given above; thus the mineral dissolution is pH-dependent. Inspection of the equilibrium equations listed above leads to the conclusion that if water is acidic, $CaCO_3$ is more likely to dissolve; in basic waters, the mineral is more likely to be formed. Because the earth's crust has enormous quantities of such minerals (see **Table 3.1.1**), the calcite dissociation reactions buffer the carbonate (CO_3^{2-}) concentrations in the ocean.

In addition to the abiotic reactions described above, biological processes play an important role in the uptake of CO_2 by ocean waters. For example, in reef environments much of the buried $CaCO_3$ is precipitated *in situ* by corals and benthic algae as the mineral aragonite. In the open ocean, most $CaCO_3$ precipitates as a mineral (calcite) in the microfossils of two groups of marine plankton: coccolithophorids and planktonic foraminifera.

Organic carbon, that is, carbon that is present in organic matter such as the cells of living organisms, is also present in the oceans. In ocean surface layers, organisms use sunlight to convert CO_2 into biomass. If these surface organisms die and sink to the deep part of the ocean before they decompose, there is a net flux of organic carbon from the atmosphere towards the deep ocean. This flux is referred to as the "**biological pump**."

These organic and the inorganic cycles are coupled. As an example of how this coupling works, consider **Figure 3.3.2**. Suppose we were to grow organisms in the sea that consume CO_2 and form biomass, then die and sink to the bottom of the ocean. This would be an "engineered" enhancement of the biological pump, and it would have two effects on the CO_2 levels of the atmosphere. The first effect is that the biological pump would remove carbon more rapidly from the surface layer of the ocean and put it into the deep ocean, allowing the surface layer to take up more CO_2 and thus decreasing atmospheric CO_2. The second, indirect effect is that in the deep ocean the organic carbon would decompose and regenerate CO_2, which in turn would increase the acidity of the deep ocean water. This acidity would be neutralized by the dissolution of $CaCO_3$. The net result would be a decrease in the global ocean's burial rate of $CaCO_3$!

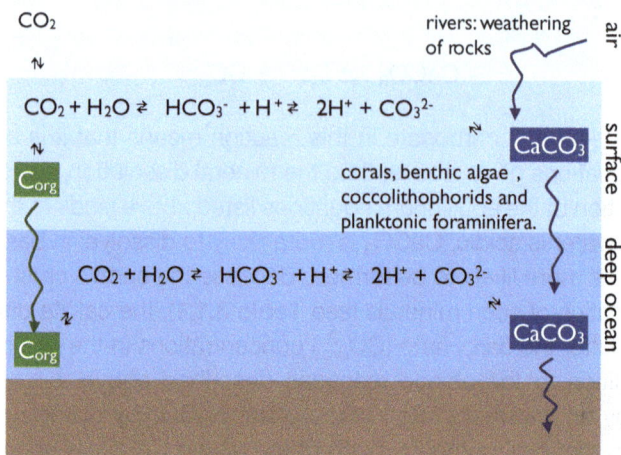

Figure 3.3.2 CO_2 chemistry in the ocean

This animation can be viewed at: http://www.worldscientific.com/worldscibooks/10.1142/p911#t=suppl

For many years these chemical equilibria have been in a steady state situation, where the flux of CO_2 leaving the ocean surface equals the flux entering the ocean surface. Furthermore, the ocean's burial rate of $CaCO_3$ equals the amount of $CaCO_3$ entering through the rivers.

Flow in the oceans

In the previous section, we discussed the CO_2 chemistry of the oceans. To fully understand the mechanism it is important to comprehend ocean flow patterns. These patterns are shown schematically in **Figure 3.3.3** and **Movie 3.3.1.** We see that water flows in a circular pattern: both in the Atlantic Ocean and in the Southern Ocean. In the deep ocean, regenerated CO_2 and regenerated nutrients follow the same pattern. An important difference between the two regions is that in the Northern Atlantic most of the surface is at a high latitude while in the Southern Ocean most of the surface is at low latitudes. In the vast low-latitude oceans of the Atlantic, the nutrients are almost completely consumed and returned to the ocean interior as sinking organic matter. However, in the warm and

Thermohaline Circulation

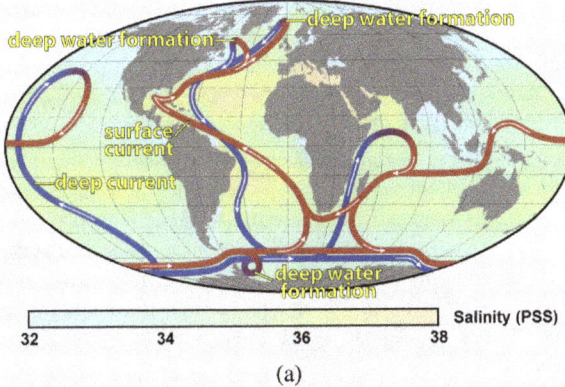

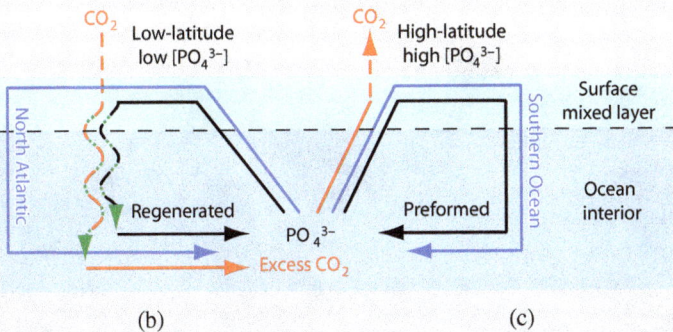

Salinity (PSS)

32   34   36   38

(a)

CO_2 Low-latitude low $[PO_4^{3-}]$

CO_2 High-latitude high $[PO_4^{3-}]$

Surface mixed layer

North Atlantic

Southern Ocean

Regenerated   Preformed   Ocean interior

PO_4^{3-}

Excess CO_2

(b)   (c)

Figure 3.3.3 Flow patterns in the ocean

(a) This map shows the pattern of thermohaline circulation (see also **Movie 3.3.1**), also known as "meridional overturning circulation." This collection of currents is responsible for the large-scale exchange of water masses in the ocean, including providing oxygen to the deep ocean. Blue paths represent deep-water currents, while red paths represent surface currents. The entire circulation pattern takes ~2000 years. *Figure by Robert Simmon, NASA. Minor modifications by Robert A. Rohde.*

(b) Symbolic diagram of the ocean's biological pump. The blue, black and orange lines show the transport of water, major nutrients (represented by phosphate), and CO_2, respectively. The North Atlantic has a high efficiency imparted to the pump by the low-latitude, low-nutrient surface regions. Nutrient-bearing subsurface water is converted into nutrient-depleted sunlit surface water. This is coupled with the complete biological assimilation of the major nutrients nitrate and phosphate in the production of particulate organic matter, which then sinks into the ocean interior where it is

(Continued)

Figure 3.3.3 (*Continued*)

decomposed to "regenerated" nutrients and excess CO_2 (CO_2 added by regeneration of organic matter), sequestering CO_2 away from the atmosphere and into the deep ocean. The nutrient-poor surface waters do not return immediately into the interior but rather become cold and thus dense. *Figure by Sigman et al.* [3.8], *reproduced with permission from Macmillan Publishers.*

(c) The loop in the Southern Ocean, however, shows the low efficiency imparted by the high-latitude, high-nutrient surface regions, currently dominated by the Southern Ocean, especially its Antarctic zone near the Antarctic margin. There, nutrient-rich and excess CO_2-rich water comes into the surface and descends again with most of its dissolved nutrient remaining (now referred to as "preformed"). In so doing, this loop releases to the atmosphere CO_2 that had been sequestered by the regenerated nutrient loop. *Figure by Sigman et al.* [3.8], *reproduced with permission from Macmillan Publishers.*

Movie 3.3.1 Thermohaline circulation

The region around latitude 60 south is the only part of the earth where the ocean can flow all the way around the world with no obstruction by land. As a result, both the surface and deep waters flow from west to east around Antarctica. This circumpolar motion links the world's oceans and allows the deep water circulation from the Atlantic to rise in the Indian and Pacific Oceans, thereby closing the surface circulation with the northward flow in the Atlantic. *Figure from NASA/Goddard Space Flight Center. This movie can be viewed at*: http://www.worldscientific.com/worldscibooks/10.1142/p911#t=suppl

buoyant surface waters of the high latitudes of the Southern Ocean the biological activity is less optimal and hence most of the nutrients are returned into the deep ocean unused. Because of this difference the biological pump is much more efficient for the Atlantic Ocean, which can sequester more CO_2 compared to the Southern Ocean. Thus, an important consequence of the difference in flow patterns is that in the North Atlantic the amount of carbon that is sequestered is much larger.

Recent theories have connected this excess to the glacial and interglacial periods. The details of this mechanism are outside the scope of this text, but we cannot resist giving a short version of this elegant theory. The hypothesis is that part of the excess CO_2 generated by the biological pump is stored in the deep part of the Southern Ocean that is not well mixed [3.8]. Because of this, there is a net flux of CO_2 from the atmosphere into this deep ocean reservoir. This drop in atmospheric CO_2 causes a temperature drop and a growth in polar ice. The reservoir of deep CO_2 is, however, finite and eventually the excess CO_2 will be released, causing a reversal of the glacial period. Many of these theories are speculations, but this build up of CO_2 in the Southern Ocean is predicted to occur on a time scale of 10^5 years, which is one of the peaks in **Figure 3.1.3** for which we have no other explanation.

Section 4

The inorganic carbon cycle

In **Figure 3.1.3** the largest peak represents fluctuations of 10,000 times the current CO_2 level with a period of about 10^9 years, which is about the age of the earth. We see that the effect of this "perturbation" has not yet fully decayed. This peak is associated with the inorganic carbon cycle, which is the most important mechanism to regulate the temperature of the earth. The fact that we need to regulate the temperature is closely related to the faint young sun paradox, described below.

Faint young sun paradox

Figure 3.4.1 compiles important events in the history of the earth. Its age is about 4.6 Ga (giga annum) and geological evidence has shown that for 4.4 Ga, liquid water has existed on the earth. This implies that the surface temperature of the earth must have been in the 0–100°C range for over 4 billion years. At first glance the presence of water on the earth this long ago appears to be inconsistent with the fact that the sun was still young, and thus much cooler, all those billions of years ago.

As a star gets older it gets denser and hence brighter. A brighter star has a higher temperature. As we have seen in the previous chapter, the amount of energy the earth receives from the sun depends on the temperature at which this radiation is emitted. A faint young star emits less energy than an otherwise identical older star. In **Figure 3.4.2** the consequence of this difference in energy is plotted for the history of the earth [3.10]. This graph shows that at the beginning of the earth the sun had about 70% of its current energy output. It also shows how this change in luminosity of the sun would be expected to influence the surface

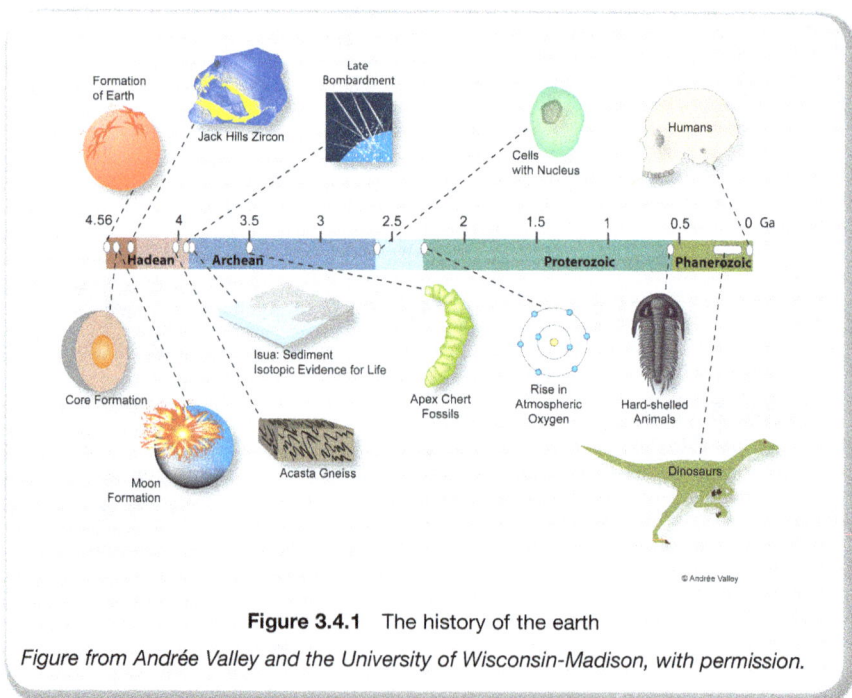

Figure 3.4.1   The history of the earth

Figure from Andrée Valley and the University of Wisconsin-Madison, with permission.

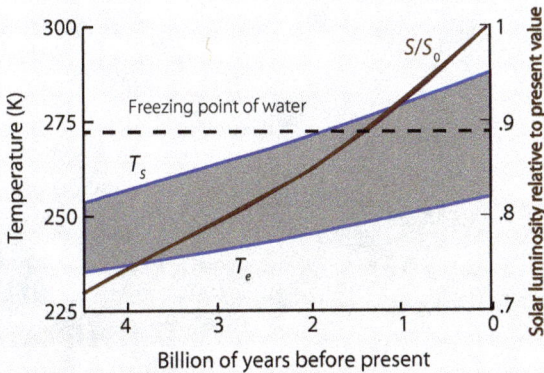

Figure 3.4.2 Solar luminosity

The solar luminosity normalized by the current luminosity as a function of time. The lower blue curve is the earth's effective radiating temperature, T_e. The upper blue curve represents the calculated mean global surface temperature, T_s. Both curves have been calculated using a model in which the atmosphere has the current (300 ppm) CO_2 concentration. *Figure based on Kasting and Catling* [3.10].

temperature of the earth. In these calculations, it is assumed that the earth has its current atmosphere. This graph predicts (erroneously) that liquid water could only be present on the earth's surface 1.5 Ga ago. The inconsistency of this prediction with experimental evidence of the existence of liquid water 4.4 Ga ago is known as the faint young sun paradox [3.11].

The hypothesis is that the greenhouse effect must have been much larger in early earth history. In fact, the amount of CO_2 must have been orders of magnitude larger at those times in order for the greenhouse effect to keep the earth's surface warm enough for liquid water. An important question is how the earth has been able to regulate CO_2 on such a massive scale. Not surprisingly, this regulation needs to involve very large amounts of carbon. The largest carbon pool is comprised of the carbonate minerals, and their central role in the carbon cycle is called the "inorganic carbon cycle."

The inorganic carbon cycle

The **inorganic carbon cycle** is illustrated in **Figure 3.4.3**. CO_2 and water form slightly acidic rain that slowly weathers rocks. The weathering reactions can be summarized as:

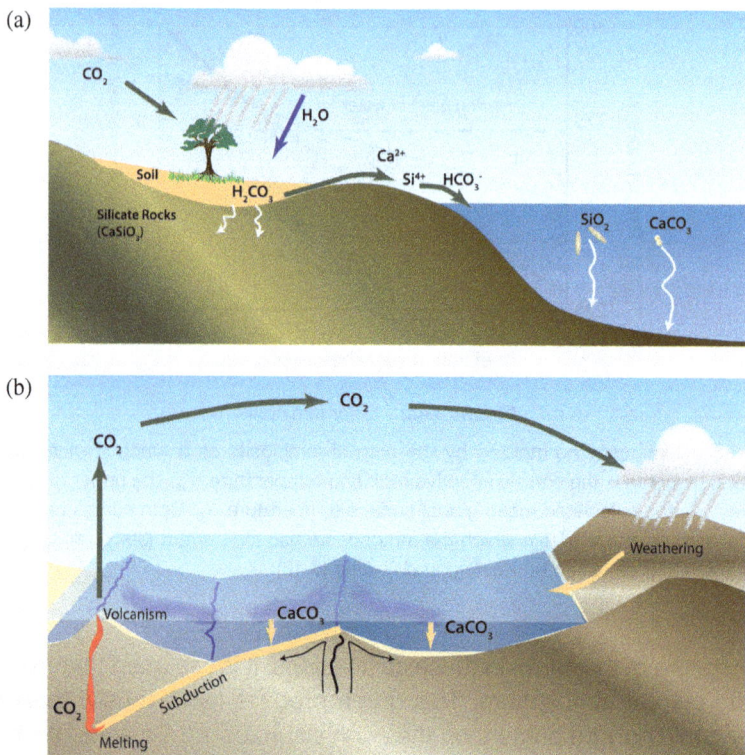

Figure 3.4.3 Inorganic carbon cycle

(a) Part of the inorganic carbon cycle: Silicates in bedrock and carbonic acid in soil are weathered on the earth's surface and transported in rivers as ions dissolved in the water. This leads to the deposition of minerals in the ocean floor as shells of ocean plankton.

(b) The figure shows the inorganic carbon cycle with subduction transporting carbon-containing minerals deep into the interior of the earth where they melt. The inorganic carbon cycle is completed when volcanism delivers decomposed minerals to the atmosphere as CO_2.

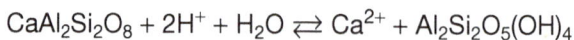

$$2CO_2 + 2H_2O \rightleftarrows 2HCO_3^- + 2H^+$$

$$CaAl_2Si_2O_8 + 2H^+ + H_2O \rightleftarrows Ca^{2+} + Al_2Si_2O_5(OH)_4$$

$$Ca^{2+} + 2HCO_3^- \rightleftarrows CaCO_3 + CO_2 + H_2O$$

The net effect of these **weathering reactions** is:

$$CO_2 + CaAl_2Si_2O_8 + 2H_2O \rightleftarrows CaCO_3 + Al_2Si_2O_5(OH)_4$$

In words, the net effect of weathering is to convert atmospheric CO_2 into carbonate minerals such as $CaCO_3$. The bicarbonate and the calcium carbonate are transported by water into the rivers and finally end up in the ocean. In the ocean, the calcium carbonates form the shells of plankton (see **Figure 3.4.4**) and a fraction settles on the ocean floor. Through this weathering reaction, 0.15 Gt C/year is transported into the oceans. As we have seen in the previous section, the exchange of CO_2 with the atmosphere is in nearly perfect equilibrium. Hence, this flux of 0.15 Gt per year would slowly drain all the CO_2 in the atmosphere if the earth did not have a mechanism to recycle these deposits.

This mechanism is the tectonic motion of the sea bottom. Because of tectonic motion, subduction takes the $CaCO_3$ deposits into the mantle, where the $CaCO_3$ decomposes and CO_2 is released. Volcanos recycle this CO_2 back into the atmosphere, thus completing the inorganic carbon cycle. As the temperature of the sun increases, rock weathering

Figure 3.4.4 Foraminifera

Foraminifera is one of the organisms that use HCO_3^- to make shells of $CaCO_3$. *Phase-contrast photomicrograph by Scott Fay.*

reactions accelerate, diluting the CO_2 concentration in the atmosphere. This feedback mechanism, i.e., a hotter sun leads to decreasing CO_2 and thus lower surface temperature, has been responsible for ensuring that the temperature of the earth's surface has remained within the narrow window that allows water to exist in its liquid phase for billions of years. Volcanos emit about 0.15 Gt C/year into the atmosphere, which is about the same quantity that the weathering reaction removes from the atmosphere. This cycle replenishes all the CO_2 in the atmosphere and oceans. But one can imagine that the typical response time of this mechanism is not very fast: the time constant is 0.5–1 million years!

For comparison, it is interesting to consider Venus and Mars (see **Figure 3.4.5**). Neither planet has water on its surface. A possible explanation for Venus is that it is so close to the sun that all the water evaporates, leaving no rain or weathering to take CO_2 out of the atmosphere.

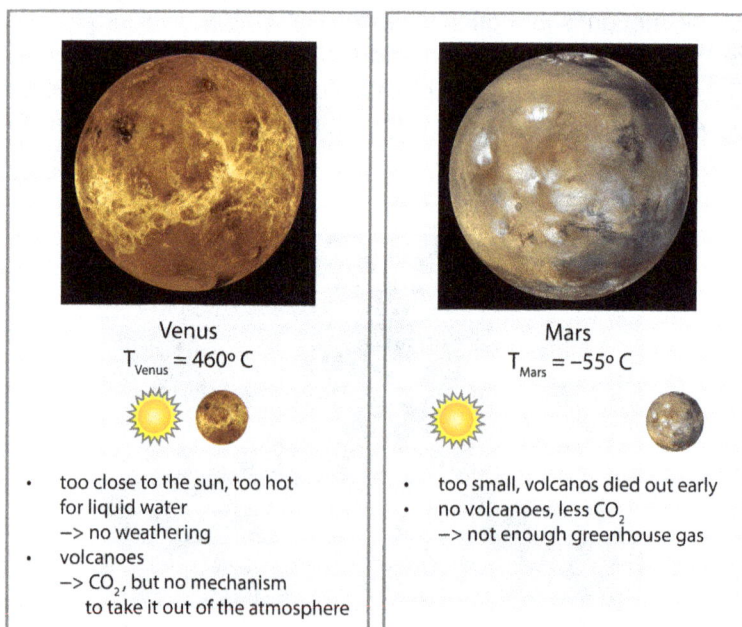

Venus
$T_{Venus} = 460° C$

- too close to the sun, too hot for liquid water
  –> no weathering
- volcanoes
  –> CO_2, but no mechanism to take it out of the atmosphere

Mars
$T_{Mars} = -55° C$

- too small, volcanos died out early
- no volcanoes, less CO_2
  –> not enough greenhouse gas

Figure 3.4.5 The atmosphere of Venus and Mars

Photo of Venus and Mars by NASA.

Because of the volcanic activity, all CO_2 ends up in the atmosphere of Venus and as a result Venus experiences a very large greenhouse effect. In this model, Venus resembles the earth in its very early existence. Mars has a different problem. It is much smaller than the earth and because of this the volcanic activity has stopped. As a result, Mars has no mechanism for CO_2 to return to the atmosphere, and hence too little greenhouse effect. Mars is thus too cold for liquid water.

Section 5

A box model for the global carbon cycle

Let us now collect all our knowledge on the interactions within the carbon reservoirs to discuss the global carbon cycle. This knowledge is summarized in our system view (**Figure 3.5.1**). We see that there is a nearly perfect equilibrium between the uptake of CO_2 by the biosphere and the release of CO_2 by the biosphere and by the decomposition of the terrestrial soils. Similarly for oceans, the uptake and release of CO_2 by the surface is in equilibrium. The inorganic carbon cycle, which creates a flux of only 0.15 Gt/year, is actually controlling the long-term CO_2 levels in the atmosphere. This global carbon cycle is an accurate description of the earth's carbon cycle until 1850.

Since 1850, humans started to use fossil fuels on a very large scale. The current carbon cycle is shown in **Figure 3.5.2**. The main difference is that by burning fossil fuels we add carbon from surface reservoirs to the atmosphere at a current rate of 9 Gt/year. The total amount of carbon that has been emitted since 1850 is about 500 Gt of carbon. If we look at this number we see that the rate at which we emit is significantly higher, by a factor of 60, than the natural rate of emission from volcanos (which is about 0.15 Gt/year). Anthropogenic emissions are depicted as the red peak in **Figure 3.1.3**, and this perturbation is too large and occurs on a time scale (100 years) too fast for natural process to remove the excess

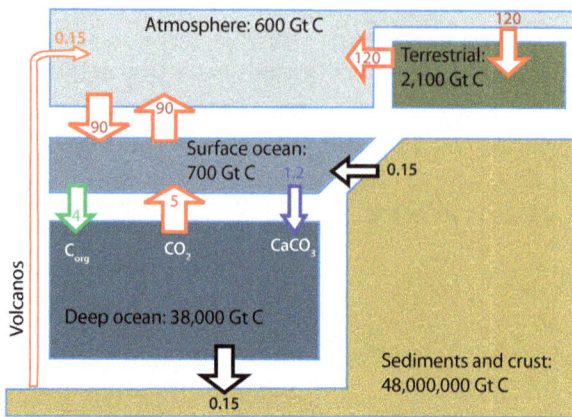

Figure 3.5.1 The box model of the carbon cycle before 1870

This animation can be viewed at: http://www.worldscientific.com/worldscibooks/ 10.1142/p911#t=suppl

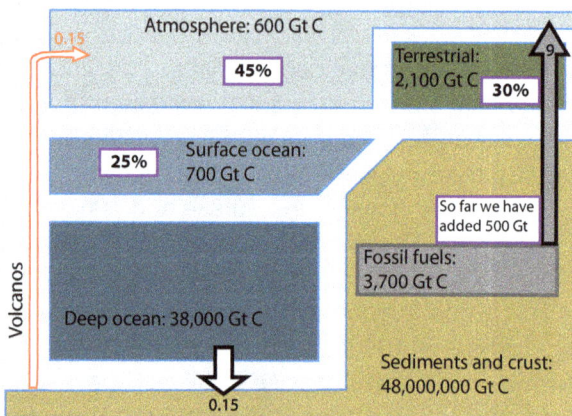

Figure 3.5.2 The box model of the carbon cycle in 2012

This animation can be viewed at: http://www.worldscientific.com/worldscibooks/ 10.1142/p911#t=suppl

of CO_2. To the best of our current understanding, some of this excess is absorbed by the oceans (~25%) and some by the biosphere (~30%). The remaining ~45% is the accumulation in the atmosphere that is responsible for the increase in CO_2 levels that we presently measure.

Section 6

The future carbon cycle

What will the carbon cycle look like in the future? In response to anthropogenic changes in atmospheric CO_2, we expect the weathering mechanism and the volcanos of the inorganic carbon cycle to eventually bring the CO_2 levels back to approximately what they were in 1850. Recall however that the time scale of this process is measured in units of 100,000 years! This is arguably relatively short on a geological time scale, but what will happen in the meantime on the 100 year time scale associated with current political, economic, and ecological forces?

Climate model predictions

To address this time scale we employ the climate models that were discussed in Chapter 2. First we need to assume specific scenarios for future carbon emissions. Climate model predictions have been made with two general scenarios [3.12]: a "moderate" and a "large" scenario. The "large" scenario assumes that most of the proven fossil fuel reserves (mainly coal) will be consumed in the coming 300 years, giving a total carbon emission of 5,000 Gt. The "moderate" scenario is one in which we use only 20% of these reserves. To put these scenarios in context, the "large" scenario assumes slightly lower emissions than the A2 SRES scenario of the IPCC (see **Figure 3.6.1**). Also, recall that the earth's natural emission from volcanos is about 0.15 Gt per year, which is a factor of 160 smaller than the maximum emissions in the "large" scenario (see **Box 2.5.1**).

Can we predict future CO_2 levels in the atmosphere based upon these emission scenarios? To answer this question, we examine the time scales that impact the carbon cycles we have discussed in this chapter; it is these processes that sequester excess CO_2 from the atmosphere.

These predictions are made with climate models. As we have seen in Chapter 2, the standard way of testing the sensitivity of the predictions is to run many different models built by different research groups around the world. The idea is that each of these models employs different assumptions, and if these models give very different predictions, then the

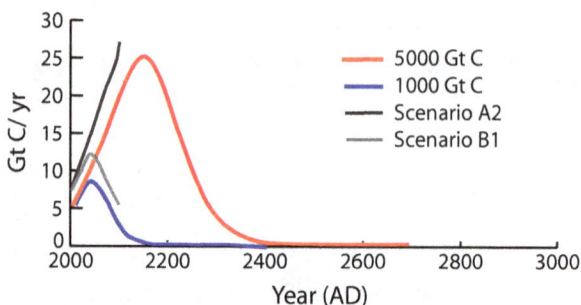

Figure 3.6.1 Future carbon emissions

The thick red line is the "large" scenario in which we emit the amount of CO_2 corresponding to most of the proven fossil fuel reserves. The thick blue line is the "moderate" scenario in which we only use 20% of these reserves. The gray lines are the SRES scenarios A2 and B1 from the IPCC (see **Box 2.5.1**). *Figure from Archer and Brovkin [3.12].*

scientific community associates considerable uncertainty with their predictions. If, on the other hand, all the models agree, then this agreement suggests that we have a good understanding of the effects and the predictions. Before looking at the actual model predictions, it is important to make some qualitative observations.

The terrestrial biosphere will take up some excess carbon as a warmer world will have a longer growing season. However, increasing the temperature will also increase the decomposition of carbon in soils. Moreover, the total amount of terrestrial biomass is simply too small to act as an efficient buffer of excess atmospheric CO_2. We therefore have to rely on the oceans for the uptake of the CO_2. It typically takes the ocean surface one year to equilibrate with the atmosphere. However, this surface layer is about 100 m deep. Once the surface layer is saturated we have to wait for the water in this layer to be mixed with the water in the deep ocean. This mixing occurs on a time scale of 100 to 1,000 years, which is the time scale of the circulation of the oceans. In **Figure 3.6.2** these time scales are illustrated.

In **Figure 3.6.2** an important assumption has been made, namely that the deep ocean is infinitely large or has an infinitely large buffering capacity. To some extent this is true. The pervasion of CO_2 in the ocean causes the pH to decrease, but the presence of $CaCO_3$ on the sea bottom acts as the buffer. The buffering reactions are slow and their capacity is limited. The

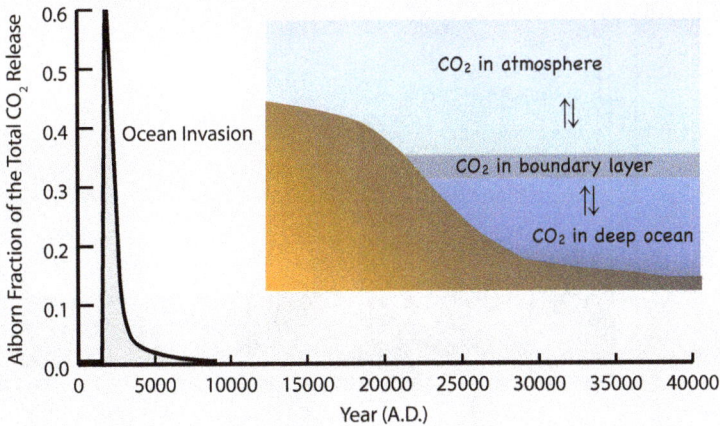

Figure 3.6.2 CO_2 in the oceans

The oceans form an important buffer to absorb the extra emitted CO_2. The equilibration of CO_2 within the boundary layer happens relatively fast (1 year). As the capacity of this boundary layer is limited, the equilibrium concentration of CO_2 in the atmosphere will stay high. Mixing of surface water with the deep ocean water, in which the concentrations of CO_2 are lower, and subsequent re-equilibration with the CO_2 in the atmosphere, allows a further decrease in the CO_2 concentration. As the mixing times of the oceans are on the order of 100 to 1,000 years, we see that it may take up to 5,000 years for atmospheric CO_2 levels to fully equilibrate with those in the oceans. *Figure adapted from Archer and Brovkin* [3.12].

final reduction in atmospheric CO_2 level has to come from the inorganic carbon cycle, i.e., through the weathering reaction of CO_2 with silicate rocks and volcanism (see Sections 3.3, 3.4), which has a time scale on the order of a hundred thousand years (see **Figure 3.6.3**) [3.12, 3.13].

Now let's return to the quantitative side of things. At present we are emitting about 9 Gt of carbon per year from fossil fuels. Of these 9 Gt, 5 Gt are taken up by the biosphere and oceans in about equal amounts. The difference, 4 Gt, is added to the atmosphere each year. In order to employ sophisticated climate models, we have to estimate the amount of carbon that will be emitted each year; the maximum atmospheric CO_2 concentration will depend on this rate. If we emit the same amount over a longer period of time the biosphere and oceans can better keep up with our emissions. The results of the different climate models for the "moderate" and "large" scenarios [3.14] are shown in **Figure 3.6.4**.

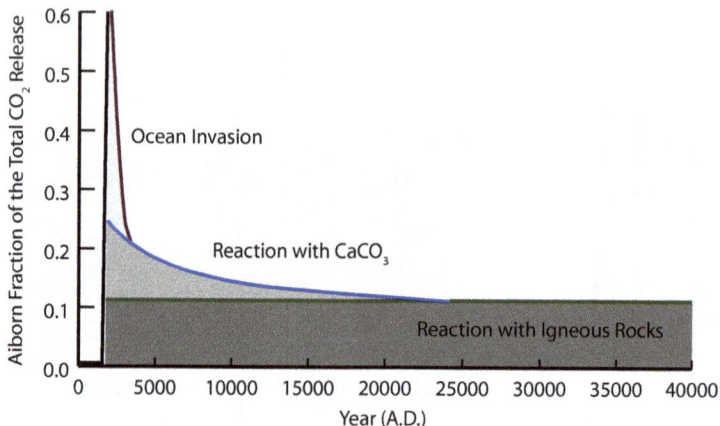

Figure 3.6.3 CO_2 absorption mechanisms

Upon the invasion of CO_2 from fossil fuels into the ocean, the acidity from the CO_2 provokes the dissolution of $CaCO_3$ from the sea floor. It takes thousands of years for the dissolution of $CaCO_3$ to restore the pH of the ocean to a natural value. Restoring the pH also replenishes the buffering ability of sea water to store more CO_2, so the airborne fraction of the fossil fuel CO_2 drops a bit further [3.13]. At the end of the neutralization stage, the atmosphere still contains more CO_2 than it held before the fossil fuel era. The rest of the CO_2 awaits reaction with igneous rocks. CO_2 is extracted from the atmosphere by these reactions and ends up on the sea floor in $CaCO_3$ deposits. This final piece of the anthropogenic CO_2 perturbation takes hundreds of millennia to subside. *Figure adapted from* [3.12].

Given these results, the next step is to address the relationship between these predicted atmospheric CO_2 levels and average temperatures on the surface of the earth. We have seen that in the early days of the earth, CO_2 levels were much higher than they are now. We will use some of the data that have been collected about these conditions to understand potential impacts of future CO_2 levels.

Paleocene-eocene thermal maximum

Figure 3.6.5 shows a compilation by the IPCC of different sources of experimental data on atmospheric CO_2 levels for the past millions of years [3.15]. In **Box 3.6.1** we introduce some of the experimental techniques that have been used to obtain these data. The data in **Figure 3.6.5**

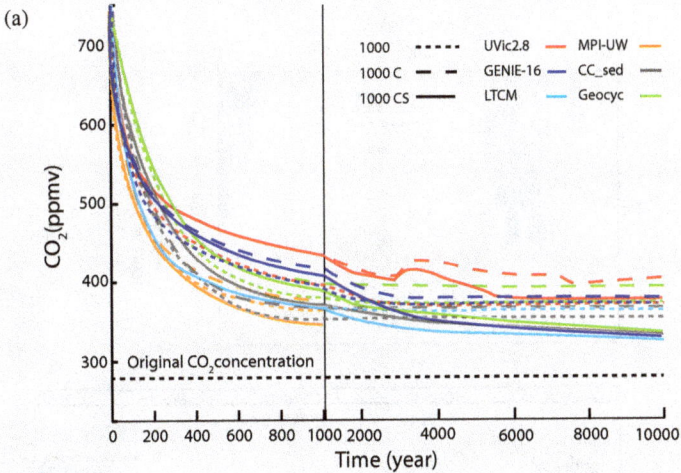

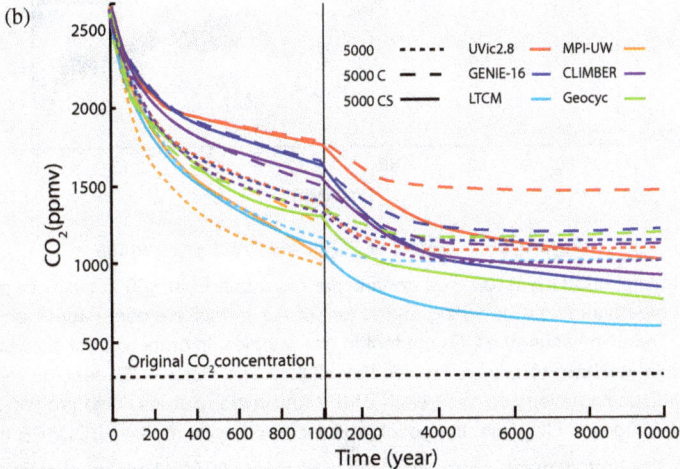

Figure 3.6.4 Predictions of the CO_2 concentration in the atmosphere from various climate models

(a) The moderate scenario.

(b) The large scenario.

The first 1,000 years are expanded, which gives the split scale. Colors denote different climate models and the line styles represent different aspects that are included in the calculations, for example, the climate feedback (C) and the climate plus sediment feedback (CS). *Figures adapted from Archer et al.* [3.14].

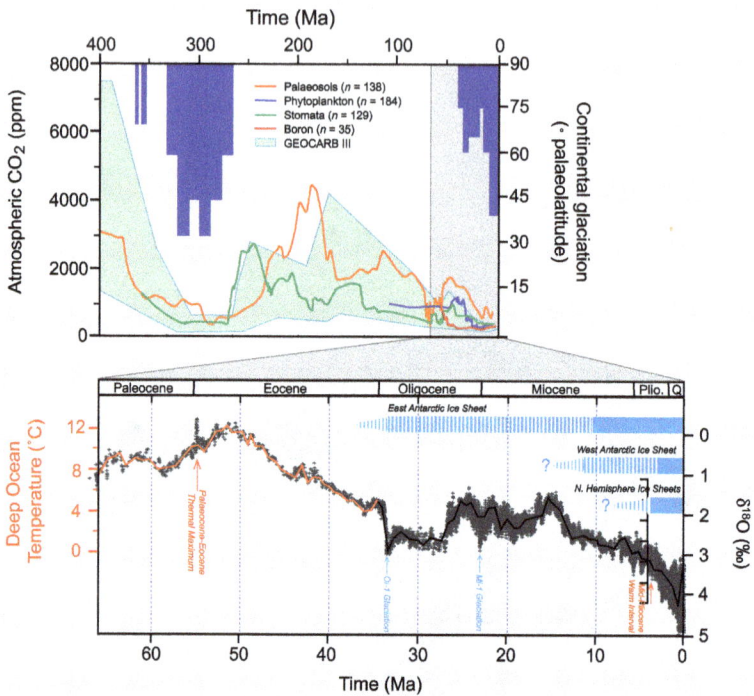

Figure 3.6.5 Historical CO_2 levels and temperatures

Top: Atmospheric CO_2 levels and continental glaciation from 400 Ma ago to present. Vertical blue bars mark the timing and extent of ice sheets. Ice core experiments were able to directly measure the CO_2 content in gas bubbles. In order to track the very early CO_2 levels, we have to rely on proxies (**Box 3.6.1**). The plotted CO_2 records represent five-point running averages from each of the four major proxies. Also plotted are the plausible ranges of CO_2 from the geochemical carbon cycle model GEOCARB III.

Bottom: The bottom graph zooms in on the most recent 60 Ma. The temperature proxies that are plotted here are the deep-sea benthic foraminifera [18]O isotope records from many different experiments. The bars on the upper right corners indicate the presence of ice sheets. If the bars are dashed, they represent periods of ephemeral ice or ice sheets smaller than those at present, while the solid bars represent ice sheets of modern or greater size. *Figures from IPCC, reproduced with permission* [3.15].

show that in the early days of the earth CO_2 levels were much higher, as discussed in Section 3.4. We see that the CO_2 levels have gone through very large fluctuations. We also see that the extent to which the earth was covered with ice correlates inversely with CO_2 levels. In fact, by inspection of this figure one can see that the CO_2 levels need to stay below 700 ppm levels to sustain ice in the polar regions.

If we look carefully at the data we see that the Paleocene and the Eocene eras are separated by a peak in the ocean temperatures. This peak is know as the **Palaeocene-Eocene Thermal Maximum (PETM)**, which took place about 52.5–55.5 million years ago. **Figure 3.6.6** shows some more detailed data for this event. **Figure 3.6.6** (top) gives the isotopic composition of ^{13}C in marine and continental records. The sudden drop indicates that the ^{13}C composition of the atmosphere had dropped dramatically, as would be expected from a massive release of organic (^{13}C–depleted) carbon into the atmosphere. At present, the origins of the PETM are not fully understood. **Figure 3.6.6** (middle) shows that during the same period the $\delta^{18}O$ level of the ocean foraminifera decreased. As this $\delta^{18}O$ level is a proxy for the ocean temperature (see **Box 3.6.1**), the ^{13}C and $\delta^{18}O$ levels are consistent with a significant increase in carbon levels (CO_2 or CH_4) in the atmosphere, and an ensuing greenhouse gas effect, causing the increase in the temperature of the oceans. It has been estimated that the total carbon release for this time period is on the order

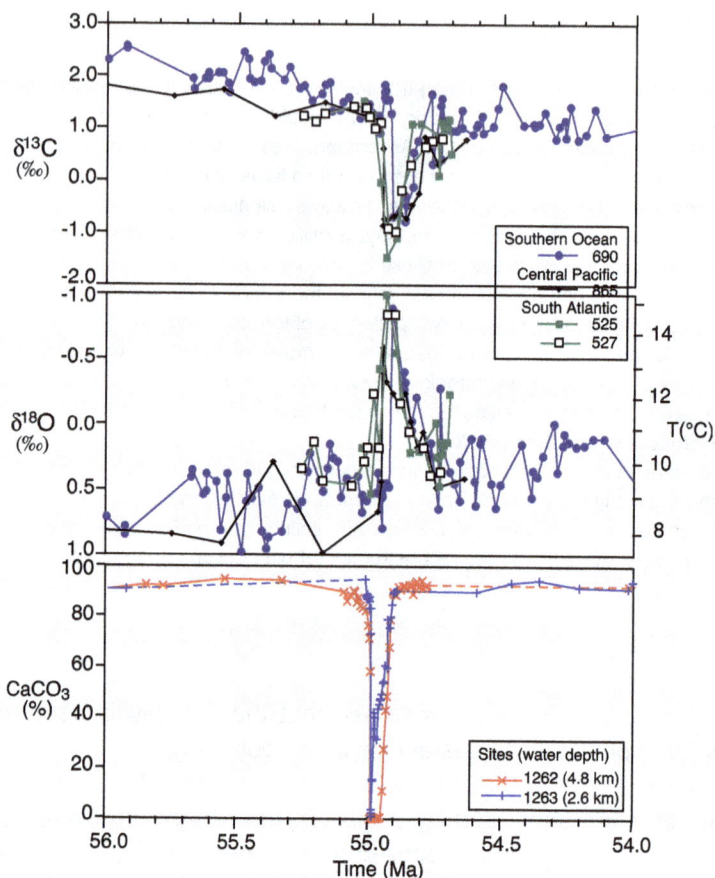

Figure 3.6.6 The palaeocene-eocene thermal maximum

Top: The foraminifer isotopic records from sites in the Antarctic, South Atlantic and Pacific.

Middle: The temperatures of the oceans as measured by the $\delta^{18}O$ proxy.

Bottom: The effect on the carbonate ($CaCO_3$) content of sediments. *Figures from IPCC, reproduced with permission* [3.15].

of 1,000 to 2,000 Gt, which is similar to the numbers in the "large" and "moderate" scenarios of the future emissions. Because of this similarity, the details of the PETM climate outcomes provide some indication of the long-term climate effects of emitting this much carbon into the atmosphere. The mass of carbon in the PETM was sufficiently large to exceed

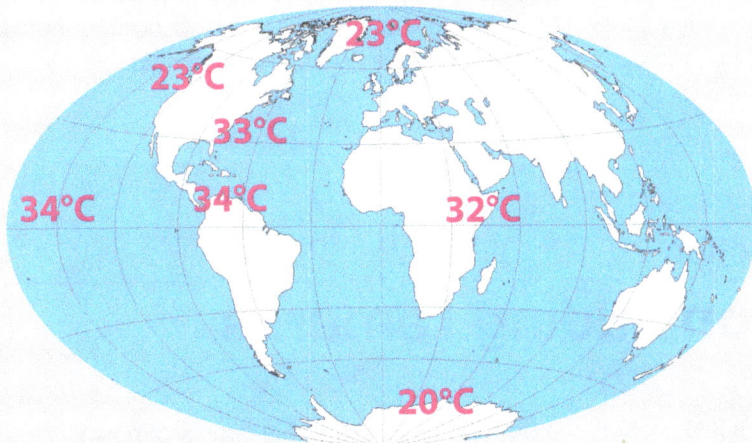

Figure 3.6.7 Average surface temperature of the earth during the peak of the PETM *The data for the temperatures are taken from* [3.16].

the buffer capacity of the ocean and lower its pH. As we have seen in the previous section, such a low pH would dissolve the carbonate minerals on the bottom of the ocean. **Figure 3.6.6** (bottom) shows that is exactly what is observed.

There are many speculations in the literature about the source of the enormous release of atmospheric carbon; it could have been methane (CH_4) from decomposition of clathrates on the sea floor, CO_2 from volcanic activity, or oxidation of sediments rich in organic matter. One of the more recent theories posits that the magnitude and timing of the PETM is related to the decomposition of soil organic carbon in circum-Arctic and Antarctic terrestrial permafrost. This massive carbon reservoir had the potential to repeatedly release thousands of petagrams (1 petagram equals 10^{15} grams) of carbon into the atmosphere-ocean system, once a long-term warming threshold had been reached just before the PETM [3.17]. The figures also show that eventually the carbon cycle of the earth was able to remove the excess of carbon. The important thing to note, however, is that it took 100,000 years before the earth recovered! The crux of the entire analysis is depicted in **Figure 3.6.7** [3.16]. This figure shows the reconstruction of the average temperature by regions at the peak of the PETM. Study this figure closely, and overlay it with your own experiences with cities, countries, and ecological resources. As scientists we like to avoid hyperbole, but in the context of our present world political and economic systems, this figure is apocalyptic.

Section 7
Concluding remarks

In the last section, we discussed an extreme example of how large-scale carbon emissions can influence the climate for many years. The importance of this example is that the amount of carbon that was emitted in the atmosphere during the PETM is of the same order of magnitude as the total carbon content of all fossil fuels we have discovered so far. The effects of emitting the corresponding CO_2 are large and will take about 100,000 years to subside.

We have also seen that the earth's carbon cycles are very complex with many feedback loops. It would be impossible to fully understand such a complex system. Thus, the predictions that we have discussed in this book contain uncertainties. These uncertainties are often the drivers for new scientific inquiry, with particular attention to those uncertainties that are largest and have the biggest impact.

If we look back 30 years, anthropogenic climate change used to be a controversial topic in the literature. One can find discussions arguing that if climate change is ongoing, one would see an increase in the CO_2 concentration in the ocean. At that time there were no experimental data to support this claim. Because of this scientific discussion between those that argued in favor of and against the existence of climate change, an experimental program was set up to measure whether such an increase actually has taken place. The experiments showed that the ocean has indeed taken up the additional CO_2. Over the years heated discussions have resulted in more innovative experiments to settle the disagreements. Little by little the most pressing questions have been settled simply by accumulating experimental evidence — not because the scientists in favor had the best arguments or the loudest voices, but because experimental data settled the issues step by step. Because of this evidence the overwhelming majority of scientists today have little doubt about the reality of climate change caused by human activity. This does not mean that uncertainties do not exist, nor does it mean that we do not need even more innovative studies of the earth's past and its chemistry.

It is difficult to touch this topic without referring to the perception by a portion of the population that climate change occurs more in the minds of scientists than in the real world. In this respect the focus of scientists on uncertainties does not help. It is important to note that while scientists debate about uncertainty, the outcome is not in debate at all. A metaphor may help. Suppose you are driving your new GPS-enabled and satellite-uplinked car when, suddenly after the next turn, you are in a very heavy fog. Suppose further that there are many scientists back at a lab looking over the data from your GPS, and they engage in a discussion as to when your car will go over a cliff. Of course, being scientists, they have models to predict when you will reach this cliff. Not surprisingly, their predictions all disagree owing to the uncertainty of the GPS system, the time to uplink data from the car, and so forth. Some argue that you will reach the cliff in 5 minutes while others argue that it will take you at least one hour to reach the cliff. What do you do?

Section 8

Review

3.1. Reading self-test

1. Which statement about the amount of carbon is NOT correct?
   a. The total amount of carbon in the atmosphere is 720 Gt
   b. Most of the carbon is in rocks
   c. There is about three times as much carbon in the biosphere as in the atmosphere
   d. The total amount of organic carbon in the ocean is 38,400 Gt

2. Which statement about the diurnal carbon cycle is NOT correct?
   a. Because of the diurnal carbon cycle, CO_2 changes with a daily frequency
   b. The diurnal carbon cycle is related to the rotation of the earth
   c. The typical fluctuation in average CO_2 levels is about 10 ppm

3. Which statement about the Vostok ice cores is NOT correct?
   a. The Vostok ice cores are named after the location of the research center in Antarctica
   b. Historical data of the temperature is obtained by measuring the concentration ratio of ^{18}O to ^{16}O
   c. The maximum CO_2 concentration observed in these ice cores is about 10 ppm less than the current levels
   d. These ice cores contain 420,000 years of history

4. Which reaction plays the most important role in buffering the pH of the ocean?
   a. $CO_2(aq) + H_2O \rightleftarrows HCO_3^- + H^+$
   b. $CO_2(g) \rightleftarrows CO_2(aq)$
   c. $HCO_3^- + H^+ \rightleftarrows CO_3^{2-} + 2H^+$
   d. $CaCO_3 \rightleftarrows Ca^{2+} + CO_3^{2-}$

5. Which statement about the inorganic carbon cycle is NOT correct?
   a. Volcanos emit 0.15 Gt of carbon per year
   b. If the temperature of the earth increases, the weathering reaction increases and hence more CO_2 reacts
   c. Without rain, the temperature of the earth would decrease
   d. Without tectonic motion, the atmosphere would become depleted of CO_2

6. The ratio of anthropogenic to natural emissions of CO_2 is
   a. 1
   b. 6
   c. 10
   d. 60
   e. 100

7. If we emit 1 Gt of CO_2, where will the CO_2 go?
   a. atmosphere 45%, biosphere 30%, and ocean 25%
   b. atmosphere 30%, biosphere 45%, and ocean 25%
   c. atmosphere 25%, biosphere 30%, and ocean 45%
   d. atmosphere 45%, biosphere 25%, and ocean 30%

8. The total amount of carbon we have emitted in the form of CO_2 since 1850 is
   a. 100 Gt
   b. 200 Gt
   c. 500 Gt
   d. 1,500 Gt

9. If we decide to use 20% of the reserves of the fossil fuels, what is the maximum CO_2 concentration predicted to be in the atmosphere?
   a. 300 ppm
   b. 700 ppm
   c. 1,000 ppm
   d. 1,500 ppm

10. What is the lowest concentration of CO_2 in the atmosphere above which there will be no ice?
    a. 500 ppm
    b. 700 ppm
    c. 900 ppm
    d. 1,100 ppm

Section 9

References

1. US Department of Energy Office of Science Genomic Science Program, March 2008. Workshop Report on Carbon Cycling and Biosequestration, DOE/SC-108, http://genomicscience.energy.gov/carboncycle/report/

2. Falkowski, P., R.J. Scholes, E. Boyle et al., 2000. The global carbon cycle: A test of our knowledge of Earth as a system. Science, **290** (5490), 291. http://dx.doi.org/10.1126/science.290.5490.291

3. Sigman, D., and E. Boyle, 2000. Glacial/interglacial variations in atmospheric carbon dioxide. Nature, **407**, 859–869. http://faculty.washington.edu/battisti/589paleo2005/Papers/SigmanBoyle2000.pdf

4. Sun, J.L., S.P. Burns, A.C. Delany et al., 2007. CO_2 transport over complex terrain. Agr. Forest Meteorol., **145** (1–2), 1. http://dx.doi.org/10.1016/J.Agrformet.2007.02.007

5. Tans, P. (NOAA/ESRL) and R. Keeling (Scripps Institution of Oceanography). http://scripps CO_2.ucsd.edu/

6. Petit, J.R., J. Jouzel, D. Raynaud et al., 1999. Climate and atmospheric history of the past 420,000 years from the Vostok ice core. Antarctica Nature, **399** (6735), 429.

7. Hays, J.D., J. Imbrie, and N.J. Shackleton, 1976. Variations in Earth's orbit — pacemaker of ice ages. Science, **194** (4270), 1121.

8. Sigman, D.M., M.P. Hain, and G.H. Haug, 2010. The polar ocean and glacial cycles in atmospheric CO_2 concentration. Nature, **466** (7302), 47. http://dx.doi.org/10.1038/nature09149

9. Honisch, B., A. Ridgwell, D.N. Schmidt et al., 2012. The geological record of ocean acidification. Science, **335** (6072), 1058. http://dx.doi.org/10.1126/Science.1208277

10. Kasting, J.F. and D. Catling, 2003. Evolution of a habitable planet. Annu. Rev. Astron. Astrophys., **41**, 429. http://dx.doi.org/10.1146/annurev.astro.41.071601.170049

11. Sagan, C. and G. Mullen, 1972. Earth and Mars — evolution of atmospheres and surface temperatures. Science, **177** (4043), 52.

12. Archer, D. and V. Brovkin, 2008. The millennial atmospheric lifetime of anthropogenic CO_2. Clim. Change, **90**, 283. http://dx.doi.org/10.1007/s10584-008-9413-1

13. Ridgwell, A. and J.C. Hargreaves, 2007. Regulation of atmospheric CO_2 by deep-sea sediments in an Earth system model. Global Biogeochemical Cycles, **21** (2), 6B 2008. http://dx.doi.org/10.1029/2006GB002764

14. Archer, D., M. Eby, V. Brovkin, *et al.*, 2009. Atmospheric lifetime of fossil fuel carbon dioxide. Annu. Rev. Earth. and Pl. Sc., **37**, 117. http://dx.doi.org/10.1146/Annurev.Earth.031208.100206

15. Climate Change, 2007. The Physical Science Basis. Working Group I Contribution to the Fourth Assessment Report of the Intergovernmental Panel on Climate Change, Figure 6.1; Figure 6.2. Cambridge, UK, and NY: Cambridge University Press.

16. Winguth, A., C. Shellito, C. Shields, and C. Winguth, 2010. Climate response at the paleocene-eocene thermal maximum to greenhouse gas forcing — a model study with CCSM3. J. Climate, **23**, 2562–2584. http://dx.doi.org/10.1175/2009JCLI3113.1

17. DeConto, R.M., S. Galeotti, M. Pagani *et al.*, 2012. Past extreme warming events linked to massive carbon release from thawing permafrost. Nature, **484** (7392), 87. http://dx.doi.org/10.1038/Nature10929

Chapter 4

Introduction to Carbon Capture

There is nothing magical about carbon capture and storage (CCS). The technology to capture and store CO_2 in geological formations is already available commercially. So why isn't carbon capture technology more widespread? Currently, the costs of separating CO_2 — in terms of required energy and capital costs, and the resulting increase in price of electricity required — are significant. If research were able to reduce the required energy and monetary costs of separation in a meaningful way, it would greatly reduce the barrier to CCS.

Section 1

Introduction

From a technological point of view, there are different ways to capture CO_2. The simplest way is to use **post-combustion carbon capture**, because it can be added to an existing power plant without having to modify the power plant itself. For this reason post-combustion capture is often the preferred economic solution, especially for older power plants.

Each type of capture technology involves a different technique for gas separation. To weigh the pros and cons of each strategy, we need to take a detailed look at how a power plant works. **Figure 4.1.1** shows a simple coal-fired power plant [4.1]. In the direct combustion step, pulverized coal is mixed with air and burned. The heat of this combustion is recovered to produce high-pressure steam, which in turn drives an electric turbine to produce electrical power. Simplified, the chemical reaction for this process can be written as:

$$fuel + O_2 \rightarrow heat + CO_2 + H_2O$$

Unfortunately, this equation is somewhat idealized. In real life, coal not only contains carbon and hydrogen, but other elements (N forming NO_x and S forming SO_x) and trace metals (e.g., Hg). These impurities have to be removed before the flue gas is vented into the atmosphere, and that process requires money and effort. Tacking CO_2 separation onto the process requires even more energy or electricity, which we assume we will take directly from the power plant itself. This energy drain is another important barrier for large-scale adoption of CCS, because the process will reduce the efficiency of a power plant (see **Box 4.1.1**).

A typical post-combustion carbon capture process is shown in **Figure 4.1.2.** There is a cost in energy associated with both the capture and the compression steps. As we will discuss in detail later, carbon capture processes involve the regeneration of the capturing materials. This regeneration is done using heat (low-pressure steam from the power plant), which detracts from the electricity production of the plant. Additionally, more energy will be required in order to prepare captured

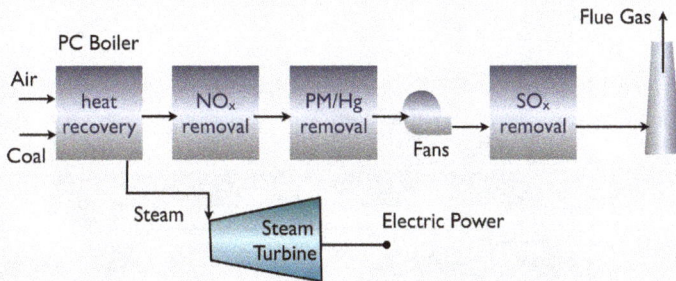

Figure 4.1.1 Coal-fired power plant

This animation can be viewed at: http://www.worldscientific.com/worldscibooks/
10.1142/p911#t=suppl

Box 4.1.1 Efficiency of a power plant

Elementary thermodynamics states that the maximum efficiency with which heat can be transferred into work is given by the Carnot Efficiency:

$$\eta_{Carnot} = \frac{T_{steam} - T_{cool}}{T_{steam}}$$

where T_{steam} is the temperature of the high pressure steam (550–600°C, or 823–873 K) and T_{cool} (~40°C, or 313 K) is the temperature of the flue gas. In addition to the theoretical Carnot efficiency, one has to take into account that a typical gas turbine has an efficiency of 75%. These are two important factors that explain why the maximum thermal efficiency of existing coal-fired power plants is about 44%.

CO_2 for geological storage because CO_2 needs to be compressed (typically to 150 bar) for easier transport. Clearly, the energy used for the capture and compression of CO_2 will reduce the efficiency of a power plant.

So to summarize, the main advantage of post-combustion carbon capture is that it can be added on as an accessory to an existing power plant without requiring the construction of an entirely new plant. The main drawbacks are the reduction of energy production for the power plant, and the fact that the size of such a separation unit can be significant — too big, even, to easily fit on the power plant site.

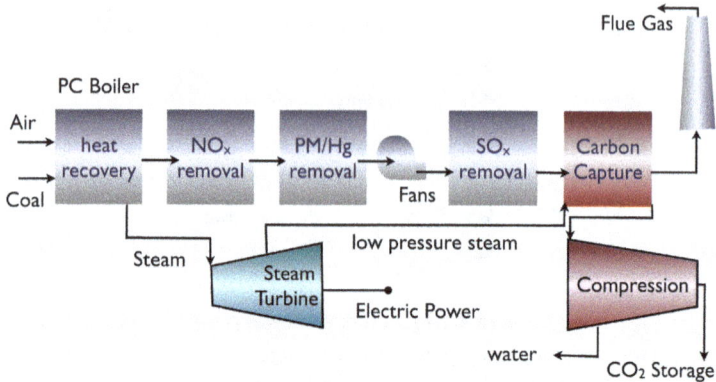

Figure 4.1.2   Coal-fired power plant with post-combustion carbon capture

This animation can be viewed at: http://www.worldscientific.com/worldscibooks/ 10.1142/p911#t=suppl

Question 4.1.1 Power plants

What is the most efficient power plant in the USA and the rest of the word? What is the average efficiency of a USA power plant? Explain the differences.

Post-combustion is not the only way to approach carbon capture. If one were willing to build an entirely new power plant, it may be possible to find more economical solutions using pre-combustion carbon capture. The idea behind **pre-combustion carbon capture** is essentially to burn coal with pure oxygen rather than air, which contains a mix of several gasses. The resulting flue gas would only contain water and carbon dioxide. Then, by condensing the water, we could easily separate CO_2 from the gas mixture. Burning coal with pure oxygen, however, means we have to find a way to efficiently separate oxygen from air. At present, this step is done using cryogenic separations, which are expensive and energy-intensive processes. So for pre-combustion methods the cost of CO_2 separation is mainly associated with the air separation step. Additional research is needed to find or develop materials that can accomplish this separation more efficiently.

There are several types of pre-combustion capture: **oxycombustion**, **integrated gasification combined cycle (IGCC)**, and **chemical looping** [4.1]. These differ in their methods of integrating the technology into a power plant, as illustrated in **Boxes 4.1.2, 4.1.3,** and **4.1.4,** respectively. From a gas separation point of view, the IGCC process involves the conversion of coal into **syngas**, a mixture of CO and H_2. The separation of these two gasses is an important step in the IGCC process. In contrast, chemical looping avoids the need for separation of oxygen from air.

An obvious question at this point is: which technology is the best? As we will see, this is a very difficult question to answer. Take the issue of cost, for example. The main difference between pre-combustion and post-combustion technologies is in construction. While post-combustion installations can be added onto existing power plants, as we saw, implementing pre-combustion carbon capture technology in an existing power plant

---

Box 4.1.2 Oxycombustion (Pre-combustion carbon capture)

In the oxycombustion process coal is burned using pure oxygen. The pure oxygen is obtained from a cryogenic air separation unit (ASU). Burning coal with pure oxygen would produce temperatures that are too high for the currently available boiler materials. The temperature is reduced by recycling part (70–80%) of the flue gasses and mixing this with pure oxygen. The resulting flue gas contains water and carbon dioxide, which can easily be separated. Pulverized coal (PC) is burned in a boiler, which in this particular case does not use selective catalytic reactors (SCR) for controlling NO_x emissions. Wet limestone is used for flue gas desulphurisation (FGD).

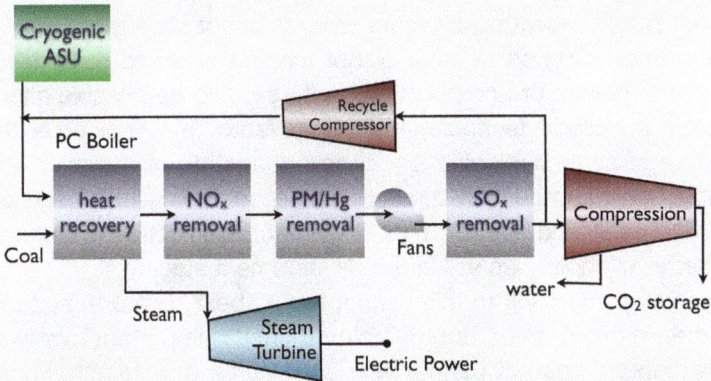

Box 4.1.3 IGCC (Pre-combustion carbon capture)

The integrated gasification combined cycle (IGCC) process relies on the conversion of coal into syngas. Syngas is a mixture of CO and H_2, and is formed by partially oxidizing coal in a coal gasifier. A water gas shift reactor, which converts water and carbon monoxide into hydrogen and carbon dioxide, is used to increase the amount of hydrogen in the mixture. Before the fuel goes into the burner, the CO_2 is separated from the H_2, and the H_2 is subsequently burned to produce the heat for generating the steam. In the IGCC process, the CO_2 separation involves a mixture of hydrogen, carbon monoxide, and carbon dioxide at high pressure.

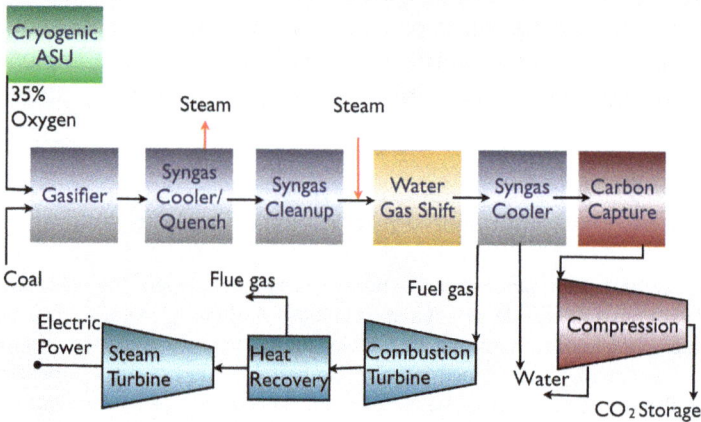

requires a significant refitting, which is not always possible for various reasons. But for new power plants, one does not have these same constraints. Depending on whether or not funding exists to build new coal-fired power plants, pre-combustion capture could be a viable alternative to post-combustion technology. Another factor to consider is that, at present, not all technologies are in an equal state of development. For example, post-combustion through amine scrubbing has been around since 1930 [4.2], and is ready to be implemented. In contrast, technologies like chemical looping are still in the development stage.

We will come back to this concept of a "best" capture method several times in this chapter, but for the moment it is important to realize that carbon capture has not been a very active area of research. For a very long time the emission of CO_2 was not seen as a real problem, so if we

Box 4.1.4 Pre-combustion carbon capture: chemical looping

The idea of chemical looping is to separate the combustion process into two separate reactors. Oxygen is taken from the air by the first reactor and then transported to the other reactor, where combustion takes place and CO_2 is produced. In this scheme, there is no mixing of N_2 with flue gas, therefore the separation only involves CO_2 and H_2O. The key question is how to separate out the combustion process. One idea is to use a metal to transport oxygen from one reactor to another. In the air (oxidizing) reactor, the metal (Me) would react with the oxygen in the air:

$$2Me + O_2 \rightarrow 2MeO$$

The metal could be transported to the fuel (reducing) reactor, where it could react with the fuel:

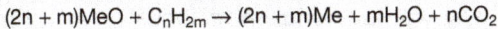

$$(2n + m)MeO + C_nH_{2m} \rightarrow (2n + m)Me + mH_2O + nCO_2$$

After the reaction, the metal would be transported back to the air reactor to close the loop.

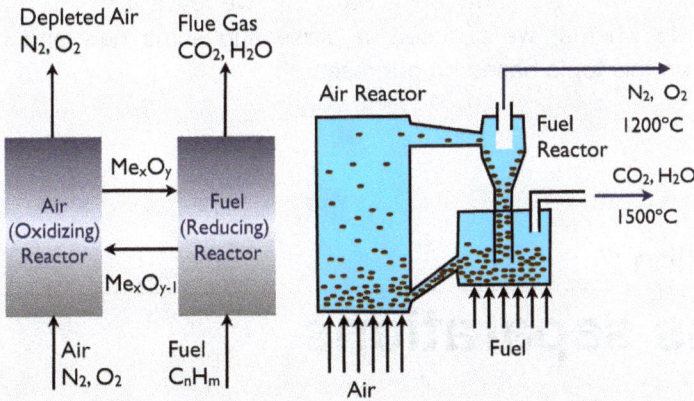

had to decide on a technology today, we would have a very limited number of options. Of the commercially available technologies that could be used for carbon capture today, most were not originally designed for this purpose. Take Bottoms' amine scrubbing process, for example. In Chapter 5, we will discuss this process in detail. Bottom developed this process for treatment of sour gas, i.e., the separation of acid gasses

(CO_2, SO_x, H_2S) from natural gas, not for separating CO_2 from flue gas. The good news is that several of these technologies can very easily be adapted for the purpose of carbon capture [4.3]. From an industry perspective, these older methods may actually be more appealing than new capture technology, because techniques like amine scrubbing are "tried and true" compared to technologies developed specifically for carbon capture that use completely new ideas and materials. As this field has become a very active area of research, however, the number of alternative technologies will likely increase rapidly in the near future.

The field of carbon capture is too large to cover completely in a single textbook, so rather than spread ourselves thin, we will focus mainly on the role of novel materials in gas separations. From our perspective, it is very tempting to make many arguments for why novel materials are the most important aspect of carbon capture. But the more accurate and honest reason why we have chosen to focus on novel materials is that two of our authors' research happens to focus on precisely this topic. Our decision to leave out many other worthy aspects of carbon capture is certainly not due to their lack of importance. Rather, it is that we want to stick to what we know best. Naturally, we leave it to the reader to conclude whether we succeed in presenting some new and original views on this topic based on our research.

Section 2

Gas separations

In the previous section, we mentioned several different types of gas separations that are used in carbon capture technologies. These included:

- CO_2 from N_2 in flue gasses
- O_2 from air (to obtain pure oxygen)
- CO_2 from the mixture of $H_2/CO/CO_2$ (the syngas in coal gasification from which the CO_2 needs to be removed)

We also touched on the separation of water and carbon dioxide, but as this separation simply involves the condensation of water, there is very little room for improving this process. The focus of this book will be on the separation of CO_2 from N_2 in flue gasses, but occasionally we will refer to the other separations listed above.

Minimum energy for separations

Before we begin discussing the different types of technologies one can use to separate CO_2 and N_2, it is interesting to take a step back and ask ourselves why it costs energy to separate gasses. Why can't we just use a Maxwell Demon (see **Box 4.2.1**) to separate CO_2 and N_2 at zero energy cost? The answer lies in thermodynamics, which not only teaches us that it costs energy to separate gasses, but also allows us to estimate the minimum energy required to carry out these separations.

Let us consider the carbon capture process shown in **Figure 4.2.1.** We assume that our flue gasses are at a given temperature T and that the mole fraction of CO_2 in our flue gasses is given by x_{flue}. If we carry out the

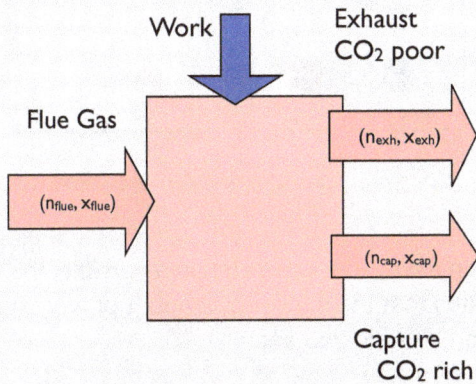

Figure 4.2.1 Carbon capture process

A flue gas, with a flux of n_{flue} [mol/s] and mole fraction CO_2 x_{flue} is separated into a CO_2 rich capture stream with flux n_{cap} and mole fraction x_{cap} and a lean exhaust stream with flux n_{exh} and mole fraction x_{exh}. The separation process costs energy as indicated by the blue arrow.

Box 4.2.1 The Maxwell Demon

James Clerk Maxwell (1831–1879) envisioned a method of separating gasses without incurring energy costs. To illustrate this idea, he described an imaginary system in which a small devil operated a gas separation unit containing hot and cold gas molecules. The demon could selectively open and close a door separating the two chambers in such a way that he would let only one type of molecule pass. Assuming that the door is frictionless, this separation would be achieved with zero energy costs. This process is illustrated here with red (hot) and blue (cold) molecules. If Maxwell's demon were successful, the net result would be for one side of the chamber to heat up and the other to cool down (a violation of the second law of thermodynamics). Maxwell's thought experiment does not just apply to heat, but to other scenarios involving gas separation. Just as systems should not spontaneously separate into hot and cold components, the second law of thermodynamics states that mixtures do not spontaneously separate into their pure components.

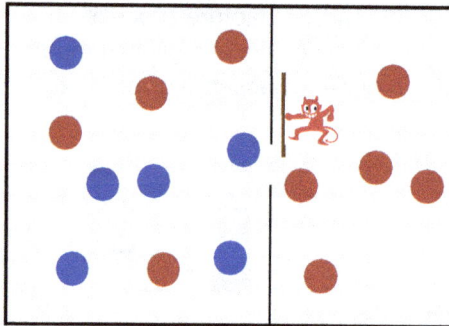

separation in such a way that each step is reversible, the first law of thermodynamics states that the minimum amount of work we need to separate one mole can be modeled as:

$$w_{min} = \Delta U^{sep} - T\Delta S^{sep}$$

We assume flue gasses behave like ideal gasses. In an ideal gas, the internal energy (U) is zero. Hence, the change in the internal energy is zero: $\Delta U^{sep}=0$. Therefore the minimum work (w_{min}) is determined by the change in entropy (ΔS^{sep}).

The entropy change is given by:

$$\Delta S^{sep} = n_{cap}s^{mix}(x_{cap}) + n_{exh}s^{mix}(x_{exh}) - n_{flue}s^{mix}(x_{flue}),$$

where s^{mix} is the molar entropy of the mixture with the given composition. We will use the convention that the total entropy of the system is written as S, while the entropy per mole is written as s. We will use a similar convention for the energy (U and u) and work (W and w). The molar **entropy of mixing** two components is given by:

$$\frac{\Delta s^{mix}}{k_B} = -x \ln x - (1-x) \ln (1-x),$$

where x is the mole fraction of one of the components. This formula can be derived from statistical thermodynamics (see **Box 4.2.2**).

Figure 4.2.1 suggests that we have 6 variables in the carbon capture process. However, not all variables can be controlled independently. As the total number of moles needs to be conserved, we have:

$$n_{flue} = n_{exh} + n_{cap},$$

Box 4.2.2 Entropy of mixing

Boltzmann has shown that the entropy S is related to the total number of configurations Ω accessible to a system:

$$S = k_B \ln \Omega$$

To illustrate how we count the total number of configurations, consider a lattice model with N_A molecules of type A (red) and N_B molecules of type B (blue). Each lattice site is occupied by exactly one molecule. If the system is separated into two compartments, there is exactly one way in which we can put the blue and red molecules on the lattice ($\Omega_A=1$ and $\Omega_B=1$). If we allow the system to mix, the total number of configurations is given by:

$$\Omega_{AB} = \frac{(N_A + N_B)!}{N_A! N_B!}$$

Box 4.2.2 (*Continued*)

The entropy of mixing is given by:

$$\Delta S^{mix} = S_{AB} - S_A - S_B = k_B \ln\left(\frac{(N_A + N_B)!}{N_A! N_B!}\right)$$

For very large number of particles we can use the Stirling approximation:

$$\ln N! \approx N \ln N - N,$$

which gives:

$$\Delta S^{mix} = k_B (N_A + N_B) \ln (N_A + N_B) - k_B N_A \ln N_A - k_B N_B \ln N_B$$

If we write for the mole fraction of component A:

$$x_A = \frac{N_A}{N_A + N_B},$$

we obtain for the entropy of the mixing:

$$\Delta s^{mix} = \frac{\Delta S^{mix}}{N} = -k_B x_A \ln x_A - k_B (1 - x_A) \ln (1 - x_A)$$

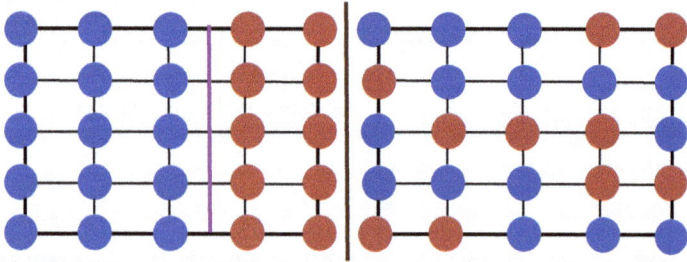

where n is the flux in number of moles per second. The number of moles of CO_2 is also conserved:

$$x_{flue} n_{flue} = x_{exh} n_{exh} + x_{cap} n_{cap}$$

As we think about how to design a separation process, it's important to realize that that we cannot control the flux of the flue gas nor the composition of the flue gas (x_{flue}) because these streams come from the power plant. We can, however, specify the concentrations in the capture stream (x_{cap}) and in the exhaust stream (x_{exh}). The corresponding follows from the mass balances:

$$n_{cap} = \frac{\left(x_{flue} - x_{exh}\right)}{\left(x_{cap} - x_{exh}\right)} n_{flue} \quad \text{and} \quad n_{exh} = \frac{\left(x_{flue} - x_{cap}\right)}{\left(x_{exh} - x_{cap}\right)} n_{flue}$$

We can now compute the entropy per mole of flue gas as a function of our two design variables, the compositions of the exhaust (x_{exh}) and the capture stream (x_{cap}):

$$\Delta s^{sep} = \frac{\Delta S^{sep}}{n_{flue}} = \frac{\left(x_{flue} - x_{exh}\right)}{\left(x_{cap} - x_{exh}\right)} s^{mix}\left(x_{cap}\right)$$
$$+ \frac{\left(x_{flue} - x_{cap}\right)}{\left(x_{exh} - x_{cap}\right)} s^{mix}\left(x_{exh}\right) - s^{mix}\left(x_{flue}\right)$$

This yields for the given flow the **minimum work** per mole of CO_2 per second:

$$W_{min} = \frac{T\Delta S^{sep}}{x_{flue} n_{flue}} = \frac{T}{x_{flue} n_{flue}} \left[\frac{\left(x_{flue} - x_{exh}\right)}{\left(x_{cap} - x_{exh}\right)} s^{mix}\left(x_{cap}\right) \right.$$
$$\left. + \frac{\left(x_{flue} - x_{cap}\right)}{\left(x_{exh} - x_{cap}\right)} s^{mix}\left(x_{exh}\right) - s^{mix}\left(x_{flue}\right) \right]$$

Figure 4.2.2 shows the effect of the initial concentration of CO_2 in the flue gas: the lower the concentration, the higher the minimum work per kg CO_2 [4.4]. This relationship explains why it is not a great idea to try to capture CO_2 directly from the air. Capturing CO_2 from the atmosphere would require up to five times the energy input compared to capturing CO_2 directly at a point source, such as a power plant (see **Figure 4.2.2**). Even among different types of power plants, there are differences in efficiency. Flue gasses from coal-fired power plants have a higher concentration of CO_2 (10–15%) compared to those from natural gas-fired power plants

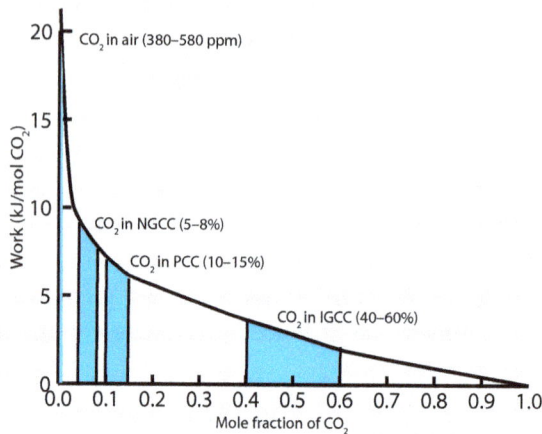

Figure 4.2.2 Minimum work to capture CO_2 as a function of the initial concentration of CO_2 in the flue gas

The figure illustrates the differences in minimum work to capture 100% of the CO_2 for flue gasses from different processes: IGCC (integrated gasification combined cycle), PCC (coal), NGCC (natural gas), and capture directly from air. *Figure based on data from* [4.4].

(5–8%). Because of this difference, carbon capture from gas-fired power plants tends to be relatively more expensive than capture from coal-fired power plants. This explains why most CCS research to date has focused on coal-fired power plants; they produce more CO_2 and are therefore better entities to first consider for capturing CO_2 more efficiently.

Another interesting way to look at the minimum energy requirement is to express it as a fraction of the total energy produced [4.5]. The average USA coal-fired power plant generates 3.43 GJ net electricity per tonne of CO_2 emitted. **Figure 4.2.3** shows that if we capture 100% of the CO_2 with a flue gas of 15% CO_2, the minimum energy needed to capture is 5.12% of the electrical energy generated by the power plant. If we capture 90% of the CO_2, this number reduces to 4.22% of the electrical energy generated by the power plant. This calculation shows that it costs 20% more energy to separate the last 10% of CO_2 from the flue gas — a disproportionate amount of energy for the outcome. For this reason most regulations do not require 100% capture, but a much more sensible 90%. For a typical reference value for a coal-fired power plant (12% CO_2), the minimum energy for a 90% separation is about 158 kJ/kg CO_2 [4.6].

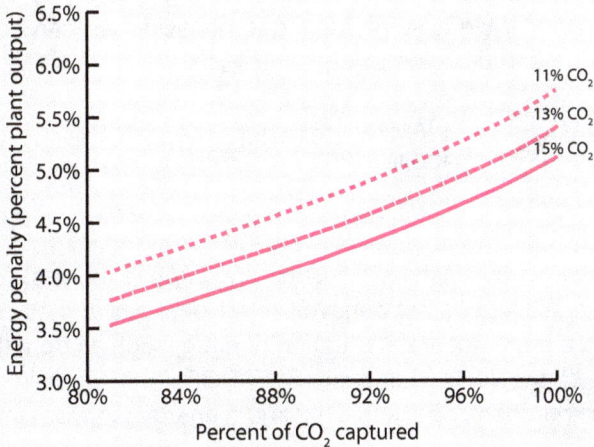

Figure 4.2.3 Energy requirement as a fraction of the energy produced

The energy penalty as a function of the percentage of CO_2 captured from the flue gas. In this calculation the separation is carried out at 40°C using different flue gas concentrations. *Figure based on the calculations given in* [4.5].

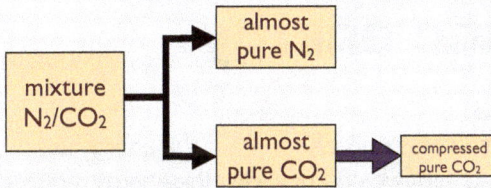

Figure 4.2.4 Separation and compression

Separation followed by compression of the capture stream.

Compression

Once the CO_2 is separated, the next step is to compress it to a pressure at which it can be stored in geological formations (see **Figure 4.2.4**). The pressure needs to be sufficiently high to ensure that CO_2 is a supercritical fluid, so we use a reference pressure of 150 bar (148 atm). The energy required to compress the CO_2 contributes significantly to the costs of carbon capture. If we assume ideal gas behavior, we can compute the energy for the compression using:

$$W_{comp} = \int_{Capture}^{Transport} p\,dV = n_{cap}RT\ln\frac{p_{transport}}{p_{capture}},$$

where n_{cap} is the amount of gas we need to compress, $p_{transport}$ and $p_{capture}$ are the pressures at which we transport and capture the gas, respectively.

However, CO_2 does not behave like an ideal gas at these conditions, so we need a more accurate equation of state, such as from the real gas properties of the NIST database REFPROP [4.7] for fluid properties. Using these numbers, one gets a minimum energy requirement of 218 kJ/ kg CO_2 [4.4].

Section 3

Parasitic energy

As we have seen in the previous sections, the energy associated with a CCS process is parsed into two components: the energy required to separate the CO_2 from the flue gas, and the energy required to compress the CO_2 for transport and storage. While minimizing the total of these two energies is a design goal, different capture processes may distribute the energy demands differently between separation and compression. To best compare different processes, then, engineers use the concept of **parasitic energy**.

We can easily formulate an equation for parasitic energy by recognizing that carbon capture consumes two forms of energy from the power plant: electricity directly from the generators, and heat in the form of the steam that derives from fuel combustion. For example, compressors that deliver 150 bar CO_2 for transport and storage will likely be powered directly by power plant electricity. The capture process itself will likely use steam to regenerate the capture medium. This diverts steam from the power cycle of the plant, which effectively imposes a parasitic load on the power plant. We can use thermodynamics to estimate the loss of electricity. First we have the Carnot efficiency (η_{final}) which gives us an upper limit of the efficiency in which heat (Q) is transferred into work. In addition, the steam turbines have an efficiency of 0.75 [4.6]. If we add these contributions, we get the following equation representing the parasitic energy of CCS for the power plant:

$$E_{par} = 0.75\eta_{final}Q + W_{comp},$$

where W_{comp} refers to the work required for compression. As an example, this equation tells us that if the steam requirements of two processes are the same, the one with the lowest compression cost will have the lowest parasitic energy and be the more favorable design.

At this point, it is important to mention that energy costs are not the only costs that matter. Energy will be an important factor in the operational cost of CCS, but building a carbon capture separation unit is an investment that can represent as much as 50% of the cost of the power plant. The chemicals necessary to run the capture process, for example, may constitute a significant fraction of the total capital cost of the new or retrofitted power plant. At the end of the day both operating and capital costs will contribute to an increase in the price of electricity. This price will be one of the most important factors in the selection of a new technology.

Researchers in the chemical and materials sciences face a special challenge when seeking to develop new chemistries for carbon capture because it is very difficult to translate fundamental research cost to manufactured cost. For example, how can we estimate the future worldwide production price of a material that is currently handmade by a PhD student at rate of 1 milligram per day, on a good day? Naturally, a material that is already used on a very large industrial scale will be orders of magnitude cheaper compared to a completely novel material. Yet if this new material could significantly reduce the *parasitic energy* of CCS compared to known technology, it could play an important role in lowering costs in the future. It is therefore the responsibility of researchers to demonstrate that new materials will result in significantly lower energy costs compared to the existing process. Hence, this metric is most commonly cited in research publications.

In the remainder of the carbon capture sections, we will discuss in greater detail the three most commonly used gas separation processes:

Liquid absorption: a liquid is used to selectively absorb CO_2.

Solid adsorption: a porous solid material is used to selectively adsorb CO_2.

Membranes: a membrane is used to separate CO_2 from the flue gas.

Section 4

Review

4.1 Reading self-test

1. What is the maximum theoretical (Carnot) efficiency of a coal-fired power plant?
   a. 70%
   b. 60%
   c. 40%
   d. 20%

2. Why is NO_x removed from the flue gas before SO_x?
   a. The temperature for NO_x removal is higher than for SO_x removal
   b. NO_x contaminates the SO_x removal process
   c. The temperature for NO_x removal is lower than for SO_x removal
   d. The order does not matter: in the alphabet N is before S

3. Which statement is correct about chemical looping?
   a. The coal is looping between two reactors
   b. It uses a solid to transport oxygen between the air and combustion parts
   c. Oxygen is separated from the air
   d. Statements B and C
   e. Statements A and C

4. Which statement is not correct about the IGCC process?
   a. IGCC = integrated gasification combined cycle
   b. Coal is converted into syngas
   c. Syngas is a mixture of CO_2 and H_2
   d. The key separation in IGCC is CO_2 and H_2

5. Which statement is not true about oxycombustion?
   a. The current technology to separate oxygen is cryogenic distillation
   b. CO_2 is added to the oxygen to lower the temperature in the boiler
   c. The theoretical efficiency of an oxycombustion plant is higher than combustion in air
   d. The most expensive part of oxycombustion is the CO_2 separation

6. Which statement about chemical looping is not correct?
   a. Chemical looping involves the transportation of solid particles from one reactor to another
   b. The solid particles "carry" the oxygen
   c. Oxygen is consumed in the reducing reaction through: $2Me + O_2 \rightarrow 2MeO$
   d. None of the above

7. Which statement about entropy is not correct?
   a. The Boltzmann relation between entropy and number of configurations is $S = k_B \ln \Omega$
   b. The entropy of mixing is higher at higher temperatures
   c. The smaller the concentration of component A in a binary mixture, the higher the entropy of mixture per molecule A

8. Which statement best describes the parasitic energy of a CCS process?
   a. The total energy needed for the separation and compression of CO_2
   b. The electric energy that a power plant cannot deliver because of the CCS process
   c. The higher the temperature of the heat needed for regeneration, the higher the parasitic energy

4.2 Parasitic energy

1. In the graph below the minimum work for separating CO_2 is given as a function of the mole fraction of CO_2. Identify the corresponding processes of CO_2 separation given by the letters A–D.

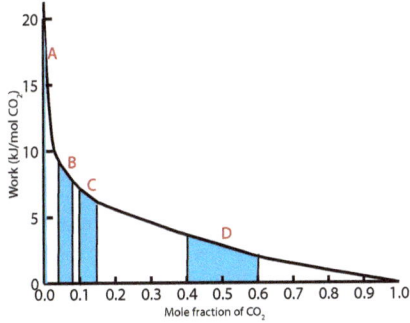

2. The energy penalty of separating CO_2 at a temperature T is given by the red curve. Now when we decrease the temperature, the blue curve gives the energy penalty at this temperature. Which one is correct?

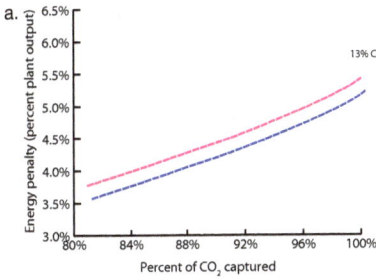

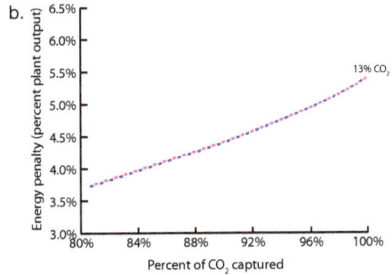

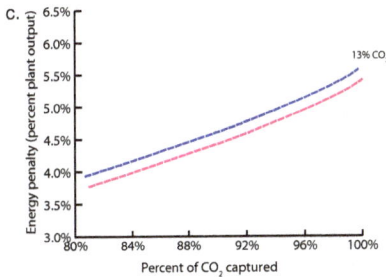

Section 5

References

1. Ciferno, J.P., J.J. Marano, and R.K. Munson, 2011. Technology Integration Challenges. Chem. Eng. Prog., **107** (8), 34.
2. Bottoms, R., 1930. Separating Acid Gases, US Patent 1,783,901.
3. Rochelle, G. T., 2009. Amine scrubbing for CO_2 capture. Science, **325** (5948), 1652. http://dx.doi.org/10.1126/science.1176731
4. Socolow, R., M. Desmond, R. Aines *et al.*, 2011. Direct air capture of CO_2 with chemicals: a technology assessment for the APS panel on public affairs. American Physical Society. http://www.aps.org/policy/reports/assessments/upload/dac2011.pdf
5. Bhown, A.S., and B.C. Freeman, 2011. Analysis and status of post-combustion carbon dioxide capture technologies. Environ. Sci. Technol., **45** (20), 8624. http://dx.doi.org/10.1021/es104291d
6. Freeman, S.A., R. Dugas, D. Van Wagener, T. Nguyen, and G.T. Rochelle, 2009. Carbon dioxide capture with concentrated, aqueous piperazine. Ener. Proc., **1** (1), 1489. http://dx.doi.org/10.1016/j.egypro.2009.01.195
7. Lemmon, E.W., M.L. Huber, and M.O. McLinden, 2010. NIST Reference Fluid Thermodynamic and Transport Properties Database (REFPROP): Version 9.0. http://www.nist.gov/srd/nist23.cfm

Chapter 5

Absorption

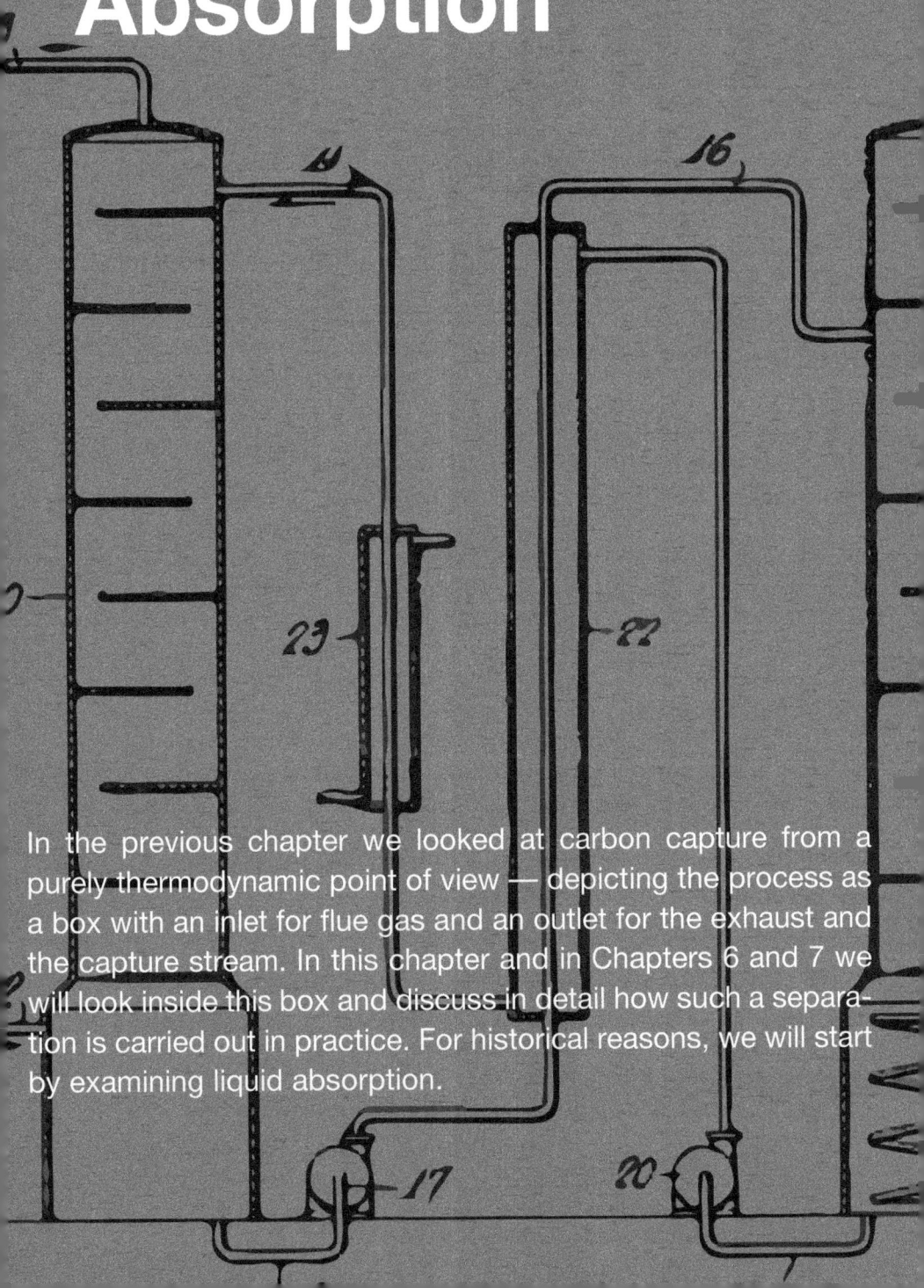

In the previous chapter we looked at carbon capture from a purely thermodynamic point of view — depicting the process as a box with an inlet for flue gas and an outlet for the exhaust and the capture stream. In this chapter and in Chapters 6 and 7 we will look inside this box and discuss in detail how such a separation is carried out in practice. For historical reasons, we will start by examining liquid absorption.

Section 1

Introduction

In 1930, Bottoms [5.1] patented a technology to clean natural gas of its impurity gasses — the amine scrubbing process, depicted in **Figure 5.1.1.** Interestingly, although this technology is nearly a century old, the current CCS version still looks very similar. As you can see, the process consists of two parts: the **absorber** and the **stripper**. The flue gas exhaust enters at the bottom of the absorber and rises through the column, which is segmented into several tray decks. Each tray consists of a flat plate with holes and a weir to manage liquid solvent flow. Liquid solvent enters at the top of the absorber and flows down the plates in the opposite direction of the flue gas. The weir "holds up" just enough solvent to ensure that the flue gas vapor bubbling up through the holes in each plate allows the solvent to absorb CO_2. As a result of the falling solvent and rising flue gas passing through many such plates, the exhaust at the top of the column will be significantly CO_2-depleted.

The solvent has a limited capacity to absorb CO_2, so once it has been in the absorber long enough to become completely saturated, it needs to be moved from the bottom of the absorber to another device where it is stripped of its CO_2, and finally returned to the top of absorber for a fresh round of absorption. The regeneration is done by the second part of the process, in the stripper. The stripper looks and acts like the opposite of an absorber. The CO_2-rich solvent from the bottom of the absorber is injected at the top of the stripper where it cascades through plates with holes and weirs. Heat is added (in the form of steam) at the bottom of the stripper so as to release CO_2 from the liquid solvent as well as boil some solvent, thereby forming a gas phase that rises through the stripper. At the top of the stripper a condenser returns the solvent, and nearly pure CO_2 is ready for subsequent cooling, compression, and transport for storage. With its CO_2 "stripped," the hot liquid solvent is transported from the bottom of the stripper and recycled back into the top of the absorber. Due to the fact that the regenerated solvent is much warmer, a heat exchanger (number 22 in **Figure 5.1.1**) is used to recover some of the heat used for the regeneration process. The entire process is a superb example of 20th century chemical engineering — moving

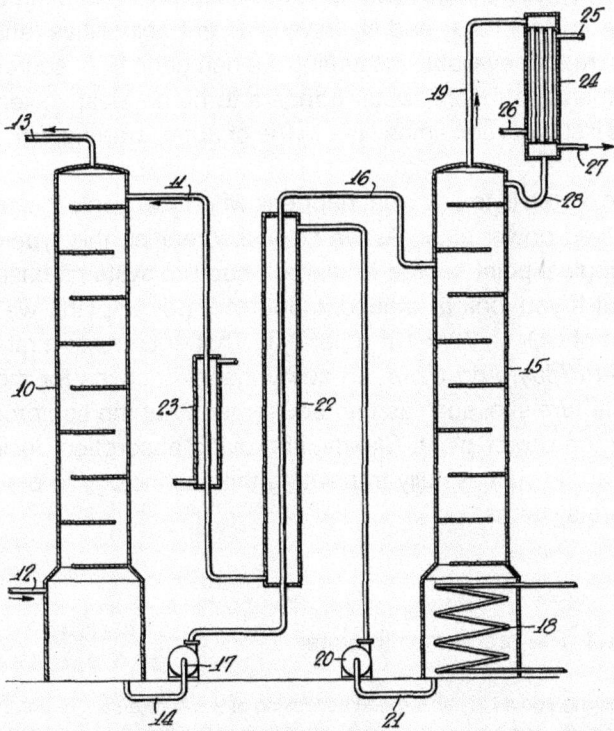

Figure 5.1.1 The carbon capture process using amine scrubbing as developed by Bottoms in 1930

Image from Bottoms [5.1].

massive quantities continuously through devices while they are subject to chemistry, thermodynamics, and mass transport between phases.

The idea of **absorption** is simple: if we have a solvent with a different solubility for CO_2 and N_2, we can design a separation process such that the exhaust and the flue gasses have the composition we desire. However, depending on the type of solvent, the size of the separation equipment can vary and energy requirements can differ. We will illustrate these ideas by designing a sample carbon capture separation using pure water as a solvent. CO_2 and N_2 have different solubilities, and water is cheap and readily available (assuming we can get it from sources like the ocean). At least in theory, water appears to be an ideal solvent. We will revisit this issue several times later in the chapter, starting with **Box 5.1.1** [5.2, 5.3].

The next step is to ask what happens when you apply this technology to a coal-fired power plant. As we discussed earlier, this type of plant is the most logical point source for carbon capture systems based on efficiency. But if you look at a generic 500 MW power plant, we're talking about a lot of CO_2 — 70×10^8 g per day. That equates to a rate of about 4,000,000 m^3/day! And there are additional restrictions for the technology at a capture site, such as the requirement that the equipment fits on the site of the power plant. Clearly, improving absorption design will be a huge factor in successfully translating this technology to practical carbon capture systems.

Box 5.1.1 The problem with water

We start talking about CO_2 absorption by looking at water as a solvent. After all, water is all around us, and it is used in many other industrial processes, including cooling, mining, fuel production, and emissions control. The reality, however, is not so simple. In addition to its use for drinking and agriculture, water is required to produce most forms of energy, including that generated by turbines, hydroelectric, and geothermal methods. Particularly when it comes to drinkable, or potable, water, there are great strains on our current global resources. Further complicating matters, energy is required to move, treat, deliver, use, and dispose of water. The inexorable relationship between water and energy generation even has a name: the **water-energy nexus** [5.2, 5.3].

The global consumption of both water and energy is on the rise. According to the World Health Organization, as much as one-sixth of the world's population does not have access to safe drinking water. Yet in most countries, more water is consumed for energy purposes than for drinking. In the USA, for example, electricity production in 2000 accounted for 39% of national freshwater withdrawals. As the population increases, our energy needs will put an even greater strain on our water resources. As we think about how to minimize the costs of absorption for carbon capture in order to make it a sustainable process, we will need to look at both water and energy as two sides of the same coin. It would not make sense for carbon capture technology to trade one form of environmental strain (CO_2) for another (H_2O).

(Continued)

Box 5.1.1 (*Continued*)

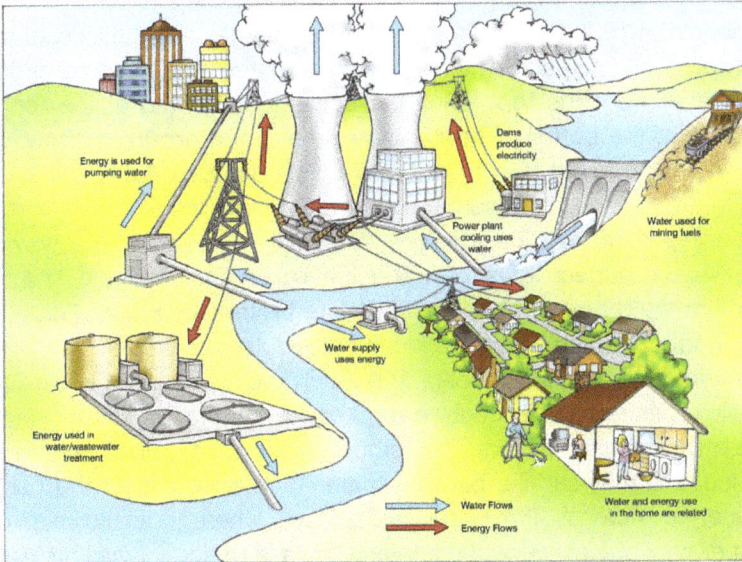

Figure from "Energy Demands on Water Resources", US DOE, 2006.

Section 2

Absorption design

Absorption equipment

Let us start out with a more detailed physical description of an absorber in order to understand how we might improve it. There are two main factors to consider: mass transfer and driving force. In the absorber, we need to transfer the CO_2 from the gas phase into the liquid solvent. As this transfer only takes place at the solvent-gas interfaces, it would make sense to try to tweak our design to increase this area as much as possible. The other important design consideration is that the driving force for CO_2 to move from the gas to the solvent is determined by the difference between the actual concentration and the equilibrium concentration. In the 1930 design, we see that the flue gas and the solvent move in opposite directions; the flue gas is injected at the bottom and removed as exhaust from the top, while the solvent is injected at the top and removed at the bottom. The importance of **countercurrent flow** compared to **concurrent flow** is further discussed in **Box 5.2.1**.

On the outside, an absorber almost universally looks like a tower. But several types of absorber designs differ on the inside in the way they increase the surface area between the liquid and the gas. The most straightforward type of absorber is a **plate tower** (see **Figure 5.2.1**). This is a cylinder in which plates are placed at regular intervals. A plate can be any metal disk with small holes that retain the liquid (see **Figure 5.2.2**), but allow gas to pass through. A **packed column** looks very similar to a plate tower from the outside, but rather than plates it is filled with material that looks somewhat like packing material. The idea is for the packing material to increase the liquid-gas surface when liquid flows over it. Another way of increasing the interface area is to spray small droplets of liquid in the column. The most important equipment in a **spray column** therefore resembles a shower head. In a **bubble column**, the area is increased by injecting small bubbles of gas in the bottom of the column.

Let's zoom in to see what's going on in one of the plate tower trays. Liquid comes down from above via a funnel and then falls down to the next plate (see **Figure 5.2.2(c)**) through the holes. In this system, we have vapor rising and liquid falling. There are several types of plates with slightly different formal geometries, such as bubble caps, to facilitate vapor/liquid mixing, but the most common form is a metal plate with the holes punched in it and a weir, or dam-like structure, that maintains a thin layer of liquid on each tray. The idea is to ensure optimal contact between the solvent and the flue gas such that the system is as close to equilibrium as possible (see **Movie 5.2.1**).

Now back to our task of designing a separation unit on the scale of 70×10^8 g of CO_2 per day of flue gas using water as a solvent. For this we need to learn a little more about absorber design. We need to address a few questions: what are the dimensions of our plate absorber? How many plates do we need to install? And how much water will we need? Of course, if we were to make a design in real life, we would need many more details. But in our case, answering these elementary questions will

Box 5.2.1 Concurrent versus countercurrent

In the design of an absorber unit we have two options: we can inject the solvent and the gas at the same place on the column, or we can inject one at the top and the other at the bottom. If both are injected in the same place, the gas and the liquid will flow in the same direction and we have *concurrent* flow; otherwise the flow is *countercurrent*. The mechanics of the problem make it easier for a gas to be injected at the bottom and the fluid at the top, as gravity naturally will cause gas to rise and liquid to fall. However, there is a more compelling reason to favor countercurrent flow.

In the countercurrent flow, the concentration of the CO_2 in the flue gas will decrease as a function of the height in the column. As the regenerated solvent is injected at the top, the concentration of CO_2 in the solvent is the lowest at the top. The concentration profile shows that the difference between the concentration of CO_2 in the liquid and gas phases remains large throughout the column. In the concurrent flow, the difference in concentration of CO_2 in the solvent (fresh) and input flue gas is large. But as you go up the absorber, the concentrations of CO_2 in the gas and solvent phase get closer, so the *difference* in concentration between the two phases gets increasingly smaller.

Why do such differences matter? A fundamental axiom of mass transport between phases is that the rate of the mass transfer between phases is proportional to the concentration differences between the two phases. Simple designs for absorber plate towers assume that in the bubbling froth above each plate the rate of mass transfer between gas and liquid phases is rapid — so rapid as to achieve thermodynamic equilibrium at each plate. By maximizing the concentration differences between the two phases in all plates in the absorber, the countercurrent design maximizes the chances that equilibrium is achieved at each plate.

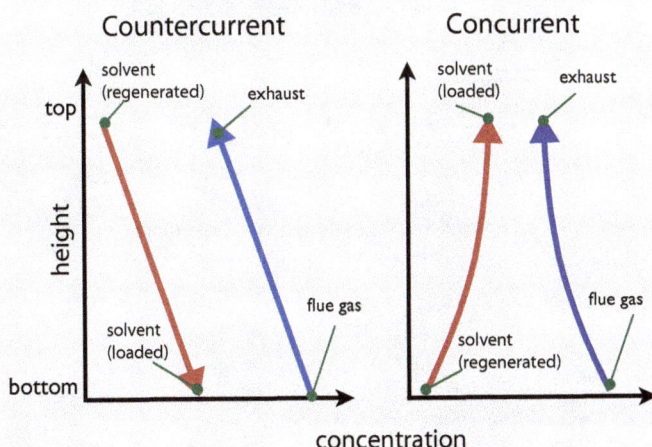

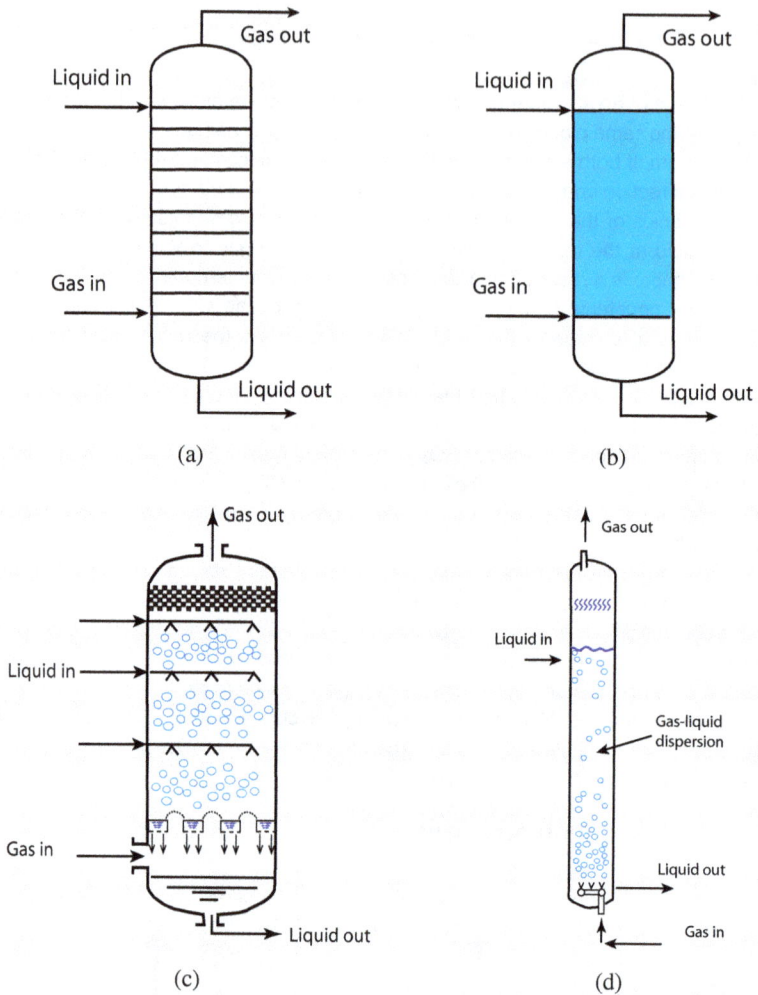

Figure 5.2.1 Different types of absorbers

(a) Plate tower.

(b) Packed columns.

(c) Spray tower.

(d) Bubble column.

(a)   (b)

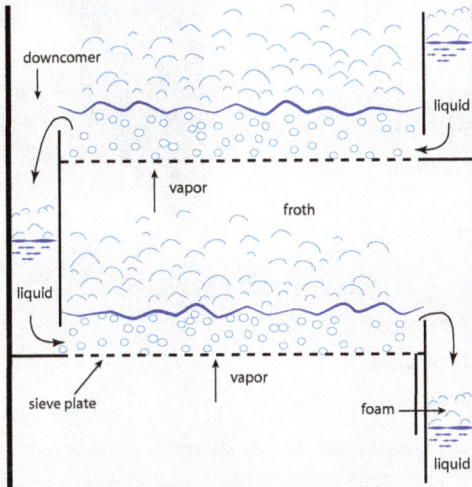

(c)

Figure 5.2.2 Different plate designs

(a) A single tray from a plate tower.

(b) A single tray from a plate tower: Each tray consists of a flat plate with holes and a weir to manage liquid solvent flow.

(c) The liquid is flowing down the column from one plate to another, while the gas is flowing through the holes in the plates.

(*Continued*)

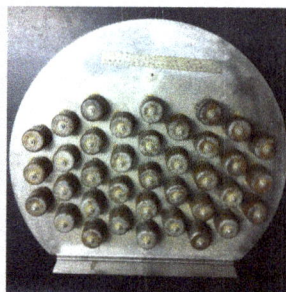

(d) (e)

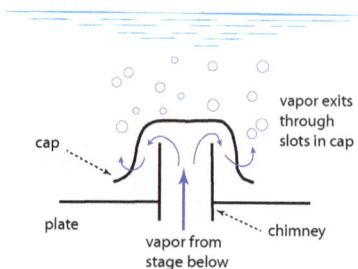

(f) (g)

Figure 5.2.2 (*Continued*)

(d) and (e) Bubble cap trays.

(f) Flow through a bubble cap.

(g) A single bubble cap.

teach us a lot about the feasibility of our separation. Along those same lines, for our purposes here we will use only the most elementary design tools, which will still give us a lot of insight into the most important parameters of the process.

Thermodynamics

At this point in our tower design we need to know, for a given temperature and partial pressure of CO_2, how much CO_2 will be absorbed in the solvent. The simplest thermodynamic model that gives us this information is Henry's law (see **Box 5.2.2**), which states that the vapor pressure of an

Movie 5.2.1 A single tray

An example of gas flowing through a plate with liquid. *This movie can be viewed at*: http://www.worldscientific.com/worldscibooks/10.1142/p911#t=suppl

ideal solution is dependent on the vapor pressure and mole fraction of each of the chemical components present in the solution, or in the formula:

$$y_{CO_2}\, p = K_{CO_2}\, x_{CO_2},$$

where p is the pressure in the column, y_{CO_2} and x_{CO_2} are the mole fractions of CO_2 in the gas and liquid phase, respectively, and the constant K is the **Henry constant** for CO_2 (see **Box 5.2.2**). We will use the common convention that x refers to a concentration in the liquid phase and y to a concentration in the gas phase. We can get the partial pressure of the gas by multiplying the pressure (p) by the mole fraction of CO_2 in the gas phase (y_{CO_2}):

$$p_{CO_2} = y_{CO_2} p$$

The constant $1/K_{CO_2}$ characterizes the solubility of CO_2 in the solvent. If a solvent has a high value of K_{CO_2}, the amount of CO_2 in the solvent will be lower compared to a solvent that has a lower value of K_{CO_2}. We can define a dimensionless constant κ:

$$\kappa = \frac{K_{CO_2}}{p}$$

We know that for water at 25°C and 1 atm, this dimensionless constant is 1,600. With this information we can form a very simple

Box 5.2.2 Henry's Law

Henry's law can be derived in several different ways. Here we start with a simple kinetic argument. We are interested in the relation between the partial pressure of CO_2 in the gas phase and the solubility in the liquid phase. If we are at equilibrium, the number of CO_2 molecules per unit of time escaping from the liquid phase into the gas phase should be equal to the number of CO_2 molecules that go from the gas phase to the liquid:

$$\Phi_{CO_2} \text{ (gas} \rightarrow \text{liquid)} = \Phi_{CO_2} \text{ (liquid} \rightarrow \text{gas)}$$

If we assume that CO_2 behaves like an ideal gas, the flux from the gas to the liquid phase is simply proportional to the number of gas molecules that collide with the surface. This number is proportional to the partial pressure of CO_2:

$$\Phi_{CO_2} \text{ (gas} \rightarrow \text{liquid)} \propto N_{CO_2} = p_{CO_2} = y_{CO_2}p$$

Also for the liquid we can assume that if the liquid behaves as an ideal solution, in which CO_2 molecules are not influenced by the presence of other molecules, the flux of CO_2 molecules escaping the liquid phase is proportional to the mole fraction:

$$\Phi_{CO_2} \text{ (liquid} \rightarrow \text{gas)} \propto x_{CO_2}$$

Equating the two fluxes gives us:

$$\frac{y_{CO_2}}{x_{CO_2}} = \kappa$$

For a real fluid, our assumptions only hold in the limit of very low pressure, which is often called the Henry regime. In this regime, we have an ideal gas and our equation can be rewritten in terms of the density of CO_2 molecules that are absorbed in a liquid and the partial pressure of CO_2:

$$\rho_{CO_2} = K_{CO_2}p_{CO_2},$$

where ρ_{CO_2} is the density of CO_2 in the liquid (number of moles per unit volume) and K_{CO_2} is the Henry coefficient of the solvent. The units of the Henry coefficient can vary. Here we have number of moles per unit volume per unit pressure, but if we express the concentration of the solvent in kg CO_2 per kg liquid, the dimension of the Henry coefficient is per unit pressure.

relation between the mole fraction of CO_2 in the solvent and in the flue gas:

$$y_{CO_2} = \kappa x_{CO_2}$$

This equation is the first of two equations that we can use to graphically describe the absorption process in a plate tower such as that shown in **Figure 5.2.1**. In particular, if we assume the flue gas and the solvent are at thermodynamic equilibrium, we can use this equation to relate the mole fractions of CO_2 in the liquid and the gas on each of the plates. This function is called the equilibrium line and is shown schematically in **Figure 5.2.3**. As we vary the design components of our multi-plate absorption column, we can assume that the equilibrium of the frothy liquid/gas mixture behind the weir can be described by a point on this line. The slope of the equilibrium line, κ, represents the physical properties of our solvent, and this simple constant gives us sufficient information on the thermodynamics of CO_2 absorption to design our separation. Of course, Henry's law only approximates the solubility. Comparison with experimental data shows that Henry's law only holds for very low concentrations of CO_2 in the solvent. A more accurate design would require taking this non-ideal behavior into account (and this is exactly what the more sophisticated design tools will do for you).

Conservation of mass

Now that we have described the equilibrium thermodynamics of our absorber, we can integrate another relationship into this system: conservation of mass. As we mentioned above, the amount of CO_2 that goes into the system must be equal to the amount of CO_2 that comes out of

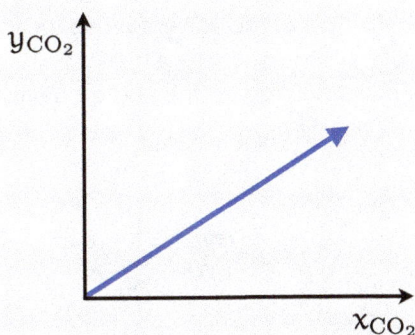

Figure 5.2.3 The equilibrium line

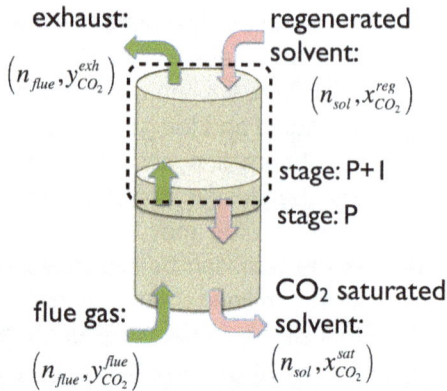

Figure 5.2.4 Mass balance over the top of the absorber

The dotted line gives the boundaries for the mass balance over the top and plate P.

the system. Let's look at mass conservation as engineers do, by drawing a system of boxes around what is entering and leaving the system.

We start at the top of an absorber (see **Figure 5.2.4**), where gas is exiting the absorption column. Even though this gas contains less CO_2 than when it entered the column, not all of the CO_2 from the original flue gas will have been absorbed. The exhaust vapor will still contain some amount of CO_2 as it exits into the atmosphere. We will represent this residual CO_2 flux as the mole fraction of CO_2 ($y_{CO_2}^{exh}$) times the flow rate n_{flue}. Entering the top of the column, we have nearly pure water, with some residual CO_2 left after the regeneration. We will represent this residual CO_2 flux as the flow of the solvent times the mole fraction of CO_2 in the solvent ($n_{sol}\, x_{CO_2}^{reg}$).

If we apply the mass balance to the entire column, we have:

$$\text{inlet} = \text{outlet},$$

where we have for the inlet:

$$\text{inlet} = n_{sol} x_{CO_2}^{reg} + n_{flue} y_{CO_2}^{flue},$$

and for the outlet:

$$\text{outlet} = n_{exh} y_{CO_2}^{exh} + n_{sol} x_{CO_2}^{sat}$$

To keep things simple, we assume that only CO_2 is transferred between the two phases and that the solvent does not evaporate. Additionally, we assume that the amount of CO_2 is relatively small, such that the gas and liquid fluxes are constant over the absorber.

Recall that both the gas and the liquid must pass through various plates, or "stages," which we will represent from the bottom up as P, P + 1, etc. So, we have vapor rising from plate 1 at the bottom of the column, and liquid descending from the top of the column.

With these assumptions, we can write a mass balance for an imaginary volume bounded by the top of the absorber and plate P + 1. The inlet streams of CO_2 are the regenerated solvent and the gas coming up from plate P:

$$\text{inlet} = n_{sol}x_{CO_2}^{reg} + n_{flue}y_{CO_2}(P)$$

The corresponding outlet streams are the exhaust at the top of the absorber and the liquid going down from plate P + 1 to P:

$$\text{outlet} = n_{flue}y_{CO_2}^{exh} + n_{sol}x_{CO_2}(P+1)$$

Equating the inlet and outlet, we deduce another relationship between the composition of the gas and liquid phases on the plates:

$$y_{CO_2}(P) = \frac{n_{sol}x_{CO_2}(P+1)}{n_{flue}} + \frac{n_{flue}y_{CO_2}^{exh} - n_{sol}x_{CO_2}^{reg}}{n_{flue}}$$

This equation defines the operating lines of the absorber (see **Figure 5.2.5**). These operating lines express the mass balance condition: the larger the solvent flow, n_{sol}, the steeper the slope of the operating line. The design of a column follows from this operating line together with the equilibrium line, for which we use:

$$y_{CO_2}(P) = \kappa\, x_{CO_2}(P)$$

McCabe-Thiele diagrams

In our column, we can control the relative amount of solvent and gas we inject in the absorber. For example, if we have a large excess of solvent, the concentration of CO_2 in the solvent will be much lower at the outlet

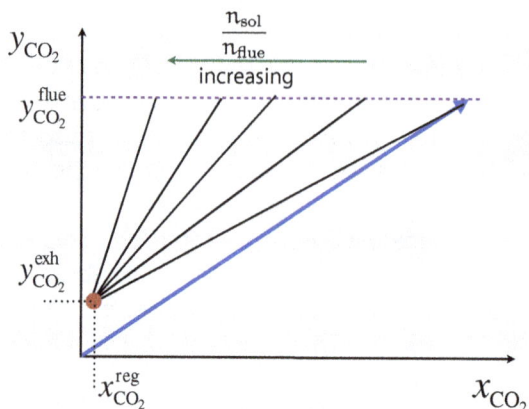

Figure 5.2.5 Equilibrium and operating lines

This figure shows Henry's law (blue) together with the mass balance equations (operating lines, black) for different ratios of gas and solvent fluxes (n_{sol}/n_{flue}). Applying the mass balance over the entire column shows that all mass balance lines have to include the point ($x_{CO_2}^{reg}$, $y_{CO_2}^{exh}$).

than in the case where we have an excess of gas. This is a simple consequence of the conservation of mass principle, and the mass balance equation above is a simple algebraic way of showing how much CO_2 comes into the system and how much CO_2 comes out of the system for a given ratio of n_{sol}/n_{flue}. This conservation principle will yield the second of the two equations we can use to graphically describe the plate tower.

From an engineering standpoint, we should note that the main design factor we can control is n_{sol}, the molar flow rate of the solvent that enters the system. Of course, we can also tune the molar flow rate of the gas (n_{flue}). For example, we can cut n_{flue} into half, but as we do need to remove 90% of the CO_2 from our flue gas (see Section 4.2), this tuning implies that we have to use two *or more* absorber/stripper reactor units. In **Figure 5.2.5**, we plotted the mass balance equation for different ratios n_{sol}/n_{flue}. We are given the initial CO_2 concentration of our flue gas entering at the bottom of the absorber, $y_{CO_2}^{flue}$. At the top of the absorber, the concentration of CO_2 in the flue gas is $y_{CO_2}^{exh}$ and the concentration of CO_2 in the regenerated solvent is $x_{CO_2}^{reg}$. By definition, every operating line will cross this point. For our design we need to specify the value of $y_{CO_2}^{exh}$, the concentration of CO_2 that we wish to achieve in the flue gas exhaust. We know from Section 4.2 that, although it would be nice to capture 100% of the CO_2 in the absorption column, it is relatively expensive to separate the last 10%. We can capture 90% of the CO_2 with reasonable efficiency, say, if our coal-fired power plant is bringing in 70×10^8 g of CO_2 per day. The mass balance will tell us the concentration of CO_2 in the solvent at different plates in the absorber and at the bottom, $x_{CO_2}^{sat}$.

Let us now start by looking at a single plate P (see **Figure 5.2.6**). Henry's law relates the solvent concentration $x_{CO_2}(P)$ to the gas concentration $y_{CO_2}(P)$ on this plate. More colloquially, Henry's law tells us that in theory all froths behind a weir are described by a point on the equilibrium line. The mass balance equation relates the solvent concentration $x_{CO_2}(P)$ in the liquid descending from a plate to the concentration in the gas ascending from the plate below, $y_{CO_2}(P-1)$. The operating line, then, assures us that mass is conserved between any two plates. A geometric construct from this graph shows the progress of CO_2 absorption as the gas moves through the absorber. CO_2 entering plate P has mole fraction $y_{CO_2}(P-1)$. It achieves thermodynamic equilibrium with the froth above plate P, then exits plate P with $y_{CO_2}(P)$. Graphically we form a triangle, which is called a stage, representing a plate in our absorber. We can

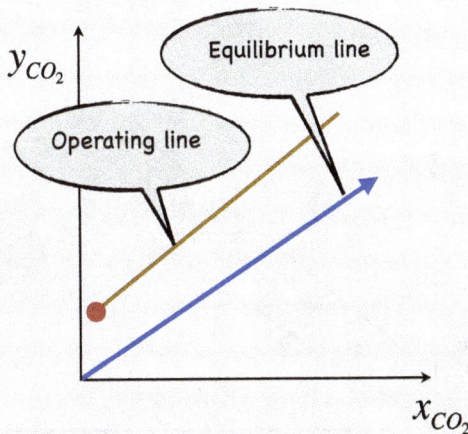

Figure 5.2.6 Mass balance and thermodynamic equilibrium on a plate

This *animation can be viewed at*: http://www.worldscientific.com/worldscibooks/ 10.1142/p911#t=suppl

"step down" between the slopes of these two lines by forming a series of triangles. Each of these triangles, or stages, represents mass conservation and thermodynamic equilibrium at one plate in our absorber.

Let us now look at a typical absorber, which is using pure water as the solvent at 25°C. Flue gas vapor from a coal-fired power plant is coming in with a CO_2 concentration $y_{CO_2}^{flue} = 0.12$. Water is exiting the absorber with some concentration of CO_2 ($x_{CO_2}^{reg}$). Let us assume a certain n_{sol}/n_{flue} ratio. We start at plate number one in the bottom of our absorber (see **Figure 5.2.7**). As we are at the bottom of the column, the gas entering the plate is the flue gas, $y_{CO_2}(1) = y_{CO_2}^{flue}$, the mass balance equation gives us the corresponding liquid concentration, $x_{CO_2}(2)$. If we assume equilibrium on each plate, the gas that is leaving this plate has a CO_2 concentration of $y_{CO_2}(2)$, according to Henry's law. We now have our first stage in the absorber. Now we can apply the mass balance and Henry's law to obtain $x_{CO_2}(3)$ and the corresponding $y_{CO_2}(3)$ — and in this way we are adding stages, each of which represent a plate. Ultimately, the stages will terminate with concentrations $x_{CO_2}^{reg}$, $y_{CO_2}^{exh}$ at the top of the absorber. This form of analysis is called a **McCabe-Thiele method,** named for the two MIT chemical engineers who developed it in the 1920s. **Figure 5.2.7** gives us an intuitive sense of how plate absorption columns operate.

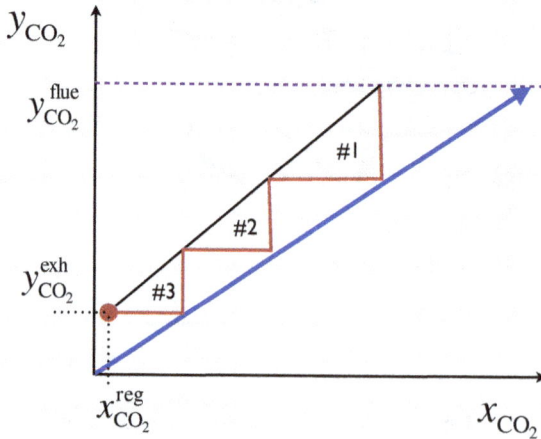

Figure 5.2.7 Absorber with three stages

The next question is: how would we use a McCabe-Thiele diagram to estimate the height of a given absorption column? One way we can think about the height of the column is by determining the required number of plates/stages, and then estimating their spacing in the column. If you were to take a ruler into an absorption column and look at the froth at each plate or tray, you would recognize that the froth height is related to the ratio n_{sol}/n_{flue}, the slope of our operating line. You would want to make sure your plates are separated far enough apart so each plate does not interfere with the froth of the tray below it. For the diameter, a trivial observation is that the smaller the height between plates, the bigger the diameter must be because all liquid and gas need to fit in the column. However, it is not as simple as this scaling. To maintain the froth and sustain the column throughput one needs to have an appropriate ratio of hole to plate areas. In addition, we can vary the size and number of holes. This is important for the hydrodynamics on the plate. Holes that are too large cause the liquid to weep through, while holes that are too small give many small "jets" and the vapor blows too fast through the liquid. Many years of experimental work have given us the empirical engineering correlations needed to find the optimal diameter and hole dimensions.

Now let's go back to the example of an absorber (**Figure 5.2.7**) for which we have determined that the separation requires three plates. How would changing the design parameters affect the number of stages in our absorber? What will happen if we decrease the ratio n_{sol}/n_{flue} by decreasing the flow of water we inject into our absorber? As we have seen in **Figure 5.2.5**, a smaller n_{sol}/n_{flue} ratio decreases the slope of our operating line, giving us a new starting point ($x_{CO_2}^{sat}$, $y_{CO_2}^{flue}$) for our McCabe-Thiele diagram. Now we need more plates in our absorber design, as we can see in **Figure 5.2.8**. Increasing the flow of water, thus increasing the slope of our operating line, on the other hand, decreases the number of plates. Why does this make sense? The distance between the operating and equilibrium lines on that graph represents the concentration gradient between the gas and liquid phases, which is the driving force for mass transfer. The bigger the separation, the bigger the driving force and the more effective the transfer process!

Now it is important to note that our analysis is based on an ideal situation. In our McCabe-Thiele analysis, we assume that the liquid and vapor flows are together long enough in time to result in thermodynamic equilibrium on each plate, but this may be wishful thinking. In practice,

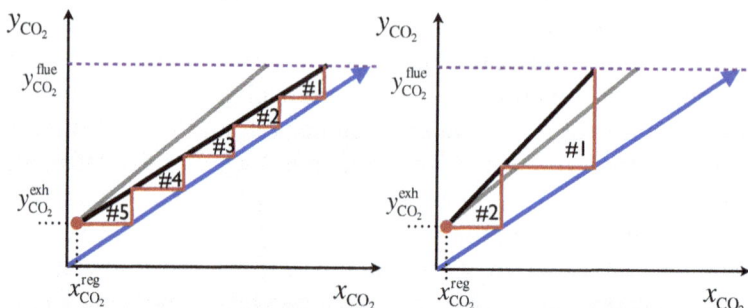

Figure 5.2.8 Effect of changing the ratio n_{sol}/n_{flue}

In the left figure, we decreased the solvent flow n_{sol} and in the right figure, this flow was increased.

the n_{sol}/n_{flue} ratio affects not just the mass-transfer relationship but the reality of what happens within each tray of the column. If the liquid flow rate is too high and the vapor flow rate is too low, for example, we will get poor contact between our liquid and vapor. In other words, the gas that comes into the bottom of the absorber and enters the froth won't be there long enough to achieve thermodynamic equilibrium. We know this is true in the case of water, for example, because mass transfer with water is actually very slow. Without enough time for full contact, each stage in our diagram won't quite reach the equilibrium line, and consequently, many more stages will be necessary to achieve the same separation.

At this point, we should also acknowledge that although we have focused our current absorber discussion on the design potential of multi-tray absorbers, the majority of absorbers use a packed design. In a packed bed, one can still use the concept of a stage, but the link to a physical plate is less direct. Nevertheless, the number of theoretical stages in a packed bed can also be used as a measure of the height of the column. In this book, we will not touch on the details of how other types of absorbers work, but the corresponding McCabe-Thiele diagrams can still give us an intuitive sense of the efficiency of these absorbers.

Based on what we now know about absorption column design, let's go back to our sample coal-fired power plant and our charge of processing 400 m^3 of flue gas per second. What is the necessary volume of water we would need to deal with that volume of gas? The answer, it turns out, is not too promising: 200,000 gallons of water per second (to give some context, a typical shower uses 7 gallons per minute, so it would take the equivalent of 2,000,000 showers to equal this volume per second). Not only is that a lot of water, we must remember from the absorber design (see **Figure 5.2.1**) where it is going: all the way to the top of the absorber column. While flue gas is coming in at the bottom of the absorption column, all that water must be pumped to the top of the column before it comes down and then over to the stripper. Making matters worse, those 200,000 gallons per second of water must also be heated in the stripper so that the CO_2 can be desorbed for storage. Some useful numbers put this in context: 200,000 gallons per second equals 27,000 cubic feet per second. This equals ten times the Rio Grande river in Texas or half of the Ganges river in India. Given all those energy requirements, it's clear this system is not practical. We need another idea. To look for a better solution, we move from the engineering perspective to the chemical perspective: how can we improve the driving force for absorption?

Section 3

Solvent Design

Introduction

In our absorption column, the effectiveness of our separation is represented by the distance between our operating and equilibrium lines. In order to drive mass transfer, we have two options: to increase the slope of the operating line (increasing n_{sol}/n_{flue}, see **Figure 5.2.8**) or to change the properties of the solvent, in order to lower κ, and hence the slope of the equilibrium line.

As shown in **Figure 5.3.1**, an increase in κ increases the number of plates; a decrease in κ makes the absorber more efficient. Recall that κ is defined as the Henry constant divided by the total pressure. The Henry constant depends on the solute, the solvent, and the temperature. One way to reduce the slope of the equilibrium line would be to raise the total pressure of the absorber, but as noted in Chapter 4, compressing large volumes of gasses takes significant amounts of energy. The natural alternative to altering pressure would be to somehow lower Henry's constant. We either have to change the solvent or add something that can help water take up CO_2 more efficiently. There are currently quite a few research opportunities in CCS that involve designing molecules to facilitate the uptake of CO_2.

Water and water+

Up until this point in the chapter, we have assumed that the absorption of CO_2 in water is a simple equilibrium between CO_2 in the gas phase and in solution. In reality, the situation is more complex. When added to water, CO_2 forms multiple ions, dissociating to become carbonic acid (H_2CO_3), bicarbonate (HCO_3^-), and carbonate (CO_3^{2-}). To understand the

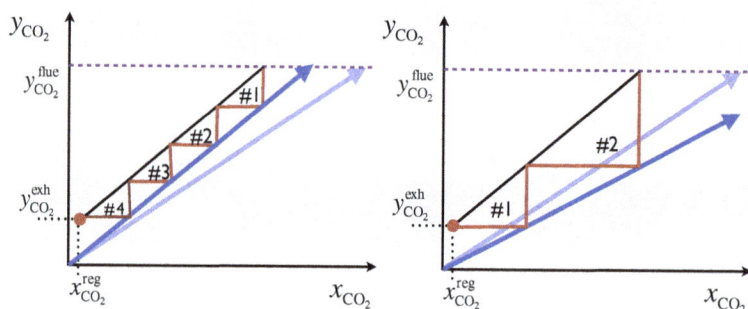

Figure 5.3.1 Effect of the solvent on the number of plates

McCabe-Thiele diagrams for different solvents: the left figure has a solvent with a higher κ and the right figure a solvent with a lower κ.

effect of these chemicals, we need to discuss the chemistry of CO_2 in water (see **Figure 5.3.2**).

CO$_2$ reacts with water to form carbonic acid (H_2CO_3). This is a relatively slow chemical reaction and therefore the rate-limiting step for the uptake of CO_2 in water. Carbonic acid further dissociates into a (hydrated) proton and a bicarbonate (HCO_3^-) ion. Bicarbonate is an acid. While this

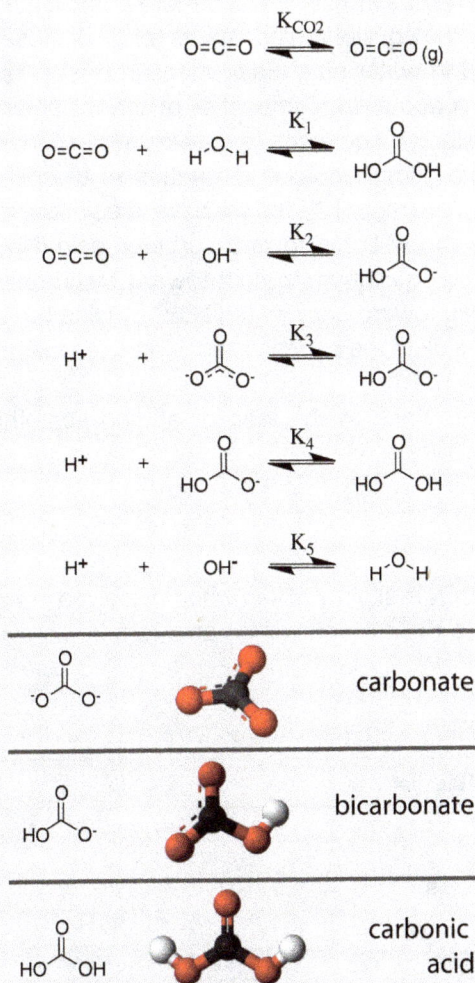

Figure 5.3.2 Chemical reactions of CO_2 in water

certainly makes the absorption of water more complicated than we previously discussed, the equilibrium constants for all of these reactions when CO_2 is dissolved in water have been measured (see **Table 5.3.1** [5.4]). One point to think about is the impact of pH on the solubility of CO_2. We can use the equilibrium equations to determine the concentrations of the various carbonate compounds as a function of the pH of the solution. The lower the pH, the more H^+ ions we have in solution, and the more the chemical equilibrium will shift toward carbonic acid (H_2CO_3). Increasing the pH causes the equilibrium to shift toward bicarbonate (HCO_3^-), and for the highest pH, toward carbonate (CO_3^{2-}). Using the equilibrium constants we can calculate the concentrations and see that this is indeed the trend (see **Figure 5.3.3**).

This brings us back to the idea of adding something to water to increase the solubility.

Bases

Remember that CO_2 is slightly acidic when dissolved in water, so with the addition of a base the solution will be able to take up even more CO_2. A

Table 5.3.1   Reaction constants of CO_2 in water (see **Figure 5.3.2**)

Reaction	K_i
1	1.15×10^{-3}
2	3.21×10^{-3} M^{-1}
3	5.95×10^{10} M^{-1}
4	2.46×10^3 M^{-1}
$\log(K_5) = -5839.48/T - 22.473 \log(T) + 61.2060 - \log(c_0)$	

Data from McCann et al. [5.4].

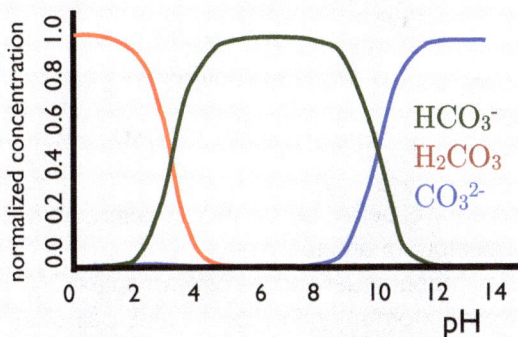

Figure 5.3.3 Concentration of the carbonate species in water as a function of the pH

good thought experiment when vetting ideas like this is to consider the extremes. What will happen to the solubility of CO_2 if we add a really strong base like sodium hydroxide (NaOH)? The nice thing about adding a strong base is that we know it will dissociate completely, meaning in this case that 1 mole of NaOH will translate to 1 mole each of aqueous Na^+ and OH^- ions. If we add a large amount of OH^- to the system, this will drive the equilibrium toward bicarbonate. This will improve the mass transfer driving force of the system, and in turn, increase the uptake of CO_2. But there's a problem with using a strong base: the absorption of CO_2 as bicarbonate is a very exothermic reaction. As a result, the process will require a significant amount of heat to desorb the CO_2 from the solvent in the regeneration step. That means more energy and more expense, so a strong base probably wouldn't be a good idea for a traditional absorber. However, this process is very efficient in capturing CO_2 even at very low concentrations, so it isn't a bad idea to use strong bases to capture CO_2 directly from air [5.5, 5.6]. In one of the next sections we will discuss the energy associated with the regeneration in more detail.

Amines

The idea that weaker bases might work for carbon capture is not exactly new. Bottoms used a similar concept in his 1930 patent, where he proposed to use an aqueous solution with amines, like **monoethanol amine (MEA)**, as a base. How amines, such as MEA, react depends on which

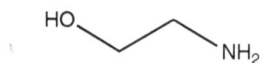

monoethanol amine

Chemical Formula: C_2H_7NO

ones we are using. To better grasp this concept, let's do some basic review on the structure of amines.

Amines are compound derivatives of ammonia (NH_3), which consists of a central nitrogen atom attached to one lone pair of electrons and up to three functional groups, or R-groups. R-groups consist of a non-hydrogen substituent, such as a methyl ($-CH_3$) or other alkane group. Amines are classified as either primary, secondary, or tertiary, depending on the number of R-groups attached to each molecule. The number and size of the R-groups that an amine has will influence the molecule's properties and ability to react with other compounds. This will be important as we decide what type of amines we can "tune" to make better solvents for carbon absorption.

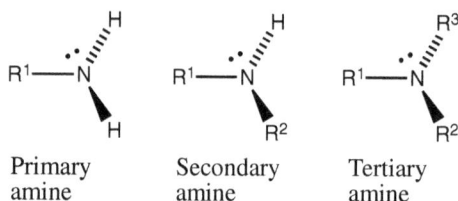

Primary amine

Secondary amine

Tertiary amine

The amines are weak bases relative to NaOH. As we have seen, 1 mole of NaOH will completely dissociate and give us 1 mole each of aqueous Na^+ and OH^- ions. For amines, we can represent equilibrium in solution as:

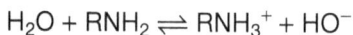

$$H_2O + RNH_2 \rightleftharpoons RNH_3^+ + HO^-$$

As an example, a 1 molar MEA solution has a pH of approximately 11.7 (compared to a pH of 14 for 1 molar NaOH). Because amines are weak bases, they influence the solubility of CO_2 by shifting the equilibrium toward the formation of carbonate ions. The process is slightly different depending on the structure of the amine. The reaction of CO_2 with primary and secondary amines, shown in **Figure 5.3.4**, involves the

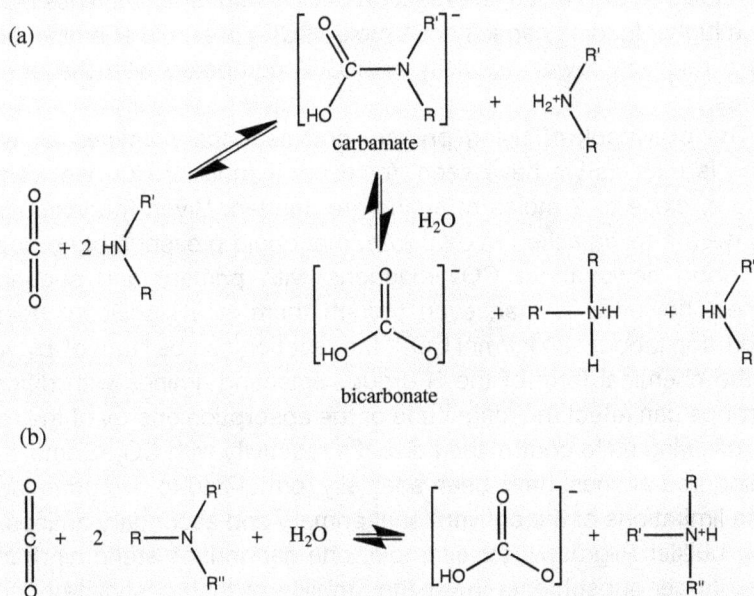

Figure 5.3.4 General reaction schemes for the chemical absorption of CO_2
Reaction scheme by (a) primary or secondary and (b) tertiary amine-containing solvents. More complete reactions schemes can be found in the literature.

formation of a carbamate bond between the carbon of the CO_2 molecule and the nitrogen atom of the amine [5.4, 5.7]. A fraction of the formed carbamate species is subsequently hydrolyzed to form bicarbonates. As a result, the CO_2 loading capacity for primary and secondary amines lies in the range 0.5–1 moles of CO_2 per mole of amine. For tertiary amines, the additional R-group gets in the way of the carbamation reaction (which does not proceed), giving a base-catalyzed hydration of CO_2 to

Question 5.3.2 Concentration of MEA

We measure that 1 molar MEA solution has a pH of approximately 11.7. What is the OH^- concentration in the solution? What is the percentage of MEA present in its un-ionized (i.e., molecular) form?

form bicarbonate. Hence, the reaction of CO_2 with tertiary amines occurs with a higher loading capacity of 1 mole of CO_2 per mole of amine, albeit with a relatively lower reactivity with CO_2 compared with the primary amines.

The drawback of using primary and secondary amines as weak bases is that, as we have seen, for every 1 mole of CO_2 we want to capture, close to 2 moles of amine are needed. Given the volumes of CO_2 we are being asked to deal with, that could present a big problem. The good news about CO_2 reactions with primary and secondary amines, however, is that, even though there is a significant heat of absorption (about -80 kJ/mol CO_2), the reaction can be "tuned" by varying the chemical form of the R-group. Selecting amines with different R-groups can affect the magnitude of the absorption energy of the reaction, allowing us to control the solvent's reactivity with CO_2. Some, such as hindered amines, have been specially formulated to overcome some of the limitations of the conventional primary and secondary amines. By using bulkier R-groups, for example, one can induce steric hindrance. These larger substituents lower the stability of their associated carbamates, allowing us to attain CO_2 loadings well in excess of 0.5 mole equivalents.

Tertiary amines form bicarbonate directly using a 1:1 molar ratio of amine to CO_2. This is a plus from a cost perspective, because we will need less solvent to process the same amount of CO_2. Also, their reaction with CO_2 is associated with a lower heat of absorption, which means less energy is needed for recovery in the stripper. Finally, tertiary amines provide the benefit of being very tunable due to the presence of three functional R-groups.

Question 5.3.3 Heat of absorption

Let's take a closer look at this number that is associated with the heat of absorption of amine solutions — 40–100 kJ/mol. To desorb the CO_2 from the solution we need to supply the heat. We do this in the stripper by boiling the solution. Amine solutions typically consist of 30% amines and 70% water. If we assume that the amine solution has the same properties as water, how much heat do we need to supply if we have a heat of absorption of 40 kJ/mol? What if we have 100 kJ/mol?

Enzymes

Now we know that tertiary amines or binary amines with steric hindrance can uptake more CO_2 than other amines. The disadvantage of these molecules is that they are much slower to react with CO_2. The reaction can be so slow, in fact, that it limits the transfer of CO_2 from the gas phase to the liquid phase. Additionally, the dissolution can be limited by the formation of bicarbonate (see **Figure 5.3.4**).

In nature, the enzyme carbonic anhydrase catalyzes the reaction of carbon dioxide and water to bicarbonate and protons, and is very adept at helping with CO_2 uptake. This reaction occurs with astonishingly fast rates in our tissues every time we exhale. Carbonic anhydrase is part of a family of enzymes that are classified as metalloenzymes due the fact that they contain a zinc atom at their center. The mechanism of the catalytic process of carbonic anhydrase enzymes is well known and it works very efficiently. One idea is that we could use a similar enzyme to increase the uptake of CO_2 in amine solutions that have an intrinsically low reaction rate. From a research point of view this is a very interesting topic, but the natural enzyme has a very poor stability under the harsh conditions of the carbon capture processes. The temperatures range from 40° to over 120°C, and high concentrations of organic amine and trace contaminants of flue gasses are not typical of a normal biological environment. To make enzymes work for carbon capture, researchers are studying the carbonic anhydrases that can be found in organisms that live in very harsh conditions, as well as looking at protein engineering techniques to create thermo-tolerant enzymes [5.8].

Ionic liquids

So far we have focused on changing the properties of water in order to improve the uptake of CO_2. A different strategy is to replace the water with a completely different solvent. In this context, there is a great deal of interest in ionic liquids (IL) [5.9]. Recall that ions have either a negative (anion) or positive (cation) charge, and that positive and negatively charged ions are attracted to each other. When they come together, ions form solid crystalline salts. These salts are very stable — to melt sodium chloride one has to heat the crystals to 801°C. In an ionic liquid, however, the cations and anions are both large, asymmetric, bulky molecules, preventing them from forming the crystalline lattice of a typical salt. As a result, these "frustrated" salts remain liquid at room temperatures. Like molten salts, they have special properties because of their ionic character. For example, ILs have relatively low vapor pressures compared to ordinary liquids. Because of charge neutrality, IL molecules can only escape in pairs from the liquid phase. As we will see in one of the next chapters, this makes it possible to use ILs as membrane materials.

Another interesting property of ionic liquids is that they are highly customizable. By changing the cation and anion, it is possible to form millions of different ionic liquids, many of which chemists have already

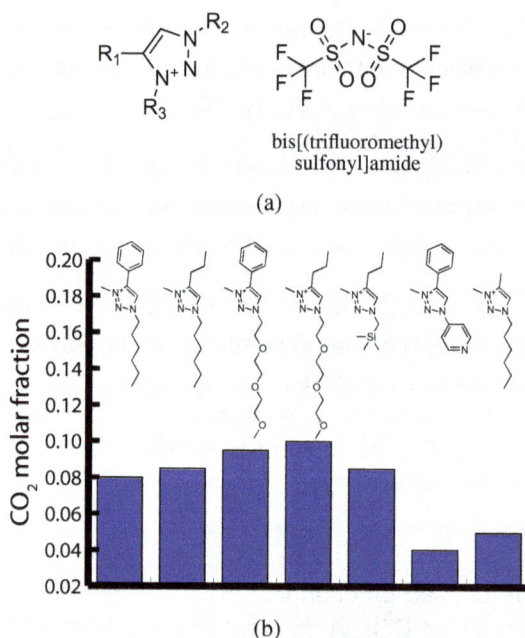

bis[(trifluoromethyl)
sulfonyl]amide

(a)

(b)

Figure 5.3.5 Triazolium-based ionic liquids

(a) Example of a triazolium-based ionic liquid. The bulky cations and anions make it very difficult for these liquids to crystalize and hence at conditions where ordinary ionic systems (salts) are solid, these materials are liquid. By varying the groups R_1, R_2, and R_3 one can make millions of different materials each having very different properties.

(b) Effect of the structure of the ionic liquid on the solubility of CO_2. *Figure based on data from* [5.10].

characterized. Take, for example, a particular cation shown in **Figure 5.3.5** with three different R-groups, and an anion with one R-group. Between these two structures, you can make thousands of different ionic liquids [5.10]. The question therefore becomes, which ionic liquid do we want to use as a solvent for carbon capture? **Figure 5.3.5** shows an example from the literature [5.10] of subtle changes in these particular functional groups. This example is related to solubility. We have the possibility of varying CO_2 solubility in, and reactivity with, ILs based on their geometry. In **Box 5.3.1** we further illustrate how ILs can be chemically functionalized to enhance the CO_2 solubility. However, at this stage we do not know whether the increased solubility will indeed lead to a better carbon capture material [5.11].

Commercial solvents

At the beginning of this chapter we mentioned that there is nothing magical about carbon capture. If we would like to add a capture unit next to our power plant, existing commercial processes are available. We will briefly touch on a few of these processes.

Box 5.3.1 Designing ionic liquids

Most ionic liquids (IL) selectively absorb CO_2. This physical solubility, however, is limited. One way of enhancing the solubility is to use an IL that can react with CO_2. As amine groups efficiently react with CO_2, an obvious choice is an amine-functionalized IL:

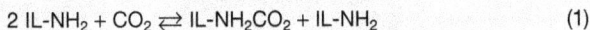

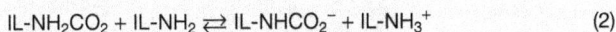

$$2 \text{ IL-NH}_2 + CO_2 \rightleftarrows \text{IL-NH}_2CO_2 + \text{IL-NH}_2 \qquad (1)$$

$$\text{IL-NH}_2CO_2 + \text{IL-NH}_2 \rightleftarrows \text{IL-NHCO}_2^- + \text{IL-NH}_3^+ \qquad (2)$$

This is the typical carbamate chemistry that we have already seen in the case of aqueous amines. Ideally one would like to suppress step (2), the formation of the zwitterion, so as to have an IL/CO_2 capture ratio of one. Ionic liquids present us with the interesting option to place the amine group on either the cation (3) or the anion (4):

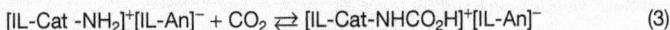

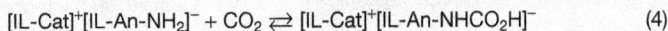

$$[\text{IL-Cat -NH}_2]^+[\text{IL-An}]^- + CO_2 \rightleftarrows [\text{IL-Cat-NHCO}_2\text{H}]^+[\text{IL-An}]^- \qquad (3)$$

$$[\text{IL-Cat}]^+[\text{IL-An-NH}_2]^- + CO_2 \rightleftarrows [\text{IL-Cat}]^+[\text{IL-An-NHCO}_2\text{H}]^- \qquad (4)$$

(Continued)

Box 5.3.1 (*Continued*)

Because the anion possesses a negative charge, it has a higher affinity for protons, and thus is much less willing to give up a proton for a reaction of the type shown in (2). Schneider, Brennecke and co-workers used this idea to synthesize amino acid based ILs with the amine functional group on the anion [5.11]. The figure shows two examples of these ILs with methioninate (Met, top structure) and prolinate (Pro, bottom structure) as amino acid-based anions and the structure known as P$_{66614}$ for the cation.

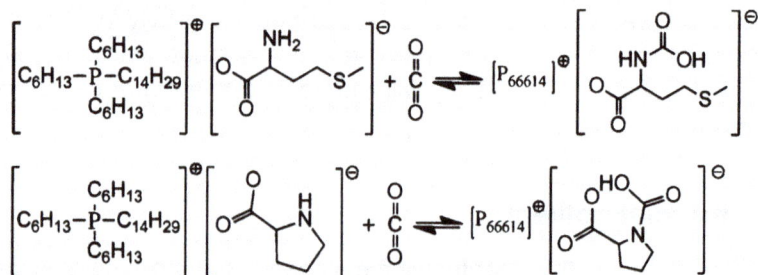

The absorption isotherms confirm that a 1:1 chemistry can indeed be obtained.

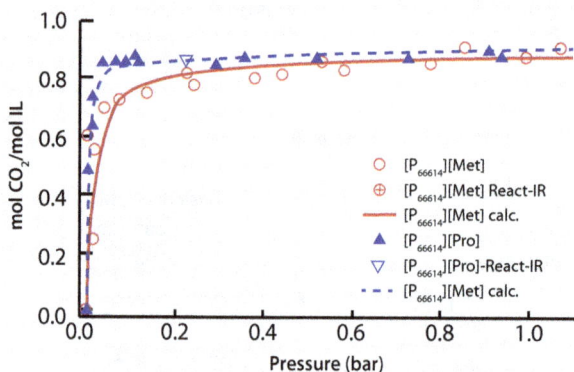

An important advantage of liquid absorption processes is that it is easy to transport fluids. However, if the viscosity of a fluid gets too high it will become increasingly difficult to pump. In practice one would like to ensure that the viscosity is below 100–200 centi-Poise (cP). One of the challenges of IL is that the viscosity increases once CO_2 is absorbed [5.9]. The reason for this viscosity increase is the formation of a hydrogen bonded network within the ionic liquid. By using steric hinderance Brennecke and co-workers were able to develop ILs in which the formation of such a network is suppressed [5.11]. *Figures reprinted with permission from Brennecke and Gurkan [5.9]. Copyright (2010) American Chemical Society.*

In the Selexol™ process the solvent, which consists of a proprietary mix of dimethyl ethers of polyethylene glycol (DEPG), absorbs CO_2 from the flue gas at a relatively high pressure. Unlike amine-based gas removal, Selexol™ uses a physical solvent that does not involve internal chemical reactions between CO_2 and the liquid. One of the main advantages of Selexol™ is that it does not use any water. The disadvantage, however, is that the solvents used have higher viscosities than water, and thus require more energy to be pumped around, have reduced mass transfer rates and tray efficiencies, and increased packing or tray requirements.

Rectisol® is another commercial process that just uses methanol as the solvent. Methanol binds CO_2 sufficiently well, but the entire system must be run at very low temperatures (between –40°C and –62°C), which necessitates expensive stainless steel refrigerated vessels.

Most of these processes were developed in the context of cleaning natural gas, and were later adopted for flue gasses. Much of the current research is focussed on developing novel solvents or processes to reduce the costs associated with carbon capture from flue gasses. One such solvent is KS-1™, a sterically hindered amine that was developed jointly by the Kansai Electric Power Company (KEPCO) and Mitsubishi Heavy Industry (MHI). KEPCO and MHI report that the regeneration energy required for KS-1™ solvents is less than that for MEA. KS-1™ solvent has been used commercially since 1999 at a steam reforming power plant at Kedah Danul Aman in Malaysia, where the process captures 160 tonnes of CO_2 per day [5.12].

Section 4

Beyond equilibrium thermodynamics

Up to this point, we have looked at solvents from a purely thermodynamic point of view. Thermodynamics tells us all about the equilibrium situation. In practice, however, a reaction never reaches true equilibrium in the froth above a plate. One should therefore view a design based on equilibrium data only as the ideal reference. In a real design, we have to correct for the fact the system is not actually in equilibrium. These corrections always make the design bigger and consequently more expensive. **Figure 5.4.1** illustrates this point. For a given gas composition, the corresponding liquid concentration of CO_2 will be lower compared to the true equilibrium concentration. As a result, a greater number of stages is required.

Next, we will try to quantify the two main processes that preclude equilibrium in the froth.

Diffusion limitation

If we have thermodynamic equilibrium, the concentrations of CO_2 in the gas and liquid phase are in equilibrium. If we know the equilibrium concentration, we can compute the flow we need to remove all the CO_2 that is coming from the flue gas. In this calculation it is assumed that equilibrium is reached instantaneously. In reality CO_2 molecules have to diffuse from the gas phase into the liquid phase. **Figure 5.4.2** shows an

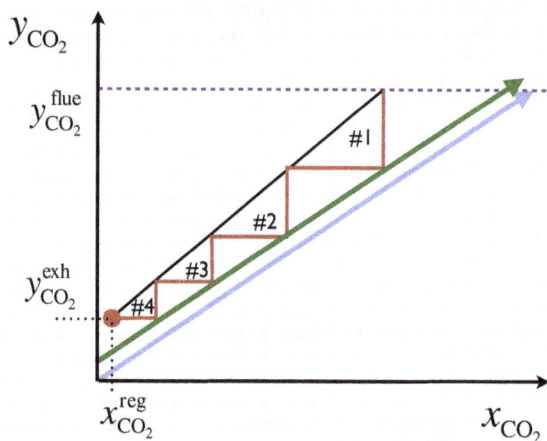

Figure 5.4.1 Number of plates for non-equilibrium

Number of plates for a system that does not reach the (blue) equilibrium concentration.

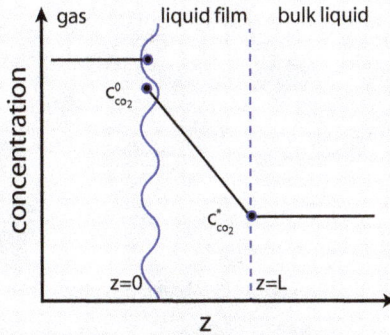

Figure 5.4.2 Concentration profile at the gas liquid interface

$C^0_{CO_2,i}$ is the equilibrium concentration in the fluid phase and $C^*_{CO_2}$ is the actual (non-equilibrium) concentration. L is the typical thickness of the interfacial layer.

approximate concentration profile at the liquid-vapor interface. In this scenario, we see that the gas phase is well mixed, so the gas concentration is constant in the gas phase. The liquid phase however is not well mixed and we see that next to the gas interface the CO_2 concentration is higher compared to the concentration in the bulk liquid. Because of this concentration gradient, CO_2 will diffuse from the interface to the bulk. This diffusion process follows **Fick's law**, which states that the flux (in molecules per unit area per unit time) is proportional to the gradient in the concentration:

$$j_{CO_2} = -D_{CO_2} \frac{dc_{CO_2}}{dz},$$

where D_{CO_2} is the diffusion coefficient of CO_2 in the liquid phase. If this diffusion coefficient is sufficiently high then the flux of CO_2 can keep up with the flux of CO_2 from the flue gas. If, however, the diffusion coefficient is small it can be a limiting factor in the process.

To quantify this effect let us consider a mass balance (**Figure 5.4.3**) on a small volume across the interfacial region:

accumulation = flow in − flow out

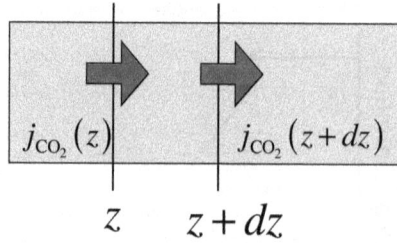

Figure 5.4.3   Mass balance

The accumulation is equal to the increase in the number of molecules in the volume ($V = dxdydz$):

$$\text{accumulation} = \frac{dc_{CO_2}}{dt} dxdydz$$

The flow in and out of the volume is given by the flux times the area ($A = dydx$):

$$\text{flow in} = j_{CO_2}(z)dydx$$

$$\text{flow out} = j_{CO_2}(z + dz)dydx$$

For the flow in minus flow out, we can write:

$$\left[j_{CO_2}(z) - j_{CO_2}(z + dz) \right]dydx = -\frac{dj_{CO_2}(z)}{dz}$$

This gives us for the mass balance:

$$\frac{dc_{CO_2}(z)}{dt} = -\frac{dj_{CO_2}(z)}{dz}$$

If we insert Fick's law into this result, we get the following differential equation for the time-dependent concentration profile:

$$\frac{dc_{CO_2}(z)}{dt} = D_{CO_2} \frac{d^2c_{CO_2}(z)}{dz^2}$$

If we assume that we have a steady state, with boundary conditions $c_{CO_2}(0) = c_{CO_2}^0$ and $c_{CO_2}(L) = c_{CO_2}^*$, where L is the thickness of the boundary layer between the gas and bulk liquid, the solution is:

$$c_{CO_2}(z) = c_{CO_2}^0 - \frac{z}{L}\left(c_{CO_2}^0 - c_{CO_2}^*\right)$$

For practical applications we are interested in the flux of CO_2 across our unit interface, which follows directly from Fick's law:

$$j_{CO_2} = \frac{D_{CO_2}}{L}\left(c_{CO_2}^0 - c_{CO_2}^*\right) = k^{eff}\left(c_{CO_2}^0 - c_{CO_2}^*\right)$$

In this equation, we have defined an effective mass transport coefficient k_{eff}. The total amount of CO_2 we have to remove is given by the total amount of flue gas. For a given solvent, we have a corresponding value for k_{eff}. Therefore in order to ensure that we can capture all the CO_2, we either need to increase the total area in the absorber, or we need to increase the driving force ($c_{CO_2}^0 - c_{CO_2}^*$). If our effective mass transfer coefficient is sufficiently large, we can allow for a very small driving force. Another interpretation of this driving force is that the actual concentration in the solvent $c_{CO_2}^*$ differs from the equilibrium concentration $c_{CO_2}^0$:

$$c_{CO_2}^* = c_{CO_2}^0 - \frac{j_{CO_2}}{k_{eff}}$$

This is the origin of the shift in the equilibrium line shown in **Figure 5.4.1**.

Chemical reaction

In our design, we assumed that absorption can be simply described with Henry's constant or solubility. But with amine solutions, we have to take into account how the chemical reaction is influencing the absorption process.

Let us assume we introduce a chemical "B" into the water. "B" reacts with CO_2 via a first order reaction:

$$B + CO_2 \rightleftarrows BCO_2$$

If we assume chemical equilibrium we have:

$$K_{eq} = \frac{[BCO_2]}{[B][CO_2]}$$

We can combine this expression with the equilibrium of CO_2 in the gas phase and in solution, as given by Henry's law:

$$(CO_2)_g \rightleftarrows CO_2$$

or,

$$K_{CO_2} = \frac{[CO_2]}{\left[(CO_2)_g\right]}$$

The solubility of CO_2 is now given by the sum of CO_2 in the liquid phase and converted to BCO_2:

$$[CO_2] + [BCO_2] = \left(1 + k_{eq}[B]\right)[CO_2],$$

which we can relate to the partial pressure by introducing an effective solubility:

$$[CO_2] + [BCO_2] = \left(1 + k_{eq}[B]\right)K_{CO_2}\left[(CO_2)_g\right] = K'p_{CO_2}$$

We see that by increasing the concentration of component B in our solution we can enhance the solubility of CO_2.

Rate limitations

As CO_2 moves from the gas to the liquid phase within an absorber, the reaction rate can limit the rate at which equilibrium is reached. Unfortunately, this complicates the way we think about mass transfer within our absorbers. The process is shown schematically in **Figure 5.4.4**.

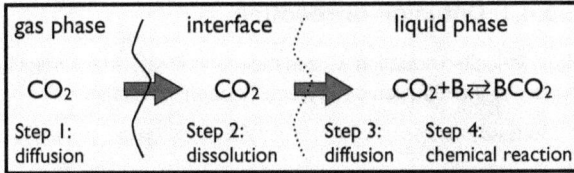

Figure 5.4.4 Mass transfer from the gas to liquid phase

As we think about the reaction rate limitation, we can use a similar approach to the one we used for diffusion limitation. If we construct another mass balance for our control volume, we get:

$$\frac{dc_{CO_2}(z)}{dt} = D_{CO_2} \frac{d^2 c_{CO_2}(z)}{dz^2}$$

Due to the fact that we now have a chemical reaction, CO_2 can disappear in our volume, according to the rate equation:

$$\frac{dc_{CO_2}(z)}{dt} = k_r c_B c_{CO_2}$$

In this rate equation, k_r is the reaction rate of CO_2 with B, which we assume to be irreversible. Substitution in our mass balance equation gives us:

$$\frac{d^2 c_{CO_2}(z)}{dz^2} = \frac{k_r c_B}{D_{CO_2}} c_{CO_2} = k' c_{CO_2},$$

with boundary conditions $c_{CO_2}(0) = c^0_{CO_2}$ and $c_{CO_2}(L) = c^*_{CO_2}$ (see **Figure 5.4.2**). This equation has as solution:

$$c_{CO_2}(z) = A \exp\left(\sqrt{k'} z\right) + B \exp\left(-\sqrt{k'} z\right)$$

Question 5.4.1 Diffusion or reaction

In the diffusion-limited equation we see the reaction rate, in the reaction-limited equation we see the diffusion coefficient. Does this make sense?

From the boundary conditions we have:

$$A = \frac{c^*_{CO_2} - c^0_{CO_2}\exp\left(-\sqrt{k'}L\right)}{\exp\left(\sqrt{k'}L\right) - \exp\left(-\sqrt{k'}L\right)} = \frac{c^*_{CO_2} - c^0_{CO_2}\exp\left(-\sqrt{k'}L\right)}{2\sinh\left(-\sqrt{k'}L\right)}$$

$$B = \frac{c^*_{CO_2} - c^0_{CO_2}\exp\left(\sqrt{k'}L\right)}{\exp\left(\sqrt{k'}L\right) - \exp\left(-\sqrt{k'}L\right)} = \frac{-c^*_{CO_2} + c^0_{CO_2}\exp\left(\sqrt{k'}L\right)}{2\sinh\left(\sqrt{k'}L\right)}$$

The flux is given by:

$$j_{CO_2}(z) = -D_{CO_2}\left[A\sqrt{k'}\exp\left(\sqrt{k'}z\right) - B\sqrt{k'}\exp\left(-\sqrt{k'}z\right)\right]$$

Of particular interest is the flux at $z = 0$:

$$j_{CO_2}(0) = -D_{CO_2}\sqrt{k'}\left[A - B\right]$$
$$= \frac{D_{CO_2}\sqrt{k'}}{\sinh\left(\sqrt{k'}L\right)}\left(c^0_{CO_2}\cosh\left(\sqrt{k'}L\right) - c^*_{CO_2}\right)$$

This equation looks complex, but we can learn a lot by looking at two limiting cases.

Let us first assume that the reaction is slow compared to diffusion:

$$\sqrt{k'}L = \sqrt{\frac{k_r c_B}{D_{CO_2}}}L \ll 1$$

We can make a Taylor expansion of the sinh and cosh:

$$\sinh\left(\sqrt{k'}L\right) = \frac{e^{L\sqrt{k'}} - e^{-L\sqrt{k'}}}{2} = L\sqrt{k'} + \frac{\left(L\sqrt{k'}\right)^3}{3!} + \cdots$$

$$\cosh\left(\sqrt{k'}L\right) = \frac{e^{L\sqrt{k'}} + e^{-L\sqrt{k'}}}{2} = 1 + \frac{\left(L\sqrt{k'}\right)^2}{2!} + \cdots$$

If we keep only the linear terms, we get the equation for a *reaction-limited* flux:

$$j_{CO_2}(0) = \frac{D_{CO_2}\sqrt{k}}{L\sqrt{k'}}\left(c_{CO_2}^0 - c_{CO_2}^*\right) = \frac{D_{CO_2}}{L}\left(c_{CO_2}^0 - c_{CO_2}^*\right)$$

We get the results of the previous section back again; the chemical reaction is too slow to have any effect.

The other limit is if the chemical reaction is fast compared to diffusion:

$$\sqrt{k'}L = \sqrt{\frac{k_r c_B}{D_{CO_2}}}L \gg 1$$

In this limit, we have the equation for *diffusion-limited* flux:

$$j_{CO_2}(0) \approx D_{CO_2}\sqrt{k'}c_{CO_2}^0 \frac{\cosh\left(\sqrt{k'}L\right)}{\sinh\left(\sqrt{k'}L\right)} \approx \sqrt{D_{CO_2}k_r c_B}\,c_{CO_2}^0$$

We see that if the chemical reactions dominate, we get a typical term related to the square root of the diffusion constant times the reaction rate constant. In the derivation of this expression, we assume that the reaction is irreversible, so we cannot relate this result to the equilibrium concentration.

For amine solutions the reaction rate is often the limiting step, in which case we have as a driving force the difference in CO_2 concentration and the concentration at the interface:

$$j_{CO_2}(0) = \frac{D_{CO_2}}{L}\left(c_{CO_2,i} - c_{CO_2}^*\right) = k_{eff}\left(c_{CO_2,i} - c_{CO_2}^*\right)$$

which shows that, similarly to diffusion limitation, reaction limitation gives an effective mass transport coefficient that causes a shift of the equilibrium line in the McCabe-Thiele diagram (**Figure 5.3.1**).

Additional constraints

Now we also consider some common-sense limitations. For example, as mentioned previously, the solvents we focus on in absorption need to be transported from the bottom of the absorber to the top of the stripper. This implies that we have to use energy to move large solvent quantities around, so naturally we would prefer a solvent with a low viscosity; a polymer gel that can take up CO_2 may sound great, but in the end it simply won't work with certain types of designs.

Absorption relies on gas molecules moving from the solvent to gas phase. CO_2 in its gas phase will reach the liquid-gas interface in the froth, where it will dissolve in the solvent with a certain rate of mass-transfer. Clever engineering can help us increase the interfacial area by several orders of magnitude, but cannot eliminate the problem of mass transport rates. Even if a potential solvent has great solubility, we will still encounter problems if its transfer is very slow. If we need to have extremely slow fluxes of gas and liquid in order to give the CO_2 molecules enough time to move from the gas phase to the liquid phase, it would completely nullify the positive effect of the solubility.

There are additional constraints specific to using amine solvents in our absorbers. So far we have looked at flue gasses as binary mixtures of CO_2 and N_2. But in reality flues gasses contain many other components, such as O_2 or SO_x. It is well known that these molecules react with amines, which causes our solvent to degenerate over time; eventually we have to change out our solvent. Amines also react with carbon steel, so we need to use expensive hydrated steel in our absorber design. Another issue is that, to the best of our knowledge, nobody has published much about the potential consequences should amines like MEA escape from these absorber/strippers, either to the air or to the water. Without knowing the toxicology of your solvent, you are at risk to create a situation like we saw with MTBE in gasoline, for example, where, if deployed on a large scale, secondary environmental effects could pose a serious problem.

Section 5

Costs of absorption

When we talk about achieving the widespread adoption of absorption techniques for carbon capture, we always have to go back to the issue of money. Costs are a significant barrier to the deployment of carbon capture. As we mentioned at the very beginning of this unit, it's possible to go out right now and order an absorber and stripper from "Absorbers-R-Us." The problem is that the energy costs associated with the technology are steep. In the design of an absorber, it's important to keep both capital and maintenance costs of the technology in mind.

Let's start with the capital costs. The price of the chemicals and bio-solvents involved in absorption are an important part of the cost equation. Earlier we touched on the immense volume of pure water solvent (200,000 gallons per second) that would be needed to process the exhaust from a standard 500 MW coal-fired power plant. Ideally, a solvent or additive should be cheap, synthesizable, and easily regenerated (in addition to being good at taking up CO_2). Aside from the solvent costs, we have to consider the capital costs of building the equipment: the absorption tower and plates.

The operating and maintenance costs of absorption (compression, pumping, heating and regeneration of solvent) are particularly significant sources of expense, and a big part of the challenge with CO_2 absorption. There is not a cheap way to perform these functions right now, and most of the breakthroughs associated with lowering these costs have to do with novel ways to siphon energy from existing sources for "free." In other words, it would be a major innovation if our absorbers could siphon off waste heat, say from a power plant. This is an area of CCS research that we hope will continue to develop.

Cost is a factor that can put stress on the relationship between the energy industry and CCS research. Imagine that you are in charge of a large department of energy research, and in this program your team needs to develop a new absorbent. Let's say there is the possibility of making new types of materials. You decide to ask your team what a sensible target would be for your project, and you get drastically different answers. Your chemist wants the best absorbent to be the one with the most beautiful chemistry. The accountant on your team wants the one with the lowest price. Who is right? And, more importantly, how do you reconcile these two points of view?

From a research standpoint, it's natural that we want to look for beautiful chemistry. Only beautiful chemistry gets published in *Science* or *Nature*, right? However, from an accountant's point of view, new chemistry and new solvents are suspicious; the chemicals look expensive, and might require new technology to implement. Therein lies the problem: if we use cost as a guiding criterion for research, we may never do anything new. If we use scientific "coolness," we may not ever be able to implement the novel technology in the real world.

There's no doubt that costs are a big part of this challenge, but they're far from the only factor. One also has to consider issues of safety, environmental impact, efficiency, sustainability, and scale. The scale we must work with is enormous. If the entire world were to function like CostCo, one would expect that the axiom "the larger the scale, the lower the price" would hold true, almost by definition. However, if the scale is very large, one might reach the limits of global supplies. This will cause the price to rise. For example, a chemist may discover exciting chemistry about a promising solvent based on tellurium, but as this is a very rare metal, the potential large-scale employment of such a material is extremely limited. This important factor of the scale and scalability of the chemistry distinguishes energy-related research from other fields of science. One of the aims of this book is to create an awareness of the importance of scale, in particular as much of the exciting new chemistry starts at the nano scale. Eventually, our new materials will need to process gigatonnes of carbon.

Section 6

Energy of absorption

As discussed earlier in this chapter, it is incumbent upon researchers to provide rational estimates for costs associated with the deployment of new materials or processes. If you have just synthesized two new materials for carbon capture, which is the best material? Eventually it will be the one with the lowest capital and operating costs. We need guidance as to how to estimate these costs.

One important reason carbon capture is so expensive is that the process requires a large amount of energy. One way of expressing the energy costs is to determine the **parasitic energy** of the carbon capture process. Minimizing this parasitic energy is a guiding principle which we will use to compare different materials. We find that this parasitic energy is based on purely physical properties of our materials, and therefore provides a good metric to compare materials. In addition, minimizing the parasitic energy will contribute to a more efficient use of fossil fuels, a concern of many researchers.

The main components for determining parasitic energy are the energy required for the separation and the energy needed for compression of CO_2 for geological storage. In Section 4.2, we used a thermodynamic analysis to estimate the minimal energy from the ideal entropy of mixing. In this section, we compare this energy with our estimates of the energy cost of an absorption process.

Parasitic energy of absorption

To estimate the parasitic energy of the absorption process, we need to make some simplifications that allow us to focus on the performance of a solvent. We will focus on the energy required to absorb and desorb CO_2, ignoring, for example, the energy costs to transport the fluids, flue gas, and other operations that cost energy. As a result, our estimate will be lower than the real energy costs. For an amine solution (MEA), for example, we know the actual energy costs are 30% higher than our calculations for a complete design [5.11]. Similarly, we assume that other solvents will add around the same percentage of energy burden onto our calculations. But when we are interested in comparing different solvents for our design, the parasitic energy remains a useful metric.

The simplest absorption process is shown in **Figure 5.6.1.** The CO_2 from the flue gas is absorbed in the solvent, which is later regenerated in the stripper. As the absorption of CO_2 releases heat, the absorption process will not cost any energy. In the stripper we need to heat the solvent so that the CO_2 is released. This heat has two contributions: the **sensible**

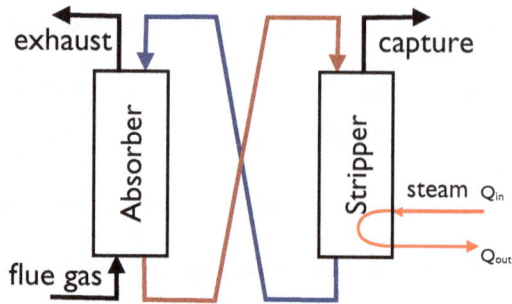

Figure 5.6.1 Simple absorption process

In an absorption process a solvent is cycled between the absorber, where it captures the CO_2, and in the stripper where the CO_2 is released through heating by steam from a power plant.

heat, which is the heat required to increase the temperature of the solvent to the desorption conditions, and the **heat of desorption**, which is the negative of the **heat of absorption**. It is useful to normalize the heat per kg CO_2, q_{tot}:

$$q_{tot} = \frac{Q_{sen} + Q_{des}}{\Delta\sigma_{CO_2}},$$

where $\Delta\sigma_{CO_2}$ is the amount of CO_2 removed in an absorption/desorption cycle, which is often referred to as the **working capacity** of the material. The sensible heat is related to the heat capacity (C_p) of the solvent:

$$Q_{sen} = C_p m_{abs} (T_{des} - T_{abs}),$$

where m_{abs} is the total amount of solvent and T_{des}, T_{abs} are the temperature of the solvent in the stripper and absorber, respectively. The desorption heat is the negative of the heat of absorption:

$$Q_{des} = \Delta h_{CO_2}\Delta\sigma_{CO_2} + \Delta h_{N_2}\Delta\sigma_{N_2},$$

where Δh_{CO_2} and Δh_{N_2} are the heats of desorption of CO_2 and N_2, respectively. In most calculations, we will assume that the absorption of N_2 is so small that the contribution of nitrogen to the heat of desorption can be ignored.

For the MEA solution, this energy is $q_{tot} = 8,776$ kJ/kg CO_2. If we compare this with the minimum energy for the separation, 158 kJ/kg CO_2 (see Section 4.2), we see that our process requires 50 times more energy than the thermodynamic minimum. A good rule of thumb is that the chemical process industry should be operating within a factor of 2–3 times the thermodynamic minimum. If we were to operate such an absorption process without modification, very little (if any) energy would be left for producing electricity. The fact that one can in practice design a relatively efficient process using MEA illustrates how clever engineering can reduce the parasitic energy to very reasonable levels.

Section 7

Optimization of an amine scrubber

(This section is based on a guest lecture given by Professor Gary Rochelle, University of Texas, Austin.)

So we've covered the fact that a simple absorption process requires enormous amounts of energy. In our calculation in the previous section we assumed that all the heat supplied to the stripper is lost. However, by using heat exchangers we can recover a large fraction of the heat (see **Figure 5.7.1**). Because we also need to compress the CO_2, integrating this compression in the optimization is important.

In this section, we will illustrate how chemical engineering tools can be used to optimize the absorption process [5.13]. This engineering approach involves selecting the optimal solvent together with a careful analysis of all steps to see how much energy each step loses compared to the theoretical minimum. **Figure 5.7.2** shows how incorporating these steps has reduced the energy costs of carbon capture.

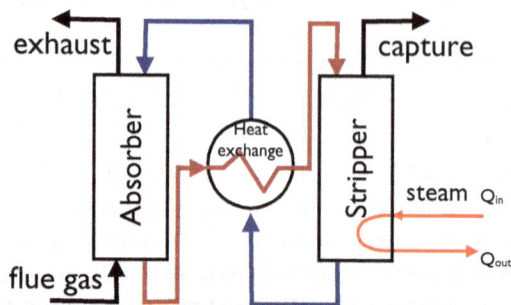

Figure 5.7.1 Simple absorption process with heat exchange

The most simple absorption process requires large amounts of energy to heat the solvent in the regeneration step. By using a heat exchanger, a significant amount of energy can be recovered.

Optimizing amine scrubbing

We learned in Section 4.2 that the minimum theoretical energy requirement for separating CO_2 from flue gas and compressing it to 150 bar is about 7% of the power plant's output. Let us borrow some typical numbers from Professor Rochelle's lecture to see how these energy requirements translate to costs. An energy cost of 20–30% of the power plant's output translates to a carbon capture cost of about $20 per tonne of CO_2 (2011 numbers). The energy cost is not the only cost associated with adding a carbon capture process to a power plant. The second major cost is the capital equipment cost for CO_2 capture, which for an 800 MW power plant will be about a billion dollars. About a third of that cost would be for the absorber, a third for the stripper, and a third for the compressor. This costs us an additional $20 to $40 per tonne CO_2. The third major cost associated with amine scrubbing results from the amine degradation. Amines can oxidize and thermally degrade, so we have to replace the solvent periodically, which adds about $1 to $5 per tonne CO_2. Environmentally, the costs are higher because these degraded amines need to be treated.

Irreversibility

The minimum energy required for a separation is an absolute thermodynamic minimum, independent of the process that is chosen. For a given

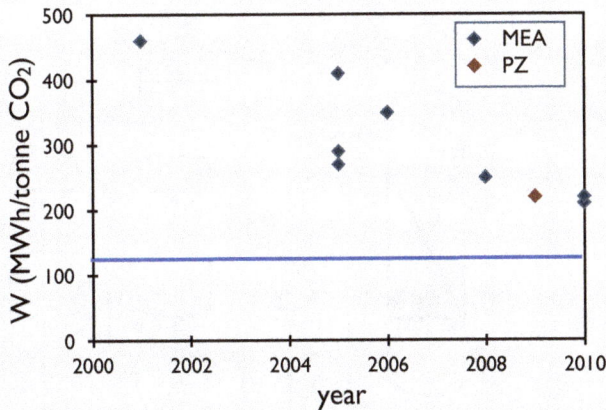

Figure 5.7.2 The parasitic energy of an amine capture process

This figure shows how different process designs and different solvents (MEA and PZ) have reduced the parasitic energy over the years. The blue line represents the thermo-dynamic minimum. *Data from Rochelle et al.* [5.13].

process we can also analyze and estimate the minimum energy required for each step in the overall process. Professor Gary T. Rochelle and co-workers at the University of Texas at Austin analyzed the amine capture process shown schematically in **Figure 5.7.3.** In this figure, we recognize the absorber and the heat exchanger. For the stripper a two stage process is used, where a first separation is achieved at a higher pressure than the second one. In the condenser, water is removed from the CO_2 gas stream before the compression stage. In this process, a flue gas with 12% CO_2 at 40°C is considered and it is assumed that 90% of the CO_2 is captured and compressed to 150 bar.

Rochelle and co-workers estimated the minimum required energy by calculating the minimum lost work in each step of the process. The idea of a lost-work analysis is that any irreversibility in the process creates additional entropy that is lost for the process. As we recall from first year thermodynamics, a truly reversible process requires infinitely small steps and infinitely small perturbations from the equilibrium configuration. In practice, "infinitely small" can be approached by building very large equipment such that the flow is so slow that the system is practically at equilibrium. However, the costs of such equipment will be prohibitively

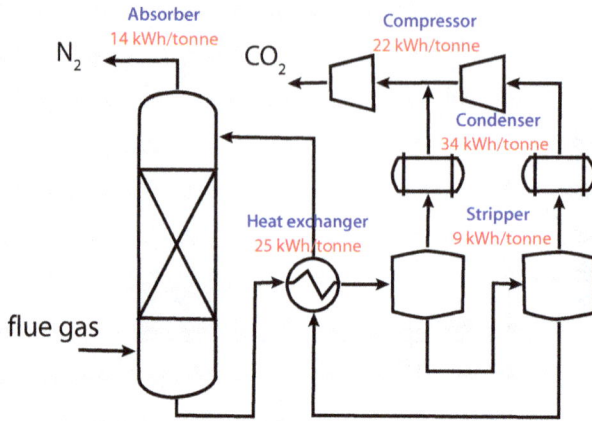

Figure 5.7.3 Irreversibility in an absorption/stripping system

Irreversibility in an absorption/stripping system where the values represent ideal lost work in kWh/tonnes CO_2. *Figure adapted from* [5.13].

large. To estimate the minimum lost work, it is therefore important to have some experience in the design of such equipment to make reasonable assumptions on the typical performance and dimensions of the equipment.

The numbers shown in **Figure 5.7.3** show how much work we lose for the different operations because our process is not reversible:

1. *The absorber*: to keep the absorber from being infinitely long and still have a reasonable mass-transfer rate, we need to have a significant driving force. We lose 14 kWh per tonne here.
2. *The heat exchanger*: a temperature gradient is the driving force for a heat exchanger. Because of the temperature gradient we lose work, typically 25 kWh per tonne.
3. *The stripper*: here we also cannot have an infinite number of stages to have an infinitely small driving force, and we lose 9 kWh per tonne.
4. *The condenser*: the condenser requires cooling water and the temperature difference gives a loss of reversibility, which equates to 34 kWh per tonne.
5. *The compressor*: the losses here are due to increases in temperature during compression, and are about 22 kWh per tonne.

The above numbers add up to an ideal work loss of 104 kWh per tonne. However, in a real-life absorber/stripper column, the actual lost work will be higher, more likely slightly above 200 kWh per tonne! If we compare this with the loss of 500 kWh/tonne of the early MEA process (see **Figure 5.7.2**), we see that careful engineering of this process has limited the loss of energy significantly. We will talk a little about how these "gains in loss" have been obtained, but for a complete discussion we refer to the work of Rochelle and his co-workers.

Let us start the discussion with the heat lost in the heat exchanger, which ideally should approach 25 kWh/per tonne. This number can be estimated by calculating the additional steam that is necessary to make up for the energy losses that occur in the exchanger. The driving force for the heat transfer is the temperature difference between the hot and the cold stream. The larger this driving force, the further the system is out of equilibrium and hence the larger the lost work. In **Box 5.2.1**, we discussed the importance of countercurrent flow. In a countercurrent flow we can operate with a nearly constant temperature difference across the exchanger (see **Figure 5.7.4**). The heat flux is given by this temperature difference, the area of the exchanger (A), and the overall heat transfer coefficient of our exchanger (λ):

$$\Phi_{\text{heat}} = \lambda A \left(T_H - T_L\right)$$

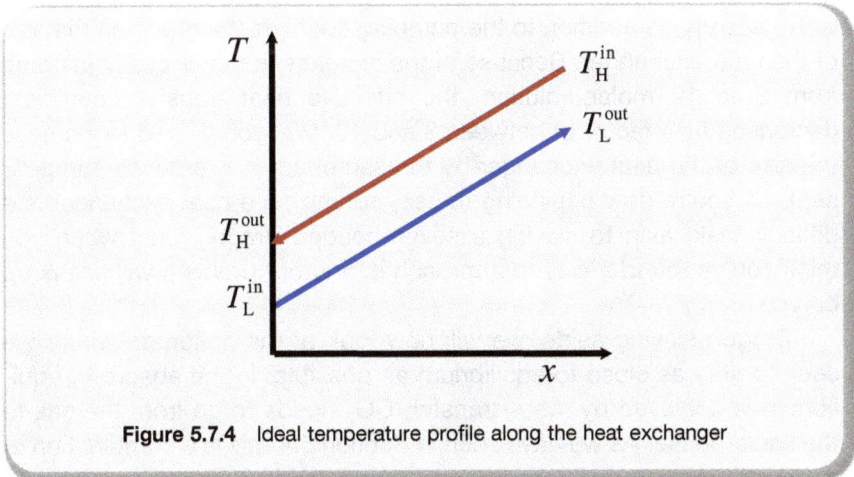

Figure 5.7.4   Ideal temperature profile along the heat exchanger

We now see that staying close to equilibrium and minimizing $(T_H - T_L)$ can be achieved by increasing the area (A) and thus the size of the heat exchanger. As heat exchangers are not cheap, we have to make the tradeoff between energy savings and capital costs. We see that because of this temperature difference, the fluid entering the stripper does not have the same temperature as the fluid leaving the stripper. This difference needs to be supplied by the steam from our power plant, which gives a lost heat per unit CO_2 that we remove:

$$W_{exc,loss} = \frac{C_p \left(T_H^{in} - T_H^{out} \right)}{\Delta\sigma_{CO_2}},$$

where C_p is the heat capacity of the solvent and $\Delta\sigma_{CO_2}$ is the working capacity of the fluid. The working capacity is defined as the amount of CO_2 removed in a cycle. We see that increasing the working capacity will decrease our losses in the exchanger. The numbers reported in **Figure 5.7.3** are converted to parasitic energy, the corresponding amount of electricity that we are unable to generate due to this lost work.

One puzzle that may have occurred to you by now is why we do not simply run our process with a higher percentage of MEA, say a 20 molar solution. It sounds like a good idea, but it turns out that such a high concentration of MEA is not soluble in water. Twelve molar solutions are also problematic, as they turn out to be too viscous. The main trouble with viscosity, in addition to the pumping costs, is the effect on the size of the heat exchanger. Because of the increase in the viscosity in going from 8 to 12 molar solution, the effective heat transfer coefficient decreases by a factor of between 3 and 10. We would need to increase the size of the heat exchanger by the same factor in order to compensate. As you're now beginning to see, optimizing a heat exchanger is a difficult task, akin to slaying a seven-headed dragon. Just when you think you've found a way to approach it, another problem will sneak up on you.

Dragon-slaying aside, we will now look at the absorber, where we seek to stay as close to equilibrium as possible. In the absorber, equilibrium is achieved by mass-transfer; CO_2 needs to go from the gas to the liquid phase. As we have seen in Section 5.4 this is a combination of

CO_2 dissolving in the solvent and subsequent reactions. The effective flux is given by the square root of the reaction kinetics and Henry coefficient:

$$j_{CO_2} = \frac{\sqrt{D_{CO_2} k \left[Am\right]^2}}{H_{CO_2}} \left(p_{CO_2}^0 - p_{CO_2}^*\right),$$

where $p_{CO_2, i}$ is the actual partial CO_2 pressure in the gas phase of the absorber and $p_{CO_2}^*$ the equilibrium partial pressure corresponding to the actual solvent loading. For a particular solvent we can define an effective mass transfer coefficient k_g':

$$j_{CO_2} = k_g' \left(p_{CO_2}^0 - p_{CO_2}^*\right)$$

Similarly, in our heat exchanger we would like to keep the driving force, $p_{CO_2}^0 - p_{CO_2}^*$, as small as possible, and selecting a solvent with the highest value of k_g' will do this. In order to select the optimal solvent, we need to know the diffusion coefficient, solubility, and reactivity.

In the compression step we lose work for two reasons. One is related to the pressure of the stripper: if we were to operate at a higher pressure, our compressor wouldn't have to do as much work. The second reason is that there is a significant amount of water vapor coming off the CO_2 stream from the stripper, and we lose work in the condensers. Interestingly, these losses can also be reduced by changing the solvent. The key factors here are the temperature and corresponding equilibrium partial pressure of CO_2. At higher temperature the vapor pressures of both CO_2 and water increase. The rate of increase depends on the heat of absorption. So by selecting a solvent with a high heat of absorption the relative amount of water in the vapor phase will decrease, and we will therefore have less water to condense per kg CO_2. Another important factor is the stability of the amine solution. If we can operate at a higher temperature at the bottom of our stripper, the pressure at the top will be larger and we save on compression costs. The lost work for compression can be computed very accurately, as this involves compressors that are well characterized.

We can quantify the lost work associated with the condensation by the amount of heat that is taken up by the cooling water used to run the condenser:

$$W_{exc,loss} = \frac{\Delta h_{H_2O}^{con} \Delta \sigma_{H_2O}}{\Delta \sigma_{CO_2}} \left(\frac{T_{cond} - T_{env}}{T_{cond}} \right),$$

where $\Delta \sigma_{H_2O}$ is the amount of water that is condensed and the Δh_{H_2O} is the heat of vaporization. The term in brackets is the Carnot factor which takes into account the efficiency at which this heat can be converted into electricity, and depends on the temperature of the condenser T_{cond} and the temperature of the environment T_{env}.

Finally, we come to the stripper process configuration. This is a typical example in which spending more on capital costs makes the process significantly more efficient: The improvements in efficiency come from increasing the stripper area and an advanced two-stage process configuration to recover the CO_2 at an elevated pressure. In the first stage, we evaporate the solution and flash off CO_2 using steam at 150°C and a system pressure of 17 bar. In that stage, half of the CO_2 evaporates at a very favorable high pressure. We then reduce the pressure of the solution to 11 bar, heat it again to 150°C with steam, and release the rest of the CO_2 at a somewhat lower pressure. This two-stage process also reduces the amount of water evaporating from the system, which, as we have seen, will save energy in the compression stage. For more details, we again refer to the work of Rochelle and his co-workers.

Solvent selection

In the previous section we have seen that in various steps of the process, the choice of the solvent has a significant effect on the efficiency of the absorption process. Let us summarize the most important properties:

1. **Working capacity:** the amount of CO_2 a solvent can capture per cycle. A high working capacity reduces the amount of solvent we have to cycle.
2. **Heat of absorption:** Solvents with higher heats of absorption tend to have larger vapor pressures at high pressure, and therefore the gas

leaving the stripper has a higher partial pressure of CO_2 and lower water content.

3. **Reactivity:** The efficiency of the absorber depends on the reactivity of the amines.

4. **Stability:** The more stable the amines, the higher the temperature at the bottom of stripper. This higher temperature translates to a higher pressure at the top of the stripper.

5. **Viscosity:** A high viscosity makes the heat exchange less efficient.

Rochelle and his co-workers evaluated different amines that are available commercially, and found that piperzine (diethylenediamine) performs significantly better than MEA on many of the points above. Changing from MEA to piperzine reduced the energy requirements significantly, as shown in **Figure 5.7.2**. The trouble with picking an amine, though, is that no single selection seems to meet all of our ideal characteristics. Our perfect solvent would have high capacity, high heat of absorption, and a high rate of CO_2 absorption. We also want a higher stripper temperature. But the amine with the best capacity is often not the same as the amine with the best heat of absorption or the best absorption rate. So it's complicated, and from a research point of view, it is interesting to see whether novel amines will be developed that outperform piperzine.

As we will see in the next few chapters, alternative technologies such as solid adsorption and membrane technologies do not have a preexisting "cousin" technology upon which to base cost and designs. Consequently, it will take us much longer to develop these methods "from scratch," so to speak. Due to these factors, it is likely that at least the first generation of carbon capture plants will be based on liquid absorption. Some critics of carbon capture research say that because liquid absorption has such a great lead time over alternative methods, it will be very unlikely that competing technologies will ever be viable. On the other hand, it is easy to point out many examples of new discoveries that have been transformative within an industry, and our research is aimed to achieve exactly this within the field of carbon capture.

A different perspective has to do with the cost of carbon capture. A liquid absorption carbon capture unit will account for something on the order of 30% of the costs of a power plant. Compared to these numbers, research is only a very tiny fraction of the costs of carbon capture adoption. In this context, it is important that we explore all possibilities for carbon capture, and that as researchers we create as many options to promote this goal as possible. Unfortunately, this process is very similar to the lottery; at the end of the day, it is likely that only one or two of these options will survive. Unlike the lottery, however, a "losing ticket" is not worthless. Even "unsuccessful" strategies for carbon capture will allow us to gain invaluable insights into the syntheses and properties of novel materials.

Section 8

Case Study: water versus amines

It is instructive to summarize this chapter by making a comparison between use of MEA and water as a solvent in a CO_2 absorption process. This case study was carried out by the UC Berkeley students, Forrest Abouelnasr, Josh Howe, Vicky Jun, and Karthish Manthiram.

Design an absorption and stripper unit that would remove 90% of the CO_2 from a coal-fired power plant (post-combustion). Be sure to clearly label the top and bottom of both units on your diagram, and state whatever assumptions you used in generating your diagrams. The analyses should demonstrate the benefits of using amines (e.g., MEA) versus water. To further illustrate the benefits of using amines, please estimate the capital cost of both a water and an MEA process.

HINT: Here are several ways to do this: you might use the same operation conditions and then compare the costs, or you could try to optimize some relevant variables for both processes to make a fair comparison. For an extra thought experiment, take the operational costs into consideration.

Coal-fired power plants in the United States produce 30% of domestic CO_2 emissions, making them an ideal target for carbon capture projects. Of the various adsorption, absorption, and membrane-based processes that have been investigated, amine scrubbing is the most developed technology for carbon capture. Amine scrubbing was invented in 1930 by R. R. Bottoms [5.1], who filed a patent on the process, which was immediately put into commercial use to separate CO_2 from hydrogen and natural gas [5.14]. With increasing concern over climate change in the 1980s, amine-scrubbing was applied for carbon capture in small-scale gas and coal-fired power plants [5.15]. Amine-scrubbing is now used in small 20 MW combustion plants, but has not yet been demonstrated in a commercial-scale plant.

Currently, aqueous monoethanolamine (MEA) is the solvent of choice for CO_2 scrubbing. Carbon dioxide reacts with MEA to primarily form a carbamate [5.16]. There are also a series of reactions involving water and CO_2 alone which lead to a small amount of CO_2 capture. For this response, we compared the solvent effectiveness in a set of absorption and stripping columns which use water alone to one that uses a 30% MEA solution in water. The effectiveness of each solvent is determined by comparing the size of the column and solvent flow rates which would be needed to complete a desired separation of CO_2 from flue gas and the associated capital cost.

Before we get started with our calculation, we defined in **Figure 5.8.1** a few key variables for our absorption and stripper column designs. We assume our analysis will be applied to a typical commercial-scale 500 MW coal-fired power plant, which produces a flue gas stream consisting of 12% CO_2 ($y_{N+1}^A = 0.12$), and a total flue gas flow rate of 9.5 x 10^7 mol/hr = 680 m^3/sec. The carrier gas flow rate entering the absorption column V_{in}^A is defined as $(1-y_{CO_2,flue})$ multiplied by the total flue gas flow rate, which gives us 8.3 x 10^7 moles per hour. We already know the absorption column is designed to remove 90% of the CO_2 in the flue gas stream. Based on that criterion and the 12% initial mixture composition, we calculate that the gas composition exiting the absorption column (y_1^A) will be approximately 0.012% CO_2. The solvent from the absorption column is sent to a stripping column, where the CO_2 is removed from the solvent by increasing the temperature, producing a relatively pure stream of CO_2. The absorption column operates at a cooler temperature (40°C) than the stripping column (120°C) to enhance the capture of CO_2 in the absorber and release of CO_2 in the stripper.

We can obtain an intuitive sense of why a 30% MEA solution would be more effective at CO_2 capture than pure water by first analyzing the

$$\left(V, y_1^A\right) \quad \left(L, x_0^A\right) \qquad \left(V, y_1^S\right) \quad \left(L, x_0^S\right)$$

stage: I   stage: I

stage: n   stage: n

stage: N   stage: N

$$\left(V, y_{N+1}^A\right) \quad \left(L, x_N^A\right) \qquad \left(V, y_{N+1}^S\right) \quad \left(L, x_N^S\right)$$

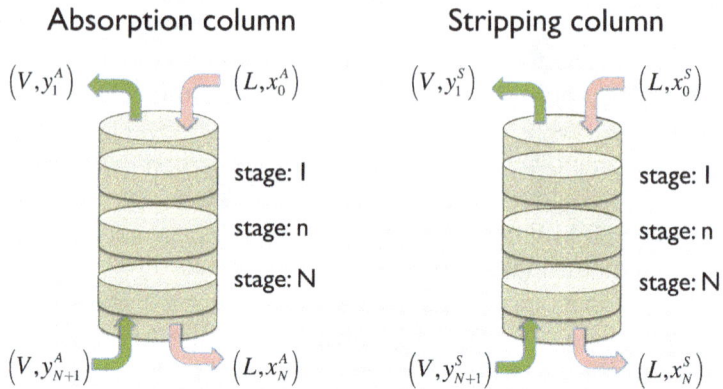

Figure 5.8.1   The absorption and stripping column

The absorption column: L = liquid flow rate, V^A, V^S = gas flow rate in the absorption/stripping unit, x,y represent the mole fraction of CO_2 in all the variables, the superscript "A" refers to the absorption column. For the stripping unit, we will use the superscript "S":

x_0^A, x_0^S = CO_2 in liquid entering the absorber/stripper

x_N^A, x_N^S = CO_2 in liquid at stage N in the absorber/stripper

y_{N+1}^A, y_{N+1}^S = CO_2 in gas entering the absorber/stripper

y_1^A, y_1^S = CO_2 in gas leaving the absorber/stripper

Note that the labeling of the stage is different from that in the previous section, here the top of the column is stage 1.

Henry constants for these two solvents. Henry constants describe the partial pressure of a solute (in this case CO_2) which is in equilibrium with a liquid; a smaller Henry constant indicates that the liquid is more effective at absorbing CO_2. At 40°C, the Henry constant for CO_2 in water is 2,397 atm; the Henry constant for CO_2 in 30% MEA is 10.1 atm [5.16]. Hence, for a given partial pressure of CO_2 in equilibrium with a liquid at 40°C, a 30% MEA solution will contain over two orders of magnitude more CO_2 compared to water alone. As a result, a 30% MEA solution can capture a significantly larger amount of CO_2.

Additionally, it is important to understand how the Henry constant changes with temperature. When the temperature of pure water is increased from 40°C to 120°C, its Henry constant increases by a factor of 5 while the Henry constant of a 30% MEA solution increases by a factor of 3237.2 with the same temperature increase. This demonstrates that the Henry constant of a 30% MEA solution is significantly more temperature sensitive, which means that temperature swings are effective at controlling whether the material absorbs or releases CO_2. This "switch-like" behavior for CO_2 absorption as a function of temperature is necessary so that we can effectively capture the CO_2 in the absorber and then subsequently release it in the stripper.

With this physical understanding of how amines change the Henry's law behavior of water, we are prepared to perform McCabe-Thiele analyses for the two systems. Going back to our absorber/stripper design diagram, one of the parameters which we may set within reasonable limits is the mole fraction of CO_2 in the solvent entering the absorption column, which is identical to that exiting the stripping column ($x_0^A = x_0^S$). We estimate this value to be 3×10^{-6} based on the equilibrium line/Henry's constant for water.

We assume that the carrier gas in the stripping column is water vapor. The water vapor which exits the re-boiler and enters the stripping column has a negligible amount of CO_2 ($y_{N+1}^S = 0$). We can now solve for the mole fraction of solute in the solvent exiting the absorption column using a mass balance. This is most easily done using mole ratios, defined as follows:

$$\overline{Y}_{N+1}^A = \frac{y_{N+1}^A}{1 - y_{N+1}^A}, \quad \overline{Y}_1^A = \frac{y_1^A}{1 - y_1^A}, \quad \text{and} \quad \overline{X}_0^A = \frac{x_0^A}{1 - x_0^A}$$

We can now solve for $\overline{X}_N$ using a mole balance:

$$\overline{X}_N^A = \frac{x_N^A}{1 - x_N^A} = \overline{X}_0^A + \frac{V}{L}\left(\overline{Y}_{N+1}^A - \overline{Y}_1^A\right)$$

From here, we can get the mole fraction of the CO_2 in both the absorber and stripper columns: $x_A^N = x_S^N = 5.35 \times 10^{-5}$. Similarly, we can construct a mole balance for the stripper column. Eventually, this yields:

$$\overline{Y}_1^S = \frac{y_1^S}{1 - y_1^S} = \overline{Y}_{N+1}^S + \frac{V}{L}\left(\overline{X}_0^S - \overline{X}_n^S\right)$$

We assume L/V is constant in our system. The L/V ratio is an adjustable design parameter. We manually adjusted the L/V ratio until we have identified a critical value which causes the operating and equilibrium lines to cross, which leads to an infinitely large number of stages. We then increased L/V by about 20% for the absorbers and decreased L/V by about 20% for the strippers, which leads to a finite number of stages. In the case of the absorber running on water, the critical L/V value was 2,500, which we increased by 20% to 3,000. In the case of the stripper running on water, the critical L/V value was about 8,800, which we reduced by 20% to about 7,040. For the 30% MEA case, we first utilized the same L/V ratios as used in the water case for purposes of comparison. Then, in a separate comparison, we adjusted the L/V ratios for the 30% MEA case to achieve the same number of trays as used in the water case.

The McCabe-Thiele diagram for using water in both the stripper and absorber is shown in **Figure 5.8.2.** In each case, we have optimized the liquid to vapor ratio (L/V) so as to minimize utility costs. We begin on the operating line and then sequentially "step-off" stages. We used a typical stage efficiency of 0.4; as a result, when we step off the operating line toward the equilibrium line, we only go 40% of the way to the equilibrium line. Given these parameters, we see that the absorber requires 33 stages and the stripper requires 27 stages.

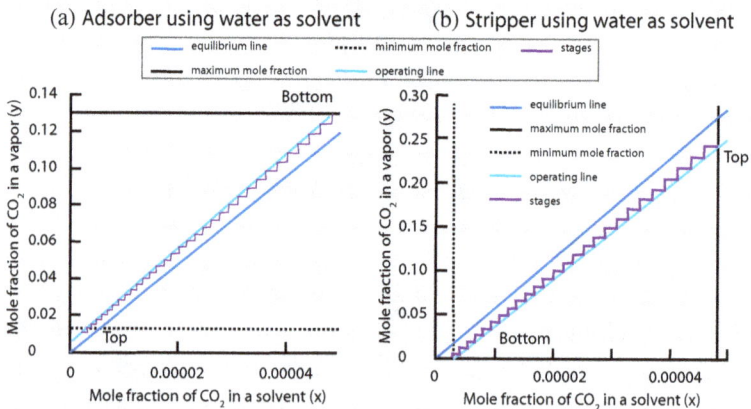

Figure 5.8.2 McCabe-thiele diagrams for water

McCabe-Thiele diagrams for (a) an absorber with 33 stages operating at an L/V ratio of 3,000 and (b) a stripper with 27 stages operating at an L/V ratio of 7,000; both units are using water as the solvent.

We now switch the solvent to 30% MEA while keeping the L/V ratio constant but changing the number of trays. We find that because 30% MEA has an extremely low Henry constant at 40°C, the equilibrium line is essentially horizontal, such that only 5 stages are required for the absorber (**Figure 5.8.3**). We find that the stripper operating on 30% MEA requires only 8 stages, since 30% MEA has an extremely high Henry constant at 120°C (see **Figure 5.8.3**). Hence, by switching from water to a 30% solution of MEA while keeping L/V constant, we reduce the number of stages from 33 to 5 for the absorber, and from 27 to 8 for the stripper.

We also consider the case in which we switch the solvent to 30% MEA while keeping the same number of trays, but change the L/V ratio to achieve the desired separation (**Figure 5.2.6**). We find that the L/V ratio for the absorber drops from 3,000 to 11 when we switch from water to 30% MEA, allowing us to reduce the solvent flow rate by a factor of 272. The L/V ratio for the stripper increases from 7,000 to 50,000, allowing us to use 7 times less vapor in the stripper upon switching from water to 30% MEA (**Figure 5.8.4**).

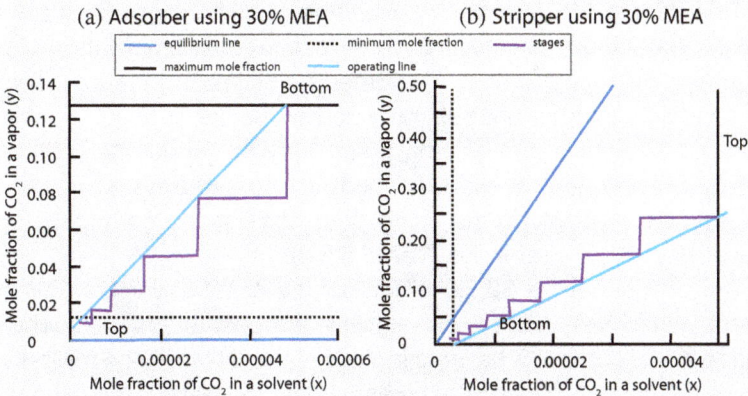

Figure 5.8.3 McCabe-Thiele diagrams for MEA (1)

McCabe-Thiele diagrams for (a) an absorber with 5 stages operating at an L/V ratio of 3,000 and (b) a stripper with 8 stages operating at an L/V ratio of 7,000; both units are using 30% MEA as the solvent. These columns are operating at the same L/V ratios as used for the water case shown in **Figure 5.8.2**. Notice that the equilibrium line in (a) has a very small slope and essentially coincides with the x-axis.

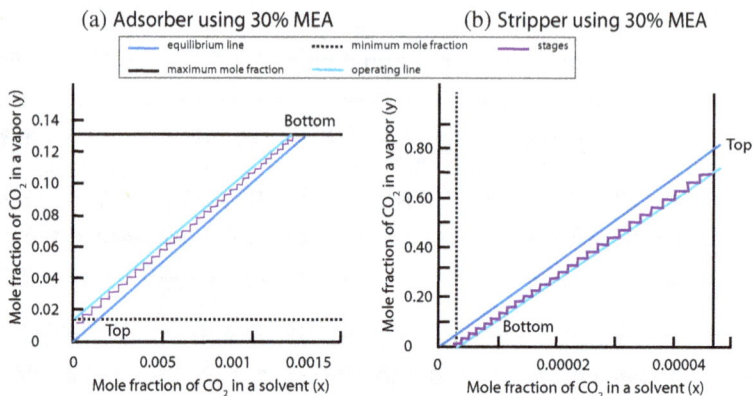

Figure 5.8.4 McCabe-Thiele diagrams for MEA (2)

McCabe-Thiele diagrams for (a) an absorber with 33 stages operating at an L/V ratio of 11 and (b) a stripper with 27 stages operating at an L/V ratio of 51,000; both units are using 30% MEA as the solvent and contain the same number of stages as used for the water case in **Figure 5.8.2**.

To estimate the economic advantage of switching from water to a 30% MEA solution, we calculated the capital cost associated with the processes described above. We started by estimating the tray diameter and spacing using the analysis provided by Seider *et al.* [5.17]. We calculated the height of the column by multiplying the number of trays by the tray spacing. We took into account the cost of sieve trays, the vessels, ladders, and platforms using cost correlations corrected for the present day (2012) cost index (**Table 5.8.1**).

The overall cost is $3.32 million for the case in which water is the solvent. When 30% MEA is the solvent, the cost is $941,775 when we operate at the same L/V as used in the water case and $2.4 million when we use the same number of trays as used in the water case. Hence, the capital cost can be reduced by a factor of 3 by switching to 30% MEA from water while keeping the same L/V conditions. The cost is driven down primarily by the fact that we use a significantly smaller number of trays with 30% MEA than water. When we use the same number of trays for both 30% MEA and water while changing the L/V ratio, we find that the capital cost drops marginally; we expect, however, that the utility

Table 5.8.1 Comparison of cost for columns using water and MEA solution

Item	Cost					
	Water		30% MEA with same L/V as water case		30% MEA with same number of trays as water case	
	Absorber	Stripper	Absorber	Stripper	Absorber	Stripper
Platforms and Ladders	$97,036	$54,128	$31,108	$25,502.3	$93,318.4	$5,964
Vessel	$37,241	$21,538	$28,080	$10,094.7	$35,583	$2,764
Trays	$2,511,373	$602,377	$576,655	$263,776	$2,259,404	$1,540
Total	$3.32 million		$941,775		$2.40 million	

cost will be significantly less with 30% MEA in this scenario since it uses significantly less liquid in the absorber and vapor in the stripper.

Our analysis demonstrates that 30% MEA significantly improves the performance of a pair of absorption and stripping columns, leading to large cost-savings. The difference in performance arises from how MEA changes the Henry's law behavior of water; specifically, MEA reduces the Henry's law constant at low temperatures and increases it at high temperatures, which allows for reversible absorption of CO_2. In order to more accurately model how MEA impacts an absorption and stripping process, it would be necessary to include more detailed kinetic effects in the model, beyond the stage efficiency that we have already incorporated. It is known that MEA has relatively poor kinetics for CO_2 absorption, which has motivated significant research to find additives which improve the kinetics. A model which contains both the thermodynamic and kinetic parameters is necessary to fully understand how MEA impacts the cost of the CO_2 capture process.

Section 9

Review

5.1. Reading self-test

1. Which statement about efficiency of an absorber is correct?
   a. If we use an infinite number of plates, the thermodynamic losses are close to zero
   b. Countercurrent flow is less efficient compared to concurrent flow in an absorber
   c. Plates are essential in an absorber to increase the contact area between the liquid and gas

2. Which statement about the equilibrium line is not correct?
   a. The smaller the slope, the better the solvent
   b. Only at low concentration is the equilibrium line a straight line
   c. If the system does not reach perfect equilibrium, the actual concentration in the solvent is lower compared to the equilibrium concentration

3. Henry's law is typically used at low partial pressures. Why?
   a. At high pressure, the gas does not behave like an ideal gas
   b. At high pressure, the number of dissolved gas molecules can become so large that the molecules will influence each other
   c. At high pressure, the molecules cannot escape from the liquid phase
   d. At high pressure, the liquid is not ideal
   e. Answers A and B
   f. Answers A, B, C and D

4. Which statement about the operating line is not correct?
   a. The slope of the operating line increases if we increase the amount of absorbent in the process
   b. All operating lines have in common the point $(x_{CO_2}^{reg}, y_{CO_2}^{exh})$
   c. The operating line gives the relation between the concentration of the fluid and the gas leaving plate P

5. At a pH of 9 what percentage of CO_2 dissolved in water is in the form of HCO_3^-?
   a. 100%
   b. 75%
   c. 50%
   d. 25%
   e. 0%

6. Which statement about the interactions of CO_2 with amines is not correct?
   a. The chemistry of CO_2 with primary amines is: CO_2:amine=1:2
   b. The chemistry of CO_2 with secondary amines is: CO_2:amine=1:1
   c. The reaction rates of tertiary amines with CO_2 are generally fast, but slower than primary or secondary ones
   d. The chemistry of CO_2 with tertiary amines is: CO_2:amine=1:1

7. Which statement about Ionic Liquids (IL) is correct?
   a. Ionic liquids have a relatively low vapor pressure
   b. Ionic liquids have a low melting temperature
   c. Ionic liquids react with CO_2 with a 1:1 chemistry if the anion is functionalized with an amine
   d. Answers A, B, and C are correct

8. Which term does not contribute to the parasitic energy of an absorption process?
   a. Compression
   b. Pumping of fluids
   c. Heating of the absorbent
   d. Cooling of the flue gas

9. Which part of an optimized amine process has the largest energetic losses?
   a. Absorber
   b. Stripper
   c. Condenser
   d. Heat exchanger
   e. Compressor

10. Which factors contribute to the selection of an absorbent?
    a. Capacity to absorb CO_2
    b. Thermal stability
    c. Corrosiveness
    d. Oxidative stability
    e. Answers A, B and C
    f. Answers A, B, C and D

5.2. Absorption

1. Assign the right names in the absorption process corresponding to the letters A–F.

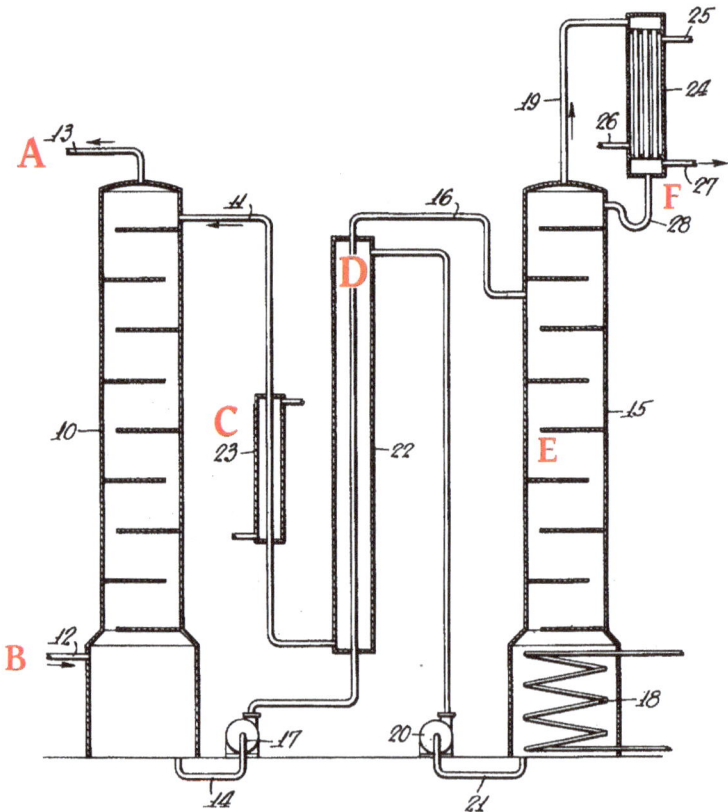

Section 10
References

1. Bottoms, R., 1930. Separating acid gases, US Patent 1,783,901.
2. Desai, S., and D.A. Klanecky, 2011. Meeting the needs of the water-energy nexus. Chem. Eng. Prog., **107** (4), 22.
3. UNICEF, 2006. Meeting the MDG Drinking Water and Sanitation Target: The Urban and Rural Challenge of the Decade. WHO and UNICEF: Geneva.
4. McCann, N., D. Phan, X.G. Wang, W. Conway, R. Burns, M. Attalla, G. Puxty, and M. Maeder, 2009. Kinetics and mechanism of carbamate formation from CO_2(aq), carbonate species, and monoethanolamine in aqueous solution. J. Phys. Chem. A, **113** (17), 5022. http://dx.doi.org/10.1021/Jp810564z
5. Keith, D.W., 2009. Why capture CO_2 from the atmosphere? Science, **325** (5948), 1654. http://dx.doi.org/10.1126/science.1175680
6. Mahmoudkhani, M., and D.W. Keith, 2009. Low-energy sodium hydroxide recovery for CO_2 capture from atmospheric air — thermodynamic analysis. Int. J. Greenh. Gas Con., **3** (4), 376. http://dx.doi.org/10.1016/J.Ijggc.2009.02.003
7. Vaidya, P.D., and E.Y. Kenig, 2007. CO_2-alkanolamine reaction kinetics: a review of recent studies. Chem. Eng. Technol., **30** (11), 1467. http://dx.doi.org/10.1002/ceat.200700268
8. Savile, C.K., and J.J. Lalonde, 2011. Biotechnology for the acceleration of carbon dioxide capture and sequestration. Curr. Opin. Biotechnol., **22** (6), 818. http://dx.doi.org/10.1016/j.copbio.2011.06.006
9. Brennecke, J.E., and B.E. Gurkan, 2010. Ionic liquids for CO_2 capture and emission reduction. J. Phys. Chem. Lett., **1** (24), 3459. http://dx.doi.org/10.1021/jz1014828
10. Nulwala, H.B., C.N. Tang, B.W. Kail, et al., 2011. Probing the structure-property relationship of regioisomeric ionic liquids with click chemistry. Green Chem., **13** (12), 3345. http://dx.doi.org/10.1039/C1gc16067b
11. Gurkan, B.E., J.C. de la Fuente, E.M. Mindrup, et al., 2010. Equimolar CO_2 absorption by anion-functionalized Ionic liquids. J. Am. Chem. Soc., **132** (7), 2116. http://dx.doi.org/10.1021/ja909305t
12. Ramezan, M., T.J. Skone, N. Nsakala, and G.N. Liljedahl, 2007. Carbon Dioxide Capture from Existing Coal-Fired Power Plants, Report No. DOE/NETL-401/110907). http://www.netl.doe.gov/energy-analyses/pubs/CO$_2$%20Retrofit%20From%20Existing%20Plants%20Revised%20November%202007.pdf

13. Rochelle, G., E. Chen, S. Freeman, D. Van Wagener, Q. Xu, and A. Voice, 2011. Aqueous piperazine as the new standard for CO(2) capture technology. Chem. Eng. J., **171** (3), 725. http://dx.doi.org/10.1016/J.Cej.2011.02.011

14. Rochelle, G.T., Amine scrubbing for CO_2 capture. Science, **325** (5948), 1652. http://dx.doi.org/10.1126/science.1176731

15. St Clair, J.H., and W.F. Simister, 1683. Process to recover CO_2 from flue-gas gets 1st large-scale tryout in Texas. Oil Gas J., **81** (7), 109.

16. Aroua, M.K., A. Benamor, and M. Z. Haji-Sulaiman, 1999. Equilibrium constant for carbamate formation from monoethanolamine and its relationship with temperature. J. Chem. Eng. Data, **44** (5), 887. http://dx.doi.org/10.1021/je980290n

17. Seider, W.D., J.D. Seader, D.R. Lewin, and W.D. Seider, 2004. Product and Process Design Principles: Synthesis, Analysis, and Evaluation. Wiley: New York.

Chapter 6

Adsorption

Only one letter separates **Ab**sorption from **Ad**sorption, yet these two methods to capture carbon dioxide differ in important ways. **Adsorption** is the process by which the surface of a solid holds on to small molecules in solution or in the gas phase. In other words, the molecules reside on the surface of the material. In **absorption**, the molecules actually enter into the material. One way these differences play out is in the development of strategies for increasing the capacity of the capture material. With **absorption**, increasing the capacity of the technology usually involves increasing the amount of capture material. In contrast, with **adsorbents**, increasing capacity involves increasing the exposed surface area of the material.

Section 1

Introduction

Adsorption has been used for many centuries. For example, in 1500 BC the Egyptians used charcoal to adsorb odorous vapors from wounds. Hindu documents dating back to 450 BC report the use of charcoal to purify water. Charcoal is one of the oldest types of adsorbents, and it is still used today in the form of "activated charcoal," also called activated carbon. In modern times, activated carbon is produced from carbonaceous sources ranging from materials as diverse as nutshells, peat, wood, coir, lignite, coal, and petroleum pitch. The raw carbon gets "activated," meaning its surface area is greatly expanded. This is achieved through various chemical and physical processes, which can yield areas as large as 500–1,500 m^2/g.

In daily life, you may have encountered adsorption in the form of the little bags of silica gel that come in the pockets of new clothes. Silica gel is very efficient at adsorbing water, and bags of it are often added to boxes in which moisture-sensitive materials are shipped. Many high school chemistry courses include an experiment that shows decolorizing a solution using a charcoal adsorber. Baking soda is often used in refrigerators to adsorb vapors emanating from stored foods. Of course, in this context we are interested in separating gasses, but it is important to realize that use of materials for adsorption is already established in industry. Examples of popular adsorption materials are shown in **Figure 6.1.1**.

A new class of solid adsorbents is comprised of nanoporous crystalline materials. These crystals have pores with diameters on the order of nanometers, which can have very large internal surface areas where adsorption can occur. As in liquid absorption, we can increase the capacity by increasing the amount of material. Hence, for the purpose of CCS, the only difference between these crystalline materials and the liquid absorbents is that one is solid and the other is liquid. So we use the term absorption when molecules are absorbed in a liquid and adsorption if they are adsorbed in a solid. This allows us to stay as close as possible to the nomenclature used in the literature.

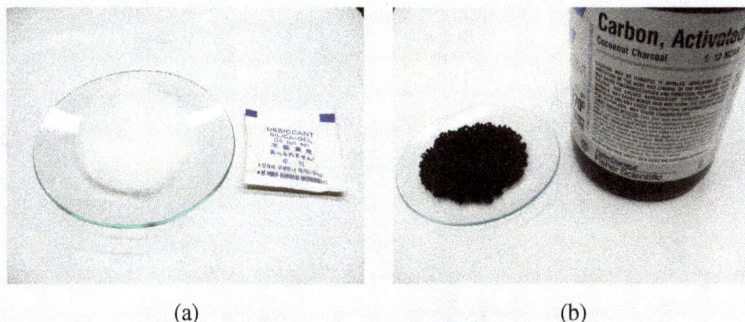

Figure 6.1.1 Adsorption materials

Examples of adsorption materials are: (a) silica (you have probably bought items with silica bags to adsorb water), and (b) activated carbon (perhaps used in your organic chemistry lab). *Photos courtesy of Joseph Chen.*

Zeolites are nanoporous materials that are used in many different applications. The basic building blocks of zeolites are corner-sharing TO_4 groups, where the T atom is usually Si, Al, or in some cases P. These tetrahedra can form different types of units, such as 6-rings, 8-rings, or 12-rings. These rings are the so-called secondary building units for different types of cylinders or cages. These cylinders or cages form a network of pores in the zeolite crystal. At present over 200 different zeolite structures are known, each of which has a similar chemical composition, but a very different pore topology (see **Figure 6.1.2**). Each zeolite structure or topology has been given a unique three letter name by the Structure Commission of the International Zeolite Association (IZA-SC).

Since the recent discovery of a new class of nanoporous materials called **Metal Organic Frameworks (MOFs),** the list of novel porous materials has grown by an order of magnitude. MOFs are metal/organic hybrid solids built from organic linkers and nodes made out of inorganic metals (or metal-containing clusters). One sample structure, MOF-5, is shown in **Figure 6.1.3**. By changing the metal and/or the linker, one can

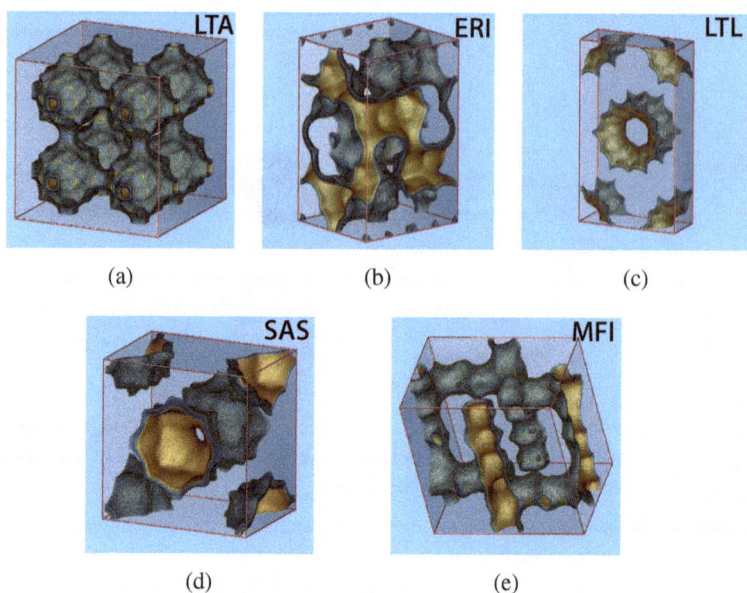

Figure 6.1.2 Some zeolite structures

(a) Computer generated representation of zeolite structure with the **LTA** topology: this material has spherical cages that are connected by narrow windows. The cages have a diameter of about 10 Å and windows of about 4.2 Å.

(b) The **ERI** topology: this material has elliptical cages that are connected by narrow windows.

(c) The **LTL** topology: this material has one-dimensional tubular channels.

(d) The **SAS** topology: this material has one-dimensional channels consisting of cavities connected with narrow windows (similar to LTA).

(e) The **MFI** topology: this material has three-dimensional intersecting channels.

Figures reproduced with permission from Beerdsen et al. [6.1]. Copyright (2006) American Chemical Society.

generate millions of different MOFs. Given the potential for so many different materials, we may ask ourselves what material will be best for carbon capture. This is the question that we will address in the present chapter.

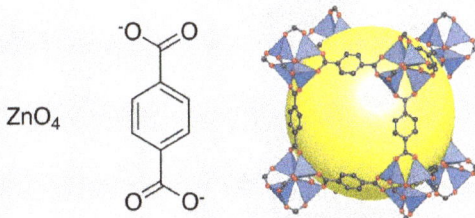

Figure 6.1.3 A metal organic framework (MOF-5)

The MOF-5 structure: this material is synthesized from the metal-containing cluster ZnO_4 with benzene dicarboxylate as the linker. The MOF-5 structure is shown as ZnO_4 tetrahedra (blue) joined by the linker (O, red, and C, black) to give an extended 3D cubic framework with interconnected pores of 8 Å aperture width and 12 Å pore diameter. The yellow sphere represents the largest sphere that can occupy the pores without coming within the van der Waals radius of the atoms in the framework. *Figures reproduced from Li et al. [6.2], with permission from Macmillan Publishers.*

Section 2

Adsorption processes

Fixed bed adsorption

Adsorption is a unit operation that can be done in a couple of different ways. The most common and easiest to think about is the so-called fixed bed, shown in **Figure 6.2.1**. In this system, a flue gas enters and flows onto a bed of solid material, which selectively adsorbs CO_2. The gas that exits the system is nearly pure N_2, and the CO_2 is retained on the bed. This process can continue until the bed is saturated, at which point CO_2 will "break through" and show up in the exhaust. At that point of saturation the bed needs to be regenerated, which is accomplished by running the process in reverse. This can be done by either increasing the temperature or by decreasing the partial pressure of CO_2, for example, by introducing a purge gas. The nearly pure CO_2 will then be recovered.

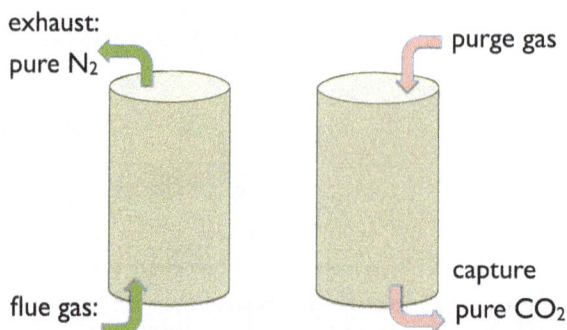

Figure 6.2.1   Fixed bed adsorber

In a fixed bed adsorber, CO_2 is captured in two steps. In the first step (left figure), the CO_2 is selectively adsorbed from the flue gas. This adsorption bed needs to be regenerated once it is saturated with CO_2. The figure on the right depicts the second regeneration step, in which pure CO_2 is produced. In this step, the bed is heated and a purge gas displaces the adsorbed CO_2.

Fluidized bed adsorption

A disadvantage of fixed bed adsorption is that it is a batch process, meaning that the process must be interrupted periodically to change the direction of the flow. In industry there are many good reasons to prefer using a continuous process. For example, a coal-fired power plant cannot be shut down just to accommodate the replacement or regeneration of an adsorption unit. Therefore, we need to run two adsorbers in parallel; one in adsorption mode and the other in regeneration mode. This is not always the most efficient way to operate an adsorption process. Fluidized bed adsorbers were developed to use solids in a continuous process configuration so that the adsorption process proceeds without any interruptions, just like the power plant. These fluidized beds are very interesting from the perspective of fluid dynamics, and a great deal more complicated from an operational point of view.

Transportation is the root of the problem. To have a continuous process, we need to move our adsorbent from one place to another. Many adsorbents are powders, which are particularly challenging to transport. Imagine, if you will, trying to move 600 pounds of loose baby powder several yards and back without spilling. It wouldn't be easy. Yet this is precisely what we are being asked to do with our adsorbent in the regeneration step. One solution is to use a fluidized bed, based on the idea that by flowing a gas through a powder, it will behave like a liquid. Such a fluidized bed allows us to bring the flow of solid particles from the bottom of the column to the top. Once the particles reach the top, they can flow to the bottom through gravity, adsorbing the CO_2 gas that flows upwards on their way down. When the particles get to the bottom, they are regenerated and ready to enter the adsorption section. **Figure 6.2.2** shows a possible implementation of this process.

An alternative to moving the solid is to continuously change the position in which we introduce the flue gas and the purge gas. In this way we mimic the movement of the solvent using a well-designed system of

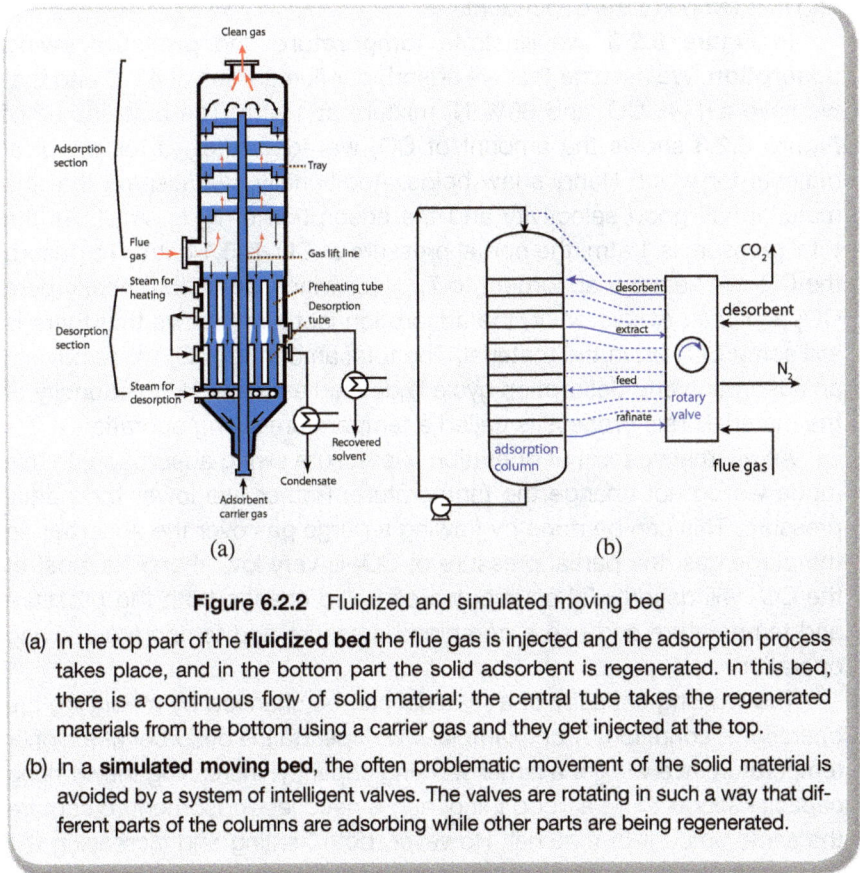

Figure 6.2.2 Fluidized and simulated moving bed

(a) In the top part of the **fluidized bed** the flue gas is injected and the adsorption process takes place, and in the bottom part the solid adsorbent is regenerated. In this bed, there is a continuous flow of solid material; the central tube takes the regenerated materials from the bottom using a carrier gas and they get injected at the top.

(b) In a **simulated moving bed**, the often problematic movement of the solid material is avoided by a system of intelligent valves. The valves are rotating in such a way that different parts of the columns are adsorbing while other parts are being regenerated.

valves. The apparatus that employs this technology is called a simulated moving bed (**Figure 6.2.2**).

Temperature and pressure swing adsorption

Let us now look at how we can operate a fixed bed adsorption column. As in absorption, the key thermodynamic property is the amount of captured CO_2, in this case the amount that adheres to the adsorbent. This is given by adsorption isotherms. We will discuss the thermodynamics of adsorption in detail in a moment. For the time being, we simply assume that the pressure of the flue gas is sufficiently low such that adsorption is proportional to the partial pressure of CO_2 (according to Henry's law). The proportionality constant is called the Henry coefficient (see **Box 5.2.2**), and it is temperature-dependent.

In **Figure 6.2.3**, we illustrate **temperature** and **pressure swing adsorption**. We assume that we adsorb our flue gasses at 40°C and that we have a 14% CO_2 and 86% N_2 mixture at 1 atm. The bottom half of **Figure 6.2.3** shows the amount of CO_2 we adsorb (σ_{CO_2}) for a typical material for which Henry's law holds. Additionally, we assume that the material has good selectivity and the adsorption of N_2 is small. As the total pressure is 1 atm, the partial pressure of CO_2 is 0.14 atm. To desorb the CO_2 we heat our adsorbent to T_{final}, for which we desorb nearly pure CO_2 at 1 atm. At this point, the adsorption isotherm shows that there is still some CO_2 left in the material. The total amount of CO_2 we remove in an adsorption and desorption cycle is defined as the working capacity of the material. This process is called a temperature swing operation.

An alternative form of operation is pressure swing adsorption. In this mode we do not change the temperature. Rather, we lower the partial pressure. This can be done by flowing a purge gas over the adsorber. In the purge gas, the partial pressure of CO_2 is very low, therefore most of the CO_2 will desorb. Of course, we can also change both the pressure and temperature and use a combined pressure and temperature swing operation.

The working capacity of a material can be changed by modifying the operational conditions. For example, if we operate the desorber at a higher temperature, we will get a larger working capacity. Increasing the working capacity seems to be a good thing, as we need less adsorbent to capture the same amount of material. However, both heating and increasing the

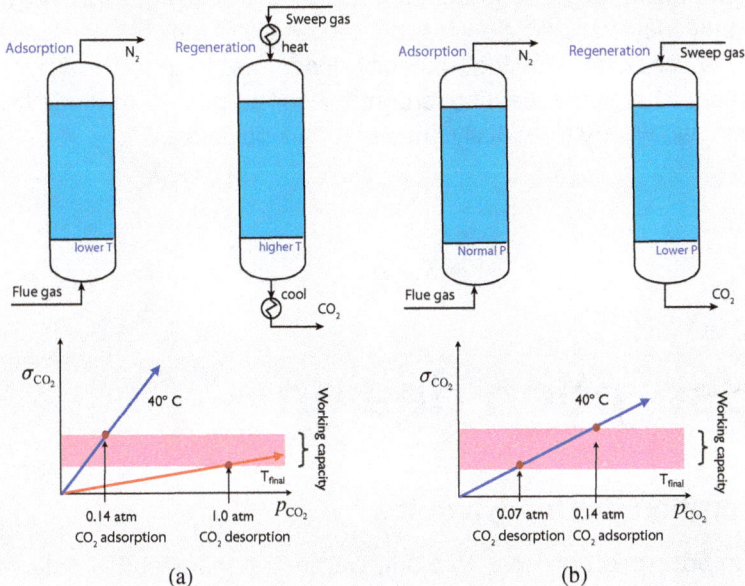

Figure 6.2.3. Temperature and pressure swing adsorption

(a) In **temperature swing adsorption**, we make use of differences in adsorption at different temperatures. The adsorption step is done at a lower temperature (1 atm, 40°C, 14% CO_2) compared to the desorption step. The isotherms give the equilibrium concentrations at adsorption and desorption conditions. The working capacity is the amount of CO_2 that is removed in one cycle.

(b) In **pressure swing adsorption**, we make use of differences in adsorption at different pressures. The adsorption step is done at a higher pressure (1 atm, 40°C, 14% CO_2) compared to the desorption step. The isotherms give the equilibrium concentrations at adsorption and desorption conditions. The working capacity is the amount of CO_2 that is removed in one cycle.

Question 6.2.1 Working capacity

Do the following changes of the operating conditions *decrease* or *increase* the working capacity?
1. Decrease the flue gas temperature.
2. Increase the flue gas pressure.
3. Decrease the pressure at desorption. (How would we do this in practice?)

pressure cost energy; so in the end, the benefit of using a material more efficiently may be completely lost by the penalty of a higher energy bill.

This issue of working capacity leads us to another important question: what is the ideal adsorbent for our purposes? In the following section, we will try to partially answer to this question.

Section 3

Adsorption design

Adsorption thermodynamics

For an adsorption process to work, we need a material that selectively takes up CO_2. The adsorption behavior of a material is described by the adsorption isotherms, which give the amount of CO_2 and N_2 adsorbed as a function of the partial pressure. As these isotherms play a central role in the design of an adsorber, we discuss in this section the underlying thermodynamics of adsorption.

A typical isotherm is shown in **Figure 6.3.1**. Initially the adsorption is proportional to the pressure, following Henry's law as described in the previous section, but as our material has a limited capacity it will require increasingly higher pressures to further increase the loading until saturation. The simplest mathematical formula that describes this behavior is:

$$\theta(p) = \frac{\sigma(p)}{\sigma_{max}} = \frac{bp}{1 + bp},$$

where p is the (partial) pressure, θ is the fractional occupancy, σ is the loading (in mol/kg adsorbent) and σ_{max} is the saturation loading. This equation is a **Langmuir isotherm**, named after Irving Langmuir, who won the Nobel prize for his work in 1932. In **Box 6.3.1**, we give a simple kinetic

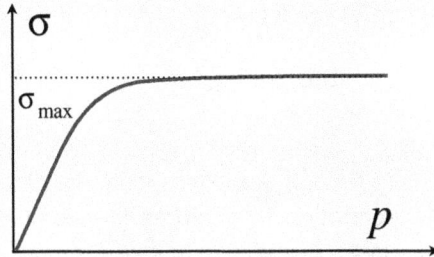

Figure 6.3.1 Langmuir isotherm

Example of a Langmuir isotherm: loading σ as a function of pressure p.

Box 6.3.1 Langmuir isotherms

We can derive the Langmuir isotherm using a kinetic argument. Suppose we have a surface with σ_{max} adsorption sites, and this surface is in contact with an ideal gas. We will have equilibrium between the surface and the gas if the number of molecules per unit time leaving the surface and going into the gas phase equals the number of molecules per unit time going in the reverse direction. If we assume that there are no interactions between the adsorption sites, the rate of molecules leaving the surface is simply proportional to the number of adsorbed molecules:

$$k\,(\text{surface} \rightarrow \text{gas}) = c_d \theta \sigma_{max}$$

The rate of molecules going in the reverse direction, from the gas to the surface, is not only proportional to the number of molecules in the gas phase (or pressure), but also to the number of empty spots on our surface:

$$k\,(\text{gas} \rightarrow \text{surface}) = c_a (1-\theta) \sigma_{max} p$$

At equilibrium these rates are equal, giving us:

$$c_d \theta \sigma_{max} = c_a (1-\theta) \sigma_{max} p,$$

with $b = c_a/c_d$, we have:

$$\theta = \frac{bp}{1+bp}$$

This is the famous Langmuir equation.

derivation of the Langmuir equation. In the limit where $p \rightarrow 0$, the adsorption is a linear function of pressure, or:

$$\sigma(p) \approx \sigma_{max} b p = H_i p,$$

where H_i is the Henry coefficient for species i. If the pressures are low, which is often the case for flue gas adsorption, we can simply extend our single component description to mixtures (see **Box 6.3.2**). In case the adsorption of both gasses is sufficiently low, we can use the ratio of the Henry coefficients to predict the selectivity of a material. The selectivity is defined as the ratio of the relative loading of the two components in the material and their relative concentrations in the gas phase, or:

$$S = \frac{\sigma_{CO_2}/\sigma_{N_2}}{p_{CO_2}/p_{N_2}} = \frac{H_{CO_2}}{H_{N_2}},$$

The other important thermodynamic relation describes the temperature dependence of the Henry coefficient. This relationship is critical as we vary the temperature of the adsorbent bed as part of our carbon capture process. The **Van't Hoff equation** (after Jacobus Henricus van't Hoff, the Dutch chemist who was the very first recipient of the Nobel Prize in chemistry) expresses the relation we need:

$$\frac{d \ln H}{dT} = \frac{\Delta h_{ads}}{RT^2},$$

where Δh_{ads} is the heat of adsorption.

Breakthrough curves

Compared to the design of an absorber, the design of either a fixed or fluidized bed is much more complicated because we have to deal with different time and length scales. The adsorption column itself is on the scale of meters, inside it is composed of irregularly packed powder, and in the powder are gaps within tiny molecular structures. So given these considerations, how do you go about trying to design the size of adsorption equipment? It turns out we can use many of the same tools that we talked about for absorption.

Box 6.3.2 Predictions of mixture isotherms

We can extend our single component Langmuir isotherm to mixtures. We use a surface with σ_{max} sites. As in the case of pure components, we assume that there are no interactions between the sites. The components A and B compete for these sites. We have equilibrium between the surface and the gas if the number of molecules per unit time leaving the surface and going into the gas phase equals the number of molecules per unit time going in the reverse direction for each of the two components A and B. If we assume no interactions between the adsorption sites, the rate of molecules leaving the surface is simply proportional to the number of adsorbed molecules:

$$k_A \left(\text{surface} \to \text{gas} \right) = c_d^A \theta_A \sigma_{max},$$

where θ_A is the fraction of sites occupied by A molecules. For component B we have:

$$k_B \left(\text{surface} \to \text{gas} \right) = c_d^B \theta_B \sigma_{max}$$

For the reverse direction, we have the molecules competing for the empty sites and for each of the components the rate of adsorption is proportional to the partial pressure:

$$k_A \left(\text{gas} \to \text{surface} \right) = c_a^A \left(1 - \theta_A - \theta_B \right) \sigma_{max} P_A$$

$$k_B \left(\text{gas} \to \text{surface} \right) = c_a^B \left(1 - \theta_A - \theta_B \right) \sigma_{max} P_B$$

Imposing equilibrium and using $b_i = c_a^i / c_d^i$ $(i = A, B)$ gives:

$$\theta_A = \left(1 - \theta_A - \theta_B \right) b_A p_A$$

$$\theta_B = \left(1 - \theta_A - \theta_B \right) b_B p_B,$$

which gives the adsorption isotherms:

$$\theta_A = \frac{b_A p_A}{1 + b_A p_A + b_B p_B}$$

$$\theta_B = \frac{b_B p_B}{1 + b_A p_A + b_B p_B}$$

The importance of this result is that one can use the experimental data of the pure component isotherms (b_i, σ_{max}^i) to predict the mixture isotherms. The assumption that molecules compete for the same sites is often too simple and one needs to use alternative approaches to predict the actual mixture isotherm. A particularly popular one is IAST (Ideal Adsorbed Solution Theory) [6.3].

Typically, the additional variables we look at are the gas velocity that comes into the bed, the amount of space the particles themselves take up, or bed voidage. In principle, it is possible to call "Adsorbers-R-Us" and ask for X pounds of our favorite solid adsorbent. But first, we would need to know how much is X? To answer this question, let us start with a fixed-bed reactor. To simplify our design we assume we have two columns; one for the adsorption and one for the regeneration. We know the flow rate of the flue gas from which we need to remove CO_2. In order to know how much material to buy, we need to know the equilibrium properties of the material, including the amount of CO_2 adsorbed per unit time, the capacity of the porous media, and the selectivity of the porous media. Furthermore, we need to understand the dynamics of the system down to the level of diffusion inside the pores. This is not an easy task.

Let us first envision how the CO_2 propagates through these devices. If we have an ideal adsorbent, equilibrium between the flue gas and the gas adsorbed in the material is established instantaneously. This equilibrium concentration is given by the adsorption isotherm, which gives the amount of CO_2 adsorbed inside the material as a function of the partial pressure. Let us assume we inject the gas in the left of the column. CO_2 adsorbs inside the material until the equilibrium concentration is reached. At this point the material cannot adsorb CO_2 anymore, so a CO_2 front will form and slowly move through the column until it has reached the right of the column. Finally, the CO_2 breaks through the adsorber and the adsorber needs to be regenerated. This behavior is illustrated in **Figure 6.3.2**.

Figure 6.3.3 shows some breakthrough curves of more realistic systems (where the CO_2 concentration in the outlet increases more gradually). Inspection of these qualitative curves informs us that steeper breakthrough curves result in more efficient use of the adsorption bed. We will now show how these breakthrough curves are calculated.

Quantitative analysis of breakthrough curves

We would like to determine the density of CO_2 as a function of the position and time in the adsorption bed. If we assume that the adsorber is a cylinder parallel to the z-direction with a uniform radial profile of adsorption, the problem has only one distance dimension, the distance down the bed. The first step is a mass balance for CO_2 over a control volume

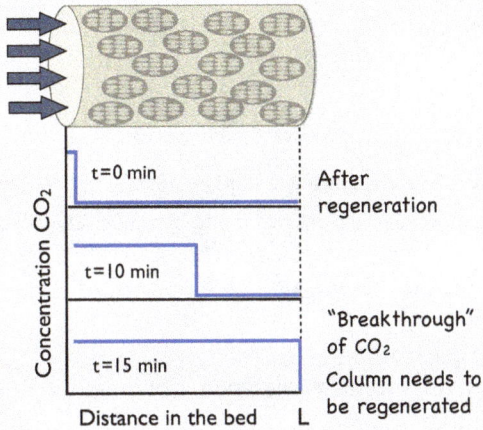

Figure 6.3.2 Adsorption bed and breakthrough times

The flue gas enters the left of the column. The blue lines give the CO_2 concentration in the flue gas as a function of its position in the bed at different time increments after the bed has been regenerated. As soon as the CO_2 front reaches the right side of the bed we reach the breakthrough, and the material needs to be regenerated again.

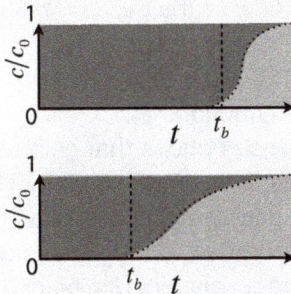

Figure 6.3.3 Experimental breakthrough curves

The concentration of an adsorbed species (like CO_2) at the outlet is divided by the concentration at the inlet c_0 as a function of time. Initially the concentration at the outlet is close to zero because all molecules are being adsorbed in the bed. After some time, the bed is becoming saturated and the concentration will increase until it equals the concentration coming in. If we have ideal adsorption, this profile is a step function. Non-ideal flow and diffusion limitations cause deviations from the ideal behavior. This non-ideal behavior decreases the efficiency of our bed, as we get too high a concentration of CO_2 earlier.

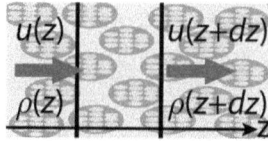

Figure 6.3.4 Mass balance

We have a gas flowing in with a velocity u and density ρ. The accumulation is in the porous material.

$dV = dxdydz$ at an arbitrary distance down the cylindrical bed (see **Figure 6.3.4**). In that volume:

$$\text{accumulation} = \text{flow in} - \text{flow out} - \text{adsorption}$$

We have included a term to account for the CO_2 that adsorbs in the adsorbent. We can write the mass balance as an equation:

$$\frac{d\rho_i}{dt} = -\varepsilon\frac{d(u\rho_i)}{dz} + \frac{\partial\rho_i}{\partial t}$$

In this equation u is the flow of the gas (m/s), ρ_i is the density of component i (mol/m^3), where i usually refers to CO_2 but could easily be another component such as N_2 or H_2O, and the term ε is the fraction of void space in the bed, or void fraction.

Let us make some assumptions that help us solve the above equation for the mass balance. First, we assume that in the adsorption bed all gasses behave as ideal gasses and that the amount of CO_2 in the gas is very small such that adsorption of CO_2 does not change the velocity u. We also assume there are no diffusion limitations for the adsorption in the pores; the CO_2 molecules in the pores are in equilibrium with the gas outside of the pores. Let us further assume that this equilibrium is in the Henry's law regime (low partial pressure of CO_2). Hence we can relate the amount adsorbed in the adsorbent, σ_{CO_2} (mol/kg adsorbent), to the partial pressure of CO_2 in the void volume using the adsorption Henry coefficient, H_{CO_2} (in units: mol/(kg adsorbent atm)):

$$\sigma_{CO_2} = H_{CO_2}p_{CO_2}$$

Note that if a CO_2 molecule gets adsorbed, it accumulates in the control volume. If ρ_A is the density of the adsorbent (kg/m³) then the total amount of adsorbent (kg) in our control volume dV is:

$$m_{ads} = (1-\varepsilon)\rho_A dV$$

We can now compute the accumulation of CO_2 due to adsorption in the adsorbent in our control volume:

$$\Delta_{CO_2} = \sigma_{CO_2} \times (1-\varepsilon)\rho_A dxdydz$$

As the adsorption is the only contribution to the accumulation, we have:

$$\frac{\partial \rho_{CO_2}}{\partial t} = \frac{d\left(\Delta_{CO_2}/dxdydz\right)}{dt} = (1-\varepsilon)\rho_A \frac{d\sigma_{CO_2}}{dt},$$

using Henry's law:

$$\frac{\partial \rho\ CO_2}{\partial t} = (1-\varepsilon)\rho_A H_{CO_2} \frac{d\rho_{CO_2}}{dt}$$

Now we can write the mass balance as:

$$\frac{1}{RT}\frac{d\rho_{CO_2}}{dt} = \frac{\varepsilon u}{RT}\frac{d\rho_{CO_2}}{dz} + (1-\varepsilon)\rho_A H_{CO_2} \frac{d\rho_{CO_2}}{dt}$$

or:

$$\left(\frac{1}{RT} - (1-\varepsilon)\rho_A H_{CO_2}\right)\frac{d\rho_{CO_2}}{dt} = \frac{\varepsilon u}{RT}\frac{d\rho_{CO_2}}{dz}$$

If we assume N_2 does not adsorb, we have the equation that we need to solve to compute the breakthrough curve, with the boundary condition, $\rho_{CO_2}(t,0) = \rho_{CO_2, \text{flue}}$. To find the solution, we guess that this solution looks like:

$$\rho_{CO_2}(z,t) = \Theta(z - Bt),$$

where $\Theta = \Theta(y)$ is an unknown function with argument $y = (z-Bt)$. Given Θ, we can write:

$$\frac{d\rho_{CO_2}}{dz} = \Theta' \quad \text{and} \quad \frac{d\rho_{CO_2}}{dt} = -B\Theta',$$

where $\Theta' = d\Theta'/dy$. Substitution of these two equations in the differential equation we need to solve gives:

$$B = -\frac{\varepsilon u}{1 + RT(1-\varepsilon)\rho_A H_{CO_2}}$$

We indeed have a solution for our equation. The boundary condition and initial condition are given as:

$$p_{CO_2}(0,t) = p_0 \quad \text{and} \quad p_{CO_2}(z,0) = 0$$

We can now calculate the breakthrough curve, without having any knowledge about the form of Θ. Surprised? Let's see how this works. For the breakthrough curve we only need to know the pressure as a function of time at position $z = L$. Why does this help us to find the solution? Let us assume we would like to know the solution at a time t' and position L. From the form of the solution and boundary condition, we have:

$$p_{CO_2}(L,t') = \Theta(L + Bt') = \Theta(0 + Bt) = p_{CO_2}(0,t) = p_0$$

Or, the solution at $z = L$, is the same as the solution at $z = 0$, at a time t', if the arguments of Θ are the same:

$$L + Bt' = 0 + Bt,$$

which tells us that we have the same solution at $z = L$ as we have at $z = 0$, not at time t but at t':

$$t' = t - \frac{1}{B}L = t + \frac{1 + RT(1-\varepsilon)\rho_A H_{CO_2}}{\varepsilon u}L$$

Therefore, if we start injecting CO_2 in the adsorber at $t = 0$, the CO_2 will simply arrive at length L a little later. The coefficient B can be interpreted as the velocity at which this front is traveling:

$$u' = \frac{\varepsilon u}{1 + RT(1-\varepsilon)\rho_A H_{CO_2}} \approx \varepsilon u \left(1 - RT(1-\varepsilon)\rho_A H_{CO_2}\right),$$

where the last step is only valid if the adsorption is small. Hence, the profile that we inject will move with an effective velocity that depends on the Henry coefficient. In addition, it will come out of the column in exactly the same shape, as is illustrated in **Figure 6.3.5**.

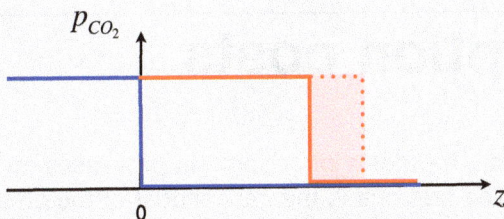

Figure 6.3.5 Propagation of the CO_2 through the column

The dotted line would have been the profile if there were no adsorption. The shaded part represents the distance the CO_2 would have travelled if there had been no adsorption of CO_2 ($H_{CO_2} = 0$).

Recall that to derive these equations we have made a few assumptions. First, we use a very simple adsorption isotherm in the limit of low pressures. For flue gasses, this linear relation is reasonable. Second, we assumed an ideal pressure and velocity profile. Lastly, we assumed that we do not take "diffusional effects" into account. In other words, we have assumed that gas moves only by convection. If a material has diffusion limitations, the effective velocity will increase as the bed is not as efficient.

The calculation we presented in this section represents an ideal case scenario. Using these equations, we can make an ideal case design for a given material. This should give us some idea about the amount of material needed for our separation. In addition, if a material performs poorly in a cost analysis using our simple model, a more realistic description will only make it worse, because the deviations from ideal behavior all hinder adsorption of CO_2.

Section 4

Adsorption costs

As in absorption, the costs of adsorption processes involve two main components. The first one is the capital cost of the equipment. As we have seen in the previous section, the size of the equipment is set by the breakthrough curves. The cost of the adsorbent also factors into the capital cost. This is more difficult to estimate, as adsorption processes for carbon capture are still in a very early stage of development. In particular, if we use a novel material it is very difficult to reliably estimate its future cost in a large-scale operation. Many of the new materials are currently synthesized in laboratories only on a very small scale. One really has to ask the question whether anything in the process would prevent us from scaling up the synthesis, say to supply the adsorbent to equip 50% of all power plants with a carbon capture adsorption process containing this material. A bottleneck for such a scale-up could be the abundance of certain elements; if the global employment of our favorite absorbent would require, say, 300% of the world reserve of this element, one would arrive at a very expensive process. The second component is the operational cost. This cost will be dominated by the energy requirement. Similarly, as in absorption processes, the energy requirements are dominated by the compression work and the energy required to regenerate the adsorbent.

Parasitic energy

As we saw in Section 5.2, we can operate our adsorber using temperature or pressure swing, or a combination of the two. These different process configurations rely on the difference between adsorption and desorption conditions to capture (and then release) CO_2, and they can vary in their methods of gas-solid contact and heat transfer. These important factors do not affect performance under equilibrium conditions, hence we will focus on equilibrium process thermodynamics for our energy analysis. The thermal energy requirement of the process is the sum of the sensible heat needed to heat the adsorption bed to the process temperature required to desorb CO_2 (Q_{sen}), and the energy needed to overcome the heat of absorption (Q_{des}). The thermal energy requirement (Q) per unit mass of CO_2 produced ($\Delta\sigma_{CO_2}$) is given by:

$$Q = \frac{Q_{sen} + Q_{des}}{\Delta\sigma_{CO_2}},$$

with:

$$Q_{sen} = C_p m_{ads} \left(T_{des} - T_{ads}\right),$$

where C_p is the specific heat capacity of the adsorbent, m_{ads} is the total mass of the adsorbent, ($T_{des} - T_{ads}$) is the temperature difference between the adsorption and desorption conditions. We also have:

$$Q_{des} = \Delta h_{CO_2}^{ads} \Delta\sigma_{CO_2}^{ads} + \Delta h_{N_2}^{ads} \Delta\sigma_{N_2}^{ads},$$

where $\Delta\sigma_i$ is the difference in mass loading for each species between the beginning and end of regeneration, and Δh_i is the heat of adsorption for each species. The loading at various conditions is calculated from the adsorption isotherms. As with our analysis for absorption, we assume that this thermal energy is supplied by diverting steam from the power cycle. Diverting steam effectively imposes a parasitic load on the power plant that can be represented as the thermal energy requirement discounted by 75%, which represents the typical efficiency of a turbine, times the Carnot efficiency η of the extracted steam. To obtain the parasitic energy, we also need to add the energy required for the compression of CO_2 to 150 bar (W_{comp}):

$$E_{par} = 0.75 \times Q \times \eta + W_{comp}$$

We see that our expression for the parasitic energy is nearly identical to the one we derived for the absorption process. Nevertheless there are important differences. In our liquid absorption process, we can use heat exchangers such that the heat we supply for the desorption is not all lost. For solid adsorption, heat exchange is much more difficult; therefore we have to assume that the heat we supply is lost. As a consequence it is much more important that we choose the material that will yield the lowest parasitic energy. At this point it is also important to emphasize that the above estimate is a lower bound on the parasitic energy; a real process involves many more operations that cost energy and the actual parasitic energy will be about 30% higher.

Optimal performance

From a research point of view an important question is: what is the ideal adsorbent? There are a few obvious criteria — the material needs to be selective toward CO_2 and the capacity needs to be reasonable. We have seen that several additional parameters, such as working capacity and heat of adsorption, come into play. This makes the selection of an optimal adsorbent complicated. For example, one might focus on research that improves the selectivity of a material for CO_2, but different materials may perform optimally at very different conditions. To address this problem, we propose to use the optimal parasitic energy as a metric to compare the performance of different materials. The idea is to determine the optimal performance, i.e., the lowest parasitic energy, for a given material and then to compare this optimum for different materials. In the remainder of this section, we provide an example of the use of parasitic energy as a metric to rank different materials for their performance in carbon capture by adsorption.

In **Figure 6.4.1** we schematically show how the working capacity of a material depends on the Henry coefficient. For materials with a small Henry coefficient, we expect a poor performance. The working capacity is small, yet the entire system needs to be heated to the desorption conditions, giving a high parasitic energy. In addition, for these materials the adsorption of CO_2 is of the same order of magnitude as the adsorption of N_2 and hence the selectivity of such a material is unusably low. Materials with a larger Henry coefficient have a significantly larger working capacity and correspondingly lower parasitic energy. This trend continues until the Henry coefficient of the material is so large that at flue gas conditions the partial pressure is too high for the CO_2 adsorption to be in the linear regime. **Figure 6.4.1** shows that at these conditions the CO_2 loading at the adsorbed state is not fully determined by a simple Henry coefficient, and that materials with the same Henry coefficient have different working capacities depending on the pore volume. **Figure 6.4.1** illustrates that at even larger Henry coefficients the adsorption of CO_2 becomes so strong that it becomes increasingly difficult to regenerate the material. From this analysis of the Henry coefficient, we hence expect to find an optimal Henry coefficient: not too big, not too small. Let us see whether we can confirm this by experiments.

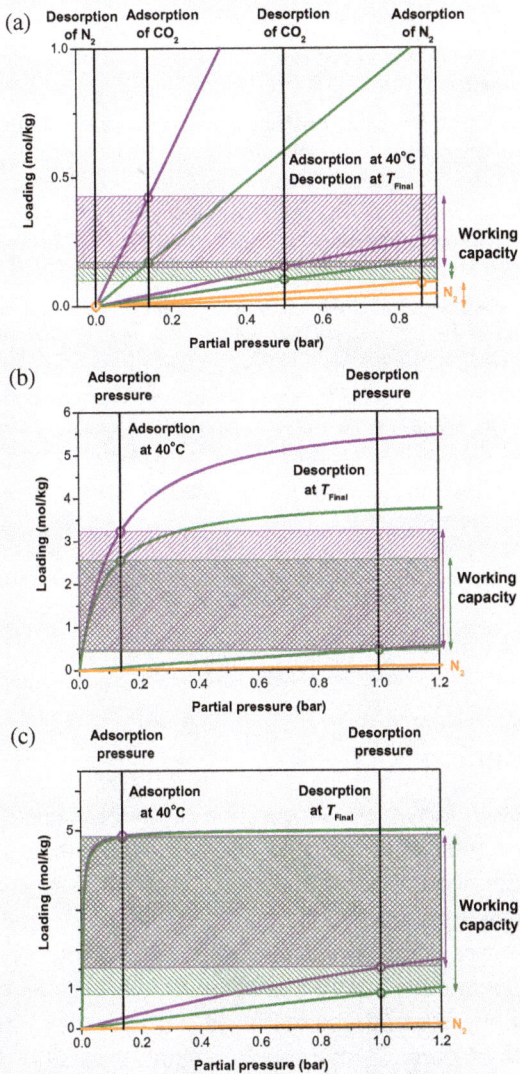

Figure 6.4.1 Adsorption in different materials

Adsorption isotherms give the loading in the zeolite as a function of the partial pressure of CO_2 (green or purple) or N_2 (orange). Adsorption is set by the flue gas conditions.

(Continued)

Zeolite screening

A possible system for which we can test our hypothesis that there is an optimal Henry coefficient would be zeolites. As we saw in Section 6.1, zeolites are comprised of basic building blocks made from corner-sharing SiO_4 units. These tetrahedral units can form different types of structures, such as 6-member rings, 8-member rings, or 12-member rings. These rings are the secondary building units of different types of cylinders or cages. The importance of these materials is that these cylinders or cages form a network of pores in the zeolite crystal. Over 200 different structures have been made, all with more or less the same chemical composition but with different pore topologies. The crystal structure and pore topologies of these materials can be found in the IZA database [6.5].

Most of the applications of these materials rely on the fact that the pores have dimensions that are comparable to the size of the molecules that can be adsorbed. Zeolite design involves selecting the optimal pore topology for a given application. As an example, zeolites are an important component of some laundry powders, where their role is to soften water. A class of zeolites can be synthesized such that some of the Si^{4+} is replaced by Al^{3+} and the charge is compensated by a cation, e.g., Na^+. Competitive adsorption makes Ca^{2+} ions adsorb much more strongly than Na^+ ions, which makes these zeolites very effective in reducing the water hardness; the calcium in the water is adsorbed in the pores of the material and washed away in the rinse cycle. Other important applications include catalysis in petrochemical applications and gas separations, where one uses the difference in adsorption in these materials to selectively retain one gas over another.

So now the question we would like to ask is what is the best zeolite for carbon capture? For this purpose, we consider materials composed of pure SiO_2. This implies that every material has the same chemical composition and only the pore topology is changing. Hence, the question we are asking is what is the optimal pore topology for carbon capture? We note that in addition to the known zeolite structures, we should include in our screening process the database of all computationally predicted zeolite structures [6.6, 6.7].

The difficulty with such a screening study is that we need the experimental adsorption isotherms to compute the parasitic energy. Unfortunately, the corresponding CO_2 and N_2 isotherms have been measured only for a very small number of zeolites. This number is too small to enable any sensible screening, but it is sufficient to develop a quantitative model that describes the interactions between gas molecules and zeolites. This model can then be used in a molecular simulation to predict the isotherms of all gasses. These simulations give a very reasonable description of the experimental isotherms (see **Figure 6.4.2**). As we can use the same model to predict the isotherms in a material with a different crystal structure, we can use these molecular simulations to predict the mixture adsorption isotherms, and for each of these materials we can estimate the optimal parasitic energy. **Figure 6.4.3** shows the optimized parasitic energy as a function of the CO_2 Henry coefficient for all known zeolite structures. **Figure 6.4.4** shows some of the structures that have an optimal parasitic energy. **Figure 6.4.3** also shows the parasitic energy for a different class of materials: **Zeolitic Imidazolate Frameworks** (ZIFs). ZIFs are a special kind of Metal Organic Frameworks (see Section 6.5) in which the size of the linker and angle between linkers are tuned in such a way that they mimic the Si—O—Si bond in zeolites and hence have similar pore topologies to zeolites.

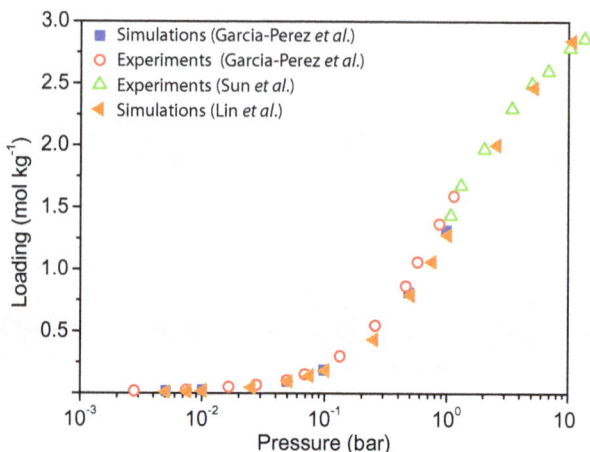

Figure 6.4.2 Comparison of experimental and simulated adsorption isotherms

Test of the reliability of the prediction through a comparison of the adsorption isotherms of CO_2 in the zeolite MFI that have been predicted using molecular simulation with the experimental.

Figure 6.4.3 indeed confirms that there is an optimal CO_2 Henry coefficient. If we have too low a Henry coefficient, the parasitic energy is high because the working capacity of the material is too low. If the Henry coefficient is too high, it takes too much energy to regenerate the material.

Another important observation is that we have a broad optimum for parasitic energy as a function of Henry coefficient. The reason for this broad minimum is that the Henry coefficient shows a strong correlation with the heat of adsorption, and the heat of adsorption has two opposing contributions to the parasitic energy. A higher heat of adsorption increases the working capacity, and while this reduces the parasitic energy, it is offset by the requirement to supply more energy to desorb CO_2, which again increases the parasitic energy.

This computational screening shows a large set of zeolite structures that have a parasitic energy well below the current MEA (monoethanolamine, see Chapter 5) technology (1,060 kJ/kg CO_2). Inspection of some of these optimal structures shows that they are diverse: we find one-, two-, or three-dimensional channel structures, cage-like topologies, and

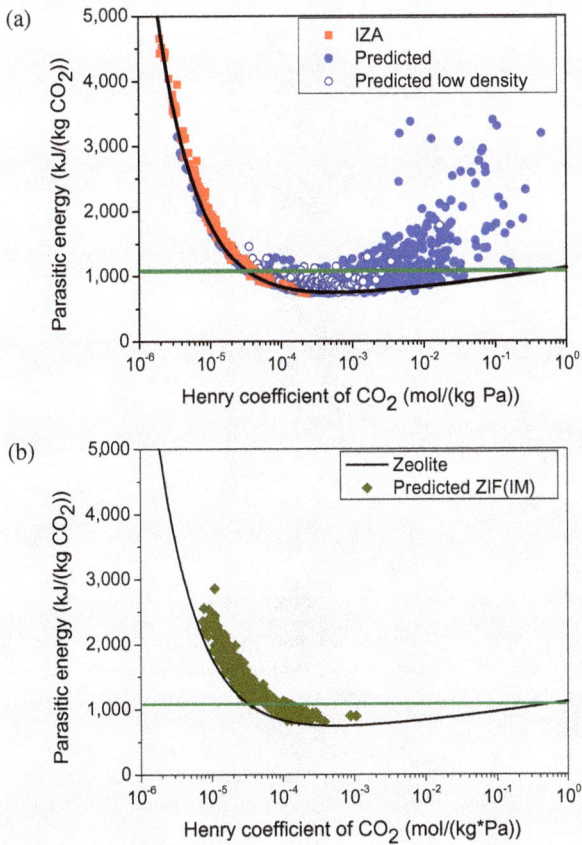

Figure 6.4.3 Parasitic energy for zeolites and ZIFs

(a) The predicted parasitic energy as a function of the Henry coefficient of CO_2 for all silica zeolite structures: the known zeolite structures (red squares) as well as the predicted structures (blue circles). The green line gives the parasitic energy of the current MEA absorption technology for comparison, and the black line is the minimal parasitic energy observed for a given value of the Henry coefficient in the all-silica structures.

(b) The predicted parasitic energy as a function of the CO_2 Henry coefficient for ZIFs is shown. The green line gives the parasitic energy of the current MEA absorption technology, for comparison, and the black line is the minimal parasitic energy calculated for a given value of the Henry coefficient in the all-silica structures. In this graph, we plotted a representative fraction of all structures.

Figures reproduced with permission from Lin et al. [6.4].

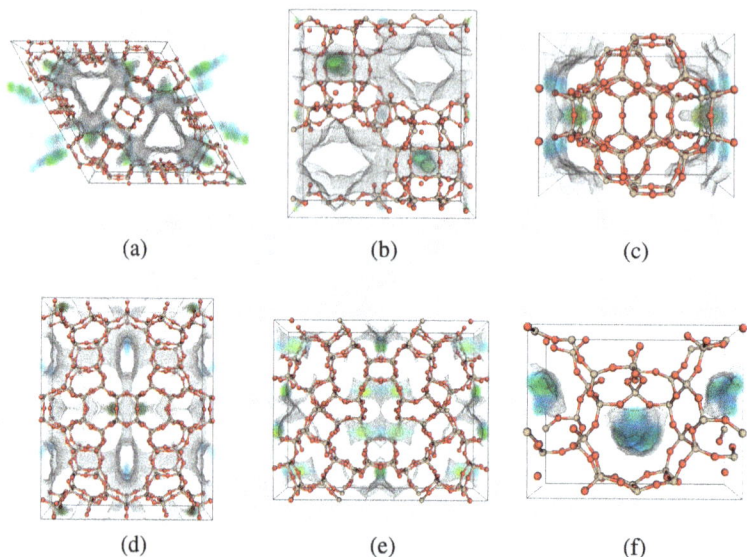

Figure 6.4.4 Examples of zeolite structures with a minimal parasitic energy

The atoms of materials are shown as ball and stick models (O — red and Si — tan). The surface gives the local free energies in the pores of the material, where cooler colors indicate the dominant CO_2 adsorption sites. *Figures reproduced with permission from Lin et al.* [6.4].

more complex geometries. To illustrate this point, **Figures 6.4.4** and **6.4.5** show some of the most diverse structures contained in the optimal zeolites and ZIFs, respectively. Such a theoretical screening does establish a theoretical limit for the minimum parasitic energy that can be achieved for this class of materials. This limit will be useful to focus experimental efforts on zeolite materials synthesis, and can be used as a reference for the consideration of new materials. Of course, an important practical question is whether we can synthesize these optimal materials in quantities sufficient for carbon capture.

At this point, it is important to mention that the above studies focused only on a CO_2/N_2 system. Flue gasses also contain water and other gasses that can influence the parasitic energy. In addition, the use of an equilibrium model assumes that gas diffusion does not play any role

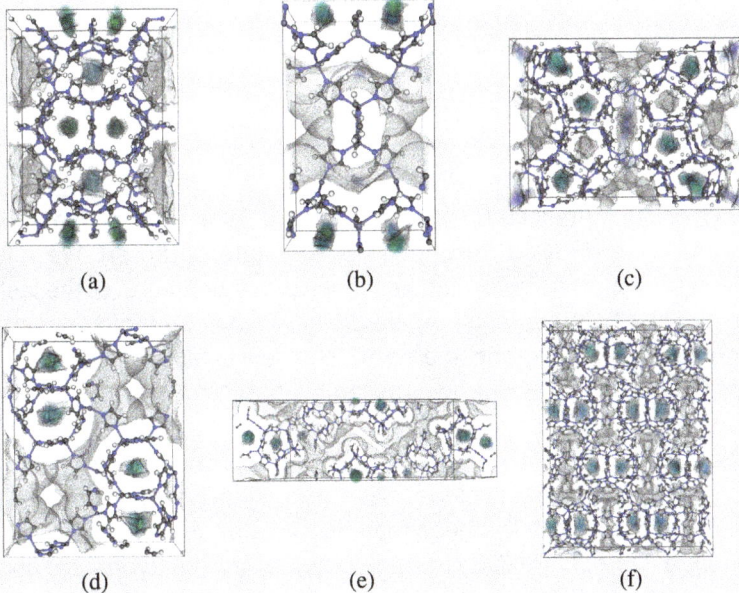

Figure 6.4.5 Examples of ZIF structures with a minimal parasitic energy

The atoms are shown as ball and stick models (Zn — blue-grey, N — blue, H — white, and C — grey). The surface gives the local free energies, where cooler colors indicate the dominant CO_2 adsorption sites. *Figures reproduced with permission from Lin et al.* [6.4].

in adsorption and desorption. As we mentioned in Section 6.3, diffusion causes a decrease of the breakthrough times and therefore a less efficient use of the bed. It is reasonable to assume that all these effects will result in an increase in parasitic energy. Hence, from a screening point of view, this implies that only for those materials that have a parasitic energy well below the current MEA technology in our simple CO_2/N_2 equilibrium model would it be worth the effort to conduct additional studies of these effects.

Section 5

Novel materials for carbon capture

Finding novel materials for carbon capture is a very active area of research. For a detailed description of the state of the art, we refer to many of the excellent review articles that have appeared over the last few years [6.8–6.12]. In this section, we give a short review of the materials that are currently being considered. This review is far from complete and is most likely biased by our own research interests. Adsorption is at present used for several commercial gas separations, and the materials that are used are the usual suspects: micro- or mesoporous inorganic and organic adsorbents like zeolites, silica gels, aluminas, or activated carbons. The applications for these materials include the separation of bulk CO_2 from a gas mixture (CO_2 from natural gas or from hydrogen) and the removal of trace CO_2 from a contaminated gas [6.13]. Because of this established practical experience with these materials in related gas separations they are the first candidates to be tested for carbon capture. More advanced materials include metal-organic frameworks (MOFs).

Physisorbents

Zeolites

We introduced zeolites previously. One of the practical limitations of zeolite research is that only a limited number of the 200 known zeolite structures are readily available. The selectivity of these structures is not sufficiently high for the all-silica structures to be attractive for carbon capture. The selectivity can be enhanced, however, by synthesizing zeolites in which some of the Si^{4+} atoms are replaced with Al^{3+} atoms and the charge deficit is compensated with a cation. With these cations the selectivity is greatly enhanced, along with the unfortunate tendency to have a strong affinity for water. As water is an important component in flue gasses, the adsorption of water limits the capacity of these materials. Zeolites are very stable materials, however, and thus are relatively easy

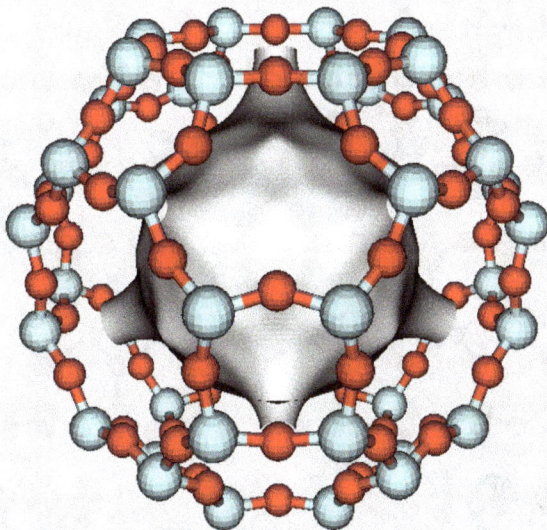

Figure 6.5.1   The zeolite LTA

Part of the LTA structure; the red spheres are oxygen atoms and the blue spheres silica atoms. The grey surface illustrates the cavity and the pore openings. *Graphic by Richard Martin. This animation can be viewed at*: http://www.worldscientific.com/worldscibooks/10.1142/p911#t=suppl

to regenerate. In **Figure 6.5.1** part of the LTA zeolite is shown, and more examples of structures were given in **Figure 6.1.2**.

Activated carbon

As we have mentioned, activated carbon is one of the oldest known adsorbents. More recent variants of carbon include carbon nanotubes [6.14] and carbon molecular sieves [6.15]. These carbon materials show a good selectivity toward CO_2, but as with the cationic zeolites, the active sites also preferentially adsorb water.

Clays

Clays are very common materials, and we will see in Section 9.2 that they play an important role in the geological sequestration of CO_2. For carbon capture, we are interested in a particular form seen in hydrotalcite-like

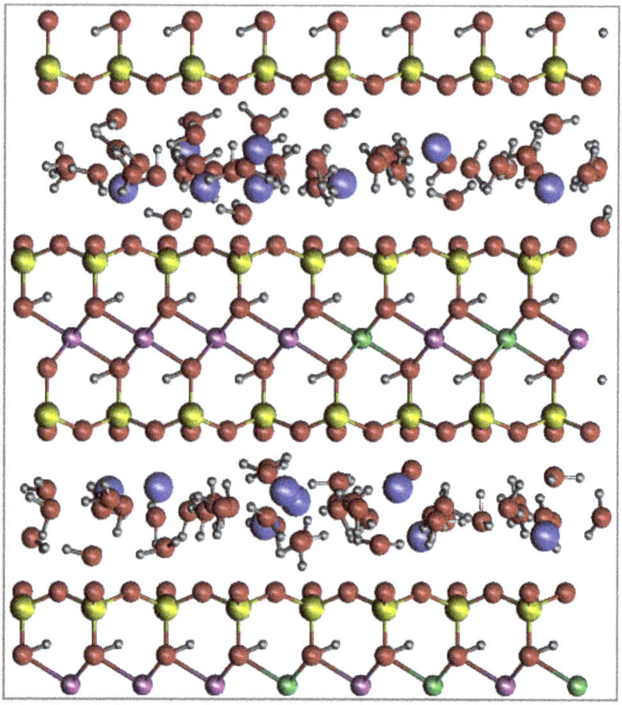

Figure 6.5.2 Molecular model of a clay

Atomistic model of sodium Montmorillonite with a monolayer of adsorbed water (O: red, H: white, Si: yellow, Na: blue, Al: purple, and Mg: green). *Figure reprinted with permission from Hensen and Smit* [6.16]. *Copyright* (2002) *American Chemical Society.*

compounds (HTlcs) (see **Figure 6.5.2**) or layered double hydroxides (LDHs). These types of clays behave as chemical bases and are used as catalysts, adsorbents, and as ion exchangers for treatment of liquid waste. The proposed mechanism of adsorption is that CO_2 forms a complex on the clay surface [6.17]. Because of this complexation chemistry, the binding of CO_2 is stronger than in comparable zeolites. One of the interesting aspects of these clays is that the capacity of CO_2 adsorption actually increases when water is co-adsorbed [6.18, 6.19].

Chemisorbents

Metal oxides

Materials that have basic groups are a natural choice for CO_2 capture, as we know from our analysis of absorption. Examples of "basic" materials include alkaline metal oxides (Na_2O, K_2O) and alkaline earth metal oxides (CaO, MgO) [6.20]. Calcium minerals are the most abundant among alkaline earth metal oxides. Limestone or dolomite are commonly found mineral forms of calcium carbonate.

These materials interact with CO_2, following the reaction:

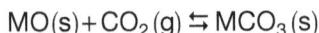

$$MO(s) + CO_2(g) \leftrightarrows MCO_3(s)$$

Heating at high temperatures (>900 K) removes the CO_2 and regenerates the metal oxide [6.21]. However, the kinetics of this reaction are slower than for zeolites or activated carbon. One interesting research direction involves additives or different metals altogether. For these materials, the energy for regeneration is generally larger compared to that for the physisorbed materials. This effect may be partially mitigated because at the higher regeneration temperatures, recovery of some of the process heat becomes easier [6.22].

Amines on supports

We have seen that amine solutions are promising solvents for absorption because they selectively bind the CO_2 (see the previous chapter). Another popular strategy for CO_2 capture is to develop materials in which these amine groups are immobilized on a solid support. One example of such materials would be amine-impregnated silicas, such as the mesoporous molecular sieve MCM-41. The CO_2 adsorption capacity is significantly increased if the material is loaded with branched polyethylenimine [6.23]. In these materials the amines are physically adsorbed, which may limit the conditions in which the material can be regenerated. An alternative strategy is to covalently bind these amines to the silica support, which prevents leaching of amines. These materials are called covalently tethered amine adsorbents [6.24, 6.25].

Amines can also be bound to organic supports. Examples of such materials include carbon-supported amines [6.26], polymer-supported amines [6.27], and solid resins [6.28].

Metal Organic Frameworks

Metal-organic frameworks (MOFs) are microporous crystalline solids which are composed of organic bridging ligands or "struts" coordinated to metal-based nodes to form three-dimensional extended networks with uniform pore diameters, typically in the range 3 to 20 Å [6.11, 6.29, 6.30]. The nodes generally consist of one or more metal ions (e.g., Al^{3+}, Cr^{3+}, Cu^{2+}, or Zn^{2+}) to which the organic bridging ligands coordinate via a specific functional group (e.g., carboxylate, pyridyl). The past 20 years have seen remarkable progress in the design, synthesis and characterization of metal-organic frameworks (MOFs) owing to their enormous structural and chemical diversity and their potential applications in gas storage, ion exchange, molecular separation and heterogeneous catalysis. This effort is motivated in part by the unique structural properties of the materials, which include: robustness, high thermal and chemical stabilities, large internal surface areas (up to $5,000 \, m^2/g$), high void volumes (55–90%) and low densities (from 0.21 to $1.00 \, g/cm^3$). Compared to zeolites they are much simpler to systematically modify in terms of pore dimensions and surface chemistry because their synthesis chemistry is based upon known organic reactions, as opposed to the hydrothermal synthesis of zeolites. In fact, from an experimental point of view finding exactly the right solvent mixture and conditions for the MOF to crystallize is often a very tedious process. For this high throughput, experimentation is used very efficiently (see **Movie 6.5.1**).

A sub-class of MOFs are zeolitic imidazolate frameworks (ZIFs). ZIFs can adopt zeolite structure types based on the replacement of: (a) tetrahedral Si^{4+} ions with tetrahedral transition metal ions such as Zn^{2+} or Co^{2+} and (b) bridging O^{2-} anions with bridging imidazolate-based ligands [6.31]. Some examples of ZIF structures are given in **Figure 6.4.5**.

Here we illustrate how this flexibility in the synthesis is used to arrive at different strategies for carbon capture: open metal sites, interpenetrated frameworks, flexible frameworks, and surface-functionalized frameworks.

Frameworks containing open metal sites

MOFs are often formed through crystallization from solution. For most MOFs the ligands fully coordinate the metal, but for some MOFs the solvent used for synthesis participates in the metal coordination. The

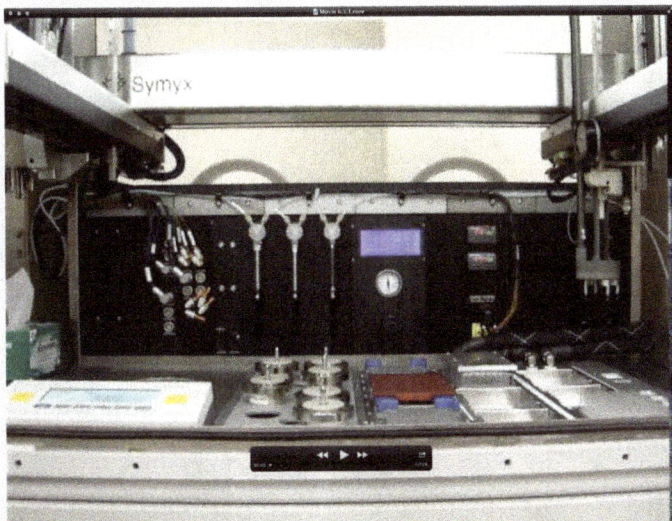

Movie 6.5.1   MOF synthesis in action

The robot is preparing different solvent mixtures and putting them in a plate with 64 small vessels in which the MOFs are formed. *Video by Professor Jeff Long, UC Berkeley. This movie can be viewed at:* http://www.worldscientific.com/worldscibooks/10.1142/ p911#t=suppl

subsequent activation of the material removes the solvent molecules and leaves an open metal site. The presence of these coordinatively unsaturated metal sites affords some robust chemistries for adsorption and the enhanced separation of gasses. **Figure 6.5.3** gives an example of such a structure.

Interpenetration as a strategy for selective adsorption

MOFs can be synthesized with many different types of linkers, and one would assume that by changing the length of the linker one could tune the pore sizes. However, for some systems the pore space can become so large that it is more favorable to form interpenetrated frameworks [6.32]. The high pressure adsorption capacity of these materials is lower than that of the corresponding non-interpenetrated structures, but at low pressure they can have a good selectivity toward CO_2.

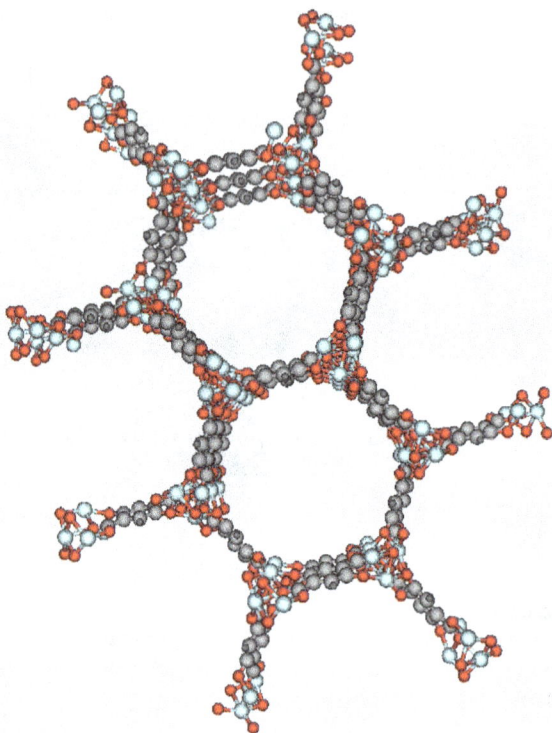

Figure 6.5.3 Example of a metal organic framework (Mg-MOF-74)

Graphic by Richard Martin. This animation can be viewed at: http://www.worldscientific. com/worldscibooks/10.1142/p911#t=suppl

Flexible frameworks

In contrast to the rigid frameworks discussed above which retain their porosity upon adsorption and desorption, flexible and dynamic frameworks collapse upon removal of guest solvent molecules, but their porous structures are restored by adsorption of gas molecules at high pressures [6.33]. The adsorption isotherms are typically characterized by a distinct step, at which point the material "opens up" as gas molecules enter the pores. In some striking cases, almost no adsorption occurs below a threshold or "gate-opening" pressure; once the threshold pressure is exceeded, the pores of the material become accessible and permit significant adsorption. The desorption isotherms are also typically characterized by hysteresis, defined as the dependence of a system on both its current and past environments. The observation of one or more distinct steps in the adsorption isotherms represents a key feature of flexible frameworks. The MIL-53 series has been extensively investigated for high pressure CO_2 and CH_4 adsorption [6.34].

Mesh-adjustable molecular sieves (MAMS) constitute another class of materials for gas separations and are based on temperature-induced gating phenomena [6.35]. MAMS represent a case where the dynamics of substituents at pore openings allow some molecules to pass but not others.

Surface-functionalized frameworks

The grafting of functional groups with a high affinity for CO_2 onto the surfaces of porous materials via ligand modification or coordination to unsaturated metal centers has been employed as a strategy to enhance the capacity and selectivity for CO_2 adsorption. This approach has analogies with other functionalized solid adsorbents such as amine-grafted silicas; however, the crystalline nature of metal-organic frameworks provides for a molecular level of control of pores, assisting in the "tuning" of frameworks for gas separations. Frameworks containing open metal sites have been selectively grafted with molecules that have a high affinity for CO_2. This grafting can be done with highly polar pyridine derivatives [6.36], and amine-functionalized ligands [6.37, 6.38].

Section 6

Amines versus MOFs

In the previous chapter, we discussed liquid absorption and in the present chapter, we have been discussing adsorption. An important practical question is: so what is the best technology? It is actually not easy to give a straightforward answer to this question. An easier question to answer is which technology would be best to use if we were to build a carbon capture plant *tomorrow*. As we have seen in the previous section, amine processes are lying on the shelves. Processes based on MOFs are in such an early stage of development that it would be impossible today to build such a process on the scale of a single power plant. To put it another way, the synthesis of new MOFs is done on the scale of a microgram per PhD student per day, while amines are already used on a commercial scale. So, if we need to build a carbon capture plant tomorrow, it will be based on amine absorption. A different question is: will the next generation of plants that capture carbon use MOFs?

This is a much debated question at carbon capture conferences. The underlying question is important: given that we have amine technologies does it make sense to develop another technology? Or, should we invest all our money in optimizing the current technologies? As you may guess, our answer is that we should do both. If we look at the costs of R&D, Development is orders of magnitude more expensive compared to Research. We need to be very critical in making the decision to develop a novel technology, but in the research stage one would like to have as many options as possible. So, yes we need to do research for the sake of the second or third generation of carbon capture technologies. We need to find the best liquid absorber, the best solid adsorber, and as we will discuss in the next section, the best membrane. At the end of the day it may very well be that only one technology will survive. In this respect, science is surprisingly similar to real life. Would it not be nice to buy only winning tickets in the lottery?

After this somewhat lengthy introduction let us ask again: can we compare amine- and MOF-based technologies? The underlying question is how to compare a mature technology with an emerging research topic on an equal footing. Sathre and Masanet [6.39] have taken the first steps to address this question. The starting point is a prospective life-cycle modeling of a CCS process. The importance of a complete **life-cycle analysis** of the process is that it includes all aspects of the process. For example, let us suppose that the degradation of material A is 1% per week and for material B 2% per week. Clearly one would argue the first is superior, but if a subsequent life-cycle analysis of the synthesis of these materials shows that it costs more than twice as much energy to synthesize material A, then material B might still be superior.

The challenge for the MOF system is to make a credible life-cycle analysis of a process that does not exist, while for the amine process real data are available. The Sathre and Masanet baseline uses the assumption that all new USA coal-fired power plants will be more efficient and equipped with CCS. The older coal-fired power plants will either be replaced by new ones or gradually retrofitted to allow for CCS such that by 2050 the full fleet will be fully equipped with CCS (see **Figure 6.6.1**).

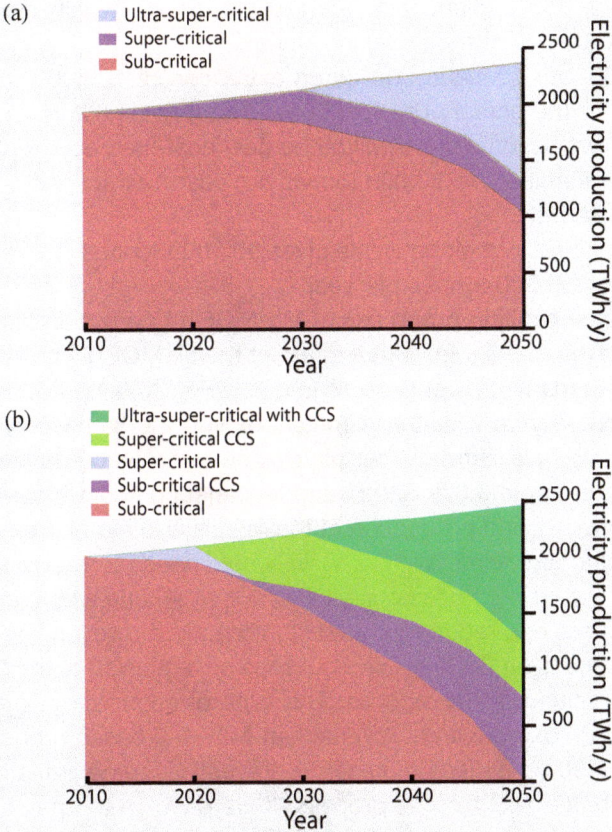

Figure 6.6.1 Scenarios for USA coal-fired power plants

(a) Scenario **without CCS**; (b) Scenario **with CCS**: Most current power plants are sub-critical, with steam pressure of around 170 bars. The current state-of-the-art is supercritical plants with steam pressure of around 250 bars. The next generation is expected to be ultra-supercritical with even higher steam pressure.

Figures adapted from Sathre and Masanet [6.39].

We assume that 90% of the flue gas is captured and we use monoethanolamine (MEA) and Mg-MOF-74 as reference materials. The molecular structure of Mg-MOF-74 is shown in **Figure 6.5.3**. For Mg-MOF-74 we know the adsorption isotherms and the working capacity. This allows us to estimate the total amount of material that one needs for the CCS scenario shown in **Figure 6.6.1**. For the total amount two numbers are important: the primary material, which is the amount needed to operate a power plant, and the recycle losses, which are typically assumed to be 5% of the recycling stream.

By 2050, our scenario implies an extraction of 120,000 tonnes per year, 50% for the primary use and 50% to supply the 5% recycling loss. After 2050, we assume there will be no new coal-fired power plants and this number stabilizes to 60,000 tonnes per year. Our scenario, however, is for the USA only.

Extrapolation to a global scale gives 900,000 tonnes per year, which stabilizes to 450,000 tonnes per year.

One of the exciting properties of Mg-MOF-74 compounds is that we can replace the metal (Mg) with a different metal. MOF-74 has been synthesized with many different metals (e.g., Ni, Fe, Co, etc). Let us assume that the compounds with these metals exhibit the same properties as those with Mg. We can now simply compute the amounts we need of each metal. In **Table 6.6.1**, Sathre and Masanet compare these numbers with the current world production of the metals and the proven reserves. We see that for some metals (Co, V) we would need more than the global reserves. If we were to base our MOF on Fe or Al, one would expect the economy of scale to reduce the cost significantly. However, if we were to base our process on Co or V, we would see an explosion of the cost once we start to scale up this process. This type of information is extremely useful to guide our research. A vanadium-MOF-74-based process would need to use 20 times less material to prevent resource limitations from creating a bottleneck for this process. If we develop the most beautiful carbon capture chemistry for Co-MOF-74, we also may want to think whether these ideas can be transferred to more abundant metals. Despite the beauty of the chemistry, if we cannot scale our materials to process gigatonnes of CO_2 it will contribute little to carbon capture.

In **Table 6.6.2**, the GHG emissions that would result from different scenarios are summarized. We see that by 2050 the net emissions will have reduced by almost 2 Gt CO_2 per year if we employ CCS. In

Table 6.6.1 Total use of metals for worldwide carbon capture using MOFs

Metal	Use for CCS in %	Total use (until 2050) in %
Fe	0.080	0.02
Al	2.2	0.2
Cu	5.6	2.3
Mn	7.0	2.3
Zn	7.5	5.9
Cr	12	12
Mg	14	1.9
Ti	23	3.6
Ni	58	19
Zr	103	36
Co	1,030	200
V	1,620	110

Suppose we implement CCS using MOFs, how will this affect the mining of metals? The first column gives the use of metals for MOFs in CCS as a percentage of the 2010 global reserves. The second column gives the total use of the metals until 2050 as a percentage of the estimated global reserves of each metal. *Data from Sathre and Masanet* [6.39].

Table 6.6.3, the different cost components are compared. This table illustrates that the bulk of the costs are from the capture process; the transport and storage only account for 10–15% of the total costs. These numbers translate into an increase in the price of electricity of about 51% by 2050. Also note that in the initial stages of the implementation the cost of CCS will be much more expensive; the first CCS operations will be the most expensive, but by the implementation of the second generation the costs will go down rapidly.

This table also shows that a process based on Mg-MOF-74 will cost $76 per tonne of avoided CO_2 emission, compared to $68 for the MEA-based process. This is a surprising result — we would have expected the amine process to be much cheaper as MOFs are expected to be more expensive to synthesize. The details in the study of Sathre and Masanet show that the costs of absorption/adsorption materials, MEA or MOF, are

Table 6.6.2 Comparison of GHG emissions in 2050 without CCS and with CCS using MEA or MOFs

	GHG emissions in 2050 (Gt CO_2 per year)			Cumulative emissions 2010–2050 (Gt CO_2)		
	No CCS	MEA CCS	MOF CCS	No CCS	MEA CCS	MOF CCS
Plant stack emissions	1.97	0.25	0.24	86.8	59.0	58.8
Coal mining and transport	0.21	0.27	0.25	9.2	10.1	9.9
Plant infrastructure	0.01	0.02	0.02	0.2	0.3	0.4
Capture media and storage	0	0.02	0.03	0	0.3	0.6
CO_2 transport and storage	0	0.03	0.03	0	0.6	0.6
Total	**2.20**	**0.59**	**0.58**	**96.2**	**70.4**	**70.2**

Estimated base-case GHG emissions in 2050 (Gt CO_2 per year) and cumulative GHG emissions from 2010 to 2050 (Gt CO_2) from the USA coal-fired power fleet without CCS and with CCS using MEA or MOF capture technology. *Data from Sathre and Masanet* [6.39].

only a fraction of the total costs. If a new material performs better, then these additional expenses can be compensated by energy savings.

Given the fact that there are very little real process data for Mg- MOF-74, the predictions of these costs estimates have many uncertainties. The uncertainties are displayed in a so-called tornado graph, **Figure 6.6.2**. This figure ranks all the parameters that are in the cost model in the order in which the uncertainty changes the cost estimate. The largest uncertainties are in the recycling time, the life span of a MOF, and the energy needed to regenerate a MOF.

The capture/regeneration cycle time is the time it takes to adsorb a bed with CO_2 and the subsequent purge. As the amount of flue that we need to process per hour is fixed, a long recycle time means that we have to operate more beds in parallel. In the calculations of Sathre and Masanet, the uncertainties in this recycling time result in the largest consequence on the cost, and further research would be needed to more precisely estimate these cycles times. In addition, these results pose some interesting questions. Will this time be faster if we operate this process as a pressure swing operation versus a temperature swing

Table 6.6.3 Cost estimate of CCS with MEA and MOF processes compared to cost without CCS in the USA

	Cost in 2050 (G$ per year)			Cumulative costs 2010–2050 (G$)		
	No CCS	MEA CCS	MOF CCS	No CCS	MEA CCS	MOF CCS
Generation (capital)	41.9	68.9	64.9	634	1131	1062
Generation (non-fuel operation)	20.5	24.6	23.6	836	901	885
Generation (fuel)	40.7 (88.4)	51.8 (112.7)	49.1 (106.8)	1,763 (2,826)	1,943 (3,178)	1,898 (3,090)
Capture (capital)	0	19.2	25.0	0	306	399
Capture (operation)	0	14.5	19.3 (19.4)	0	230	309 (311)
CO_2 transport and storage	0	22.9 (24.0)	23.1 (24.0)	0	359 (372)	362 (373)
Total	**103.1 (150.8)**	**202.0 (263.8)**	**205.0 (263.7)**	**3,232 (4,296)**	**4,871 (6,118)**	**4,915 (6,120)**

Estimated base-case annual cost in 2050 (G$ per year) and cumulative cost from 2010 through 2050 (G$) for the USA coal-fired power fleet without CCS and with MEA- or MOF-based CCS, with reference coal cost. Costs with high coal cost projections are shown in parentheses. *Data from Sathre and Masanet* [6.39].

Parameter	Units	GHG mitigation costs ($ per t CO_2 avoided) (low / high)
Capture/regeneration cycle time	minutes	30 / 90
Mass ratio, solvent/MOF	ration(mass/mass)	0 / 200
Solvent recycling rate	percent recycled	98 / 75
Life span of MOF material	number of cycles	12000 / 4000
MOF regeneration equavalent energy	MJe/tCO2	200 / 600
Capital cost of capture systems, MOF/MEA	ration ($/$)	1.0 / 2.0
Solvent production costs	$/t	400 / 1300
MOF working capacity	weight	14 / 24
Learning rate, capital cost, CO2 capture	-	0.17 / 0.06
Coal transport distance	Skm	400 / 1600
Average lenght of feeder pipeline	km	50 / 150
Capital costs of flue gas cleaning, MOF/MEA	ratio ($/$)	1.0 / 1.5
Average length of trunk pipeline	km	100 / 300
CO2 transport cost	$/tCO2	4.5 / 7.6
Metal production cost	$/t	260 / 1000
O&M cost of flue gas cleaning, MOF/MEA	ratio ($/$)	1.0 / 1.5
CO2 injection costst	$/tCO2	4.0 / 6.6
Depth of injection well	m	800 / 2000
Learning rate, maximum installed capacity	GW	150 / 50
Re-compression needed	yes or no	no / yes
MOF synthesis cost	$/t	400 / 3240
Learning rate, minimum installed capacity	GW	5 / 10
Learning rate, O&M cost, CO2 capture	-	0.3 / 0.1
MOF reaction yield	percent	100 / 70
Learning rate, CO2 transport	-	0.09 / 0.03
Learning rate, CO2 injection	-	0.09 / 0.03
Auxilliary load of capture system, MOF/MEA	ration MJ/MJ	1.0 / 1.5
Bed utilization factor	percent	100 / 90
Solvent production GHG	tCO2e/t	1.2 / 4.0
O&M cost of capture system, MOF/MEA	ratio ($/$)	1.0 / 1.5
Learning rate, captical cost, CO2 compression	-	0.1 / 0
Organic ligand production cost	$/t	1000 / 1800
Metal production GHG	tCO2e/t	0.1 / 3.0

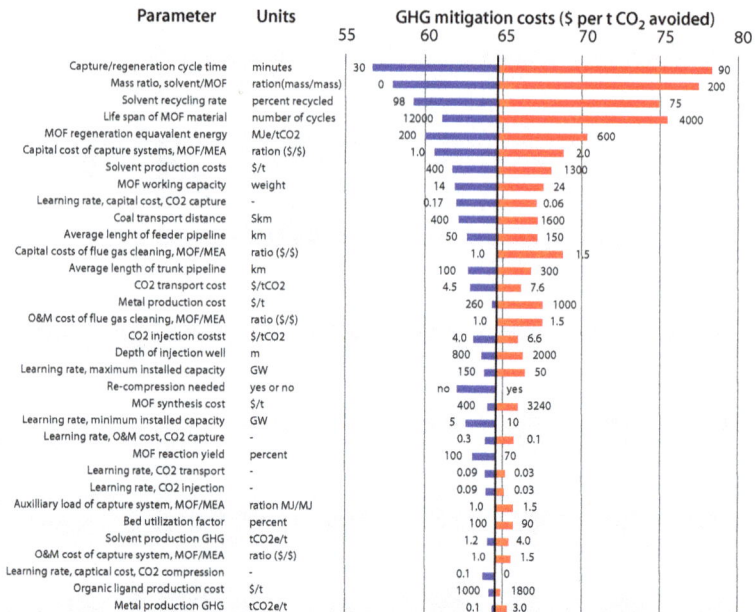

Figure 6.6.2 Tornado plot of the costs

Change in estimated GHG mitigation cost due to variation of individual parameters between low and high estimates. *Figure adapted from Sathre and Masanet* [6.39].

Question 6.6.1   MEA, MOF and coal price

Table 6.6.3 shows that the amine technology is cheaper for the reference coal price but the MOF technology is cheaper for the higher coal price. Explain why.

operation, as was assumed in these calculations? Or, is there an optimal combination of the temperature and pressure swing?

Another important process parameter is the life span of the MOF material. How many adsorption/regeneration cycles can be carried out before the material degenerates? These calculations show that if we would be able to expand the life span from 8,000 to 12,000 cycles, the cost would decrease significantly. If, on the other hand, contaminants in flue gas would decrease this to 4,000 cycles, the cost would increase significantly.

Section 7

Review

6.1. Reading self-test

1. Which of the following statements regarding adsorption is correct?
   a. Because adsorption uses a solid material and not a liquid, adsorption can only be a batch process
   b. The larger the working capacity of an adsorbent, the smaller the adsorption column
   c. Because we are working with a solid material, heat management is simple.
   d. There are many commercial examples of CO_2 separations by adsorption

2. Which statement is correct for a Langmuir isotherm $\frac{\sigma}{\sigma_0} = \frac{bp}{1+bp}$?
   a. b is equal to the Henry coefficient
   b. The underlying model assumes independent adsorption sites
   c. σ_0 gives the maximum loading
   d. Answers A and B
   e. Answers A, B and C
   f. Answers B and C

3. Which statement for the parasitic energy for adsorption is correct?
   a. The parasitic energy estimates the increase in the price of electricity
   b. The parasitic energy weighs all energies required for carbon capture and sequestration equally
   c. The Carnot efficiency tells us that the higher the temperature of the steam, the more efficient the separation
   d. None of the above

4. Which statement is correct concerning Metal Organic Frameworks (MOFs)?
   a. ZIFs form a special class of MOFs
   b. Because of the covalent metal ligand bond, MOFs are intrinsically more stable than zeolites
   c. As MOFs are a new materials at present, there are far fewer MOF structures published than zeolite structures
   d. Answers A and C

5. Which statement regarding the working capacity is not correct?
   a. The working capacity increases if we decrease the temperature of the flue gas
   b. The working capacity increases if we increase the pressure of the flue gas
   c. The working capacity increases if we increase the temperature at which we desorb the gas
   d. The working capacity increases if we increase the pressure at which we desorb the gas

6. Which statement regarding breakthrough times (t_b) is not correct? We assume a column with length L, a gas front moving with uniform velocity u, and void fraction ε.
   a. If our column is empty (no absorbent), the breakthrough time is $t_b = L/u$.
   b. If our material does not adsorb the gas, the breakthrough time is $t_b = L/(1-\varepsilon)u$
   c. If our gas has a concentration ρ and our material a Henry coefficient H: $t_b = L(1+(1-\varepsilon)\rho RTH)/\varepsilon u$

7. Which metal would be the most expensive if MOF-based carbon capture were to be implemented on a global scale?
   a. Iron
   b. Zinc
   c. Manganese
   d. Copper
   e. Vanadium
   f. Titanium

8. Which statement about the optimal zeolite structures for carbon capture is correct?
   a. It has a minimal parasitic energy
   b. It has a heat of adsorption that is not too low and not too high
   c. It has an Henry coefficient that is not too high and not too low
   d It has a maximum number of adsorption sites for CO_2
   e. Answers A, B, and C
   f. Answers A, B, C, and D

9. Which factor contributes the most to the uncertainty of the costs of carbon capture with MOFs?
   a. MOF synthesis costs
   b. MOF regeneration energy
   c. Metal production costs
   d. Organic ligand production costs

10. What is the largest source of GHG emissions in generating electricity with a plant with CCS?
    a. The residual emission of the power plant
    b. Coal mining and transport
    c. Capture materials production
    d. CO_2 transport and storage

Section 8

References

1. Beerdsen, E., D. Dubbledam, and B. Smit, 2006. Loading dependence of the diffusion coefficient of methane in nanoporous materials. J. Phys. Chem. B, **110** (45), 22754–22772. http://pubs.acs.org/doi/abs/10.1021/jp0641278
2. Li, H., M. Eddaoudi, M. O'Keeffe, and O.M. Yaghi, 1999. Design and synthesis of an exceptionally stable and highly porous metal-organic framework. Nature, **402**, 276–279. http://dx.doi.org/doi:10.1038/46248
3. Myers, A.L. and J.M. Prausnitz. 1965. Thermodynamics of mixed gas adsorption. AIChE J., **11** (1), 121.
4. Lin, L.C., A.H. Berger, R.L. Martin, et al., 2012. In silico screening of carbon-capture materials. Nature Materials, **11**, 633–641. http://dx.doi.org/doi:10.1038/nmat3336
5. International Zeolite Association (IZA), 2011. www.iza-structure.org/databases
6. Deem, M.W., R. Pophale, P.A. Cheeseman, and D.J. Earl, 2009. Computational discovery of new zeolite-like materials. J. Phys. Chem. C, **113** (51), 21353. http://dx.doi.org/10.1021/jp906984z
7. Pophale, R., P.A. Cheeseman, and M.W. Deem, 2011. A database of new zeolite-like materials. Phys. Chem. Chem. Phys., **13** (27), 12407. http://dx.doi.org/10.1039/c0cp02255a
8. Choi, S., J.H. Drese, and C.W. Jones, 2009. Adsorbent materials for carbon dioxide capture from large anthropogenic point sources. ChemSusChem, **2** (9), 796. http://dx.doi.org/10.1002/cssc.200900036
9. D'Alessandro, D.M., B. Smit, and J.R. Long, 2010. Carbon dioxide capture: prospects for new materials. Angew. Chem. Int. Edit., **49** (35), 6058. http://dx.doi.org/10.1002/anie.201000431
10. Sumida, K., D.L. Rogow, J.A. Mason, *et al.*, 2012. Carbon dioxide capture in Metal–Organic Frameworks. Chem. Rev., **112** (2), 724. http://dx.doi.org/10.1021/cr2003272
11. Li, J.-R., J. Sculley, and H.-C. Zhou, 2012. Metal–Organic Frameworks for separations. Chem. Rev., **112** (2), 869. http://dx.doi.org/10.1021/cr200190s
12. Wang, Q.A., J.Z. Luo, Z.Y. Zhong, and A. Borgna, 2011. CO(2) capture by solid adsorbents and their applications: current status and new trends. Energy Environ. Sci., **4** (1), 42. http://dx.doi.org/10.1039/c0ee00064g
13. Lee, K.B., M.G. Beaver, H.S. Caram, and S. Sircar, 2008. Reversible chemisorbents for carbon dioxide and their potential applications. Ind. Eng. Chem. Res., **47**, 8048.

14. Zhao, J.J., A. Buldum, J. Han, and J.P. Lu, 2002. Gas molecule adsorption in carbon nanotubes and nanotube bundles. Nanotechnology, **13** (2), 195.

15. Nakashima, M., S. Shimada, M. Inagaki, and T.A. Centeno, 1995. On the adsorption of CO_2 by molecular-sieve carbons — volumetric and gravimetric studies. Carbon, **33** (9), 1301.

16. Hensen, E.J.M. and B. Smit, 2002. Why Clays Swell. J. Phys. Chem. B, **106** (49), 12664–12667. http://dx.doi.org/doi:10.1021/jp0264883

17. Lee, K.B., A. Verdooren, H.S. Caram, and S. Sircar, 2007. Chemisorption of carbon dioxide on potassium-carbonate- promoted hydrotalcite. J. Colloid. Interf. Sci., **308** (1), 30. http://dx.doi.org/10.1016/J.Jcis.2006.11.011

18. Yong, Z., V. Mata, and A.E. Rodrigues, 2002. Adsorption of carbon dioxide at high temperature — a review. Sep. and Pur. Tech., **26** (2), 195.

19. Yong, Z. and A.E. Rodrigues, 2002. Hydrotalcite-like compounds as adsorbents for carbon dioxide. Energ. Convers. Manage., **43** (14), 1865.

20. Wang, S.P., S.L. Yan, X.B. Ma, and J.L. Gong, 2011. Recent advances in capture of carbon dioxide using alkali-metal-based oxides. Energy Environ. Sci., **4** (10), 3805. http://dx.doi.org/10.1039/c1ee01116b

21. Shimizu, T., T. Hirama, H. Hosoda, K. Kitano, M. Inagaki, and K. Tejima, 1999. A twin fluid-bed reactor for removal of CO_2 from combustion processes. Chem. Eng. Res. Des., **77** (A1), 62. http://dx.doi.org/10.1205/026387699525882

22. Abanades, J.C., E.J. Anthony, D.Y. Lu, C. Salvador, and D. Alvarez, 2004. Capture of CO_2 from combustion gases in a fluidized bed of CaO. Aiche J., **50** (7), 1614. http://dx.doi.org/10.1002/aic.10132

23. Xu, X., C. Song, B.G. Miller, and A.W. Scaroni, 2005. Adsorption separation of carbon dioxide from flue gas of natural gas-fired boiler by a novel nanoporous "molecular basket" adsorbent. Fuel Process Tech., **86** (14–15), 1457.

24. Hiyoshi, N., D.K. Yogo, and T. Yashima, 2004. Adsorption of carbon dioxide on amine modified SBA-15 in the presence of water vapor. Chem. Lett., **33**, 510. http://dx.doi.org/10.1246/cl.2004.510

25. Tsuda, T., T. Fujiwara, Y. Taketani, and T. Saegusa, 1992. Amino silica-gels acting as a carbon-dioxide absorbent. Chem. Lett. **1992** (11), 2161.

26. Plaza, M.G., C. Pevida, A. Arenillas, F. Rubiera, and J.J. Pis, 2007. CO_2 capture by adsorption with nitrogen enriched carbons. Fuel, **86** (14), 2204. http://dx.doi.org/10.1016/J.Fuel.2007.06.001

27. Satyapal, S., T. Filburn, J. Trela, and J. Strange, 2001. Performance and properties of a solid amine sorbent for carbon dioxide removal in space life support applications. Energ. Fuel, **15** (2), 250.

28. Drage, T.C., A. Arenillas, K.M. Smith, C. Pevida, S. Piippo, and C.E. Snape, 2007. Preparation of carbon dioxide adsorbents from the chemical activation of urea-formaldehyde and melamine-formaldehyde resins. Fuel, **86** (1–2), 22. http://dx.doi.org/10.1016/J.Fuel.2006.07.003

29. Tranchemontagne, D.J., Z. Ni, M. O'Keeffe, and O.M. Yaghi, 2008. Review: Recticular chemsitry of metal-organic polyhedral. Angew. Chem. Int. Ed., **47**, 2.

30. O'Keeffe, M., and O.M. Yaghi, 2012. Deconstructing the crystal structures of metal–organic frameworks and related materials into their underlying nets. Chem. Rev., **112** (2), 675. http://dx.doi.org/10.1021/cr200205j

31. Phan, A., C.J. Doonan, F.J. Uribe-Romo, C.B. Knobler, M. O'Keeffe, and O.M. Yaghi, 2010. Synthesis, structure, and carbon dioxide capture properties of zeolitic imidazolate frameworks. Acc. Chem. Res., **43** (1), 58. http://dx.doi.org/10.1021/ar900116g

32. Eddaoudi, M., D.B. Moler, H. Li, *et al.*, 2001. Review: Modular chemistry: Secondary building units as a basis for the design of highly porous and robust metal–organic carboxylate frameworks. Acc. Chem. Res., **34** (4), 319.

33. Horike, S., S. Shimomura, and S. Kitagawa, 2009. Soft porous crystals. Nature Chem., **1** (9), 695.

34. Bourrelly, S., P.L. Llewellyn, C. Serre, F. Millange, T. Loiseau, and G. Ferey, 2005. Different adsorption behaviors of methane and carbon dioxide in the isotypic nanoporous metal terephthalates MIL-53 and MIL-47. J. Am. Chem. Soc., **127** (39), 13519. http://dx.doi.org/10.1021/Ja054668v

35. Ma, S., D. Sun, X.-S. Wang, and H.-C. Zhou, 2007. A mesh-adjustable molecular sieve for general use in gas separation. Angew. Chem. Int. Ed., **46** (14), 2458.

36. Bae, Y.S., O.K. Farha, J.T. Hupp, and R.Q. Snurr, 2009. Enhancement of CO_2/N_2 selectivity in a metal-organic framework by cavity modification. J. Mater. Chem., **19** (15), 2131. http://dx.doi.org/10.1039/B900390h

37. Couck, S., J.F.M. Denayer, G.V. Baron, T. Remy, J. Gascon, and F. Kapteijn, 2009. An amine-functionalized MIL-53 with large separation power for CO_2 and CH_4. J. Am. Chem. Soc., **131**, 6326.

38. Demessence, A., D.M. D'Alessandro, M.L. Foo, and J.R. Long, 2009. Strong CO_2 binding in a water-stable, triazolate-bridged metal–organic framework functionalized with ethylenediamine. J. Am. Chem. Soc., **131** (25), 8784. http://dx.doi.org/10.1021/ja903411w

39. Sathre, R. and E. Masanet, 2013. Prospective life-cycle modeling of a carbon capture and storage system using metal-organic frameworks for CO_2 capture. RSC Adv., **3**, 4964–4975. http://dx.doi.org/10.1039/C3RA40265G

Chapter 7

Membranes

One can also use membranes to separate gasses. Normally membranes operate using a large pressure difference, but flue gas is at the "end of the pipe," and thus the exhaust pressure is very close to atmospheric pressure. Compressing flue gas is very expensive, so membrane separation does not look very attractive for carbon capture. We show that one can nevertheless arrive at promising process options by immersing oneself into the physics and chemistry of the membrane, then using that knowledge in clever process design schemes.

Section 1

Introduction

In the previous chapters we have discussed absorption and adsorption as mechanisms to capture CO_2. These technologies are based on differences in solubility or adsorption of the different components of the flue gas in a solvent or solid material, respectively. An alternative technology is to use a membrane. Membranes are based on the difference in permeabilities of different gasses in a material. The higher the permeability the higher the flux, the number of molecules that flow through the membrane per unit time. Hence, if these permeabilities are different enough and sufficiently large we can separate the components of the flue gas by passing them through a membrane. The driving force for membrane separation is the difference in pressure, or more accurately, the difference in chemical potential between the two sides of the membrane.

In this chapter, we aim to discuss two aspects of membrane separations. The first aspect is engineering. For many years people have argued it is useless to consider membranes to separate flue gasses. So why a chapter on membranes? The answer is: very clever engineering.

A second aspect concerns how to design an ideal membrane. We show that understanding membranes requires knowledge of adsorption and diffusion. Transport through a membrane requires molecules to adsorb and diffuse through the material. For this understanding, we go to the molecular level and see how diffusion and adsorption depend on the interactions with the materials. This will require us to obtain some understanding of how entropy, energy, and chemical potential are related to the chemical structure of a material.

Membrane separations are used in several applications (see **Figure 7.1.1**). Reverse osmosis membranes exclude salt from sea water, which is becoming an important technology in places that lack potable water. Dialysis is another common application of membranes. Dialysis is based on a membrane system that is designed to remove salts from the blood stream.

Another application of membranes is in filtration. Here the membrane acts as a sieve; depending on the size of the objects we would like to separate, we select a certain well-defined pore diameter for the sieve.

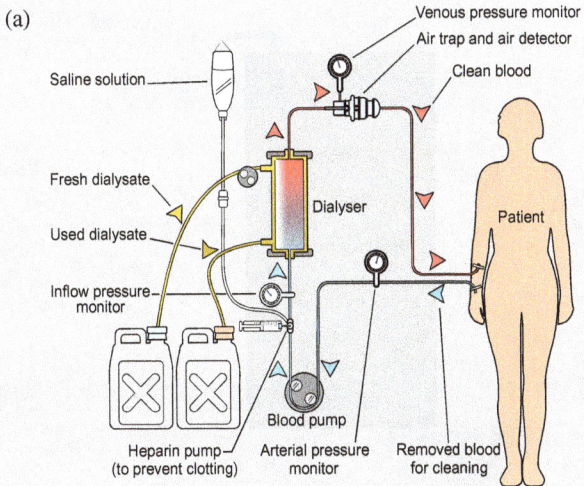

Figure 7.1.1 Examples of membrane separation

(a) Dialysis. *Image by Yassine Mrabet (2008) from WikiCommons:* http://en.wikipedia.org/wiki/File:Hemodialysis-en.svg

(b) Water desalination plant. *Image by James Grellier (2010) from WikiCommons:* http://en.wikipedia.org/wiki/File:Reverse_osmosis_desalination_plant.JPG

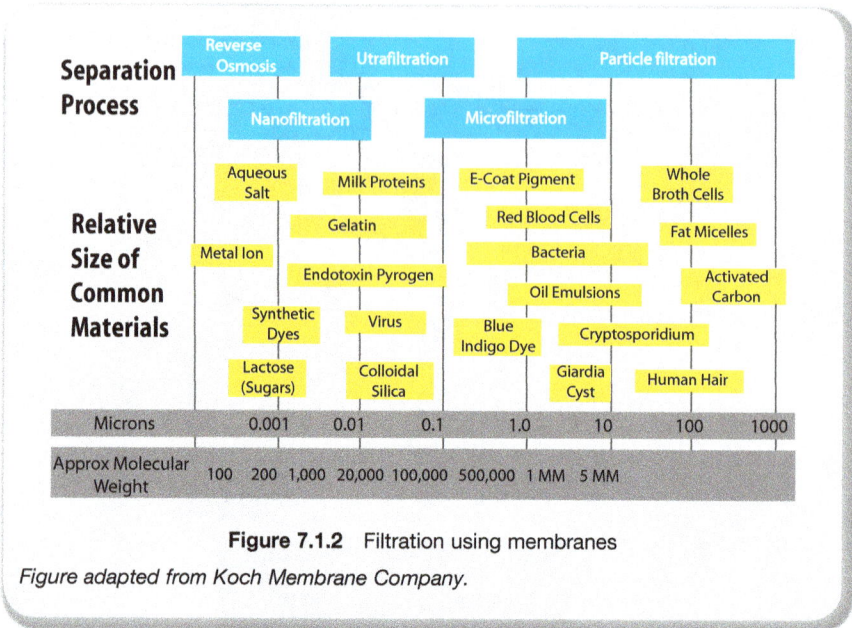

Figure 7.1.2 Filtration using membranes

Figure adapted from Koch Membrane Company.

Figure 7.1.2 illustrates the types of separations that are carried out by filtration. We see that through microfiltration we can remove bacteria from water. However, if we would like to use this sieving mechanism for gas separation, we will need to make materials with very narrow pores. In fact, if we were to plot the molecular weights of the gasses we are interested in separating on **Figure 7.1.2**, we would see they are outside the scale of "common" materials as defined in this graphic.

Membranes are often made of materials that are common and not too expensive, such as rubber, polycarbonate, or polyimide. While not exactly cheap either, they are common and accessible and can be made in large quantities. A significant number of applied technologies presently use membrane processes for large volumes of fluids. It is not too much of a stretch of the imagination to think that membrane separation could become part of CCS.

Membrane separation

Before continuing this section, it is important to discuss in some detail how membrane separation works. We have already mentioned filtration. Filtration is based on sieving; molecules or particles that are too big will not pass through the pores of a membrane. If filtration were the only mechanism, we would not be able to separate two types of molecules if both can pass through the membrane. As we will show, to separate two gasses, a difference in permeation — the ease with which molecules pass through a membrane — is sufficient to achieve a separation.

Simple membrane

Let us look at the membrane shown schematically in **Figure 7.2.1**. In this figure, our membrane is a piece of material through which there is a flow.

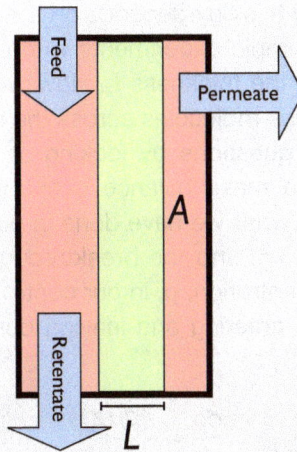

Figure 7.2.1 Simple membrane

A simple membrane with a thickness L and area A. The feed splits into two streams, one going through the membrane, which is called the permeate, and the retentate.

We have the **feed gas** containing different components, and this feed splits into two parts: the **permeate**, which is the part that goes through the membrane, and the **retentate**, which is the part that does not go through the membrane. The idea is that if there is a difference in the ease with which components in the gas pass through the membrane, these components will get separated between the permeate and the retentate. For example, for the permeate gas mixture to be enriched in CO_2, the CO_2 molecules must flow more easily through the membrane than the other flue gas components (i.e., N_2); the retentate will consequently be depleted of CO_2.

Permeation and permeability

Before we consider an actual separation, let us look in more detail at a pure component. Suppose we have a feed with a given pressure, p_R, and on the other side of the membrane we maintain a pressure, p_P (see **Figure 7.2.2**). Because of this pressure difference, the molecules will flow through the material. If we continuously remove the permeate and maintain a constant flow of the feed, the system will reach a steady state. We would like to know the flux of molecules (mol/s) through our membrane and examine how this flux depends on the properties of our material. We can do some simple experiments with our membrane, such as change the area A and the thickness L. How will this change the flux? What is the density of gas molecules across the membrane?

We address these questions by looking at a small volume of the material and applying a mass balance over this control volume (see **Figure 7.2.3**), similar to what we have done in our calculations for diffusion limitation (Section 5.4) and the breakthrough curves (Section 6.3). The change of the concentration, ρ, in our control volume is equal to the difference of the flux, j, entering and leaving our volume. In differential form, this reads:

$$\frac{d\rho}{dt} = -\frac{dj(z)}{dz}$$

The flux is related to the concentration gradient and the diffusion coefficient (Fick's law):

$$j = -D\frac{d\rho}{dz}$$

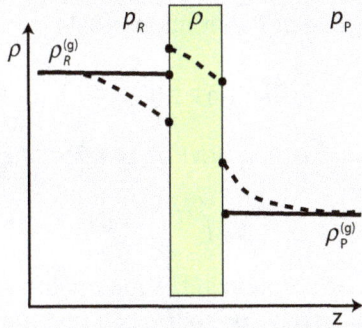

Figure 7.2.2 Concentration profile

The concentration profile of a pure component across a membrane; the dotted lines are the real profiles, while the solid lines assume that we have perfect mixing on the two sides of the membrane. The concentration ρ (in moles per unit volume) has a superscript "(g)" if it is in the gas phase. The pressure on the retentate and permeate sides are given by p_R and p_P, respectively.

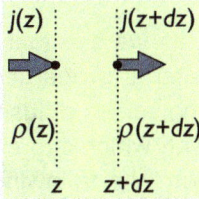

Figure 7.2.3 Mass Balance

Mass balance over part of the membrane: $\rho(z)$ is the concentration (mol/volume) of the gas molecules across the membrane and $j(z)$ is the flux [mol/(sec unit area)].

If we use Fick's equation in the expression for the mass balance, we get the following differential equation for the time-dependent concentration profile:

$$\frac{d\rho}{dt} = D\frac{d^2\rho}{dz^2}$$

If we assume that we have a steady state, such that the time derivative is equal to zero, the concentration depends only on the position and we can solve the differential equation. If we use as boundary conditions $\rho(0) = \rho_0$ and $\rho(L) = \rho_L$, the concentration profile is given by:

$$\rho(z) = \rho_0 + \frac{z}{L}(\rho_L - \rho_0),$$

and the corresponding flux can now be obtained from Fick's equation:

$$j = -\frac{D}{L}(\rho_L - \rho_0)$$

If we assume that the adsorption in the membrane is in the Henry regime (see Section 6.3) and assume perfect mixing on both sides of the membrane, we can relate the concentrations at $z = 0$ and $z = L$ to the pressure of the gas on the retentate and permeate side (see **Figure 7.2.2**), respectively:

$$\rho_0 = H p_R \quad \text{and} \quad \rho_L = H p_P$$

Substitution of these Henry coefficients into the expression for the flux gives:

$$j = \frac{DH}{L}(p_R - p_P)$$

This is an important result as we see that the flux through the membrane depends on two material properties: the diffusion coefficient, which expresses whether molecules diffuse fast or slowly in the material, and the Henry coefficient, which expresses whether the molecules adsorb or not. Compare with the following: cats can swim relatively fast, yet the flux of cats through a river is very small as their affinity with water is so low.

Experimentally, it is often easier to measure the overall transport through a material as a function of the concentration differences rather than measuring the adsorption and diffusion separately. To use this information, we define two new quantities: the **permeability** P and **permeance** P'. The permeability relates the flux to the concentration difference and depends on the thickness of our membrane. Permeance P' is a material property independent of the thickness:

$$j = \frac{P'}{L}(p_R - p_P) = P(p_R - p_P)$$

At this point it is important to mention that in our derivation above we made several assumptions (ideal gas law, Henry regime, ideal mixing, and a diffusion coefficient that is independent of the concentration). If these assumptions hold, we can express the permeance of a material for a given gas as a simple product of the **Henry coefficient** and the **diffusion coefficient**:

$$P' = DH$$

The more general case is more complex, as both the adsorption and diffusion depend on the concentration. The most important conclusion — that the experimental permeance of a material is a combination of diffusion and adsorption — is independent of these details.

That was the easy part, now the tricky part — units. The above equations have been derived using a particular selection of units. We have used for the flux mol/m^2s, but it is equally possible to write the equation in terms of kg/m^2s or m^3/m^2s, similarly for the Henry coefficient, we use ((moles gas)/mole material atm), but the Henry coefficient can also be expressed in moles of gas per kg material per atm. Finally, for the diffusion coefficient we have D in m^2/s.

If we combine the terms, we can get the units for the permeance:

$$\frac{(\text{moles of gas})}{(\text{volume of membrane}) \times (\text{unit pressure})} \times \frac{(\text{unit of length})^2}{(\text{unit time})}$$

$$= \frac{(\text{moles of gas}) \times (\text{unit of length})}{(\text{unit area}) \times (\text{unit time}) \times (\text{unit pressure})}$$

Note that we use moles per second as the flux. In most of the engineering literature, the flux is defined in volume per second.

At the time when the field of membrane separations was developing, people were naturally determining the permeance of materials that were readily available at that time. As with many other fields, researchers introduced a unit (which was later named a "**Barrer**" after one of the pioneers in the field, Richard Barrer) that took a value on the order of one for most of the available materials. Here is the definition of 1 Barrer:

$$1\,\text{Barrer} = \frac{10^{-10}(\text{cm}^3 \text{ gas STP})(\text{cm thickness})}{(\text{cm}^2 \text{ membrane area})\sec(\text{cm Hg pressure})}$$

Table 7.2.1 Permeabilities of gasses for some common polymer membrane materials

Polymer	H_2	He	CH_4	N_2	O_2	CO_2
Silicone	940	560	1,370	440	930	4,600
Natural rubber	49	30	29	8.7	24	134
Polycarbonate	—	14	0.28	0.26	1.5	6.5
Polyimide	2.3	—	0.007	0.018	0.13	0.41

Permeabilities given in Barrer at 25–30°C.

Here "cm^3 gas (STP)" represents the quantity of gas that would take up 1 cm^3 at standard temperature and pressure (STP) as calculated via the ideal gas law, i.e., the molar volume. The "cm thickness" represents the thickness of the material whose permeability is being measured, and "cm^2 membrane area" is the surface area of that material. The conversion to SI units is 1 Barrer = 3.348 × 10^{-19} kmol m/(m^2 s Pa). In **Table 7.2.1**, the permeabilities of some common materials are shown.

Let us now consider a mixture of two components. For flue gasses we can assume that the pressure is sufficiently low that adsorption in the membrane is still in the Henry regime. In this case the two components permeate independently and we can use the same formula to describe the permeation of each component. We do need to replace the pressure by the partial pressure of the corresponding component, and we use the permeabilities for the pure components. Also, here we again stress that the more general case is more complex, the molecules do interact in the membranes, and understanding the effects of these interactions on the permeation of the two components is important in the design of a real membrane.

Diffusion mechanisms

In one of the following sections, we will look in much more detail at the molecular aspects of membranes. In this section, we give a brief overview of the different types of diffusion mechanisms one can find in the literature (see **Figure 7.2.4**).

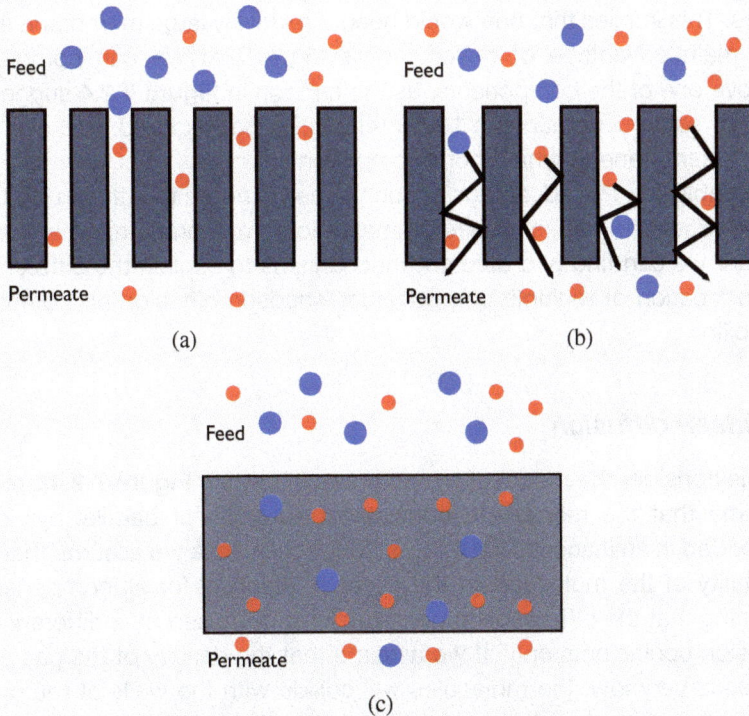

Figure 7.2.4 Diffusion mechanisms

(a) **Molecular sieving**: the pores are sufficiently small that only one component can pass.

(b) **Knudsen diffusion**: molecules with smaller mass have a higher average velocity and move faster through the membrane.

(c) **Diffusion solubility**: separation is due to a difference in the Henry coefficient and diffusion coefficient of the two components in the membrane.

Molecular sieving

The simplest concept of a membrane is a molecule sieve (see **Figure 7.2.4 (a)**); molecules that are bigger than the pore diameter of the sieve will not pass through. From a practical point of view, the main disadvantage is that one needs to use relatively narrow pores. Also, for those molecules that do pass through the membrane, the flux can be relatively small compared to the flux through membranes that have larger pores. This implies that one would need a relatively large membrane area. The main advantage of molecular sieving is that we can completely remove one of the components, as the cartoon in **Figure 7.2.4** suggests.

For some separations a 100% removal is not required, and one can use a membrane in which both components can pass but with differing permeabilities. The advantage of such a membrane is that the fluxes tend to be larger and hence require a smaller total membrane area. In the literature we can find two different mechanisms to explain the differences in permeation of multiple components: Knudsen diffusion and diffusion solubility.

Knudsen diffusion

Let us consider the model of a membrane (shown in **Figure 7.2.4(b)**). We assume that the membrane consists of an array of parallel cylinders imbedded in an inaccessible matrix. For convenience we assume that the solubility of the molecules in the pores is identical for all components, meaning that the difference in permeabilities is caused by a difference in diffusion coefficients only. If we assume that the density of the gas molecules is very low, the molecules will collide with the walls of the cylinders more often than they will collide with other molecules. We further assume that with every wall collision the molecules bounce back with a velocity taken from a Maxwell distribution, determined by the temperature of the membrane. As the kinetic energy of the gas molecules is related to this temperature, we see that molecules with a large mass will have a lower average velocity. Hence, the diffusion coefficient will depend on the (square root of) the mass of the particles. In Knudsen diffusion, the diffusion coefficient will also depend on the pore diameter, and this is the only membrane material property that plays a role.

Diffusion solubility

If we are not in one of the extreme cases that either one of the molecules does not adsorb (molecular sieving) or that Knudsen diffusion holds (low pressures), we have a material in which all the components dissolve and diffuse, see **Figure 7.2.4(c)**. The resulting permeability is then the product of the solubility and the diffusivity, and a separation process can be achieved to the extent that one or both of these properties are different for the different components of the feed gasses.

Membrane separation

In the previous sections, we have seen that a material can have a different permeability for two different gasses. This difference causes the fluxes of these gasses to be different, which is the basis for a separation. Let us now look with more detail into such a separation. Consider the membrane shown in **Figure 7.2.5**. We have two components, N_2 and CO_2. We assume that the permeability of the material is not influenced by the presence of the other component. The only difference from the pure component case is that the driving force is not the total pressure difference between the retentate and permeate, but the partial pressure difference:

$$j_{CO_2} = \frac{P'_{CO_2}}{L}\left(p_{CO_2,R} - p_{CO_2,P}\right) \quad \text{and} \quad j_{N_2} = \frac{P'_{N_2}}{L}\left(p_{N_2,R} - p_{N_2,P}\right),$$

where P'_i is the permeance of the material for component i. The partial pressure on the retentate side follows from:

$$p_{i,R} = x_{i,R}p_R,$$

where $x_{i,R}$ is the mole fraction of component i on the retentate side and p_R is the total pressure. An equivalent expression holds for the permeate side.

Let us now apply these mixture expressions to our membrane. **Figure 7.2.1** shows that the feed splits into a retentate and a permeate. If we carry out a mass balance over our membrane, we have for the fluxes (mol/sec):

$$\Phi_F = \Phi_R + \Phi_P = Aj_R + Aj_P$$

The **stage cut** θ, is defined as the fraction of the feed that flows through the membrane:

$$\theta = \frac{j_P}{j_F}$$

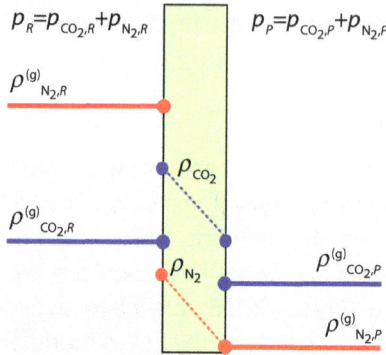

Figure 7.2.5 Mixture
Concentration profiles of a gas mixture across a membrane.

Note that $\theta \leqq 1$. We can now write a mass balance for both CO_2 and N_2, but as we have a binary mixture only one is independent because total mass is conserved:

$$j_F x_{CO_2,F} = j_R x_{CO_2,R} + j_P x_{CO_2,P}$$

$$j_F x_{CO_2,F} = j_F \left((1-\theta) x_{CO_2,R} + \theta x_{CO_2,P} \right)$$

If we know the permeability of our material and the thickness (L) of our membrane, we can use the result of the previous section to relate the flux per unit area through the membrane to the partial pressures of the two components on both sides of the membrane:

$$\theta j_F x_{CO_2,P} = \frac{P'_{CO_2}}{L} \left(p_R x_{CO_2,R} - p_P x_{CO_2,P} \right)$$

$$\theta j_F x_{N_2,P} = \frac{P'_{N_2}}{L} \left(p_R x_{N_2,R} - p_P x_{N_2,P} \right)$$

These two equations, together with the mass balance equations, are characterized by the known values of the feed $x_{CO_2,F}$, the flux of the feed ϕ_F, and the pressures on the permeate and retentate sides. We seek expressions for the stage cut, θ, the total area (A) and thickness (L) of our membrane, as well as the compositions of the retentate and permeate

Table 7.2.2 Designing a membrane

	CO₂	N₂	
Knowns			
Feed flux	Φ_F		Flue gas from the power plant
Composition of feed	$x_{CO_2,F}$	$1-x_{CO_2,F}$	
Pressure of retentate	p_R		Our purchased pumps and compressors
Pressure of permeate	p_P		
Permeance of CO₂	P'_{CO_2}	P'_{N_2}	Material property of our membrane
Unknowns			
Stage cut	$\Theta = \Phi_P/\Phi_F$		Design or optimization parameters
Area of the membrane	A		
Thickness of the membrane	L		
Composition of permeate	$x_{CO_2,P}$	$1-x_{CO_2,P}$	Using the ratio α we can eliminate an unknown
Composition of retentate	$x_{CO_2,R}$	$1-x_{CO_2,R}$	

($x_{CO_2,R}$ and $x_{CO_2,P}$, respectively). Hence, with three equations and five unknowns, in the next section we show how to further specify two of these unknowns in our design (see **Table 7.2.2**).

In the remainder of this chapter, we will use the "**ideal separation factor.**" The separation factor for a binary mixture is defined as:

$$\alpha_{CO_2,N_2} = \frac{x_{CO_2,P}/x_{N_2,P}}{x_{CO_2,R}/x_{N_2,R}}$$

The numerator of this equation may be evaluated using equations given above.

The maximum separation can be obtained if we use vacuum on the permeate side. Thus we define an ideal separation factor α^*, if we assume that the permeate pressure is zero:

$$\alpha^*_{CO_2,N_2} = \frac{\left(\dfrac{P'_{CO_2} A}{L} p_{CO_2,R}\right) \Big/ \left(\dfrac{P'_{N_2} A}{L} p_{N_2,R}\right)}{x_{CO_2,R} / x_{N_2,R}} = \frac{P'_{CO_2}}{P'_{N_2}} = \frac{H_{CO_2} D_{CO_2}}{H_{N_2} D_{N_2}}$$

This equation shows that the ideal separation factor is given by the ratio of the permeabilities of the material.

Section 3

Separating flue gasses

Let us focus on the separation of flue gasses. We will discover that the low concentration of CO_2 and the enormous flow rates make this design a challenging task. We will follow the analysis of Merkel et al. [7.1]. It is worth noting that membranes are also used for other separations. Examples include the separation of H_2 and CO_2, which is important in the context of the IGCC process, and for the separation of CH_4 and CO_2. We refer to the literature for details on these applications. In the remainder of this section, we focus on the separation of CO_2 from flue gasses in coal-fired power plants.

Single-stage separation

To better understand the performance of a membrane for flue gas applications, let us consider the design of a simple one-stage separation shown in **Figure 7.3.1**. **Table 7.3.1** lists some typical values we need for our design of a flue gas separation system. To illustrate a design issue, we assume that we can change the permeability of nitrogen in our

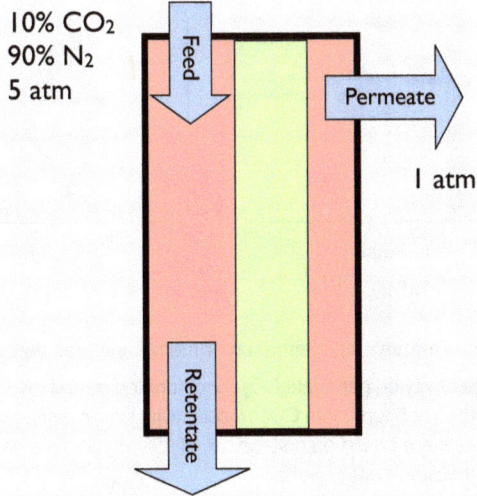

10% CO$_2$
90% N$_2$
5 atm

Feed

Permeate

1 atm

Retentate

Figure 7.3.1   Single-stage membrane for carbon capture

Table 7.3.1   Typical conditions for a membrane separation of flue gas

p_F	5 atm
p_P	1 atm
Φ_F	500 m^3/s
$x_{CO_2,F}$	0.1
P'_{CO_2}	1,000 gpu

p_F — pressure of the flue gas, p_P — pressure of the permeate, P'_{CO_2} — permeance of the membrane where gpu = $(10^{-6}$ cm^3(STP)/cm^2 sec cm-Hg), $x_{CO_2,F}$ — flue gas composition, and Φ_F — emissions of flue gas.

membrane independently. For example, we can increase the selectivity of our membrane by decreasing the N$_2$ permeability of our membrane while keeping the permeability of CO$_2$ fixed.

As we saw in the previous section, the values in **Table 7.3.1** together with the selectivity fully specify our membrane and we can use the

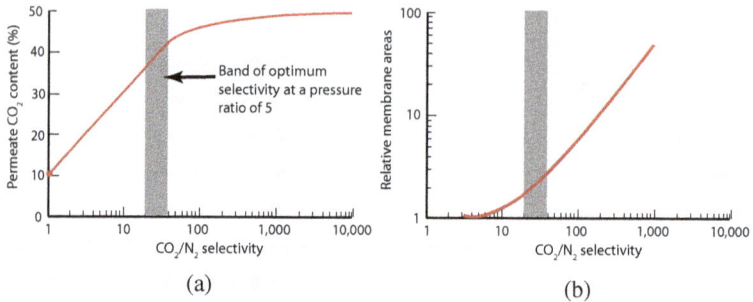

Figure 7.3.2 Permeate concentration or membrane area versus selectivity

Effect of CO_2/N_2 selectivity on permeate CO_2 concentration (a) and relative membrane area (b). For a pressure ratio of 5, optimum CO_2/N_2 selectivity falls in a selectivity band between 20 and 40. The results are based on a stage cut of 1%. *Figures are redrawn from Merkel et al. [7.1], with permission from Elsevier.*

equations derived previously to compute the area and compositions of the permeate and retentate.

Merkel *et al.* [7.1] carried out these calculations, and the results are shown in **Figure 7.3.2**, where the red curves are the solutions to the equations and the hatched regions indicate those process conditions that are economically available with commercial compressors and pumps.

We see a few remarkable conclusions. First we need to stress that our intuition for membrane separation is very poor. If we improve the selectivity of our membrane while keeping the permeance for CO_2 constant, we could in theory have a membrane that is nearly 100% selective. Hence, our intuition would tell us that this is the perfect membrane for which we can expect a permeate that is nearly 100% pure with the smallest membrane area. **Figure 7.3.2** shows that we are wrong on both accounts:

- the maximum purity we can obtain is only 50%.
- the *largest* membrane area is found for the membrane with the highest selectivity.

Clearly, we have to analyze our membrane separation in much more detail to understand why our intuition is very wrong.

We have seen that the design of a membrane follows from the mass balance:

$$j_F x_{CO_2,F} = j_R x_{CO_2,R} + j_P x_{CO_2,P},$$

$$j_F x_{CO_2,F} = j_F \left((1-\theta) x_{CO_2,R} + \theta x_{CO_2,P} \right),$$

where θ is the stage cut (j_P/j_F), and from the fact that all CO_2 and N_2 molecules have to pass the membrane:

$$\theta j_F x_{CO_2,P} = \frac{P'_{CO_2}}{L} \left(p_R x_{CO_2,R} - p_P x_{CO_2,P} \right),$$

$$\theta j_F x_{N_2,P} = \frac{P'_{N_2}}{L} \left(p_R x_{N_2,R} - p_P x_{N_2,P} \right),$$

where P'_i is the permeance of component i, p_R is the pressure on the retentate side, and p_P is the pressure on the permeate side.

These equations, together with the definition of the selectivity in terms of the ideal selectivity, can be solved numerically. **Figure 7.3.2** gives the solution for the values in **Table 7.3.1** and a stage cut of 1%.

If we focus on the limit $\theta \to 0$, the stage cut is so small that we can assume that the composition and pressure of the retentate is equal to the composition and pressure of the flue gas. This simplifies our calculations significantly:

$$\theta j_F x_{CO_2,P} = \frac{P'_{CO_2}}{L} \left(p_F x_{CO_2,F} - p_P x_{CO_2,P} \right),$$

$$\theta j_F x_{N_2,P} = \frac{P'_{N_2}}{L} \left(p_F x_{N_2,F} - p_P x_{N_2,P} \right) = \frac{P'_{CO_2}}{\alpha^* L} \left(p_F x_{N_2,F} - p_P x_{N_2,P} \right),$$

where α^* is the ideal selectivity of our membrane. We can use this equation in the mass balance for CO_2, to obtain:

$$x_{CO_2,P} = \frac{\left(P'_{CO_2}/L \right) p_F x_{CO_2,F}}{\left(P'_{CO_2}/L \right) p_P + j_P} = \frac{\left[\left(P'_{CO_2}/L \right)/(\theta j_F) \right] p_F x_{CO_2,F}}{\left[\left(P'_{CO_2}/L \right)/(\theta j_F) \right] p_P + 1} = \frac{f_A p_F x_{CO_2,F}}{f_A p_P + 1},$$

where we have defined the factor f_A:

$$f_A = \frac{P'_{CO_2}}{L\theta j_F}$$

And for N_2 we have, using the mass balance in the flue gas:

$$x_{N_2,P} = \frac{f_A p_F \left(1 - x_{CO_2,F}\right)}{f_A p_P + \alpha^*}$$

As CO_2 and N_2 are the only components, the sum of these mole fractions should add up to 1 in the permeate:

$$x_{CO_2,P} + x_{N_2,P} = \frac{x_{CO_2,F} f_A p_F}{f_A p_P + 1} + \frac{\left(1 - x_{CO_2,F}\right) f_A p_F}{f_A p_P + \alpha^*} = 1$$

This equation can be rearranged to give a quadratic equation in f_A:

$$\left(p_F p_P - p_P^2\right) f_A^2 + \left(-\left(\alpha^* + 1\right) p_P + \left(1 + \left(\alpha^* - 1\right) x_{CO_2,F}\right) p_F\right) f_A - \alpha^* = 0$$

For $\alpha^* = 1$ this equation has as solution:

$$f_A^{(1)} = \frac{1}{p_F - p_P}$$

If we use this as a reference, we can define the relative area:

$$\frac{A}{A^{(1)}} = \frac{f_A}{f_A^{(1)}}$$

These equations can be solved easily. In **Question 7.3.1** you are asked to compare the approximate solution with the numerical solution for non-zero stage cut as is shown in **Figure 7.3.2**.

We cannot obtain a concentration of CO_2 in our permeate higher than 0.5. In fact, this result is even *independent* of the performance (i.e., permeance) of our membrane. To better understand this, let us look at the

Question 7.3.1 Membrane area

Compare the approximate solution of the expression of the relative membrane area with the full solution shown in **Figure 7.3.2**.

driving force for transport across the membrane. This driving force is the difference in partial pressure between the feed and permeate, or:

$$p_F x_{CO_2,F} - p_P x_{CO_2,P} \geq 0$$

Hence, the maximum concentration in the permeate is set by:

$$x_{CO_2,P}(max) = \frac{p_F x_{CO_2,F}}{p_P}$$

In our example we have used $p_F/p_P = 5$ atm and $x_{CO_2,F} = 0.1$. Hence, to obtain a higher purity we need to increase the pressure of our feed. Importantly, this is *independent* of the performance of our membrane. Irrespective of the permeation or selectivity, we will *never* obtain a higher purity than the limits set by the flue gas concentration and the pressure difference.

This pressure ratio issue has important practical consequences. On a laboratory scale it is easy to achieve any reasonable pressure range, but our choices are very limited in coal-fired power plant applications considering we need to compress on the order of 500 m^3 gas per second. This requires very special equipment, and to compress the gas above 10 atm on the feed side or below 0.1 atm on the permeate side would require most of the energy that is produced by the power plant! If we assume that we are to use compressors that are readily available, we have to work with a pressure ratio (the feed pressure divided by the permeate pressure) of 5 in order to ensure that most of the energy of our power plant is not used for the compression or for pulling a vacuum. As we have used this limit in our example, we can conclude that our single-stage membrane separation will never give us the required purity.

The second question was related to the membrane area. Our calculations show that if we increase the selectivity while keeping the CO$_2$ permeation constant, we obtain the counterintuitive result that the required area of our membrane increases. We would have expected to see that if we make our membrane more selective we can get by with using a smaller membrane. To understand this result we have to realize that in our membrane the pressures at the permeate and feed are fixed. The pressure ratio limit requires that the maximum CO_2 concentration in the feed be 50%. Hence, the remainder must be N_2. All these nitrogen molecules need to pass the membrane! The fact that we need to obey the mass balance implies that if we decrease the permeability of our membrane for nitrogen, we have to make a much larger membrane to keep up with the flux of nitrogen.

With this background information, we can now better understand that there exists an optimal selectivity. If we increase the selectivity, the concentration of CO_2 in our permeate increases. However, once we have reached a selectivity above 100, we see that a further increase of the selectivity by, say, a factor of two, has only a very small effect on the concentration of CO_2 in our permeate, but does increase the membrane area by almost a factor of two. The shaded area in **Figure 7.3.2** gives the range for which we have an optimal separation.

Making membranes

The manufacture of industrial scale membrane separation units is a mature industry. Asymmetric hollow fiber membranes are one common design in industries such as dialysis and water purification. The ideal membrane unit has a high selectivity and a high permeability with as thin a membrane as is physically possible. However, such an extremely thin membrane can be fragile. The main idea of an asymmetric membrane is to deposit the active membrane material on a support that distributes the mechanical stresses associated with the operation. The support itself is often a polymer with a porous microstructure that affords little resistance to gas or fluid flow, yet gives the active membrane mechanical stability. The resulting membrane construct is what we call an asymmetric membrane (see **Figure 7.3.3**) [7.2].

The other attribute of membrane separations is the ability to create a very large surface area with a compact device. The technology that has been developed to achieve this uses a technique in which one simultaneously co-extrudes the selective layer and the support layer. With this co-extrusion process, one can create hollow fibers. These fibers can be

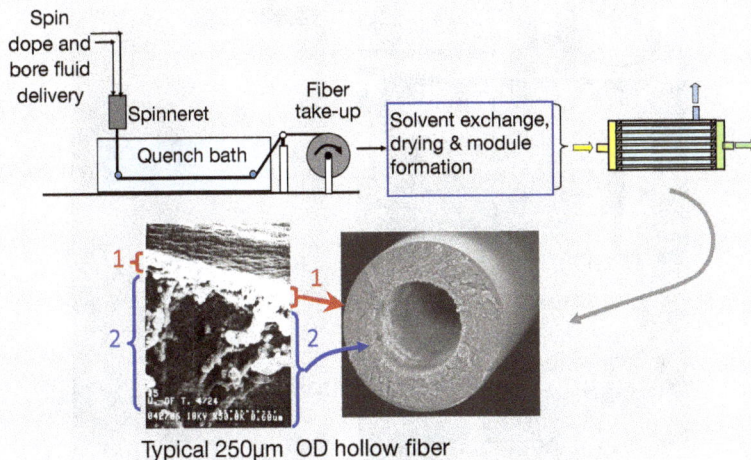

Typical 250µm OD hollow fiber

Figure 7.3.3 Hollow fiber membranes

Schematic process to make asymmetric hollow fibers. The starting point is homogeneous polymer solution (or dope) consisting of a polymer, a solvent, and additives. This
solution and the bore fluid are fed to the spinneret. The co-extruded solution is immediately immersed in a quench bath, the fibers are collected on the take-up bobbin and
the solvent is removed. The fibers are then assembled into modules that can contain as
much as 10^6 fibers with an area of 10,000 m^2/m^3. The bottom-left figure shows an electron microscope picture of a detail of the hollow fiber of the selective layer (1) and support layer (2), which can be simultaneously co-extruded to form a hollow fiber
(bottom-right picture), which makes this process more cost-effective (i.e., about $20/m^2$,
2012 price). *Figure adapted from Koros and Lively [7.2], with permission from John Wiley
and Sons.*

bundled into modules containing 10^6 fibers with an area per volume ratio
of 10,000 m^2/m^3 [7.2] (see **Figure 7.3.4**).

Given a hollow fiber, one can use different designs for the membrane
units. These designs differ in their flow patterns. We saw in Section 5.2
that countercurrent flow gives the best separation. In our hollow fiber
membrane unit, we can achieve a countercurrent situation as is illustrated in **Figure 7.3.5**.

Recently, polymer chemists have been able to synthesize hybrid
polymer-solid adsorbent materials in which these membranes contain
zeolites [7.3] or Metal Organic Frameworks [7.4].

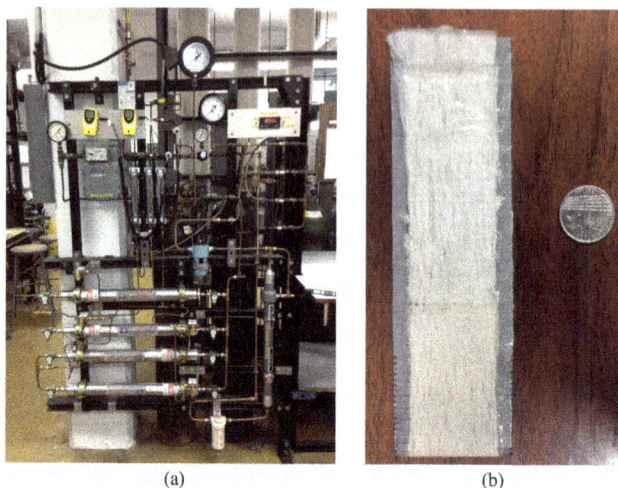

(a)   (b)

Figure 7.3.4 Membrane unit in action

(a) Membrane Separation Unit in Berkeley's Chemical Engineering undergraduate laboratory. The cylinders are five hollow fiber membrane units installed to separate N_2 from air.

(b) Photo of the interior of the cylinder containing the hollow fibers. The cylinder has been cut lengthwise. The quarter provides a size reference.

Improving membrane performance

In the previous analysis we have simply assumed that we have a ratio of feed to permeate pressure of 5. An important question is whether it makes a difference to obtain this factor by compressing the flue gas or by drawing a vacuum on the permeate side. Normally one would argue that compression is easier and one can recover part of the energy using a turbo expander. Drawing a vacuum requires more expensive equipment and recovering part of the energy is not easy. However, in making the comparison it is important to realize that we have to compress all the flue gas, while the drawing of a vacuum only requires us to pump the gas that is passing through the membrane. Merkel *et al.* [7.1] quantified the differences, and the results are shown in **Figure 7.3.6 (a) and (b)**. To put these numbers into perspective, 5 atm does not seem to be a very large pressure, but the volume that needs to be compressed is enormous. The required 114.7 MW of energy to compress the flue gas is 20% of the

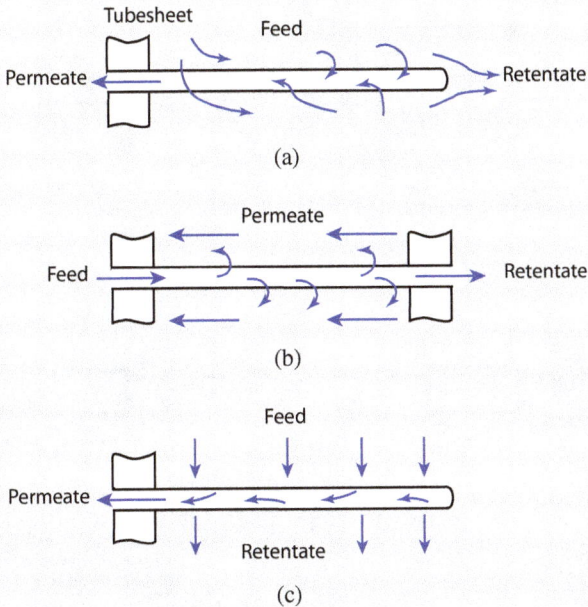

Figure 7.3.5 Membrane designs

In each figure, the membrane is a hollow fiber. In (a) and (c), the feed is introduced on the outside of the fiber (shellside feed) and in (b) the feed is inside (tubeside feed). For a shellside feed, one side of the fiber is closed and the other side (where the permeate comes out) is glued to the tubesheet. The two flow patterns are countercurrent (a, b) or crossflow (c) depending on which side of the unit the retentate is removed from. For the tubeside feed, the flow is countercurrent if the permeate is removed at the same side as the feed enters.

electricity produced by our power plant. These numbers explain why we cannot increase the feed pressure to much higher values. We also see that drawing a vacuum requires significantly less energy.

As mentioned before, there are limits to the ultimate vacuum pressure we can achieve with available technology. Vacuum pumps can reach 0.2 atm, but lower pressures would require capital costs that are far too expensive.

Figure 7.3.6 (c) shows the results for a membrane with a crossflow configuration, which is at present the standard for commercial applications. Similarly to what we saw in Section 5.2 for absorbers, operating

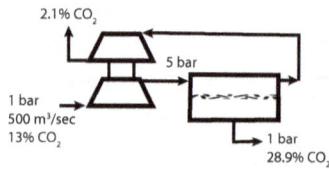

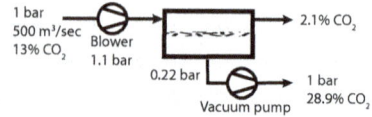

Compressor	151.3 MW
Turboexpander	(−36.6 MW)
Total energy	114.7 MW
Membrane area	2.4×10⁶ m²

(a)

Blower	7.2 MW
Vacuum pump	56.3 MW
Total energy	63.5 MW
Membrane area	11×10⁶ m²

(b)

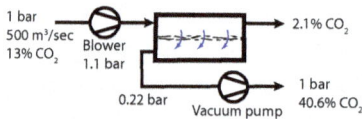

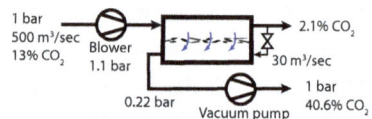

Blower	7.2 MW
Vacuum pump	38.8 MW
Total energy	46.0 MW
Membrane area	6.8×10⁶ m²

(c)

Blower	7.2 MW
Vacuum pump	39.2 MW
Total energy	46.4 MW
Membrane area	4.3×10⁶ m²

(d)

Figure 7.3.6 Different designs to separate flue gas

(a) In this design, the driving force for the separation is created by compressing the flue gas. By using a turboexpander, we can recover part of the compression energy.

(b) In this design, a driving force for the separation is created by pulling vacuum on the permeate side. We only have to pull vacuum for the gas that passes through the membrane.

(c) We can make a more efficient use of the membrane if we use a counterflow configuration. These configurations are not very common commercially.

(d) Configuration in which we use part of the product as a sweep flow. As the partial pressure of CO_2 in the retentate is very low, this has a very small effect on the driving force, but as not all the N_2 has to pass the membrane this significantly reduces the area of the membrane.

membranes in a countercurrent configuration is more efficient. Merkel *et al.* [7.1] have shown that indeed the total energy can be reduced to 46 MW for a countercurrent configuration (see **Figure 7.3.6 (c) and (d)**). Such membrane configurations are described in the patent literature, and this application would benefit from the further development of these types of systems (see **Figure 7.3.5**).

The conclusion we gather from **Figure 7.3.6 (a)** through **(c)** is that the pressure ratio that can be reached in practical terms places serious limits on the separation one can achieve. For flue gasses from a coal-fired power plant the maximum separation is about 50%. Conservation of mass is most unforgiving: all the CO_2 and all the N_2 have to pass through the membrane. If we increase the selectivity of the membrane by decreasing the permeation of N_2, we have to increase our area to maintain the flux of N_2. Changing from using compression on the flue side to using vacuum on the permeate side or changing the membrane design does not solve this pressure ratio issue. One solution is to use two or more membranes. This, however, would require compressing the flue gas or pulling a vacuum twice. The energy requirement of such a two-stage process would require a too large fraction of the electricity production!

Now we can do some magic. Let us look at the design in **Figure 7.3.6 (d)**. Here we use some of the retentate (6%), at a reduced pressure and feed it back to the permeate side of the membrane as a sweep gas. Why would this be a good idea? This looks like mixing a waste stream with the product! However, if we run the numbers we see a *reduction* in the total area of the membrane required. How can this be? If we look at the design of **Figure 7.3.6 (c)**, we have a permeate of about 40% CO_2 and hence 60% N_2. Recall that a constraint in determining the area required for membranes with a high selectivity is that all the N_2 has to pass through the membrane. If we use the retentate as a sweep gas, we have a second source of N_2 and the N_2 mass balance can now be obeyed with a much smaller membrane area. If we examine the design in **Figure 7.3.6 (d)**, we see that we have 2–3% CO_2 at the end of the counterflow module; this very small partial pressure and hence concentration of CO_2 does not affect the driving force, but now not all the N_2 has to pass through the membrane.

Separating flue gasses

In the previous section, we discussed some membrane design configurations. The main conclusion is that at present the real bottleneck in the efficiency of operation is not the material, but the energy required to operate the compressors or vacuum pumps. Given the enormous volume of flue gas we need to process we can at best obtain a pressure ratio of 5, and hence even for the best membrane materials the concentration of CO_2 in the permeate will be maximally 50–60%, and thus will fail to reach the desired purity. To achieve greater purity one would need two stages, and hence two compression steps, which would require too large a fraction of the produced electricity. Therefore, for a long time the logical conclusion was that research to develop novel membranes for carbon capture does not make much sense. The work of Baker and co-workers [7.1] completely changed this conventional wisdom by demonstrating that one can achieve the desired purity by invoking sound process design principles. A hint was already given in the previous section by the demonstration that we can increase the efficiency of the membrane by using an N_2-rich sweep gas. Baker and co-workers [7.1] proposed the scheme shown in **Figure 7.3.7**, in which two membrane units are used. The main innovation is to use air as a sweep gas in membrane unit II and burn the coal with CO_2-enriched air. The net result is that the flue gas has a significantly higher concentration of CO_2 and, given the pressure ratio constraint of 5, this higher concentration of CO_2 in the feed gives a higher concentration in the permeate.

Given that we now have a membrane separation process that can be used for flue gas separations, it is interesting to look at how the performance of the membrane depends on its material properties. **Figure 7.3.8** shows how the capture costs depend on the selectivity and permeance of the material. The costs are a combination of the energy requirements

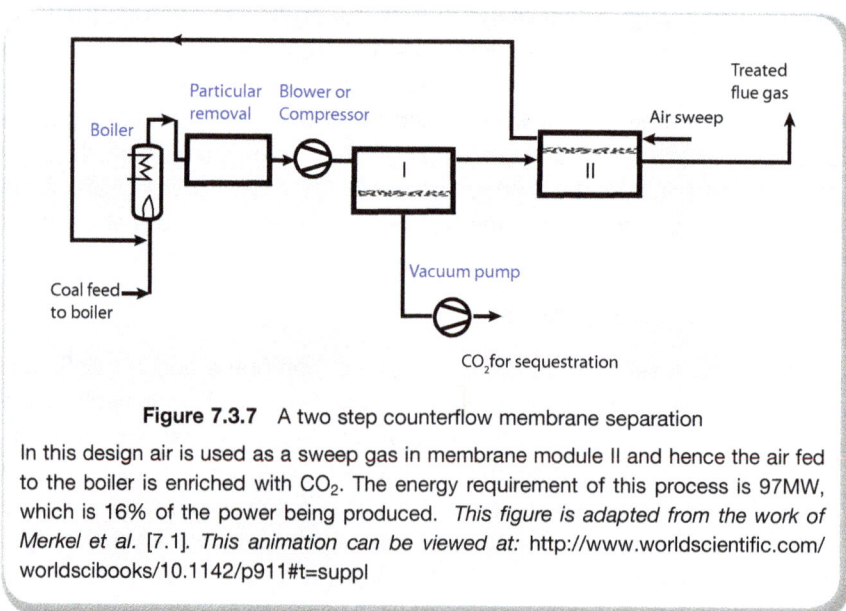

Figure 7.3.7   A two step counterflow membrane separation

In this design air is used as a sweep gas in membrane module II and hence the air fed to the boiler is enriched with CO_2. The energy requirement of this process is 97MW, which is 16% of the power being produced. *This figure is adapted from the work of Merkel et al. [7.1]. This animation can be viewed at:* http://www.worldscientific.com/worldscibooks/10.1142/p911#t=suppl

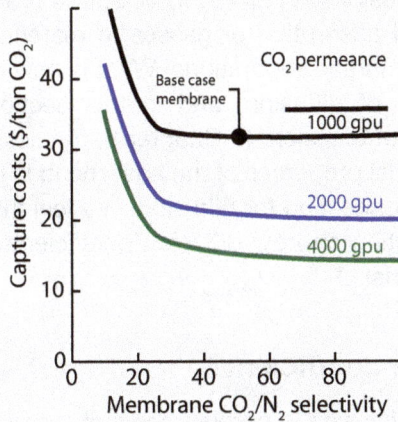

Figure 7.3.8 Selectivity versus costs

Effect of membrane CO_2/N_2 selectivity on the cost of capturing 90% of the CO_2 in flue gas for membranes with a CO_2 permeance of 1,000, 2,000, and 4,000 gpu at a fixed pressure ratio of 5.5. *Figure based on data from* [7.1].

and the capital costs. Clearly, we now see that finding materials with an optimal selectivity and high CO_2 permeance will have a significant effect on the carbon capture costs.

Section 4

Microscopic aspects: diffusion

Until now we have characterized our membranes as materials with given permeabilities. In this section and the next one, we will try to glean some molecular level understanding of the factors that determine the permeability of a material. As we have seen, the permeability is a product of adsorption and diffusion properties. In Chapters 5 and 6 we discussed the absorption and adsorption of gasses in materials. The same concepts apply for membrane separations. What is new in membrane separations is the role of diffusion, and in this section we discuss the molecular aspects of diffusion. In Chapter 6, the solubility of a gas was related to the material properties of the adsorbent; in this chapter we aim to make a similar connection for diffusion. We feel that it is important to develop some intuition on how diffusion coefficients are related to the structure of a material.

Which diffusion coefficient?

Before discussing diffusion in porous media, it is instructive to start with diffusion in the bulk. Most people who start reading the literature about diffusion get confused because the concept of a diffusion coefficient has many meanings, and these differences in meaning result in numerical values for diffusion coefficients that can differ by orders of magnitude. So, the first thing we will do is introduce these diffusion coefficients. We have to warn you that you will read, and hopefully understand, a little more about diffusion coefficients than you strictly need for carbon capture. In the literature one can find three distinct types of diffusion coefficients:

- The **Fick diffusion coefficient**: this is the diffusion coefficient that is associated with transport of mass caused by a difference in concentration. This is the diffusion coefficient we use in practical applications.
- The **Maxwell-Stefan** (*or* **Darken-corrected**, *or* **collective**) **diffusion coefficient**: this is the diffusion coefficient that relates transport of mass to a gradient in the chemical potential. This is a more fundamental way of describing the diffusion coefficient and typically follows from a molecular simulation. If we know how the concentration is related to the chemical potential, we can easily convert this diffusion coefficient into the Fick diffusion coefficient.
- The **self-diffusion coefficient**: this coefficient characterizes the diffusion of a single molecule in a fluid of identical molecules. This type of diffusion at the molecular level can be measured by labeling some of the molecules (e.g., by using Nuclear Magnetic Resonance spectroscopy (NMR)) or by molecular dynamics simulations.

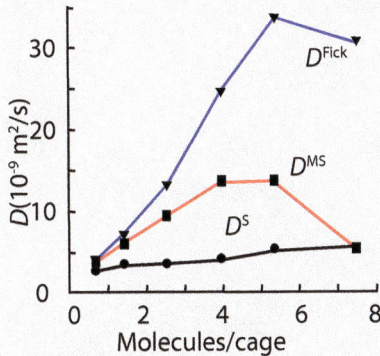

Figure 7.4.1 The different diffusion coefficients of hydrogen

Diffusion coefficients of hydrogen in the zeolite FAU. The top curve is the Fick diffusion coefficient (D^{Fick}), followed by the Maxwell-Stefan diffusion coefficient (D^{MS}), and finally the self-diffusion coefficient (D^{S}). *Figure based on data from Jobic et al.* [7.5].

Figure 7.4.1 shows experimental data for hydrogen diffusing in the zeolite FAU. For this system Jobic and coworkers [7.5] simultaneously measured the Fick diffusion coefficient and the self-diffusion coefficient using quasi elastic neutron scattering (QENS) [7.6]. We see that the three diffusion coefficients are indeed different. There are two important observations. In the limit of very low loading, the three different diffusion coefficient converge to the same value. We will demonstrate that this holds for all systems. In addition, we see that if we increase the loading, the Fick diffusion coefficient increases. This is counterintuitive! It seems to suggest that at rush hour in New York Central Station people will diffuse faster! At the end of this section you should be able to explain what is going on here.

Let us now consider these three different diffusion coefficients and show how they are related. Let's start with Fick's definition of the diffusion coefficient. From a practical point of view this is the most important definition as it directly relates to mass transport. Suppose that we take a material and apply a concentration gradient; we will observe a flux (in molecules per unit area per unit time, see **Figure 7.4.2**). If the concentration gradient is not too large, this flux is proportional to the concentration gradient driving force, or:

$$j = -D^{Fick} \frac{d\rho}{dz},$$

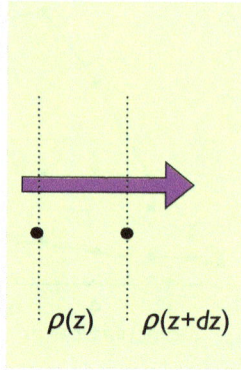

Figure 7.4.2 Flux through part of the membrane

where ρ is the number of molecules per unit volume. Only in exceptional cases is the Fick diffusion coefficient constant. More typically the diffusion coefficient will depend on the temperature and concentration (loading) of guest molecules in the material.

Onsager argued (see **Box 7.4.1**) that the fundamental driving force for diffusion is not a concentration gradient, but rather a gradient in the chemical potential, μ, or:

$$j = -\frac{L}{RT}\frac{d\mu}{dz} = -D^{MS}\frac{d\mu}{dz},$$

where L is the so-called Onsager coefficient and D^{MS} the Maxwell-Stefan (or Darken-corrected, or collective) diffusion coefficient. The concentration of the adsorbed molecules in the material and the chemical potential of these molecules are not independent quantities. This relation is given by the thermodynamic coefficient Γ at isothermal conditions:

$$\Gamma = \left(\frac{\partial\mu}{\partial\rho}\right)_T$$

With this coefficient, we can write an expression that relates the flux to the chemical potential gradient:

$$j = -D^{Fick}\frac{d\rho}{dz} = -\frac{D^{Fick}}{\Gamma}\left(\Gamma\frac{d\rho}{dz}\right) = -D^{MS}\frac{d\mu}{dz},$$

Box 7.4.1 Concentration or chemical potential

The idea of Fick diffusion is intuitive: molecules flow from a high to a low concentration. Diffusion is, however, not always as simple as this. For example, the figure shows the adsorption of molecules in a porous material. In such a system, we can have a gas in equilibrium with a dense fluid inside the pores. If we increase the pressure, molecules flow from a low concentration, the gas phase, to a high concentration in the pores.

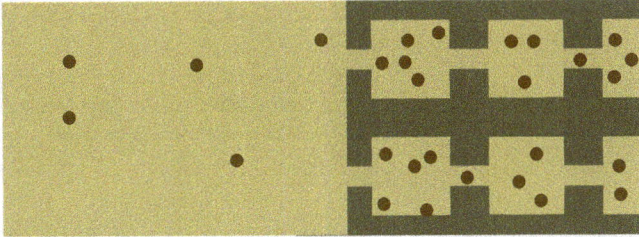

Molecules diffusing from a low density gas phase to a high density one in the pores.

Let us describe the same experiment in terms of the chemical potential. The chemical potential has an energetic contribution and an entropy contribution. As the adsorbed particles have to move in a smaller volume, adsorption causes the entropy to decrease. The interactions of the particles with the walls inside the pores, however, can compensate for this entropy loss and the chemical potential can be lower in the adsorbed phase than the gas phase. According to Onsager, molecules will flow toward a lower chemical potential, which in our example corresponds to molecules flowing from a low to high concentration!

Is the conclusion that using a Fick diffusion coefficient is incorrect? No, we can always use Fick's law as a definition of the Fick diffusion coefficient. But, we now understand why we would get for our adsorption case a negative diffusion coefficient.

where D^{MS} is the Maxwell-Stefan diffusion coefficient, which is related to the Fick diffusion coefficient by:

$$D^{MS} = \frac{D^{Fick}}{\Gamma}$$

This thermodynamic coefficient Γ can be calculated from the experimental isotherm (see **Box 7.4.2**).

Box 7.4.2 Thermodynamic coefficient

Let us calculate the thermodynamic coefficient Γ for a material of which the adsorption can be described with a Langmuir isotherm:

$$\rho = \rho_0 \frac{bp}{1+pb},$$

where p is the pressure of the gas phase and ρ_0 is the maximum loading in moles per unit volume. We have in this equation a relation between the pressure of the gas phase and the number of adsorbed molecules per unit volume. For the thermodynamic coefficient, however, we need to know the relation between the loading and the chemical potential of the gas phase. If we assume that the bulk gas pressure is sufficiently low, we can consider the gas phase to be an ideal gas. For an ideal gas, the relation between the chemical potential and pressure is given by

$$\mu = \mu^0 + RT \ln(p/RT),$$

which gives the Langmuir isotherm in terms of the chemical potential as a function of loading:

$$\mu = \mu^0 + RT \ln\left[\frac{1}{bRT}\left(\frac{\rho}{\rho_0 - \rho}\right)\right],$$

and we obtain for the thermodynamic coefficient:

$$\Gamma = RT \frac{\rho_0}{\rho_0 - \rho}$$

This gives for the relation between the Maxwell-Stefan and the Fick diffusion coefficient:

$$D^{\text{Fick}} = RT \frac{\rho_0}{\rho_0 - \rho} D^{\text{MS}}$$

This equation shows that if the loading in the material approaches the maximum loading $\rho \to \rho_0$, the Fick diffusion coefficient becomes infinitely large. The reason is illustrated in the figure. If our material is completely saturated and we add one more molecule, another molecule needs to come out instantaneously, which corresponds to an infinitely large diffusion coefficient!

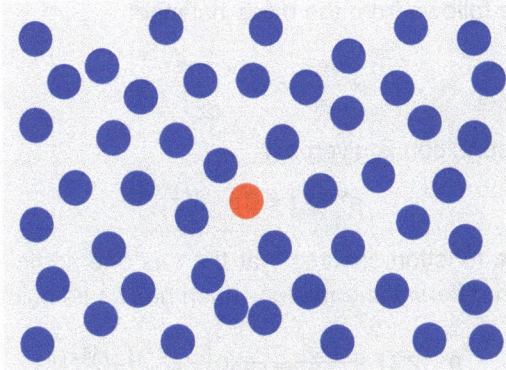

Figure 7.4.3 Labeled particle in a solvent

Now let us look at a different diffusion coefficient that is not directly defined by looking at the transport of mass. For this we label a molecule in our fluid (see **Figure 7.4.3**). This label is something that does not affect the properties of the molecule, but allows us to follow this particular molecule. If we could follow the movement of this molecule, we would observe that because of collisions with the other molecules it exhibits some kind of Brownian motion. If we could do this experiment many times, we could determine the probability of finding this particle at a distance z from the initial point. We can calculate this probability from our mass balance equation:

$$\frac{d\rho^*}{dt} = -\frac{dj}{dz}$$

where ρ^* is the density (in number of particles per unit volume) of our labeled particle.

We assume that the flux is given by Fick's law:

$$j = -D^s \frac{d\rho^*}{dz}$$

The diffusion coefficient we have introduced here is the self-diffusion coefficient. Of importance is that all particles are identical except that we tag one of them, and hence define the concentration of the tagged particle. The time dependence of this concentration, or density, of the tagged particle follows from the mass balance:

$$\frac{d\rho^*}{dt} = D^s \frac{d^2\rho^*}{dz^2},$$

and the initial condition is given by:

$$\rho^* (z, t = 0) = \delta(z),$$

where the delta function ensures that there is one particle at $z = 0$. This is a well-known differential equation which has as its solution:

$$\rho^* (z,t) = \frac{1}{\sqrt{4\pi D^s t}} \exp\left[-z^2 / \left(4D^s t\right)\right]$$

This result looks a bit strange, as we have only one particle: how can the density be less than one? The way to interpret this result is to suppose that we repeat the experiment many times and at each time t the particle will be in a different position. The expression for the density gives the probability of finding a particle at time t at position z. With this equation we can compute the average position of a particle at time t:

$$\langle z(t) \rangle = \int_{-\infty}^{\infty} z\rho^* (z,t)dz = \frac{1}{\sqrt{4\pi D^s t}} \int_{-\infty}^{\infty} z \exp\left[-z^2 / \left(4D^s t\right)\right]dz = 0,$$

which is intuitively clear as there is no gradient in the concentration and hence the particle has an equal probability of going right or left. Hence, the average will be zero. The second moment, or mean squared displacement, does not disappear:

$$\langle z^2(t) \rangle = \int_{-\infty}^{\infty} z^2\rho^* (z,t)dz = \frac{1}{\sqrt{4\pi D^s t}} \int_{-\infty}^{\infty} z^2 \exp\left[-z^2 / \left(4D^s t\right)\right]dz = 2D^s t$$

This equation shows that by measuring the mean squared displacement of a labelled particle one can get the diffusion coefficient. Because the integral in the above equation extends from $-\infty$ to $+\infty$, the simple proportionality of mean squared displacement to the diffusion coefficient assumes many displacements, i.e., over long time scales:

$$D^s = \lim_{t \to \infty} \frac{1}{2t} \langle (z(t) - z(0))^2 \rangle$$

As our particle is moving in a fluid that is composed of identical particles, this diffusion coefficient is called the self-diffusion coefficient. Interestingly, one can label a particle experimentally using NMR spectroscopy methods, and thus NMR has become the principal method used to measure the self-diffusion coefficient. Also, in a molecular simulation one can follow a single particle and from the mean squared displacement get the self-diffusion coefficient directly from simulations.

To derive D^{MS}, the Maxwell-Stefan diffusion coefficient, rather than label a single particle, we monitor the movement of the center of mass of the entire system. The center of mass is defined as:

$$z_{cm}(t) = \frac{1}{mN}\sum_{i=1}^{N} mz_i(t) = \frac{1}{N}\sum_{i=1}^{N} z_i(t),$$

where m is the mass of a single molecule and N is the total number of molecules. The Maxwell-Stefan diffusion coefficient is related to the mean squared displacement of the center of mass by:

$$D^{MS} = \lim_{t\to\infty}\frac{1}{2t}\left\langle\frac{1}{N}\left(\sum_i z_i(t) - \sum_i z_i(0)\right)^2\right\rangle$$

Let us look at the term in brackets:

$$\left\langle\left(\sum_i z_i(t) - \sum_i z_i(0)\right)^2\right\rangle = \left\langle(z_1(t) - z_1(0))\left[(z_1(t) - z_1(0)) + (z_2(t) - z_2(0)) + \cdots\right]\right.$$

$$+ (z_2(t) - z_2(0))\left[(z_1(t) - z_1(0)) + (z_2(t) - z_2(0)) + \cdots\right] + \cdots\right\rangle$$

$$= \left\langle\sum_i(z_i(t) - z_i(0))^2\right\rangle + \left\langle\sum_i\sum_{j\neq i}(z_i(t) - z_i(0))(z_j(t) - z_j(0))\right\rangle$$

We see that the first term gives the expression for the self-diffusion coefficient. The second term accounts for the correlations between the molecules. If the molecules i and j do not influence each other this term will be zero. In other words, correlations between molecules cause the Maxwell-Stefan coefficient to differ from the self-diffusion coefficient. Such correlations are far more likely at high concentrations of diffusing species. At very low concentrations the thermodynamic factor, which relates the Maxwell-Stefan diffusion coefficient to the Fick diffusion coefficient, is one. Hence, at the limit of very low concentrations all three diffusion coefficients are identical.

Random walk

A different way of looking at diffusion is to assume that because of collisions our molecule is performing a random walk. If we assume a random walk on a cubic lattice, if our molecule jumps from one lattice point to another, we can define the mean squared displacement as (see **Figure 7.4.4**):

$$\left[\mathbf{z}(N) - \mathbf{z}(0)\right]^2 = \left(\sum_{i=1}^{N} \mathbf{l}_i\right)^2,$$

where $\mathbf{l}_i$ is a random step on the lattice. If we now take an average over many random walks:

$$\left\langle\left[\mathbf{z}(N) - \mathbf{z}(0)\right]^2\right\rangle = \left\langle\mathbf{l}_1\mathbf{l}_1\right\rangle + \left\langle\mathbf{l}_1\mathbf{l}_2\right\rangle + \ldots \left\langle\mathbf{l}_2\mathbf{l}_1\right\rangle + \left\langle\mathbf{l}_2\mathbf{l}_2\right\rangle + \cdots$$

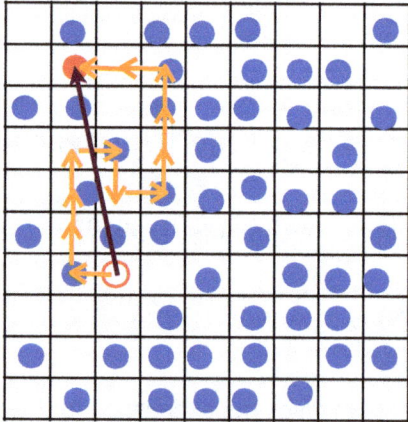

Figure 7.4.4 Diffusion as a random walk

One can envision the motion of a labelled particle (red) as a random walk on a lattice. The orange arrows give the random step $\mathbf{l}_i$ and the black arrow shows the displacement after N steps.

Because a random walk by definition has uncorrelated sequential steps, we have:

$$\langle l_i l_i \rangle = a^2 \quad \text{and} \quad \langle l_i l_j \rangle_{i \neq j} = 0$$

This gives the mean squared displacement:

$$\left\langle \left[z(N) - z(0) \right]^2 \right\rangle = Na^2,$$

where N is the number of steps in our random walk and a is the distance between two neighboring lattice points. This number of steps is equal to the hopping rate k (in number of steps per unit time) times the time t, or:

$$\left\langle \left[z(N) - z(0) \right]^2 \right\rangle = kta^2$$

A comparison with the relation of the diffusion coefficient and the mean squared displacement shows that we can relate our self-diffusion coefficient to the hopping rate, or:

$$D^S = \frac{1}{2} ka^2$$

This is a very useful relation as we can now relate the hopping of a molecule from one site to another to the diffusion coefficient.

Now, looking at the system where we have many molecules, we can tag a particle and let that particle hop with a hopping rate k. If the particles do not see each other, we can use this relation to obtain both the self- and Maxwell-Stefan diffusion coefficients from the hopping rate. However, when the particles interact we see differences. One example comes from the assumption that the particles only successfully hop to an open lattice site. As a consequence, the effective hopping rate decreases with increased loading. We observe that the self-diffusion coefficient decreases as a function of the fraction of occupied sites, θ:

$$D^S = D_0^S (1 - \theta)$$

It is interesting to look at the situation where we have a high loading. If molecule i successfully jumps, then the molecules surrounding the site that is left empty by molecule i will have a significantly higher probability of jumping. This is the type of correlation that we can ignore in the limit of very low loading. In addition, if a particle jumps it leaves a vacancy by definition. Hence, at high loading the next hop has a high probability that the particle will jump back to its original position. This correlation is much more important for the self-diffusion coefficient than for the collective diffusion coefficient, and explains why the self-diffusion coefficient is always lower.

Diffusion in porous media

Let us now go back to **Figure 7.4.1**, our experimental system of hydrogen diffusing in the zeolite FAU. The data shown in this figure nicely demonstrate the difference between the various diffusion coefficients. We also see that in the limit of low loading the three diffusion coefficients are identical.

We can now also understand the increase in the diffusion coefficient as a function of loading. Intuitively, one would argue that as we increase the number of molecules in the pores we restrict the movement of those molecules. This is indeed what we observe in bulk. What makes the difference in a porous medium?

Let us first look at the thermodynamic factor. In **Box 7.4.2** we have shown that if the adsorption can be described with a Langmuir isotherm and the gas phase approximated by an ideal gas, we have:

$$\Gamma = RT \frac{\rho_0}{\rho_0 - \rho}$$

What we see is that this coefficient gets very large for high loading. The thermodynamic factor approaches infinity at the maximum loading. Indeed, toward the maximum loading one needs to increase the chemical potential by an infinite amount to increase the concentration. Hence, at these conditions a small concentration gradient gives a very large thermodynamic factor Γ. Or, in more physical terms, if a pore is completely filled with molecules, the addition of yet another molecule at one end of the pore causes a molecule at the other end to leave (see figure in **Box 7.4.2**). As the transport diffusion coefficient is related to the flux, it does not matter which particle leaves the pore, and we observe an infinitely large diffusion coefficient. If we remove this thermodynamic contribution from our diffusion coefficient, we obtain the Maxwell-Stefan diffusion coefficient, which gives a more true representation of the mobility of the molecules. In many practical systems the Maxwell-Stefan diffusion coefficient is far less dependent on the concentration. Therefore, many engineering applications simply assume that this diffusion coefficient is independent of the concentration.

The above illustrates what is known as the "Darken" assumption, namely, that the corrected diffusion coefficient is assumed to be independent of the loading. It is important to mention that it was not Darken who made this assumption. Darken realized that in many cases the corrected diffusion coefficient does depend on loading. For a historical note on these aspects, see the article by Reyes *et al*. [7.7]. Nevertheless, in many engineering applications this assumption has been widely used. Indeed, from a practical point it is very convenient as it implies that it is sufficient to have knowledge of either the self-diffusion coefficient at low loading or a single transport diffusion coefficient plus the complete adsorption isotherm to estimate the transport diffusion coefficient for all loadings.

Section 5
Materials: polymers

In the previous sections, we have seen that the engineering design plays an essential role in defining whether or not a membrane separation for CO_2 is feasible. Given the right design, we again have a target for materials research: to make a material with a high permeability and a sufficient selectivity. In this section, we will look in more detail at the molecular aspects of the design of that material. We start our discussion with polymer membranes. In the next section, we will discuss nanoporous materials.

Polymer membranes

Polymer membranes are typical examples of systems in which permeation is ruled by the solution-diffusion mechanism. **Figure 7.5.1** shows a caricature of a disordered polymer film. The film has a typical distribution

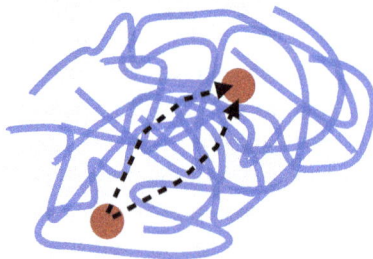

Figure 7.5.1 Diffusion in a polymer membrane

Example of a polymer (blue) with a gas molecule (red) hopping from one cavity to another. The black arrows indicate different paths.

of holes in which the gas can dissolve (solubility) and the gas molecules can hop between one cavity and another (diffusion). Depending on the chemical make-up of the polymer films, these cavities can be dynamic; they form or disappear depending on the movement of polymer segments. Most materials will have a distribution of different cavities and hence of hopping rates.

We have seen that the ideal membrane for separation has a high selectivity and a high permeability. Robeson [7.8,7.9] discovered that polymeric materials with a high selectivity have a low permeability, while materials with a high permeability have a low selectivity. This behavior is illustrated in **Figure 7.5.2**. In this "**Robeson plot**", the selectivity for CO_2/N_2 mixtures is plotted as a function of the CO_2 permeation. This plot shows that for a wide range of materials, the upper bound in selectivity is an approximately linear function of the permeation, as shown by the black line. In the simple model we described previously, we saw that we can increase the permeability in two ways: change the diffusion coefficients or change the solubility. It turns out that the solubility of gasses in most of the available polymeric materials does not vary significantly. The diffusion coefficients, on the other hand, do change if we change the material. However, the materials with the highest diffusion coefficients tend to have more open structures; as a consequence all adsorbed molecules move faster and one loses selectivity. Because of this effect, we see that materials with a high permeability also have a low selectivity. This is the explanation for the upper bound shown in **Figure 7.5.2**.

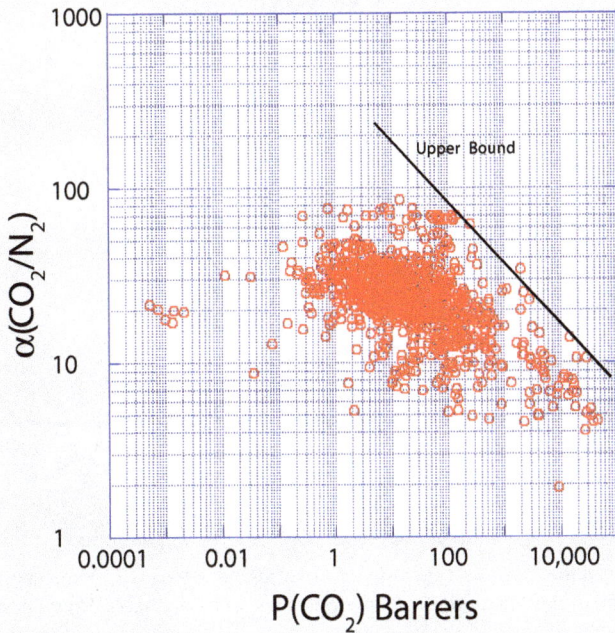

Figure 7.5.2 Robeson plot for CO_2/N_2

The CO_2/N_2 selectivity of a membrane versus CO_2 permeation. Each dot represents a different material. If the permeation is large the selectivity is typically poor. The solid line is the upper bound found in polymers experimentally. *Figure reprinted from Robeson* [7.9], *with permission from Elsevier*.

To gain more insight into these Robeson plots it is interesting to look at the plot for the separation of CO_2 and methane (see **Figure 7.5.3**). In this plot we see that glassy polymers tend to have higher selectivities that are closer to the upper bound. Glassy polymers are clearly performing better [7.10]. Why?

First let us look at some simple properties of polymers. **Figure 7.5.4** shows a diagram of the specific volume (volume/mass) of a polymer as a function of temperature. Consider this plot carefully. The "occupied volume" represents the space occupied by the atoms that make up the polymer. This volume only weakly increases with temperature (the hotter

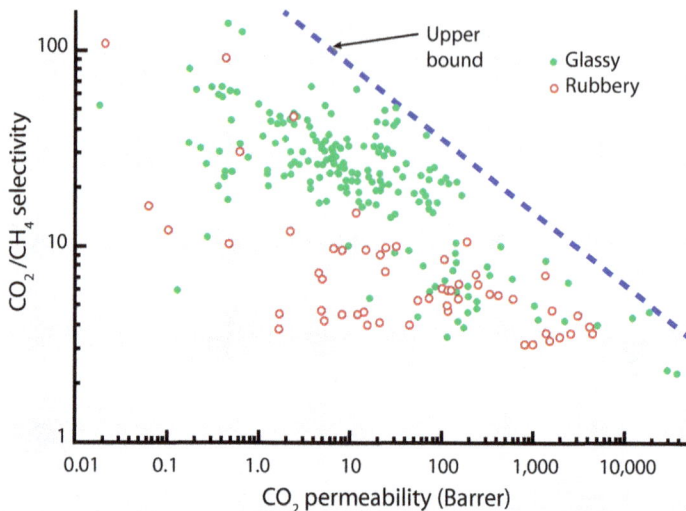

Figure 7.5.3 Robeson plot: the CO_2/CH_4 selectivity

Robeson plot: the CO_2/CH_4 selectivity of a membrane versus CO_2 permeation for glassy polymers (closed symbol) and rubbery polymers (open symbol). *Figure adapted from* [7.10].

the polymer, the more its atoms rattle around their equilibrium position in space). Because of the space between polymer chains, there is an additional "free volume" available as cavities for adsorbates. The nature of adsorbate interactions with this free volume depends on whether or not the polymer is "rubbery" or "brittle." These differences in a polymer's mechanical character are characterized by a "glass transition temperature." Below this temperature polymers are glass-like, which means that they are brittle and break. Above that temperature, the polymers are rubbery. (Imagine a bowl of cooked spaghetti, and the noodles are very wobbly and active; each individual noodle can mark out a space much larger than the noodle itself because of its motion.) In this case, the chains of the rubbery polymer are rearranging on a short time scale, compared to adsorption and diffusion. As a consequence, as CO_2 goes into the material it looks as if it is adsorbing into a material that can, like a liquid, equilibrate. This means it appears to obey Henry's law in that the concentration of adsorbed CO_2 is a linear function of partial pressure. If

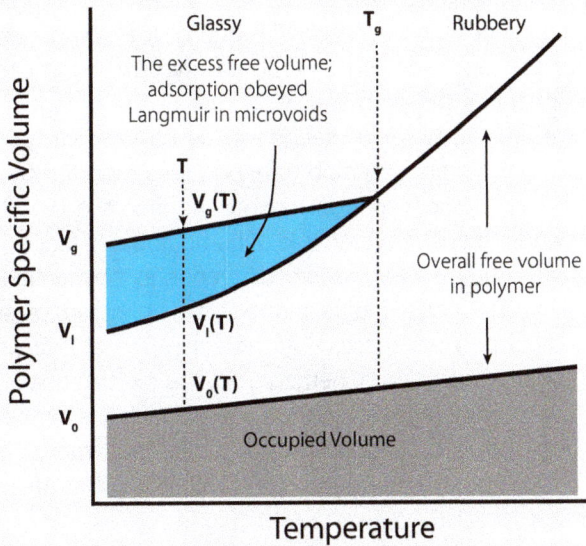

Figure 7.5.4 Phase diagram of a typical polymer

Figure adapted from [7.10]. *This animation can be viewed at:* http://www.worldscientific. com/worldscibooks/10.1142/p911#t=suppl

we drop the temperature below the glass transition temperature, the individual polymer chains are now locked in place, but there is an excess volume present. This excess volume represents micropores inside the material that exist because the previously moving chains are no longer taking up that extra room. Adsorption of CO_2 in that medium looks very much like adsorption in a porous medium. Glassy polymers "push" the limits of the Robeson plot because they appear to the adsorbing molecule as a porous material. A theoretical basis for this upper bound is given by Freeman [7.11].

To further break the boundaries of the Robeson limit researchers are looking at materials that have the potential for molecular control of adsorbate solubility by way of *"facilitated transport."*

Facilitated transport

In materials in which transport is governed by the diffusion solubility, we have limited opportunities to change the solubility of adsorbates. When considering absorption, we realized that one method to improve solubility was to use a solvent that can react with one of the components. We can utilize a similar concept to design a membrane system. In this case, a chemical reaction enhances the permeability of one of the components without decreasing the selectivity. This mechanism is called facilitated transport. One of the first articles on facilitated transport was reported by Scholander in the context of his discoveries for oxygen transport in blood [7.12]. The work of Scholander focused on the role of hemoglobin to facilitate the transport of oxygen. He showed a marked difference in permeation between a blood plasma and a plasma plus hemoglobin. In blood plasma without hemoglobin, oxygen is transported by the conventional diffusion solubility mechanism. With the addition of hemoglobin, the oxygen has an alternative route to be transported, which significantly enhances the permeation.

Let us look at the mechanism of facilitated transport in detail. **Figure 7.5.5** illustrates a membrane in which a substrate molecule S selectively reacts with CO_2:

$$CO_2 + S \rightleftharpoons [SCO_2]^*$$

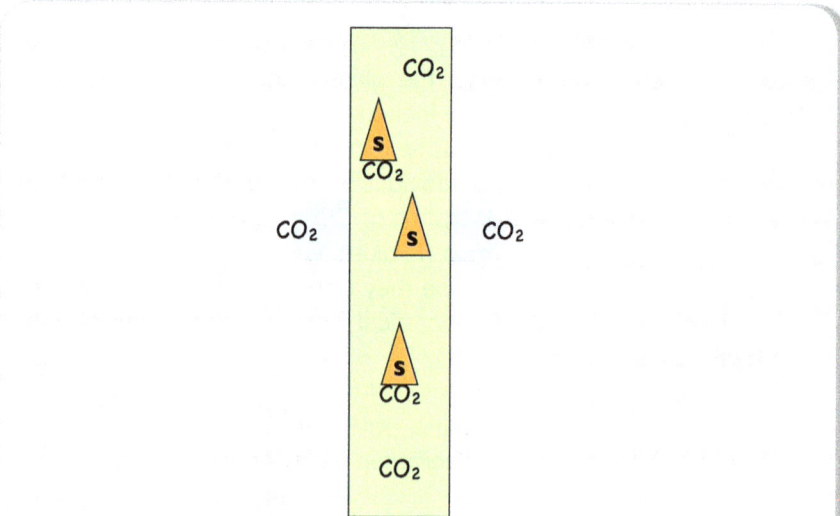

Figure 7.5.5 Facilitated transport

Membrane in which we have a substrate S that reacts with CO_2. This allows CO_2 to be transported via solubility and with the substrate as a carrier.

If we have such a chemical reaction in our membrane, we need to change our mass balance to include the consumption of CO_2 due to the reaction:

$$\text{accumulation} = \text{flux in} - \text{flux out} - \text{chemical reaction}$$

$$\frac{dc_{CO_2}}{dt} = D_{CO_2}\frac{d^2c_{CO_2}}{dz^2} - r_{SCO_2}* = 0,$$

where r_{SCO_2} is the reaction rate associated with the formation of the complex $[SCO_2]*$. As before, we assume steady state conditions. Previously we only had a mass balance for CO_2, now we have to include a mass balance for S and for the complex $[SCO_2]*$:

$$D_S\frac{d^2c_S}{dz^2} - r_{SCO_2}* = 0$$

$$D_{SCO_2}*\frac{d^2c_{SCO_2}*}{dz^2} + r_{SCO_2}* = 0$$

If we assume that the reaction rate is much faster than the diffusion time, the system is in equilibrium:

$$c_{SCO_2}* = k_S c_{CO_2} c_S$$

We can also assume that the total concentration of substrate molecules c_{S^T}, with and without CO_2, is independent of the position in the membrane:

$$c_{S^T}(z) = c_{SCO_2}*(z) + c_S(z) = c_{S^T}$$

or:

$$c_{SCO_2}*(z) = \frac{k_S c_{S^T} c_{CO_2}(z)}{1 + k_S c_{CO_2}(z)}$$

If we further assume that all diffusion coefficients are equal $D^0 = D_{CO_2} = D_{SCO_2}* = D_S$, we can add the two differential equations that involve CO_2:

$$D^0\left[\frac{d^2 c_{CO_2}(z)}{dz^2} + \frac{d^2}{dz^2}\left(\frac{k_S c_{S^T} c_{CO_2}(z)}{1 + k_S c_{CO_2}(z)}\right)\right] = 0$$

This equation has a simple solution:

$$c_{CO_2}(z) + \frac{k_S c_{S^T} c_{CO_2}(z)}{1 + k_S c_{CO_2}(z)} = Az + B,$$

with boundary conditions: $c_{CO_2}(L) = 0$ and $c_{CO_2}(0) = c_{CO_2,R}$. If we assume the adsorption is in the Henry regime, $(c_{CO_2,R} = H_{CO_2} p_{CO_2,R})$ we get:

$$c_{CO_2}(z) + \frac{k_S c_{S^T} c_{CO_2}(z)}{1 + k_S c_{CO_2}(z)} = -\left(H_{CO_2} p_{CO_2,R} + \frac{k_S c_{S^T} H_{CO_2} p_{CO_2,R}}{1 + k_S H_{CO_2} p_{CO_2,R}} \right)\left(\frac{z}{L} - 1 \right)$$

This gives for the flux of CO_2 across the membrane:

$$j_{CO_2} + j_{SCO_2^*} = \frac{D^0 H_{CO_2}}{L}\left[p_{CO_2,R} + \left(\frac{k_S c_{S^T} H_{CO_2} p_{CO_2,R}}{1 + k_S H_{CO_2} p_{CO_2,R}} \right) \right]$$

It is instructive to consider the two limiting cases. At low partial pressures of CO_2 we have:

$$j_{CO_2} + j_{SCO_2^*} = \left(1 + k_S c_{S^T}\right) \frac{D^0 H_{CO_2}}{L} p_{CO_2,R},$$

where we see that the substrate enhances the transport in proportion to the concentration of S. At high pressures, we obtain:

$$j_{CO_2} + j_{SCO_2^*} = \frac{D^0}{L}\left(H_{CO_2} p_{CO_2,R} + c_S \right),$$

at these conditions each substrate molecule has a CO_2 molecule and hence the enhancement is limited by the concentration of substrate molecules.

It is interesting to compare these results with the original experiments of Scholander for oxygen transport in blood (**Figure 7.5.6**). This figure gives the transport as a fixed difference between the retentate and the permeate side. In our equations, we have assumed that the pressure on the permeate side is zero; taking a non-zero version makes the equations slightly more complicated. We can see for the N_2 transport that the flux is independent of the total pressure; it only depends on the pressure difference. For the oxygen, however, we see that if we increase the pressure on the retentate side we will, at a given pressure, saturate the hemoglobin and the transport will become independent of the pressure.

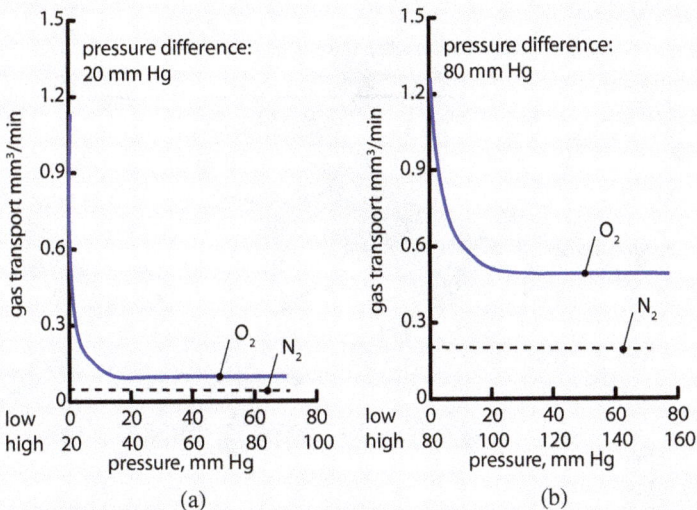

Figure 7.5.6 Facilitated transport of oxygen in blood

Transport of oxygen and nitrogen across a "membrane" of blood with hemoglobin. In this experiment, the pressure difference between the retentate and permeate sides is kept constant (left, 20 mm Hg and right, 80 mm Hg), while the total pressure on the retentate side is increased. *Figure adapted from Scholander* [7.12].

For flue gasses, we need to replace the hemoglobin with a component that selectively binds CO_2. Amines are obvious candidates. Of practical importance is how to make a CO_2-reactive membrane. The early solutions were to use a reactive liquid inside the pores [7.10]. Using an immobilized phase inside a membrane has as main disadvantage that it may leak out or evaporate from the membrane, which will slowly deteriorate the performance of the membrane. An alternative strategy is to have the polymer of the membrane functionalized as illustrated in **Figure 7.5.7**.

The figure shows a system in which there is water. Without water one needs two amine groups: one amine interacts with CO_2 to form a zwitterion.

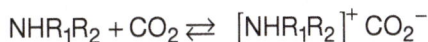

$$NHR_1R_2 + CO_2 \rightleftarrows [NHR_1R_2]^+ CO_2^-$$

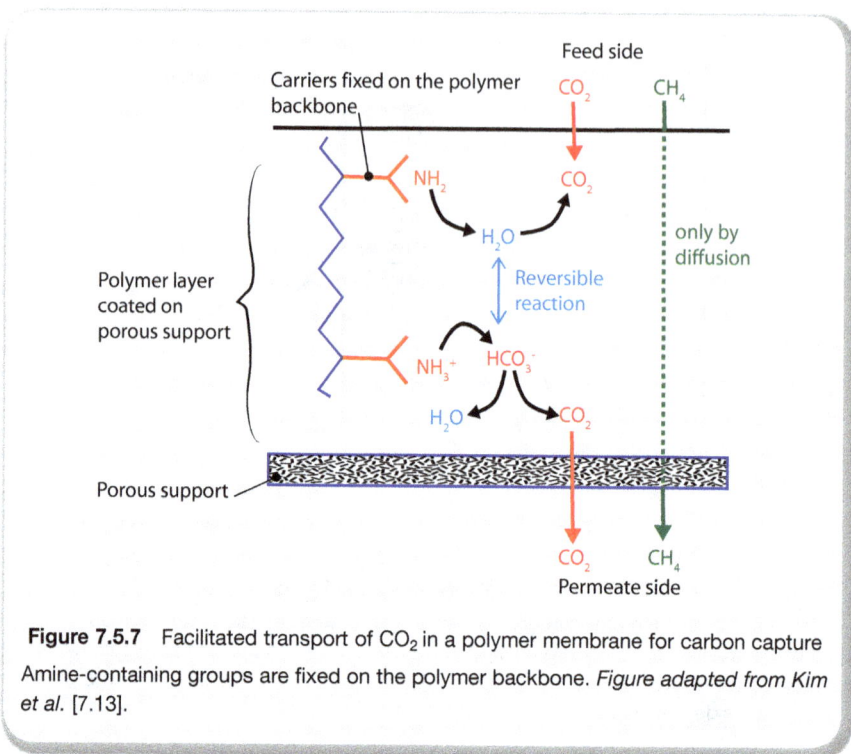

Figure 7.5.7 Facilitated transport of CO_2 in a polymer membrane for carbon capture Amine-containing groups are fixed on the polymer backbone. *Figure adapted from Kim et al. [7.13].*

This zwitterion reacts with another amine, which takes the proton:

$$[NHR_1R_2]^+ CO_2^- + NHR_1R_2 \rightleftarrows [NR_1R_2]CO_2^- + NH_2R_1R_2^+$$

The CO_2^- group can move along the backbone of the polymer through hops from one amine group to the other (like a bucket brigade). As this hopping of the CO_2^- group is on top of the diffusion of CO_2 in the gas phase, we obtain an enhancement of the permeation. Since neither CH_4 nor N_2 interacts with the amine group, we also see an enhancement of the selectivity. If in addition to the amine groups we also have water in our membrane, we can have CO_2 reacting with the amine groups and now we can form HCO_3^- and NH_3^+. Because HCO_3^- can diffuse in the water phase, it will generally have a higher diffusion coefficient compared to the dry amines.

Several examples of these amines have been published (e.g., [7.8, 7.14, and 7.15]). **Box 7.5.1** gives an example of a membrane for CO_2 separations that uses facilitated transport.

Box 7.5.1 Membranes with facilitated transport

One of the experimental strategies to overcome the Robeson upper limit in polymer materials is to use facilitated transport. We have seen that the idea of facilitated transport is that the CO_2 in the gas mixture reacts with the membrane. Because of the chemical reaction, we enhance the CO_2 solubility. If these reacted CO_2 molecules have similar diffusion coefficients, our material will have a significantly better permeability and selectivity. We have seen that one of the reasons polymer materials have this upper limit is that the CO_2 solubility varies only very little from one material to another. This completely changes if CO_2 reacts.

(a)

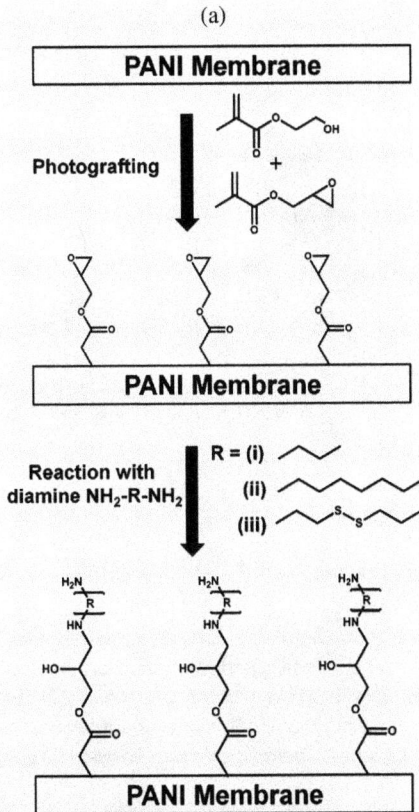

(a) Scheme of functionalization of the polyaniline membrane first photografted with glycidyl methacrylate and 2-hydroxyethyl methacrylate and then reacted with diamines. *Figure reproduced with permission from Elsevier [7.15].*

(Continued)

Box 7.5.1 (*Continued*)

(a) illustrates the method developed by Blinova and Svec [7.15], which uses a polyaniline (PANI) as the base material. The first step is to photograft glycidyl methacrylate and 2-hydroxyethyl methacrylate, and a subsequent step is to react these films with different types of diamines. (b) shows that because of this facilitated transport, these materials have a performance that is better than the Robeson upper limit.

(b)

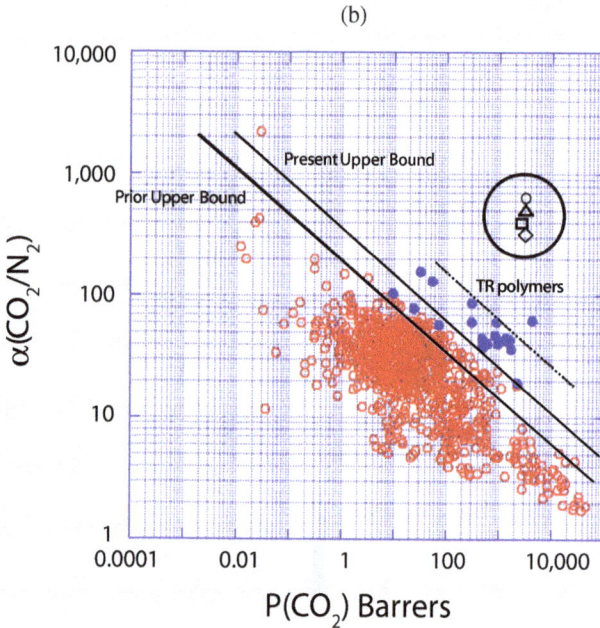

(b) Robeson's plot for separation of CO_2/CH_4 using membranes. Experimental points within the circle represent permeability and separation factors determined for polyaniline membranes photografted with glycidyl methacrylate and 2-hydroxyethyl methacrylate reacted with ethylene-diamine (diamond), cystamine (triangle), hexamethylenediamine (square). *Figure reproduced with permission from Elsevier* [7.15].

Section 6

Materials: nanoporous materials

We have seen that polymer membrane research is focused on developing strategies to beat the Robeson upper limit. In this section, we will see that for nanoporous materials the chemistry is fundamentally different.

Nanoporous membranes

Instead of starting with an overview of experimental data, we will look at a model membrane in order to develop some intuition about how to design a nanoporous membrane material. **Figure 7.6.1** shows a small part of a model membrane consisting of nanoporous cavities that are connected by narrow windows. The figure shows three cavities. The dimensions of these cavities and windows are of the same order as the sizes of the adsorbed molecules (0.5–2 nm). An ideal membrane will be a perfect unit crystal with a large surface area (parallel to the x,y plane) and thickness (in the z direction) of a few micrometers. In this crystal, the cavities form channels that run parallel to the surface (z-direction).

We assume that we can change the size of the cavities, the size of the window region, and the corresponding energies of interaction between the adsorbate and the walls. In this section, we show how we can tune the permeation of our model membrane. For example, if we change the diameter of the windows (L_{wy}) how will this change the adsorption and diffusion properties? Or, to change the interaction energies of our gas molecules with the walls of the material, do we need to change the interactions in the cavities (U_c) or in the windows (U_w)?

In previous sections, we have shown that the permeability of our material is the product of adsorption and diffusion. At low loadings the adsorption can be computed from the Henry coefficient. We have also shown that at these low loadings the self-, Maxwell-Stefan, and Fick diffusion coefficients are the same. We will assume that at flue gas conditions the loading is sufficiently low that we can obtain the number of adsorbed gas molecules from the Henry coefficient and that the

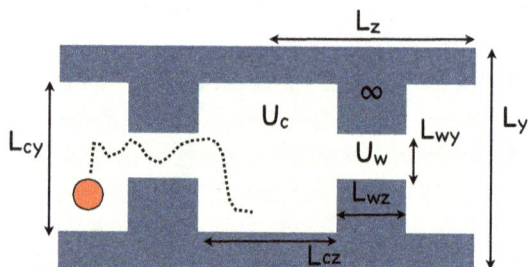

Figure 7.6.1   Microscopic model of a membrane

A small part of a simple microscopic model of a nanoporous membrane; the green area is accessible to the molecules. The membrane channels consists of cavities (of size L_{cy} × L_{cz}) that are separated by windows (of diameter L_{wy}). In the cavities the energy of the molecule is U_c and in the windows U_w. The dark blue shading indicates membrane molecules and we assume gas molecules do not occupy that space.

self-diffusion coefficient is a reasonable approximation of the Fick diffusion coefficient. At this point, we emphasize that at real flue gas conditions these assumptions may not hold true for all materials.

We introduced this model in order to analytically compute simple expressions for both the Henry coefficient and the self-diffusion coefficient for a simple (spherical) gas molecule. The formulae we derive allow us to compute the permeation and permeation selectivity. Hence, this model allows us to develop some intuition about what a Robeson plot would look like for our nanoporous materials. However, by now you must have developed a healthy skepticism about our intuition regarding membrane behavior!

Adsorption

We saw in the adsorption section that we can compute the Henry coefficient of adsorbed molecules by randomly inserting a molecule and computing its average energy:

$$H = \frac{1}{k_B T} \langle e^{-U/k_B T} \rangle_{random}$$

This formula is used in molecular simulations to compute the (excess) chemical potential µ, which is closely related to the Henry coefficient [7.16]. To apply this formula to our model membrane we have to know the energy of a molecule in a pore. We assume that our pore is structureless and that the energy is simply U_c if the molecule is in the cavity and U_w in the window region. The energy would be infinitely large if we were to place a molecule in a position occupied by the membrane molecules, and hence we only get a contribution from the cavity and window regions.

To compute the Henry coefficient for the model shown in **Figure 7.6.1**, we can envision inserting a gas particle at random in either the cavity or the window. As the contribution to the Henry coefficient is proportional to the volumes of these two regions, we have:

$$H = \frac{1}{k_B T}\left[\frac{V_c}{V}e^{-U_c/k_B T} + \frac{V_w}{V}e^{-U_w/k_B T} + \frac{V - V_c - V_w}{V}0\right],$$

where V is the volume of a unit cell ($L_y \times L_y \times L_z$), V_c is the volume of the cavity ($L_{cy} \times L_{cz}$), and V_w is the volume of the window region ($L_{wz} \times L_{wy}$).

From this formula, we can obtain some important insights. Let us assume that the energies in the cavity and in the window region are both zero: $U_c = U_w = 0$. Then our equation for the Henry coefficient reads:

$$H^{U=0} = \frac{1}{k_B T}\left[\frac{V_c + V_w}{V}\right]$$

At this point, you may wonder why is this interesting. From your thermodynamic course, you may recall that equilibrium is the state for which the free energy takes the minimum value. For a system in which we have a fixed volume, number of particles, and temperature, we have to look at the Helmholtz free energy, which is:

$$A = U - TS$$

Question 7.6.1 Henry coefficient and temperature

Is the statement that the Henry coefficient is dominated by the energy of the cavity valid at all temperatures? Assume that the energy in the cavity is −70 kJ/mol and on top of the barrier −30 kJ/mol. What if these energies are −35 kJ/mol and −30 kJ/mol? Hint: plot the contributions as a function of temperature.

If $U_c = U_w = 0$ the Helmholtz free energy has only entropic contributions, and hence the Henry coefficient has only entropic contributions. An adsorbed molecule loses entropy because the volume in which the particle can move $(V_c + V_w)$ is smaller than the corresponding volume of the gas phase (V). The density of adsorbed particles is given by:

$$\rho^{(U=0)} = \frac{P}{k_B T} \frac{V_c + V_w}{V} = \frac{V_c + V_w}{V} \rho^{gas}$$

If we want a higher density in our pores, we therefore need to compensate for this entropy effect with an energetic component: the interactions with the wall, $U_c < 0$.

To simplify our calculation we assume that the energy of a molecule in the cavity is lower compared to a molecule in a narrow window $(U_c < U_w)$, which gives for the Henry coefficient:

$$H = \frac{1}{k_B T} \left[\frac{V_c}{V} e^{-U_c / k_B T} + \frac{V_w}{V} e^{-U_w / k_B T} \right] \approx \frac{1}{k_B T} \left[\frac{V_c}{V} e^{-U_c / k_B T} \right]$$

Because of the exponent in the energy we see that we only need a small energy difference for the Henry coefficient to be dominated by the lowest energy, which we assume is the cavity (see also **Question 7.6.1**).

Diffusion

We can also make an estimate of the diffusion coefficient using these free energy calculations. The Henry coefficient is related to the (excess) chemical potential, which is for a pure component equal to the (Gibbs) free energy (F) per particle. The trick here is to use the same random insertion formula, but now to compute our free energy as a function of the position of the gas molecule in the channel of the membrane. This can be done by inserting a molecule at a random position and computing the test particle's energy as a function of the position in the z direction:

$$\exp\left[-F(z) / k_B T \right] = \left\langle e^{-U / k_B T} \delta(z' - z) \right\rangle_{random},$$

where the delta function expresses that only those particles that are randomly inserted at the position $z' = z$ (but can differ in x and y coordinates)

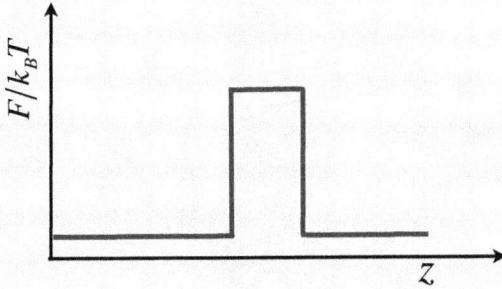

Figure 7.6.2 Free energy profile in the membrane

The free energy as a function of the position of a molecule in the pore for our model membrane, where the step represents a window between cavities.

contribute to the free energy F at position z. We can see that this free energy takes two values:

$$\exp\left[-F(z)/k_BT\right] = \begin{cases} \dfrac{L_{cy}}{L_y}\,e^{-U_c/k_BT} & z \in \text{cavity} \\[2ex] \dfrac{L_{wy}}{L_y}\,e^{-U_w/k_BT} & z \in \text{window} \end{cases}$$

In **Figure 7.6.2**, we plot this free energy as a function of the position for our model membrane. We see that the window forms a barrier to the diffusion of our adsorbed molecule. If this free energy barrier is sufficiently high we see that a molecule will be in a cavity most of the time and only once in a while hop from one cavity to another. This hopping is equivalent to a random walk on a lattice for which we have shown how to compute a diffusion coefficient in the previous section.

The important point is that if we have a situation in which hopping from one cage to another is a rare event, we can make a very good estimate of the hopping rate using the transition state theory [7.16]. In **Box 7.6.1**, we show that this theory gives the hopping rate:

$$k\left(\text{cavity}_1 \to \text{cavity}_2\right) = \frac{1}{2}|v_a| \frac{L_{wy}\exp\left[-U_w/k_BT\right]}{L_{cz}^2\exp\left[-U_c/k_BT\right]}$$

Figure 7.6.3   Changes in material properties of our model membrane

Changes in the pore structure: the left figure is our starting material, which is also represented with red dotted lines in the right figures. In our model, we can either change the cage size (top figures) or the size of the windows connecting the cages (bottom figures).

Box 7.6.1   Transition state theory

This theory assumes that the probability of finding a molecule at a given position follows from the equilibrium distribution. Statistical thermodynamics tells us that this distribution can be computed from the free energy:

$$P(z)dz = \frac{\exp\left[-F(z)/k_B T\right]dz}{\int_{\text{cavity}} \exp\left[-F(z)/k_B T\right]dz}$$

Hence, we can compute the probability that a particle reaches the top of the barrier, given this particle was in the cavity:

$$P(z^*)dz = \frac{\exp\left[-F(z^*)/k_B T\right]dz}{\int_{\text{cavity}} \exp\left[-F(z)/k_B T\right]dz}$$

where z^* is the transition state, which is the maximum of the free energy. To obtain the hopping rate we simply assume that our particle on top of the barrier is in equilibrium with its environment and hence the velocity of a molecule is what it would be in a bulk medium at the same temperature, i.e., it is described by the Maxwell velocity distribution. Moreover, with this distribution, 50% of the time the molecule has a velocity in the positive direction and 50% in the negative direction. Transition state theory simply assumes that if a molecule has a positive velocity it

(Continued)

Box 7.6.1 (*Continued*)

will reach another cavity and if it has a negative velocity it will fall back to the cavity from which it came. Combining the probability to be on top of the barrier with the positive average velocity, $|v_a|$, of the Maxwell distribution gives us a hopping rate:

$$k(\text{cavity}_1 \rightarrow \text{cavity}_2) = \frac{1}{2}\left(\frac{8k_BT}{\pi m}\right)^{1/2} \frac{\exp\left[-F(z^*)/k_BT\right]}{\int_{\text{cavity}} \exp\left[-F(z)/k_BT\right]dz}$$

Combining all these approximations, we get the hopping rate for our model membrane:

$$k(\text{cavity}_1 \rightarrow \text{cavity}_2) = \frac{1}{2}|v_a|\frac{L_{wy}\,\exp\left[-U_w/k_BT\right]}{L_{cz}^2\,\exp\left[-U_C/k_BT\right]}$$

From the hopping rate we can get the self-diffusion coefficient, using our random walk model:

$$D^S = \frac{1}{4}|v_a|\frac{L_{wy}\,\exp\left[-U_w/k_BT\right]}{L_{cz}^2\,\exp\left[-U_C/k_BT\right]}L_z^2$$

From this hopping rate, we can get the self-diffusion coefficient using our random walk model, derived in Section 7.4:

$$D^S = \frac{1}{4}|v_a|\frac{L_{wy}\,\exp\left[-U_w/k_BT\right]}{L_{cz}^2\,\exp\left[-U_C/k_BT\right]}L_z^2$$

Permeability

To get the permeance of our membrane we combine the resulting expression for the diffusion coefficient with the expression for the Henry coefficient:

$$H = \frac{1}{k_BT}\left[\frac{V_C}{V}e^{-U_C/k_BT} + \frac{V_w}{V}e^{-U_w/k_BT}\right] \approx \frac{L_{cz}^2}{L_yL_zk_BT}e^{-U_c/k_BT}$$

This gives for the permeance of our membrane:

$$P' = D^SH \approx \left[\frac{1}{4}|v_a|\frac{L_{wy}\,\exp\left[-U_w/k_BT\right]}{L_{cz}^2\,\exp\left[-U_C/k_BT\right]}L_z^2\right]\left[\frac{L_{cz}^2}{L_yL_zk_BT}e^{\left[-U_C/k_BT\right]}\right]$$

$$= \frac{1}{4} \frac{|v_a|}{k_B T} \frac{L_{wz} L_z}{L_y} \exp\left[-U_w / k_B T\right]$$

In this calculation, we have assumed that the concentration in the pores is sufficiently low that in our expression for the permeance the transport diffusion coefficient can be approximated with the self-diffusion coefficient.

Let us now use this result to see how changes in the pore geometry of our material will influence the permeation. **Figure 7.6.3** illustrates some of the changes we can make in the geometry. We can change the size of our cavity or the diameter of the window, and we assume that these changes do not influence the energies. Volume changes influence the entropy of our molecules. For example, if we increase the volume of the cavity, the entropy of our adsorbed molecules increases and hence the Henry coefficient increases. For the diffusion coefficient, an increase in the entropy of a molecule in the cavity corresponds to a decrease in the free energy. This decrease, however, increases the barrier to diffusion and we see a decrease in the diffusion coefficient. The opposite will happen if we decrease the volume of the cavity; the Henry coefficient will decrease and the diffusion coefficient will increase.

Let us now focus on changing the geometry of the window. As the window region contributes little to the Henry coefficient, changing the window volume will predominantly affect the diffusion coefficient. Making the window diameter smaller will increase the free energy and hence we will see a decrease of the diffusion coefficient. Similarly, increasing the window diameter will increase the diffusion coefficient.

The results are summarized in **Figure 7.6.4**. We see that only changes in the window diameter will influence the permeation, whereas for our model changes in the cavity volume will cause a change in the Henry coefficient that is compensated by a change in the diffusion coefficient in an opposite direction. We can use these results to compute the permeability of our materials.

At this point, it is important to note that we have created a very simple model based on many assumptions. For example, it is difficult to envision how we would change the volume of a cavity without changing the energy in the cavity at the same time. Therefore, in our model we see an exact compensation, which is unlikely in real materials. Despite these limitations, our model does illustrate why some changes in the material have significant effects on the diffusion coefficients and Henry coefficients, but not on the permeability of the material.

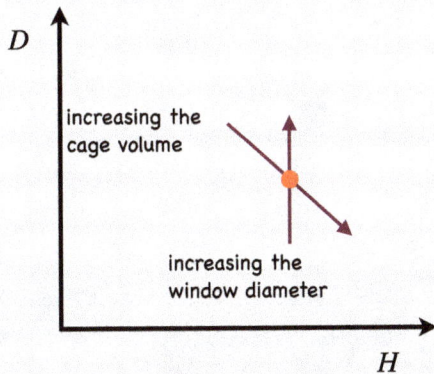

Figure 7.6.4 Effects of changes in the pore structure on the transport and thermodynamic properties

In our model membrane, we can change the volume of our cavity and the diameter of the window separating the cavities. Changing the window diameter has little effect on the adsorption but a large effect on the diffusion coefficient. Changing the cavity changes both the diffusion and the Henry coefficients.

Let us now use this model to study the effect of changes in the chemistry. In our model, the chemistry is manifested in the energy terms U_c and U_w. Let us assume that we can modify the interactions of our adsorbed molecules with the wall of our pores (see **Figure 7.6.5**). These changes will affect the permeation of the material.

For example, suppose we modify the chemistry such that the pores are more attractive for our molecules. This would correspond to making U_c more negative, which in turn would increase the Henry coefficient. However, if the energy in the cavity is lower, the free energy barrier for hopping from one cage to another increases. Hence, the diffusion coefficient will decrease and, as the formula for the permeation shows, these two effects will cancel. The net result is that *the permeation will not change significantly if we modify the chemistry of the cages*.

We can also try to modify the chemistry of the window. As the probability of finding a molecule on top of a free energy barrier is small, the expression for the Henry coefficient shows that these window regions do not contribute significantly to the Henry coefficient. So changes in the chemistry of the window region will not change the Henry coefficient. For the diffusion coefficient, on the other hand, changes to the barrier are

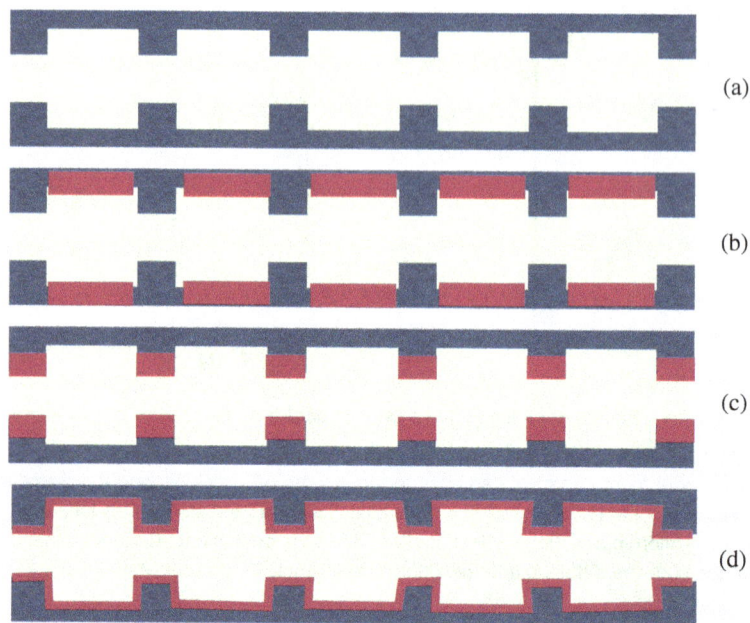

Figure 7.6.5 Changing the chemistry of our model membrane

Chemical modifications of the membrane: (a) the reference material, (b) changes in the chemistry in the cavities, (c) changes to the pore window, and (d) changes to both the cavities and the windows. The red lines indicate that the interactions of an adsorbed molecule have locally changed.

important; increasing the barrier interaction energy will decrease the diffusion coefficient. So here we see that we can tune the permeation of our material by controlling the diffusion coefficient.

In practical situations, we would most likely change the chemistry of both the cavity and the window region at the same time. If this chemistry could modify the energy by, say, adding a constant term to both the window energy, $U_w \rightarrow U_w + \Delta U$, and the cavity energy, $U_c \rightarrow U_c + \Delta U$, we see that this would not influence the diffusion coefficient but would change the Henry coefficient. That is, by changing both the interactions in the window region and the cavity, we can tune the Henry coefficient and hence the permeation (see **Figure 7.6.6**). Of course, this would require a very delicate control of the chemistry. It is exactly this type of molecular control which is the topic of modern synthetic chemistry applied to membrane separations.

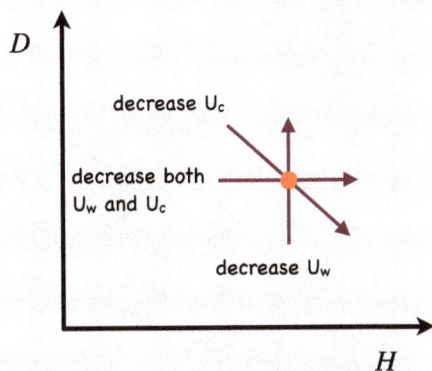

Figure 7.6.6 Effects of changes in the chemistry on the transport and thermodynamic properties of a membrane

Effects of changes in the chemistry on the diffusion coefficient and Henry coefficient. A decrease of U_c corresponds to changing the interactions in the cavities **(Figure 7.6.5b)** and a decrease of U_w corresponds to changing the interactions in the window region **(Figure 7.6.5c)**.

At this point, it is important to note that we did not make any assumptions about the types of molecules that are adsorbed. If we change the interactions with the walls these changes may have different effects on, say, CO_2 compared to N_2. For most materials, CO_2 will have a higher permeability, and our reference material will have a selectivity that is larger than one. We assume that we have such control of the chemistry that our changes will mostly affect the interactions with CO_2 and only minimally affect the interactions with N_2. We have seen that to change the permeation in our model membrane, we can modify the interactions for the entire material or only for the window region.

If we make the barrier more positive, the diffusion coefficient will decrease. This will also decrease the selectivity and CO_2 permeation. Decreasing the energy of the barrier gives an increase of our diffusion coefficient and hence the permeation. If we change the chemistry of both the cavity and the window, we can increase or decrease the Henry coefficient. Both the selectivity and the CO_2 permeation will increase if we increase the Henry coefficient and similarly both will decrease if we decrease the Henry coefficient. In our model we can change the size of the cavities or the dimensions of the window. This would give us a new reference material for

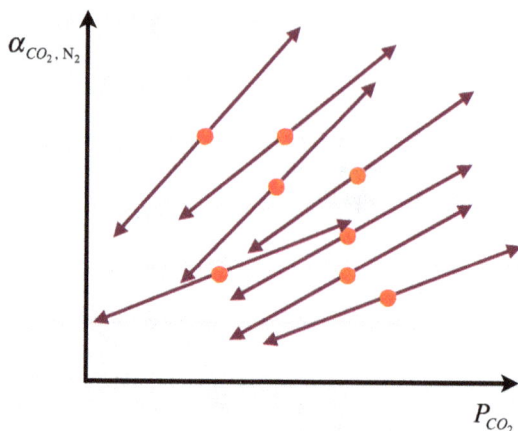

Figure 7.6.7 Selectivity versus permeation for a porous material

The effect of changes in the chemistry on the selectivity α and permeation. Each dot represents a model membrane with a different pore structure (larger cavities, smaller windows, etc.). The arrows indicate how the performance of the materials would change if we change the interactions with the walls.

which we could make the same chemical modifications. The results of this simple analysis are shown in **Figure 7.6.7**. Our model predicts a very different Robeson plot than those found for polymers in the literature (see, for example **Figure 7.5.2**). Our simple analysis suggests that nanoporous materials do not exhibit the Robeson upper bound.

"Experimental" nanoporous membranes

In the previous section, we developed a simple theory that predicts that nanoporous materials will not exhibit a Robeson upper bound. Robeson's observation of that upper bound was based on the analysis of a large number of different polymer membranes. The synthesis of membranes comprised of nanoporous materials has only been developed very recently. For example, Tsapatsis [7.17] and his co-workers have been able to synthesize membranes using zeolites. How the pore topology of these zeolites influences the performance of the membrane is an interesting question. As the number of structures is so small, we have to rely on molecular simulations to predict the permeability and selectivity. In Chapter 6, we showed

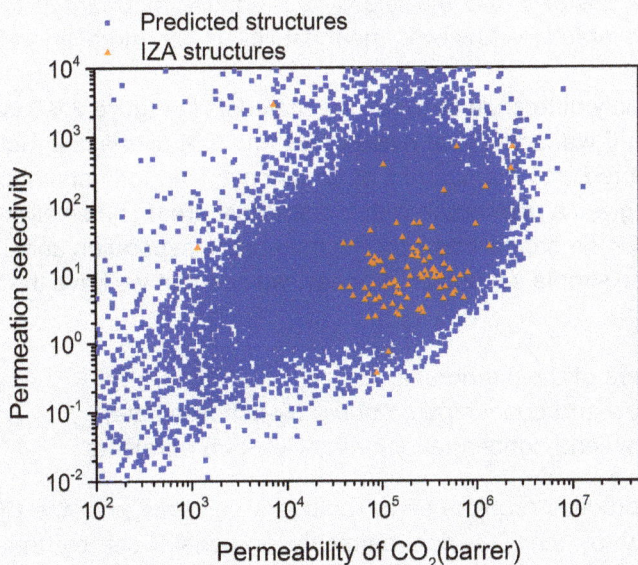

Figure 7.6.8 Robeson plot for zeolite structures

Robeson plot showing the CO_2/N_2 permeation selectivity as a function of the CO_2 permeation for over 80,000 predicted zeolite structures (blue) and known zeolite structures (red triangles). *Figure reproduced with permission from Kim et al.* [7.18].

how molecular simulations for some of the same materials can be used to find the optimal material for adsorption. We can use the same set of materials and predict the diffusion coefficient and Henry coefficient. **Figure 7.6.8** gives the corresponding Robeson plot for over 80,000 different zeolite structures. Comparison with the Robeson plot for polymer materials (see e.g., **Figure 7.5.2**) shows that nanoporous materials indeed behave differently; we do not observe the characteristic upper bound that is found in polymer materials. The simulation results are surprisingly similar to what we have predicted with our simple model (see **Figure 7.6.7**).

Selecting the best membrane

The Robeson plots are normally used to rank different materials, the ideal material having a high selectivity and high permeability. This approach suggests, however, that the selectivity is equally important as the permeability. In practice, however, permeability is far more important than selectivity.

The only difference between the materials in **Figure 7.6.8** is the pore structure. If we, somewhat naively, assume that all these structures can be converted into membranes at equal cost, the optimal material is the one that gives a membrane with the smallest area. The smaller the area, the smaller the capital costs for the membrane separation unit. If we now look at the simple separation process; we see that we have the following variables:

- The area of the membrane;
- The flow and concentration of the N_2-rich retentate;
- The flow and concentration of the CO_2-rich permeate.

The process requires us to remove a percentage of the CO_2. Also, the purity of the CO_2-rich permeate is usually set by the process parameters.

To illustrate how one can find the optimal material, we assume that we remove a given fraction of the CO_2 with a given purity. Let us consider a simple membrane separation of CO_2 and N_2 as was shown in **Figure 7.2.1**. Assume that our material has a given CO_2 permeation and permeation selectivity α_{CO_2,N_2}. In addition, the concentration of CO_2 in the feed as well as the total flue gas per second that needs to be separated is known. If we assume that we use the design given in **Figure 7.3.7**, in which air is used as a sweep gas and this CO_2-enriched sweep gas is subsequently used in the boiler, we have a flue gas with a higher CO_2 concentration (~25%) than is typical for a coal-fired power plant. Recall that in the optimized process the use of a second membrane ensures that 90% of the CO_2 will be removed from the flue gas and that the purity required for sequestration is reached.

In the design, we are looking for the zeolite structure that will give us the smallest membrane area. In Section 7.3, we saw that if we assume that all membranes have the same thickness, for a given permeation selectivity the area follows from the mass balances for CO_2 and N_2. The best material in this analysis is the one with the highest permeation for which the permeation selectivity is sufficiently high that it meets the required purity.

The status of CO_2 capture

The capture of CO_2 from gas streams is an area of active research. In Chapter 1, we introduced the vast scale on which capture technologies must operate in order to ameliorate the climate-changing effects of CO_2 in the atmosphere. This scale considerably constrains the engineering and materials design for CCS.

The engineering design of *absorption processes* proceeds along lines familiar to students of chemical engineering. Carbon dioxide is absorbed into a working fluid in a tower that contains plates or packing that bring the offending gas into thermodynamic equilibrium with the working fluid that captures the CO_2. The fluid, now laden with CO_2, is further processed in another tower (plate or packed bed) whereby the fluid is stripped of CO_2, yielding both a CO_2-rich gas stream ready for compression and delivery to a pipeline and a regenerated working fluid that is recycled back to the absorber. The analysis of absorber-stripper towers is made complicated by the rates of mass transfer within and between the gas and fluid streams, as well as by complex chemical reactions (and/or equilibria) that occur when CO_2 reacts with the working fluid. Finally, we see considerable opportunity for new research in the chemical design of fluids that capture CO_2 effectively, yet are compatible and sustainable for use in absorbers and strippers. Absorption is a mature industry, and thus this strategy for CCS has the advantage of a pre-existing industry (largely from natural gas treatment).

The *adsorption of CO_2 onto solid surfaces* is an emerging area of active research. This subject has already been enabled by other process industries, though the scale of carbon capture is daunting. Chemists and engineers have been using the uptake of specified gasses onto surfaces for decades, and the capture and regeneration of such gasses in these processes, either by systematically varying temperature or pressure ("temperature swing" or "pressure swing," respectively), are well known. The critical concept is that of the isotherm — a measure or calculation of the amount of gas on the surface of the working solid (adsorbent) as a function of the pressure of the gas over the solid. Engineers have used such isotherms to analyze the adsorption profiles, in both space and time, of CO_2 on the solids as a function of time on-stream, thereby calculating process conditions and operations. With such analysis and design schemes at hand, it can be seen that, in principle, adsorption can be cost competitive with, or even superior to, absorption. Adsorption also has the benefit of being more sustainable in the sense that water use and environmental impact (via emissions) are considerably less than for absorption. For adsorption to become a dominant player in CCS, however, considerable research must be done to imagine, synthesize, and deploy crafted nanoporous materials for CCS.

Membrane separation is also a mature chemical technology, most often associated with water desalination. The engineering design of membrane separation processes has similarities to absorption processes in the sense that the device is operated in a countercurrent arrangement so as to maximize the driving forces for transfer of CO_2 from the feed stream to the exhaust stream. A detailed analysis of these operations, though, reveals that both the absorption of CO_2 into the membrane material, as well as the diffusion of CO_2 through the membrane, must be crafted in order to address the scale of CCS. There is considerable research activity throughout the world in the design of new types of membranes that overcome limitations on diffusion solubility mechanisms. These new designs employ a sophisticated molecular-level understanding of the kinetics and thermodynamics of small molecules in solid materials. No less exciting are the emerging process schemes that greatly minimize the compression and/or vacuum requirements of membranes for CO_2 separations. The synergism of the molecular view of membrane absorption and diffusion with the (macroscopic) process analysis is thrilling.

The deployment of carbon capture awaits the political will to make it happen. In the meantime, power plants and their operating companies have several different technologies that may be employed for CCS, and the ultimate choice for which technology is used will likely depend on local constraints or opportunities. In the meantime, fundamental, process, and demonstration scale research is needed to make these technologies the most cost effective and sustainable.

Section 7

Review

7.1. Reading self-test

1. To separate viruses, one would need a membrane with pores of a diameter (in microns):
   a. 100–1,000
   b. 10–100
   c. 1–10
   d. 0.1–1
   e. 0.01–0.1

2. Which statement is not correct for a one stage membrane separation in which the permeation of one component (A) is fixed and the other (B) can be tuned independently?
   a. The higher the selectivity, the larger the membrane
   b. The maximum achievable concentration of component A in the permeate is independent of the quality of the membrane
   c. The optimal membrane selectivity is a trade-off between purity and the required pressure

3. Simple membrane separation — assign the correct name corresponding to the letters A–C in the blue arrows.

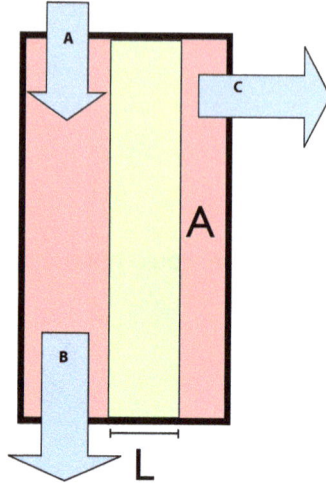

4. Match the correct energy for the separation to the corresponding letters:

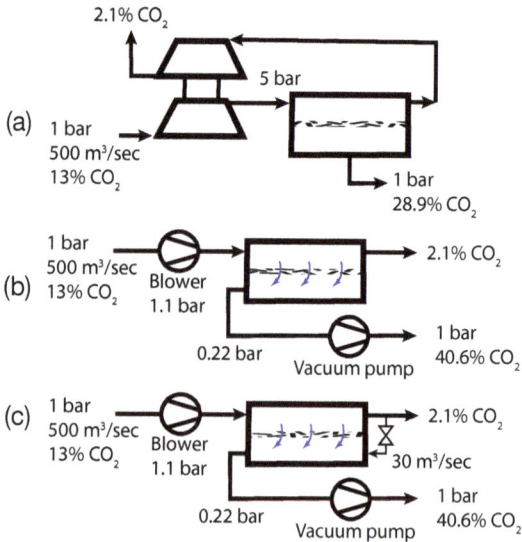

(a) 1 bar
500 m³/sec
13% CO₂

2.1% CO₂
5 bar
1 bar
28.9% CO₂

(b) 1 bar
500 m³/sec
13% CO₂
Blower
1.1 bar

2.1% CO₂

0.22 bar
Vacuum pump
1 bar
40.6% CO₂

(c) 1 bar
500 m³/sec
13% CO₂
Blower
1.1 bar

2.1% CO₂
30 m³/sec

0.22 bar
Vacuum pump
1 bar
40.6% CO₂

____ 114.7 MW
____ 46.4 MW
____ 46.0 MW

5. Which statement about flue gas separations using membranes is not correct?
   a. For a single stage membrane, the pressure ratio is so large that too much energy is required for compression
   b. The partial pressure of CO_2 in air is so low that it is an ideal sweep gas for a membrane
   c. A higher concentration of CO_2 in the air for the burner requires retrofitting of the power plant

6. Which statement on diffusion coefficients is not correct?
   a. The Maxwell-Stefan diffusion coefficient uses the chemical potential as a driving force
   b. The Fick diffusion coefficient uses the concentration as a driving force
   c. In the limit of zero loading in a porous material, the self-diffusion coefficient is identical to the Fick diffusion coefficient
   d. NMR can be used to measure the Fick diffusion coefficient

7. Which statement about polymer membranes is not correct?
   a. The solubility of gasses in membranes can vary significantly
   b. The higher the permeability of a gas, the lower the selectivity
   c. Rubbery polymers have lower selectivity compared to glassy polymers

8. Which statement on facilitated transport in membranes is NOT correct?
   a. Facilitated transport of oxygen in blood uses hemoglobin as a carrier
   b. Amine solutions show strong facilitated transport
   c. Ionic liquids with facilitated transport are of limited use because the liquid evaporates

9. Which pore has the highest diffusion coefficient (given that the interactions with the materials are the same)?

a.

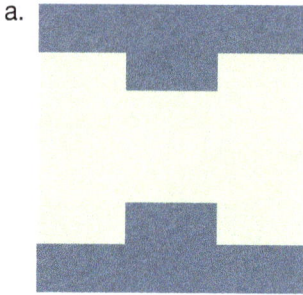

b.

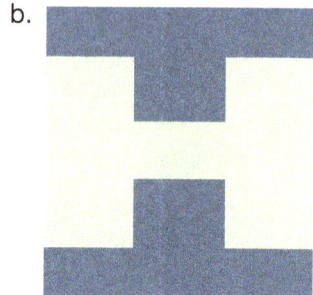

c.

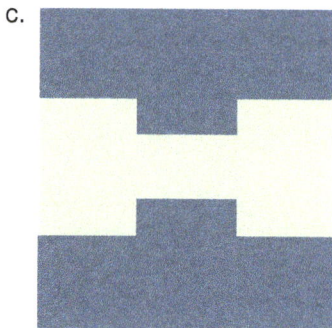

d.

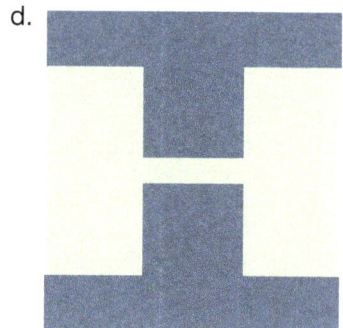

10. Which statement about nanoporous membranes is not correct?
   a. By changing the chemistry of the pores, one can tune the solubility and diffusion coefficient independently
   b. The solubility of gasses in these materials can vary significantly
   c. The diffusion coefficients can be changed by tuning the pore diameter

Section 8

References

1. Merkel, T.C., H.Q. Lin, X.T. Wei, and R. Baker, 2010. Power plant post-combustion carbon dioxide capture: An opportunity for membranes. J. Membrane Sci., **359** (1–2), 126. http://dx.doi.org/10.1016/J.Memsci.2009.10.041
2. Koros, W.J. and R.P. Lively, 2012. Water and beyond: Expanding the spectrum of large-scale energy efficient separation processes. Aiche J., **58** (9), 2624. http://dx.doi.org/10.1002/aic.13888
3. Husain, S. and W.J. Koros, 2007. Mixed matrix hollow fiber membranes made with modified HSSZ-13 zeolite in polyetherimide polymer matrix for gas separation. J. Membrane Sci., **288** (1–2), 195. http://dx.doi.org/10.1016/J.Memsci.2006.11.016
4. Perez, E.V., K.J. Balkus, J.P. Ferraris, and I.H. Musselman, 2009. Mixed-matrix membranes containing MOF-5 for gas separations. J. Membrane Sci., **328** (1–2), 165. http://dx.doi.org/10.1016/J.Memsci.2008.12.006
5. Jobic, H., J. Karger, and M. Bee, 1999. Simultaneous measurement of self- and transport diffusivities in zeolites. Phys. Rev. Lett., **82** (21), 4260.
6. Baerlocher, C., W.M. Meier, and D.H. Olson, 2001. Atlas of Zeolite Framework Types. Elsevier: Amsterdam.
7. Reyes, S.C., J.H. Sinfelt, and G.J. DeMartin, 2000. Diffusion in porous solids: The parallel contribution of gas and surface diffusion processes in pores extending from the mesoporous region into the microporous region. J. Phys. Chem. B, **104** (24), 5750.
8. Robeson, L.M., 1991. Correlation of separation factor versus permeability for polymeric membranes. J. Membrane Sci., **62** (2), 165.
9. Robeson, L.M., 2008. The upper bound revisited. 2008. J. Membr. Sci., **320** (1–2), 390. http://dx.doi.org/10.1016/j.memsci.2008.04.030
10. Scholes, C.A., S.E. Kentish, and G.W. Stevens, 2008. Carbon dioxide separation through polymeric membrane systems for flue gas applications. Rec. Patents Chem. Eng., **1** (1), 52.
11. Freeman, B.D., 1999. Basis of permeability/selectivity tradeoff relations in polymeric gas separation membranes. Macromolecules, **32** (2), 375. http://dx.doi.org/10.1021/ ma9814548
12. Scholander, P.F., 1960. Oxygen transport through hemoglobin solutions. Science, **131** (3400), 585. http://dx.doi.org/10.1126/science.131.3400.585
13. Kim, T.J., B.A. Li, and M.B. Hagg, 2004. Novel fixed-site-carrier polyvinylamine membrane for carbon dioxide capture. J. Polym. Sci. Pol. Phys., **42** (23), 4326. http://dx.doi.org/10.1002/ Polb.20282

14. Yamaguchi, T., L.M. Boetje, C.A. Koval, R.D. Noble, and C.N. Bowman, 1995. Transport-properties of carbon-dioxide through amine functionalized carrier membranes. Ind. Eng. Chem. Res., **34** (11), 4071. http://dx.doi.org/10.1021/Ie00038a049

15. Blinova, N.V. and F. Svec, 2012. Functionalized polyaniline-based composite membranes with vastly improved performance for separation of carbon dioxide from methane. J. Membrane Sci., **423**, 514. http://dx.doi.org/10.1016/J.Memsci.2012.09.003

16. Frenkel, D. and B. Smit, 2002. Understanding Molecular Simulations: from Algorithms to Applications, 2nd ed. Academic Press: San Diego.

17. Tsapatsis, M., 2011. Toward high-throughput zeolite membranes. Science, **334** (6057), 767. http://dx.doi.org/10.1126/science.1205957

18. Kim, J., M. Abouelnasr, L.-C. Lin, and B. Smit, 2013. Large-scale screening of zeolite structures for CO_2 membrane separations. J. Am. Chem. Soc., **135** (20), 7545–7552.

Chapter 8

Introduction to Geological Sequestration

Once we have separated the CO_2 from the flue gas, we need to ensure that the CO_2 is permanently stored. In this and the following two chapters, we discuss geological sequestration, which is at present the most promising option to ensure that the captured CO_2 is not released into the atmosphere.

Introduction

We have discussed ways to capture CO_2 from flue gasses using absorption, adsorption, and membrane processes. In this chapter and the following two chapters, we address the question of what to do with the CO_2 we have captured. Let us start by recalling two questions we raised in the introduction. It is not by accident that these are the questions most frequently posed to anyone working in carbon capture and sequestration research:

- Why don't we put the flue gas directly into selected geological formations? Capturing carbon is very expensive! Could the cost of carbon capture be avoided by injecting all flue gas directly into the subsurface?
- Is it safe to store CO_2 in geological formations? Will giant bubbles of CO_2 not become a threat to our environment?

Transporting CO_2 over large distances and injecting it into geological formations is something we know how to do. We have carried out these processes for around 40 years for the purpose of enhancing oil recovery through injection of CO_2 into oil fields with declining primary productivity. This process is known as CO_2-**enhanced oil recovery** or CO_2-EOR. While the quantities of CO_2 transported and injected are considerable for the individual oil fields involved, the amounts are small when compared to the scale one would need to significantly affect global CO_2 emissions. So, we will need to find alternative geological formations once we have saturated all current oil and gas fields that would benefit from CO_2 injections.

The scale of CCS

Let us start by recalling some results from the first chapters. In CCS, we are dealing with very large numbers. For example, our medium size 500 MW coal-fired power plant emits about 400 m^3/s of flue gas containing about 12% of CO_2 (volume). This gives us each year about 2.6 million tonnes of CO_2. If we assume a lifetime of 50 years, we have to sequester 140 million tonnes of CO_2 coming from 6.3×10^{11} m^3 of flue gas at each power plant.

The first idea would be to eliminate the capture process and inject the flue gasses directly. Let us assume that just below the power plant we have an ideal geological formation in which we can inject our flue gas. We will assume that this ideal geological formation has the shape of a disk with a thickness of 10 m. The question now is: what will be the radius of our disk filled with CO_2 after 50 years of injection?

Before answering this question we have to know a bit more about the geological conditions. To compute the density of our gas, we need to know the temperature and the pressure in the target formation at our injection site. The deeper our injection zone, the higher the temperature. The geothermal gradient can range from 15–30°C/km depending on the geothermal activity of the region. Pressure also increases with depth according to the hydrostatic pressure gradient (about 0.1 bar/m). If we assume that we inject the flue gas at a 2 km depth in a region of small thermal activity, and we know that at ambient conditions the density of air is 1.3 kg/m^3, then the average density of air in the storage formation (200 bar, 60°C) is 200 kg/m^3. We see that at storage conditions, the volume of our flue gas decreases to 4 x 10^9 m^3.

We now have to know a little more about the geological formations. The formations in which we would inject are very similar to those that store natural gas. In the next chapter, we will discuss in more detail what these formations look like, but for our calculation the most important factor is the porosity: if we inject our flue gas, what percentage of the rock volume will be occupied by the gas? We assume optimistically that the porosity is about 40% and that our injection accesses 50% of the available pore volume. This gives us a capacity of 20% of the total volume of our formation. Hence, we need for our total 50 years of flue gas production a geological formation with a total volume of 2 x 10^{10} m^3. If we assume that we inject into a disk of 10 m thickness, this gives us a radius of 25 km.

Let us now compare this number with the situation in which we capture the CO_2. Because we concentrate the CO_2 and thus decrease the nitrogen fraction in our capture stream, we decrease the volume by a factor of 7.7. In addition, the density of CO_2 at flue gas conditions is 1.7 kg/m^3, and 2 km below the surface it is 725 kg/m^3 (200 bar, 60°C, see **Figure 8.1.1**). Unlike air, CO_2 is supercritical at these conditions and hence has a much larger density. The effective volume of the geological formation that we need for sequestering the captured CO_2 decreases by an additional factor of nearly 4, to 7 x 10^8 m^3, which would give us a radius of 5 km.

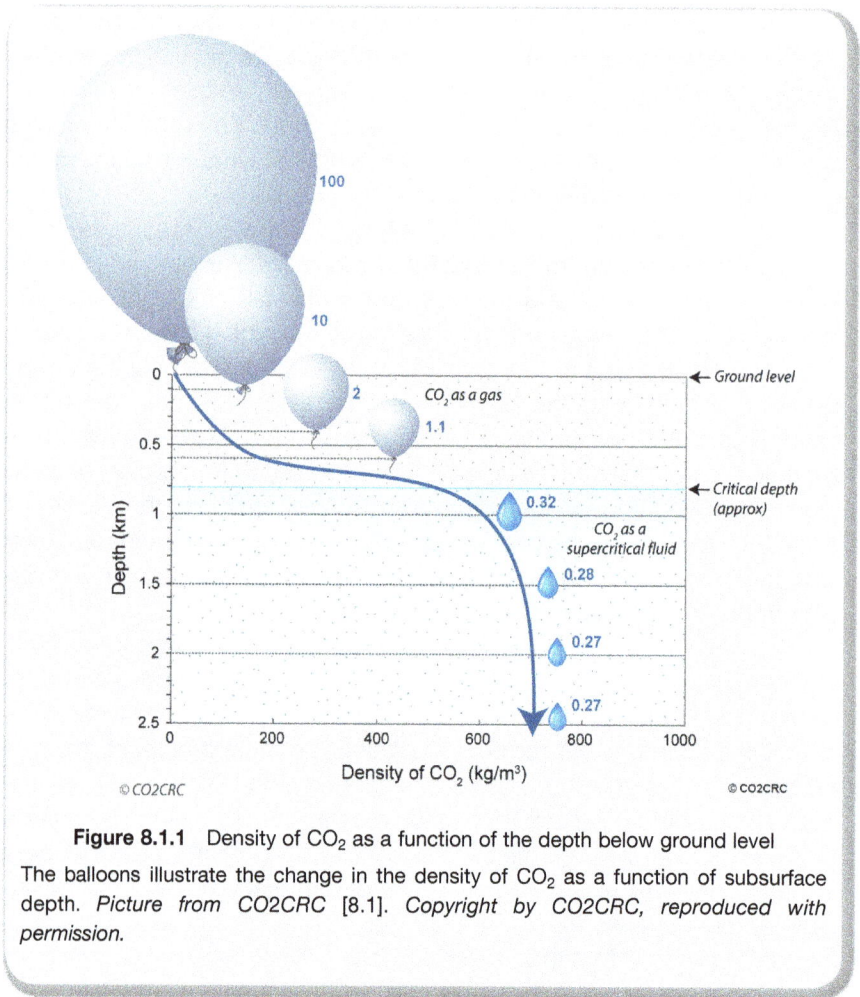

Figure 8.1.1   Density of CO_2 as a function of the depth below ground level

The balloons illustrate the change in the density of CO_2 as a function of subsurface depth. *Picture from CO2CRC [8.1]. Copyright by CO2CRC, reproduced with permission.*

In **Figure 8.1.2**, the difference between sequestering all the flue gas or just CO_2 is illustrated. This simple calculation illustrates why it is important to capture the CO_2; not only does it reduce the total amount of gas that needs to be sequestered but, unlike flue gas, nearly pure CO_2 is supercritical at sequestration depths and therefore has a much higher density compared to flue gas at the same conditions. This effect of depth on the density of CO_2 is illustrated in **Figure 8.1.1**; indeed a significant gain in density can be obtained if we sequester CO_2 below 1.5 km depths.

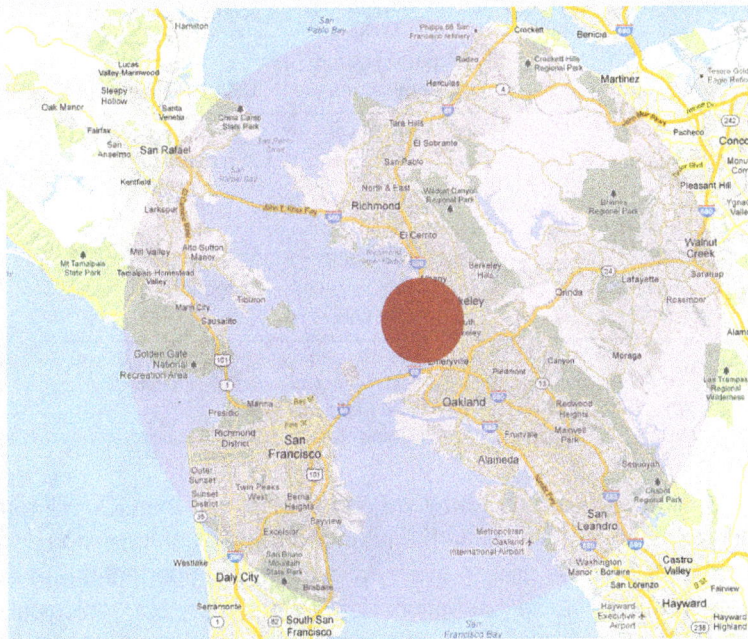

Figure 8.1.2 Storage of flue gasses with and without carbon capture

Geological formation necessary for sequestering flue gas from a medium sized coal-fired power plant located in Berkeley and operating for 50 years; with carbon capture (red circle with a 5 km radius) and without separating CO_2 (pink circle with a 25 km radius). *Map data from Google.*

The above discussion was for a single medium-size coal-fired power plant. If we envision applying CCS to all coal-fired power plants, we get very large numbers. These numbers are so large that some people are quite skeptical about the feasibility of CCS. It is therefore important to put these numbers into perspective. In the production of oil, water is currently used to replace the oil that has been pumped out of the ground. In **Table 8.1.1**, the total volume of water injected as part of oil production is compared to the total volume of CO_2 point source emissions for the USA. Perhaps surprisingly, the total volume of CO_2 we need to sequester is only slightly larger than the amount of water we are already injecting as part of our oil and gas production. However, nobody has injected such large volumes of CO_2 into geological formations and all our science is

Table 8.1.1 Sequestration of CO_2 versus H_2O

	USA point source CO_2 emissions	USA H_2O injection
Mass	2.4 Gt CO_2/year	3 Gt H_2O/year
Volume	3.4 Gm^3 CO_2/year	3.0 Gm^3 H_2O/year
Volume ratio relative to CO_2	1.0	0.9
Significance	displaces existing fluid, buoyant relative to existing fluid	replacement of produced oil and water

Comparison of the CO_2 that can be captured from point sources (**left column**) with the amount of water that is injected in geological formations in the context of oil and gas production (**right column**).

focused on the question of how we can ensure that the CO_2 will remain sequestered indefinitely within the intended geological formation.

In this chapter, we give an overview of the most important issues associated with geological sequestration: how we can carry out geological carbon sequestration efficiently, how we can mitigate related impacts, and how we can ensure that the injected CO_2 will remain permanently sequestered, irrespective of whether the CO_2 molecules like it or not (see **Movie 8.1.1**).

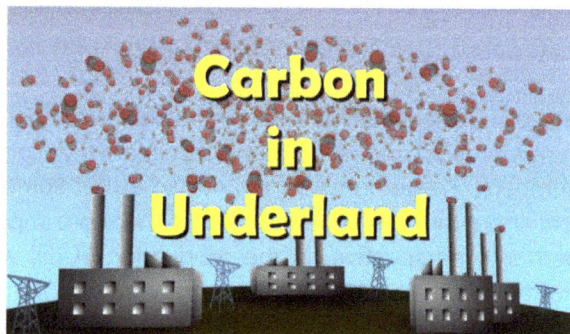

Movie 8.1.1 Carbon in Underland

Movie produced by Sergi Molins and Jennifer Cappuccio for the Center for Nanoscale Control of Geologic CO_2 (NCGC) at the Lawrence Berkeley National Laboratory (LBNL). It can be viewed at: http://www.worldscientific.com/worldscibooks/10.1142/p911#t=suppl

Section 2

Trapping mechanisms

The aim of injecting CO_2 into geological formations is to keep it from entering the atmosphere. We inject the CO_2 as a supercritical fluid. The density of this fluid is lower than the density of most of the fluids that exist in these formations, and elementary laws of physics tell us that **buoyancy** effects will tend to drive the CO_2 plume upward. We therefore need to select a geological formation capable of trapping the CO_2, and we need to examine the likelihood of escape of the injected CO_2. There is much experimental evidence that we indeed can trap gasses in geological formations. For example, we have confidence in the feasibility of trapping CO_2 in depleted natural gas reservoirs, because natural gas was trapped in these geological formations for millions of years by some of the very same mechanisms that are known to trap CO_2. In addition, there are places on earth where CO_2 is produced by decarbonation of carbonate rocks (e.g., limestone and dolostone) and becomes trapped in geological formations in the form of natural CO_2 reservoirs (e.g., Bravo Dome, NM; Jackson Dome, MS; and Sheep Mountain, CO). These CO_2 sources are used for most of the CO_2-enhanced oil recovery that is carried out in the USA.

Let us visualize what happens when we inject CO_2 into a geological formation. **Figure 8.2.1** shows the plume of injected CO_2. As the injection proceeds, the brine that occupied the pore space before the injection will be displaced and initially we will see a large plume of CO_2 around the injection site. Because the density of supercritical CO_2 is lower than that of the surrounding fluids, buoyancy effects will cause the plume to move upward. As long as the CO_2 remains in this large plume, we rely on the geological formation above the injection layer, a so-called **caprock** or top seal, to prevent the CO_2 from escaping the storage formation. Caprock is a fine-textured rock that has a very limited permeability. Hence, during the initial phase we rely on **structural trapping** or **stratigraphic trapping** of the CO_2 by this caprock. If, because of buoyancy effects, the plume moves through the storage formation, some of the CO_2 will become immobilized as small disconnected bubbles in the pores of the formation (see **Movie 8.2.1**).

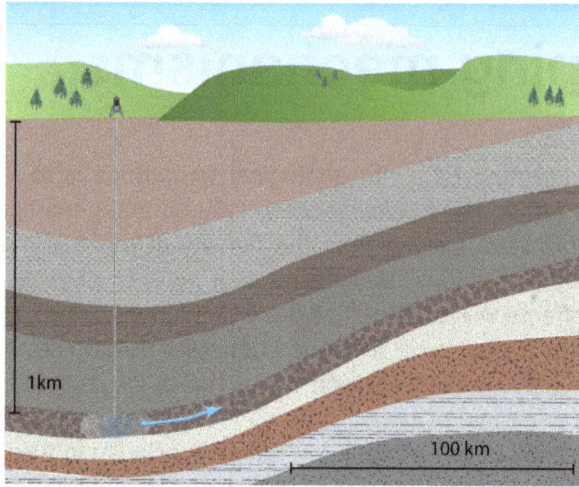

Figure 8.2.1 Injecting CO_2 below a caprock formation

Flow of the CO_2 plume; the brown layer above the injection site is a caprock which is impermeable for CO_2. Because of the lower density of the CO_2, the plume will slowly migrate to the highest point in the formation. *Figure based on information provided by CO2CRC.*

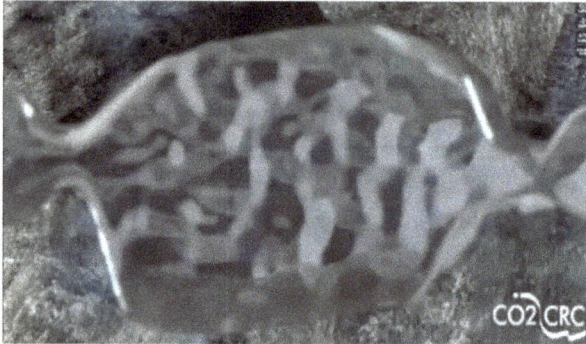

Movie 8.2.1 CO_2 residual trapping simulation

Residual gas trapping occurs when a small amount of CO_2 becomes disconnected or "snaps off" from the CO_2 plume as the CO_2 moves through the porous rock. The CO_2 is stored in the pores in tiny bubbles, trapped by surface tension. The CO_2 can't move out of the pore space and remains fixed underground. *Movie from the CRC for Greenhouse Gas Technologies (CO2CRC) [8.2]. It can be viewed at:* http://www.worldscientific.com/worldscibooks/10.1142/p911#t=suppl

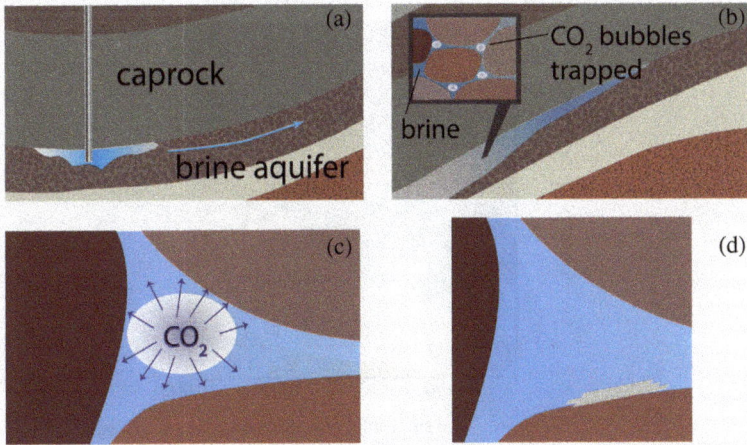

Figure 8.2.2 Mechanisms of CO_2 trapping

(a) **Stratigraphic trapping:** the injection of CO_2 displaces the brine from the aquifer, leaving a large plume of supercritical CO_2. This supercritical CO_2 is less dense than brine, so because of buoyancy effects it will move toward the highest point of the aquifer. The caprock prevents the CO_2 from moving out of the storage formation.

(b) **Residual trapping:** at the end of the CO_2 injection, as the CO_2 moves to the highest point of the aquifer, the brine flows back into the trailing edge of the plume. As the brine wets the rock, it preferentially fills the smallest pores and pore throats in the rock, and CO_2 bubbles become trapped by capillary forces in the larger pores.

(c) **Solubility trapping:** the droplets of CO_2 that are residually trapped will slowly dissolve in the surrounding brine. The dissolution of CO_2 in brine generates carbonic acid. The coexistence of supercritical CO_2, brine, and carbonate minerals buffers the pH of the brine to about 5.

(d) **Mineral trapping:** the low pH of CO_2-equilibrated brine causes the weathering of silicate mineral grains. The weathering reactions of certain silicate minerals release divalent metals (Ca, Mg, Fe). These ions combine with CO_2 to form carbonate minerals (such as limestone, $CaCO_3$), the most stable state of carbon.

This trapped amount will depend on time but may constitute roughly 20% of the total amount of injected CO_2 (**Figure 8.2.3**). In this **capillary trapping** or **residual trapping**, part of the CO_2 becomes immobilized as small bubbles at the trailing edge of the mobile CO_2 plume, while the plume itself continues to migrate to the highest point in the formation. Eventually, CO_2 will dissolve in the brine, a process known as **solubility trapping**. Finally, the dissolved CO_2 will react with minerals such as feldspars to liberate cations (e.g., Mg^{2+}, Fe^{2+}, Ca^{2+}), and

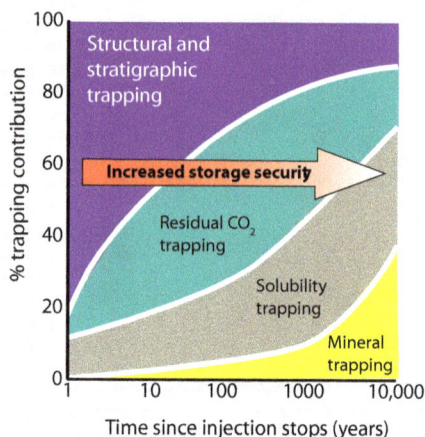

Figure 8.2.3 Trapping mechanisms as a function of time

Figure adapted from Benson and Cole [8.3].

these cations then become available to react with CO_3^{2-} in the aqueous phase to form carbonate minerals. As carbonates are the thermodynamically most stable form of carbon, **mineral trapping** is the final fate of our injected CO_2. You may have noticed that the chemistry of CO_2 sequestration presented here (CO_2 dissolution in water, the weathering of silicate minerals, and the eventual formation of carbonate minerals) is the same as the chemistry of the natural CO_2 cycle discussed in Sections 3.3 and 3.4.

The sequential trapping mechanisms are illustrated in **Figure 8.2.2.** It is important to realize that each of these mechanisms has its own time scale. For example, mineralization may take thousands of years. Another important point is that for each subsequent step in this trapping sequence, it becomes increasingly difficult for CO_2 to escape, and hence the likelihood of the CO_2 escaping decreases as time goes on [8.3]. This evolution of the storage security is illustrated in **Figure 8.2.3.** In the next chapter, we will discuss the physics and the time scales associated with these mechanisms in more detail.

Section 3

Selection of geological sites

In the previous section, we assumed that a site suitable for geological carbon sequestration is simply a site with pore space into which CO_2 can be injected. The conditions we imposed were that it should be sufficiently deep that CO_2 will be supercritical and that the formations should be below a caprock that prevents the CO_2 from escaping. In **Figure 8.3.1,** a few different types of geological formations are shown schematically [8.1]. They include depleted oil and gas reservoirs, oil reservoirs undergoing enhanced oil recovery, deep saline water-saturated reservoir rocks, and unminable coal seams. We discuss these different types of formations below.

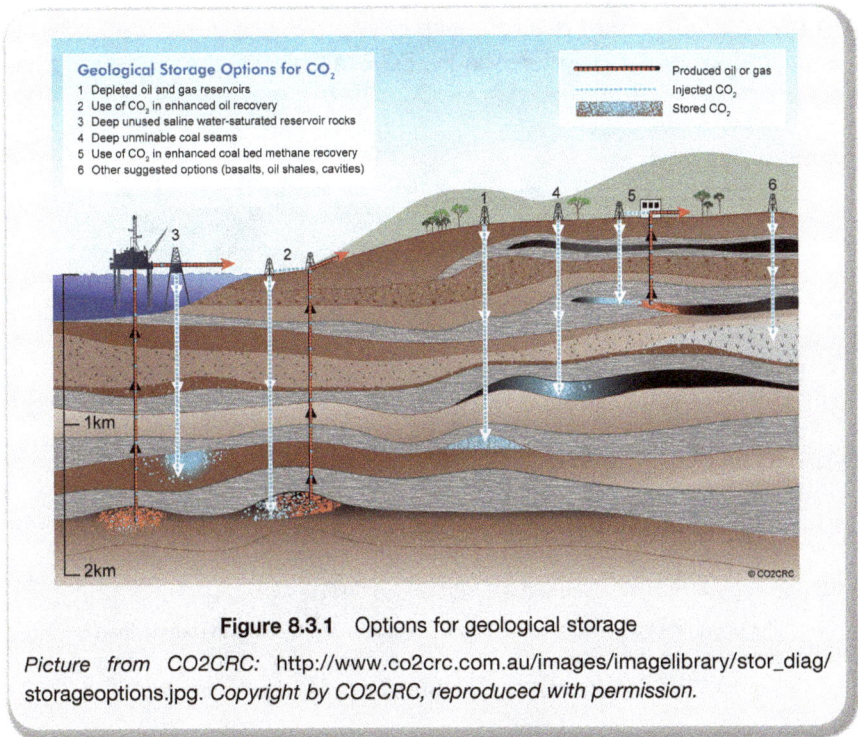

Figure 8.3.1 Options for geological storage

Picture from CO2CRC: http://www.co2crc.com.au/images/imagelibrary/stor_diag/storageoptions.jpg. *Copyright by CO2CRC, reproduced with permission.*

Depleted oil and gas reservoirs

The fact that natural gas has been trapped in these reservoirs for millions of years suggests that they can trap buoyant CO_2. In addition, data collected from wells and seismic surveys usually provide excellent knowledge about these sites. A disadvantage of these sites is that, because of the gas and oil production, wells have been drilled in the caprock, so one would need to be absolutely sure that these old wells are adequately plugged.

Enhanced oil and gas recovery

Under certain conditions, CO_2 can dissolve in oil, reducing its viscosity and density and thereby increasing its mobility through the pore space. This phenomenon has been used in the oil industry for many years to increase the fraction of oil recovered from an oil field, a process known as CO_2-enhanced oil recovery (CO_2-EOR) (see **Figure 8.3.2**). In fact, **Figure 8.3.3** shows some of the pipelines that have been built in the USA to transport CO_2 from natural reservoirs to oil fields for the purpose of enhanced oil recovery. Interestingly, EOR research in the oil industry has

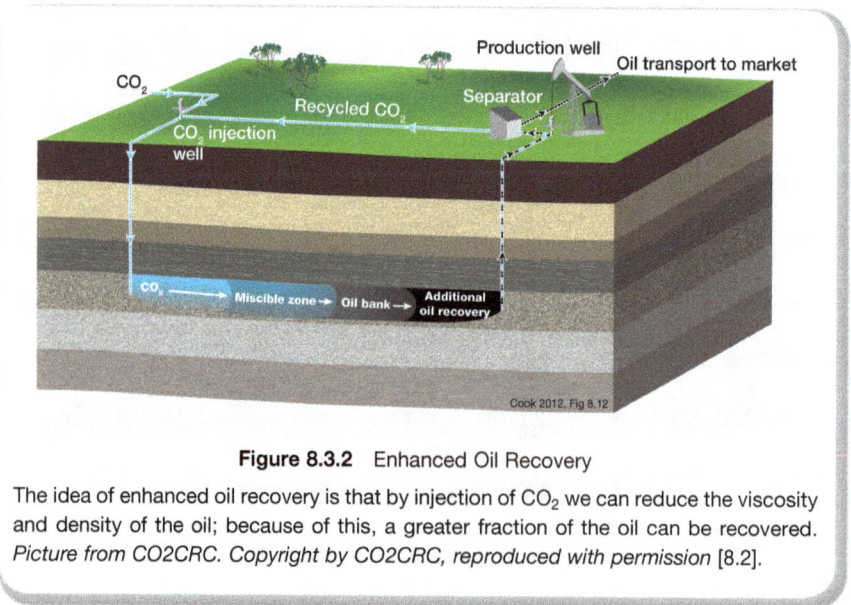

Figure 8.3.2   Enhanced Oil Recovery

The idea of enhanced oil recovery is that by injection of CO_2 we can reduce the viscosity and density of the oil; because of this, a greater fraction of the oil can be recovered. Picture from CO2CRC. Copyright by CO2CRC, reproduced with permission [8.2].

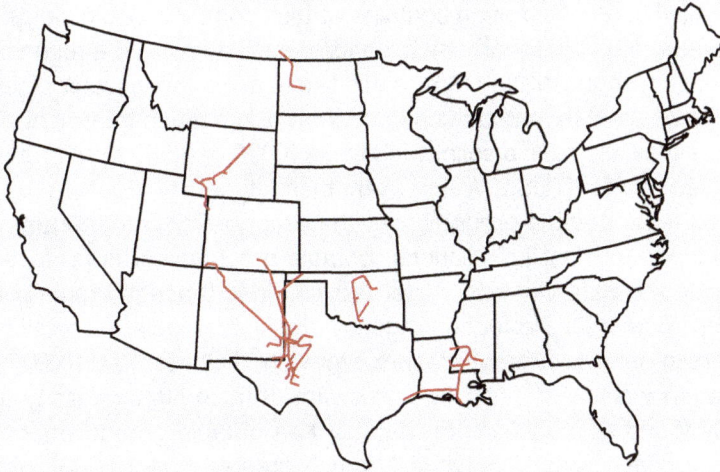

Figure 8.3.3 Pipelines in the USA for transporting CO_2 for enhanced oil recovery

Figure based on information from [8.4].

focused for many years on ensuring that as little CO_2 as possible is used, because producing and transporting CO_2 is expensive. The importance of enhanced oil recovery is that it demonstrates that we have the technology to transport CO_2 over large distances and inject it into geological formations [8.4]. Of course, in the context of CCS one would like to maximize CO_2 utilization! But even if one could maximize the amount of CO_2 "lost" in the reservoir, the total use of CO_2 for enhanced oil recovery would still be a small fraction of the total amount of CO_2 we would like to sequester in the context of CCS. CO_2 can also be used to enhance the production of natural gas. In this application CO_2 is used to re-pressurize the gas fields.

Deep saline aquifers

An **aquifer** is defined as a formation of permeable rocks saturated with water, with a degree of permeability that allows water withdrawal through wells. If an aquifer allows water withdrawal, then it will also allow injection of fluids. At present those deep aquifers not considered usable for drinking water are already used for disposal of waste water, disposal of acid gas (e.g., mixtures of H_2S and CO_2), and seasonal natural gas storage [8.5, 8.6]. If such a formation contains various combinations of fluids other than water, e.g., hydrocarbons (oil and/or gas), it is called a reservoir.

For CO_2 sequestration, one would like to avoid those water reservoirs that can be used for human consumption without much treatment. These aquifers typically have a salinity less than 3,000 or 4,000 ppm (mg/l) total dissolved solids (TDS). Water from these types of aquifers is often referred to as potable groundwater (the exact definition depends on the jurisdiction). In most jurisdictions, potable groundwater is protected. In the USA, the Safe Drinking Water Act provides the legal framework for the Environmental Protection Agency (EPA) to regulate injection into aquifers in order to protect groundwater with TDS less than 10,000 ppm. Formation water is defined as water with a much higher salinity and is usually found at much greater depths than potable groundwater. Deep saline aquifers, with TDS typically much greater than 10,000 ppm, are being considered for CO_2 storage.

Deep unminable coal seams

Fractures give coal formations some permeability. If one were to magnify these regions one would see a very large number of micropores into which gas molecules can adsorb. Experiments have shown that these pores can adsorb many gasses and often they contain large quantities of methane. Interestingly, CO_2 adsorbs more strongly than methane and hence the injection of CO_2 will displace the methane. In unminable coal formations, we can recover the methane while permanently storing the CO_2. This process is called Enhanced Coalbed Methane (ECBM) recovery.

Capacity

An important question is whether we have enough capacity in geological formations to store all the CO_2 we produce. From a practical point of view, it is also important that these formations be "conveniently" distributed over the planet. **Table 8.3.1** shows the global capacity of the different formations in which CO_2 can be sequestered [8.5, 8.6]. If we compare these capacities with the annual production of CO_2 of approximately 31 Gt per year, we see that even according to the lowest estimated storage capcity, we have sufficient capacity to store all our CO_2 for at least 50 years.

Table 8.3.1 Summary of storage capacity

Reservoir type	Lower estimate of storage capacity (Gt CO_2)	Higher estimate of storage capacity (Gt CO_2)
Oil and gas fields	675	900
Unminable coal seams (ECBM)	3–15	200
Deep saline formations	1,000	Uncertain, but possibly 10^6

Data from Metz et al. [8.7].

The distribution of the various formations over the planet is shown in **Figure 8.3.4.** This figure shows a reasonable distribution over all continents of sedimentary basins (the type of geological structure where one finds aquifers, hydrocarbon reservoirs, and coal seams). However, one would need to make a more detailed regional assessment to see how the availability of formations matches the location of point sources that emit the CO_2 we wish to sequester. In Chapter 10, we will discuss the capacity aspects in more detail.

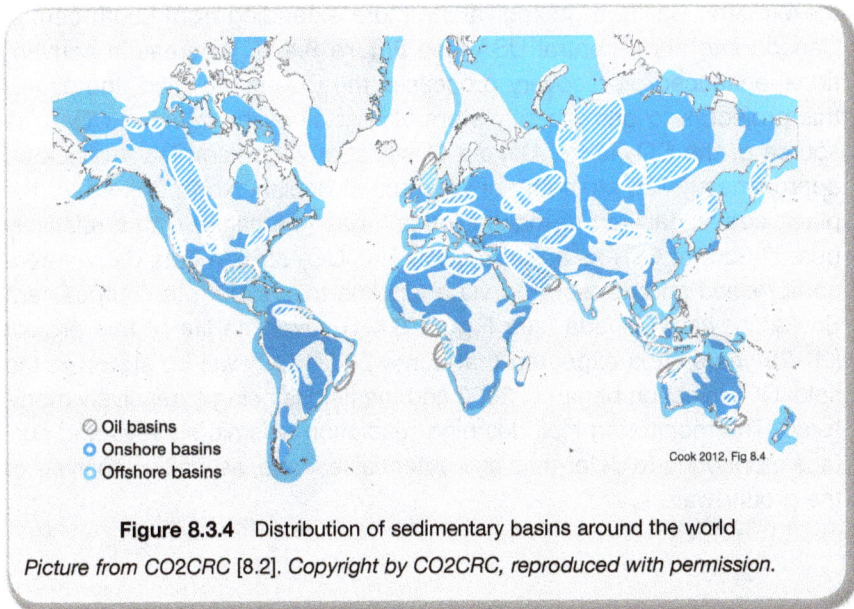

Oil basins
Onshore basins
Offshore basins

Cook 2012, Fig 8.4

Figure 8.3.4 Distribution of sedimentary basins around the world

Picture from CO2CRC [8.2]. Copyright by CO2CRC, reproduced with permission.

Section 4

Examples of sequestration projects

At present, several large scale projects store CO_2 around the world. These projects are summarized in **Tables 8.4.1** and **8.4.2.** They have a combined capture and storage capacity of approximately 36 Mtpa, equivalent to the emissions of more than seven million cars per year and roughly equivalent to the current annual emissions of Singapore or New Zealand [8.8]. Of course, compared to the total CO_2 emissions, this is a modest amount. These projects are, however, important because they help us to obtain the necessary experience in injecting CO_2 and monitoring the injected CO_2 plume.

Weyburn

The Weyburn CO_2-enhanced oil recovery (CO_2-EOR) project is located in the Williston Basin, a geological structure extending from south-central Canada into north-central USA (see **Figure 8.4.1**). Whereas in conventional enhanced oil recovery processes the CO_2 is recycled, the aim of this project is to permanently store almost all of the injected CO_2. The source of the CO_2 is the Dakota Gasification Company facility, located approximately 325 km south of Weyburn in Beulah, North Dakota. At the plant, coal is gasified to make synthetic gas (methane), with a relatively pure stream of CO_2 as a by-product. This CO_2 stream gets dehydrated, compressed and transported via a pipeline to Weyburn in southeastern Saskatchewan, Canada (see **Figure 8.4.1**). Over the life of this project (20–25 years), it is expected that some 20 Mt CO_2 will be stored in the field. CO_2 injection began in 2000 and the field is being extensively monitored. The monitoring includes high-resolution seismic surveys and surface monitoring to determine any potential leakage, as well as analysis of the groundwater.

Table 8.4.1 Examples of CCS projects that are in operation (2012)

Name	Country	Capture type	Volume CO_2 (Mtonne/year)	Storage type	Date of operation
Val Verde Gas Plants	United States	Pre-combustion (gas processing)	1.3 Mtpa	EOR	1972
Enid Fertilizer CO_2- EOR Project	United States	Pre-combustion (fertiliser)	0.68 Mtpa	EOR	1982
Shute Creek Gas Processing Facility	United States	Pre-combustion (gas processing)	7 Mtpa	EOR	1986
Sleipner CO_2 Injection	Norway	Pre-combustion (gas processing)	1 Mtpa (+0.2 Mtpa in construction)	Deep saline formation	1996
Great Plains Synfuel Plant and Weyburn–Midale Project	United States/ Canada	Pre-combustion (synfuels)	3 Mtpa	EOR	2000
In Salah CO_2 Injection	Algeria	Pre-combustion (gas processing)	1 Mtpa	Deep saline formation	2004
Snøhvit CO_2 Injection	Norway	Pre-combustion (gas processing)	0.7 Mtpa	Deep saline formation	2008
Century Plan	United States	Pre-combustion (gas processing)	5 Mtpa (+ 3.5 Mtpa in construction)	EOR	2010

EOR stands for enhanced oil recovery. *Data from Global CCS Institute* [8.8].

Table 8.4.2 Examples of CCS projects that are in the execution phase (2012)

Name	Country	Capture type	Volume CO_2 (Mtonne/year)	Storage type	Date of operation
Air Products Steam Methane Reformer EOR Project	United States	Post-combustion (hydrogen production)	1 Mtpa	EOR	2012
Lost Cabin Gas Plant	United States	Pre-combustion (gas processing)	1 Mtpa	EOR	2012
Illinois Industrial CCS Project	United States	Industrial separation (ethanol)	1 Mtpa	Deep saline formation	2013
ACTL with Agrium CO_2 Stream	Canada	Pre-combustion (fertiliser)	0.59 Mtpa	EOR	2014
Boundary Dam Integrated CCS Demonstration Project	Canada	Post-combustion (power generation)	1 Mtpa	EOR	2014
Kemper County IGCC Project	United States	Pre-combustion (power generation)	3.5 Mtpa	EOR	2014
Gorgon Carbon Dioxide Injection Project	Australia	Pre-combustion (gas processing)	3.4–4.1 Mtpa	Deep saline formation	2015
Quest	Canada	Pre-combustion (hydrogen production)	1.08 Mtpa	Deep saline formation	2015

EOR stands for enhanced oil recovery. *Data from Global CCS Institute* [8.8].

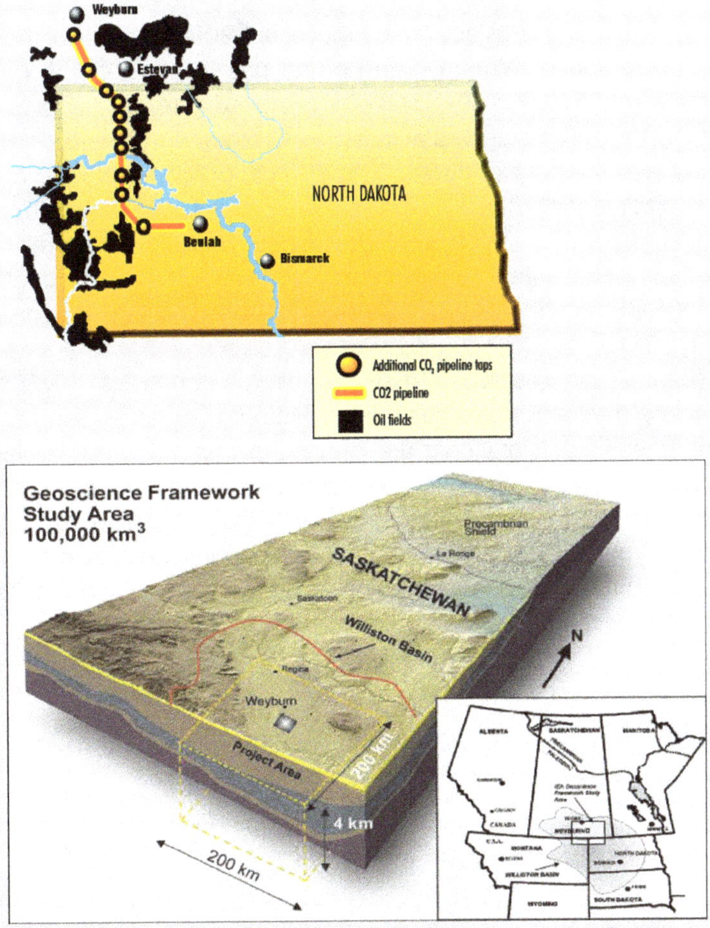

Figure 8.4.1 Weyburn CO_2-enhanced oil recovery project

(a) The CO_2 is captured from a coal gasification facility located in Buelah, North Dakota. The gas is compressed to a liquid phase and transported via a 320 km pipeline to the Weyburn and Midale oil fields for injection. *Image provided by Dakota Gasification Company, reproduced with permission:* http://ptrc.ca/+pub/image/NDKmap.jpg

(b) Cenovus' Weyburn field and Apache's Midale field are located in southeast Saskatchewan, Canada. This project represents the first time that an anthropogenic source of CO_2 has been used for enhanced oil recovery. *Image © Department of Natural Resources Canada, all rights reserved, from:* http://geoscan.nrcan.gc.ca/starweb/geoscan/servlet.starweb?path=geoscan/downloade.web&search1=R=214968

In Salah

The In Salah Gas Project (see **Figure 8.4.2**) is located in the central Saharan region of Algeria. The natural gas in surrounding gas fields contains up to 10% CO_2, which is removed through absorption with amines, the technology discussed in Chapter 5. This is required for natural gas to meet commercial specifications. The unique aspect of this project is that the CO_2 was re-injected into long horizontal wells at a depth of 1,800 m, and annually up to 1.2 Mt CO_2 was stored. Carbon dioxide injection started in April 2004. An abundance of caution and lack of CCS incentive led to the cessation of CO_2 injection for the project in 2011.

Natural gas reservoirs may contain as much as 60–70% CO_2. Extracting methane from these fields without re-injecting the CO_2 would make these fields significant greenhouse gas emitters. As we have seen for carbon capture, the equipment for these gas separations can be very large. This poses few problems for gas fields in the Sahara, but is very difficult to implement for off-shore production. Therefore, finding technologies that can do the separation using far less space presents a very interesting research topic.

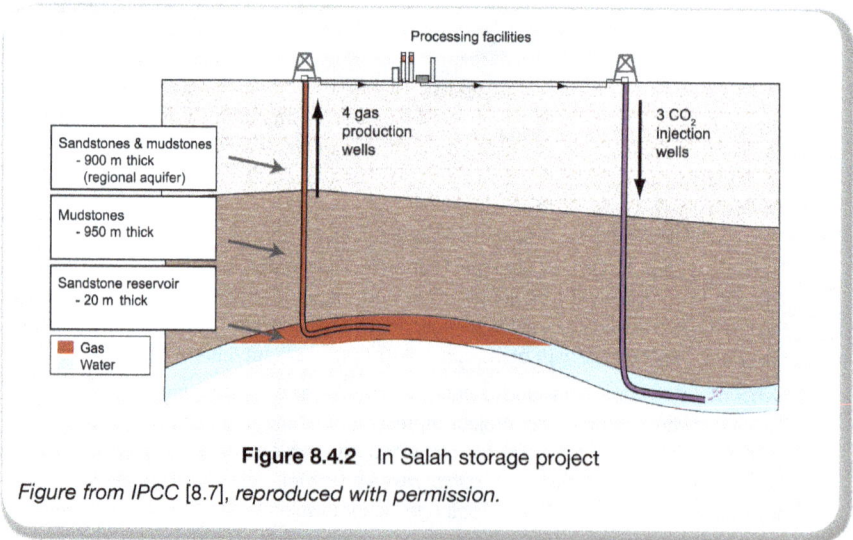

Figure 8.4.2   In Salah storage project

Figure from IPCC [8.7], reproduced with permission.

Sleipner

The Sleipner project, illustrated in **Figure 8.4.3,** is located in the North Sea about 250 km off the coast of Norway. This is the first commercial scale project in which CO_2 is injected into a saline formation. Similar to the In Salah project, about 9% CO_2 is produced with the natural gas. The saline formation into which the CO_2 is injected is a brine-saturated unconsolidated sandstone about 800–1,000 m below the sea floor. The formation also contains secondary thin shale layers, which influence the internal movement of injected CO_2. The CO_2 injection started in 1996 and by 2013 more than 11 Mt CO_2 have been sequestered. The aim of this project is to sequester 20 Mt CO_2 in total. The saline formation has a very large storage capacity (1–10 Gt CO_2) [8.6].

The fate and transport of the CO_2 plume in the storage formation has been monitored successfully by seismic time-lapse surveys. These surveys provide experimental evidence that the caprock is an effective seal that prevents CO_2 migration out of the storage formation. In 2007 (after

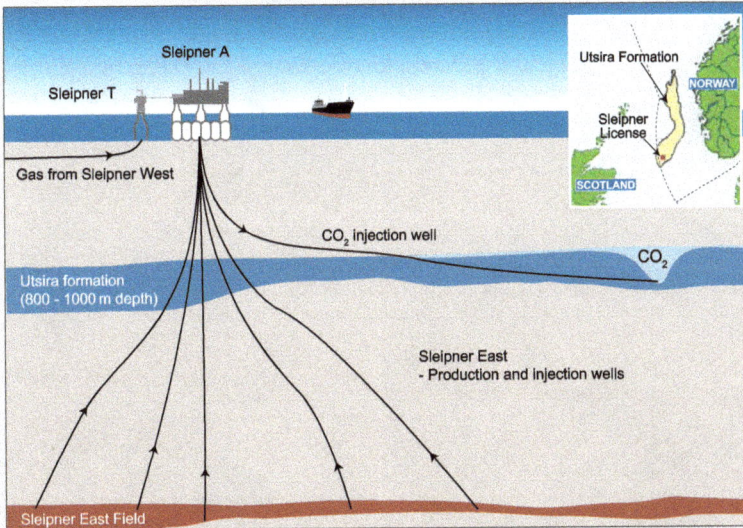

Figure 8.4.3 The Sleipner CO_2 storage project

Figure from IPCC [8.7], reproduced with permission.

10 years), the footprint of the plume at Sleipner extended over an area of approximately 5 km^2.

Summary

In this chapter, we have introduced some of the key concepts in the geological sequestration of CO_2. We find that separation of CO_2 from flue gas is essential and that there are four trapping mechanisms for keeping CO_2 sequestered for geological time scales: stratigraphic or structural, residual or capillary, solubility, and mineral trapping. Finally, we introduced the types of geological formations that can be used to effectively store CO_2, and provided specific examples of a few sites that demonstrate the feasibility. In the following two chapters, we will discuss the fundamental and practical aspects of geological carbon sequestration.

Section 5

Review

8.1. Reading self-test

1. How much flue gas does a medium size coal-fired power plant emit over its 50-year life?
   a. 150,000 cubic miles
   b. 15,000 cubic miles
   c. 1,500 cubic miles
   d. 150 cubic miles

2. What is the approximate density of CO_2 at a depth of 2 km?
   a. 0.70 kg/m^3
   b. 7 kg/m^3
   c. 70 kg/m^3
   d. 700 kg/m^3

3. The typical depth of injection of CO_2 for geological sequestration is
   a. 100 m
   b. 500 m
   c. 2,000 m
   d. 5,000 m

4. Which drawing illustrates residual CO_2 trapping?

a.
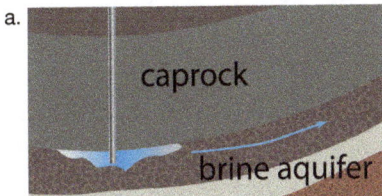

caprock

brine aquifer

b.
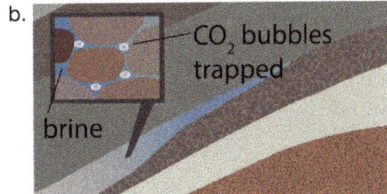

CO$_2$ bubbles trapped

brine

c.

d.

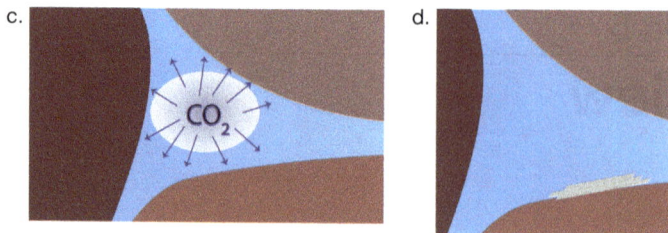

5. Mineral trapping of CO_2 occurs on a time scale of
   a. 1 year
   b. 10 years
   c. 100 years
   d. 1,000 years

6. How many years of current CO_2 emissions can maximally be stored in oil and gas fields?
   a. 1
   b. 10
   c. 30
   d. 100

Section 6

References

1. CO2CRC, 2008. Storage Capacity Estimation, Site Selection and Characterisation for CO_2 Storage Projects. Cooperative Research Centre for Greenhouse Gas Technologies, Canberra. CO2CRC Report No. RPT08-1001. http://www.CO2crc.com.au/dls/pubs/08-1001_final.pdf
2. CO2CRC, Cooperative Research Centre for Greenhouse Gas Technologies. www.CO2crc.com.au/imagelibrary
3. Benson, S.M. and D.R. Cole, 2008. CO_2 sequestration in deep sedimentary formations. Elements, **4** (5), 325. http://dx.doi.org/10.2113/gselements.4.5.325
4. Department of Engineering and Public Policy, Carnegie Mellon University, 2009. Carbon Capture and Sequestration: Framing the Issues for Regulation: An Interim Report from the CCS Reg. Project. Dept. of Engineering and Public Policy, Carnegie Mellon University: Pittsburgh. http://s3.amazonaws.com/zanran_storage/www.aiche.org/ContentPages/29660479.pdf
5. Bachu, S., D. Bonijoly, J. Bradshaw, et al., 2007. CO_2 storage capacity estimation: Methodology and gaps. Int. J. Green. Gas Con., **1** (4), 430. http://dx.doi.org/10.1016/S1750-5836(07)00086-2
6. Bradshaw, J., S. Bachu, D. Bonijoly, et al., 2007. CO_2 storage capacity estimation: Issues and development of standards. Int. J. Greenh. Gas Con., **1** (1), 62. http://dx.doi.org/10.1016/S1750-5836(07)00027-8
7. Metz, B., O. Davidson, H. De Coninck, M. Loos and L. Meyer, 2005. IPCC Special Report on Carbon Dioxide Capture and Storage. Canada: Cambridge University Press. (In this chapter, we have used Figures 5.4 and 5.5.) http://www.ipcc.ch/pdf/special-reports/srccs/srccs_wholereport.pdf
8. Global CCS Institute, 2012. The Global Status of CCS: 2012, Report No. 1838-9473. http://cdn.globalccsinstitute.com/sites/default/files/publications/47936/global-status-ccs-2012.pdf

Chapter 9

Fluids and Rocks

Geological formations contain enormous volumes of pore space, but most of this space is located in pores with diameters of nanometers to tens of micrometers. What happens when CO_2 is injected into such tiny pores?

Section 1

Introduction

In Chapter 8, we provided an overview of geological CO_2 sequestration. The idea is to select our geological formations with care so we can ensure that the CO_2 stays at those locations as we planned. At the molecular level, we need to understand how CO_2 is interacting with rocks. With the broad outline and overview of geological carbon sequestration presented in the last chapter as a foundation, we are in a position now to address key issues and questions surrounding the physical and chemical processes related to sequestration:

- Within the wide variety of geological formations on earth, what are the characteristics of the formations that are most promising for sequestering CO_2, i.e., which rock formations ensure that the CO_2 does not reach the surface?
- What will be the physical properties of CO_2 (e.g., density and viscosity) within these formations at a given depth?
- How will CO_2 flow within these formations over time?

  Some of these questions are very similar to those discussed in the carbon capture chapters. A CO_2 molecule may not notice the difference between adsorption in an adsorbent material versus in a rock, or whether it diffuses through a membrane or within a geological formation. Indeed as we will see, the concepts to describe transport in a geological formation are similar to those we have previously described — similar, but not identical. In our design of a membrane or adsorbent, we could safely assume that the small structure for which we obtained a molecular level understanding is representative of the entire mass of material in our separation unit at the power plant. In geological sequestration, however, our CO_2 plume can stretch tens of kilometers around the injection site. On this scale, assuming the earth is a perfect, homogeneous substance without any defects does not make any sense. Over such a large scale, the subsurface will be heterogeneous and we need to understand how the heterogeneities influence the behavior of the CO_2 plume. We will also see that at such a large scale a molecular description becomes less useful and we have to use a more coarse-grained description.

Before we can discuss the physical and chemical processes that govern geological sequestration, we have to go back to the basics: we need to refresh our understanding of different rock types, how geological formations are formed, and the properties of fluids in these formations.

Section 2

Rocks

In the last chapter, we saw that different options are available for geological storage of CO_2. Of these options, deep saline water-saturated reservoir rocks (deep saline aquifers) have the largest capacity, which is one of the reasons we focus on these formations in this chapter.

In **Figure 9.2.1**, we illustrate the injection of CO_2 in a saline aquifer. CO_2 is typically injected at a depth of 800 meters or deeper. In selecting a storage formation, we need to ensure that it has a sufficiently large rock pore volume at this depth to store the amount of CO_2 we are planning to inject. We need to ensure that this pore volume is connected so that the CO_2 can access the entire aquifer from a single or a small number of

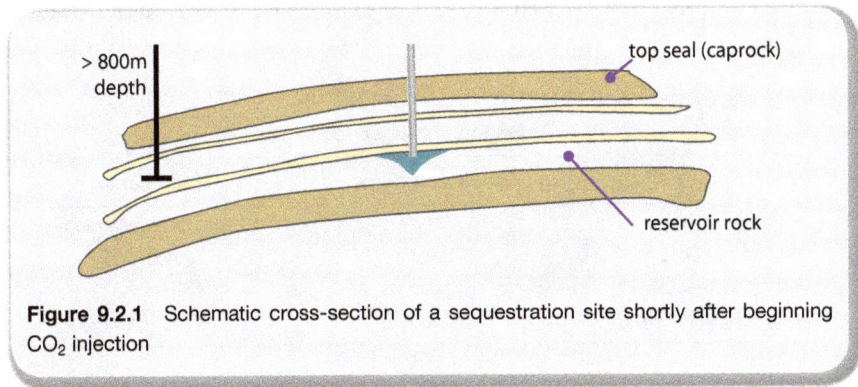

Figure 9.2.1 Schematic cross-section of a sequestration site shortly after beginning CO_2 injection

injection wells. We also need to ensure that the aquifer is insulated from the atmosphere by a seal (so-called caprock). This seal needs to prevent CO_2 from moving upward over an area that can span a hundred square kilometers. Finally, we need to be able to access the storage formation, e.g., through deep injection wells.

Geology

Before starting our geology review section, let us think a few moments about our requirements. We need an aquifer with a large pore volume in which CO_2 can flow easily, and on top of this aquifer we need a seal in which CO_2 does not flow at all. How is it possible that nature created at the same location such different types of materials?

Clastic sedimentary rocks

Aquifers are often found in **clastic sedimentary rocks**. Clastic sedimentary rocks are formed from eroded fragments of older rocks. These rock fragments are transported by water or wind, then deposited (in the process of sedimentation). The grains of these sediments are subsequently consolidated into a rock (in the process of lithification).

The different clastic sedimentary rocks are classified according to their grain size (see **Table 9.2.1**). If the rock were composed only of grains stacked together against one another, it would not have any strength and would disaggregate like sand on a sand dune. Many sandstones are in fact poorly consolidated; they readily disintegrate into constituent grains and cannot be quarried or used as building or paving stones. Other sandstones are very hard by virtue of being held together with other natural

Table 9.2.1 Clastic rocks

Grain size	Sediment name	Rock name
Coarse (>2 mm)	Gravel	Conglomerate/ breccia
Medium (0.06–2 mm)	Sand	Sandstone
Fine (0.008–0.06 mm)	Mud	Siltstone
Very fine (<0.008 mm)	Mud	Shale

minerals that bond the grains tightly, giving them strength and resistance to erosion.

Figure 9.2.2 shows examples of sandstone and shale and in **Table 9.2.2** the properties of these rocks are compared. Sandstones have a large porosity and high permeability, which makes these rocks suitable for CO_2 sequestration. Shales, on the other hand, have a lower porosity and a much lower permeability, which makes shale an ideal seal or caprock.

Table 9.2.2 shows that the different clastic rocks can also be characterized by differences in their mineral composition [9.1]. **Figure 9.2.3** shows their structures and **Box 9.2.1** gives some more details on their structures. These rocks are mixtures of quartz grains, other tectosilicates (primarily feldspars), phyllosilicates (such as mica, smectite, chlorite, and kaolinite), and carbonates. Sandstones tend to be dominated by quartz minerals, but shales by phyllosilicates.

Geological formations

We have discussed the clastic rocks that can be found in geological formations. Let us now discuss how these different clastic rock regions

(a)   (b)

Figure 9.2.2 Sandstone and shale

(a) Sandstone. *Image by Dave Waters, Oxford University.*

(b) Shale. *Reproduced with permission from Paleontological Research Institution, Ithaca, New York.*

Table 9.2.2 Typical properties of sandstone aquifers and shale seals from the USA Gold Coast

	Sandstone	Shale
Porosity (ϕ)	0.15 to 0.35	0.06 to 0.15
Permeability	High	Low
Mineralogy		
Quartz (vol. %)	58	19
Feldspars (vol. %)	28	10
Phyllosilicates (vol. %)	12	58
Carbonates (vol. %)	2	11
Organic matter (vol. %)	0	2

Typical values of the porosity ϕ and mineralogy in sandstone aquifers and shale seals. The mineralogy data are representative of sandstone-shale sequences from the USA's Gulf Coast. *Data from Xu et al.* [9.1].

were formed. This will provide some insight into how nature has provided us with sites that are ideal for geological carbon sequestration.

Long-term tectonic processes acting on the earth's crust throughout geological time have resulted in regions of large vertical uplift (mountains and highlands) from which sediments are eroded and large vertical subsidence (valleys and basins) in which sediments are deposited. These processes have created thick sedimentary basins where vast volumes of sedimentary rocks have formed.

The local conditions prevailing during sedimentation are referred to as the depositional environment. Depositional environments over space and time can be highly variable, ranging from vigorous fluvial systems (e.g., rivers), to gentle tidal systems (e.g., estuaries), to off-shore marine environments (continental shelves, submarine canyons), to subaerial arid and windy environments (alluvial fans and dunes). Coarse-grained sediments are formed during periods of high-energy water or wind transport, while fine-grained sediments are formed by sedimentation from low-energy fluvial or marine systems.

Repeated cycles of climate and sea-level change during the filling of sedimentary basins have resulted in periodic variations in the depositional environment such that recurring cycles of deposition of coarse

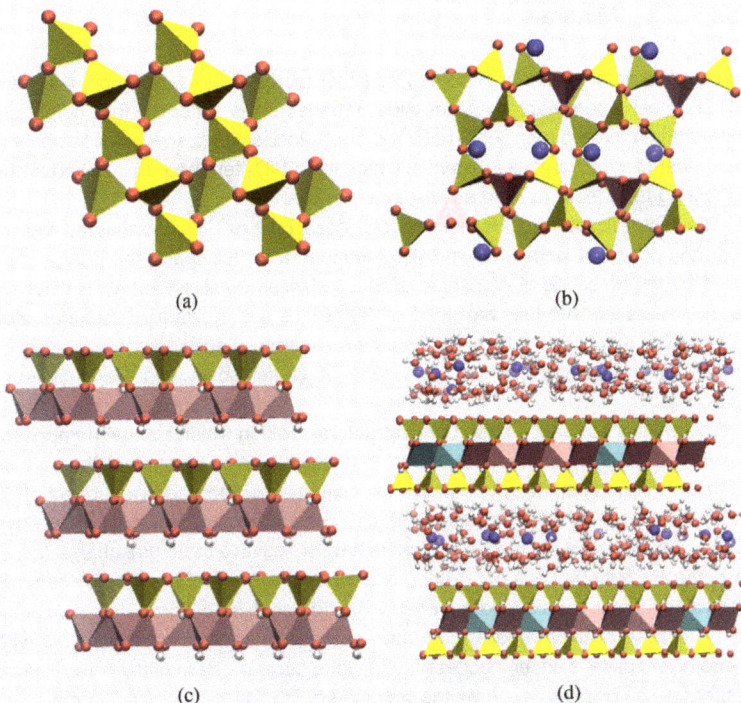

Figure 9.2.3 Structure of minerals

(a) Tectosilicates: quartz (SiO_2). SiO_4 tetrahedra are shown in yellow, O atoms in red.

(b) Tectosilicates: albite (Na-feldspar: $NaSi_3AlO_8$). SiO_4 tetrahedra are shown in yellow, AlO_6 octahedra in pink, O atoms in red, and Na atoms in blue.

(c) Phyllosilicates: kaolinite [$Si_4Al_4O_{10}(OH)_8$]. SiO_4 tetrahedra are shown in yellow, AlO_6 octahedra in pink, O atoms in red, H atoms in white. The sheets are held together by hydrogen bonding.

(d) Phyllosilicates: smectite [$C_x(Si,Al)_8(AlFeMg)_4O_{20}(OH)_4 \cdot nH_2O$, where C is the interlayer cation that balances the negative structural charge of the clay sheet, in this case Na^+. SiO_4 tetrahedra are shown in yellow, AlO_6 octahedra in pink, MgO_6 octahedra in green, O atoms in red, H atoms in white, Na atoms in blue. As smectites are swelling clay minerals, the interlayer space contains a variable number of water molecules (in this case, two statistical water monolayers; therefore, the figure shows the two-layer hydrate of Na-smectite).

sediments followed by fine sediments are commonly preserved in sedimentary rock sequences. **Figure 9.2.4** illustrates the layered structure of sediments that are formed by these periodic variations.

Box 9.2.1 Minerals

Silicate minerals are broadly classified as either tectosilicates (also known as framework silicates) where each SiO_4 (or AlO_4) tetrahedron is linked to four neighboring tetrahedra in a three-dimensional network, or phyllosilicates (also known as sheet silicates, most of which are clay minerals) where each SiO_4 tetrahedron is linked to three neighboring tetrahedra to form a two-dimensional sheet.

The tectosilicates include quartz, composed only of SiO_4 tetrahedra, and feldspars, composed of both SiO_4 and AlO_4 tetrahedra. In feldspars, the charge deficit caused by the presence of Al instead of Si is balanced by structural K^+, Na^+, or Ca^{2+} ions. Feldspars that are fully balanced by these ions are known as K-feldspar, albite, and anorthite, respectively. Most feldspars are either solid solutions of K-feldspar and albite (or *alkali feldspar*) or solid solutions of Na-feldspars and Ca-feldspars (or *plagioclase*).

The phyllosilicates are based on a structural motif in which a sheet comprised of AlO_6, FeO_6 or MgO_6 octahedra is bound to a sheet of SiO_4 tetrahedra (the so-called *1:1 structure* found particularly in kaolinite) or sandwiched between two sheets of SiO_4 tetrahedra (the so-called *2:1 structure* of mica, illite, smectite, and most other phyllosilicates). Most 2:1 structure phyllosilicates have a negative structural charge (caused by isomorphic substitutions of tetrahedral Si or octahedral Al, Fe, or Mg by cations of lower valence) that is balanced by loosely bound interlayer cations (Na^+, K^+, Ca^{2+}, or Mg^{2+}). Because of their lamellar structure, phyllosilicates tend to form stacks where the interlayer space is either inaccessible to water (as in the case of kaolinite, illite, and mica) or can be propped open by the solvation of interlayer cations (as in the case of smectite, a *swelling clay*).

Several observations relevant to carbon sequestration can be made regarding the silicate minerals in **Figure 9.2.3**. First, the weathering rates of silicate minerals tend to decrease with increasing Si:O ratio (hence among the tectosilicates weathering rates increase in the order quartz < alkali feldspar < plagioclase) and increasing connectivity of the SiO_2 network (hence tectosilicates tend to weather more slowly than phyllosilicates). Second, most phyllosilicates (except mica) are clay minerals and they have very small particle sizes. For example, smectite crystals are flake-shaped nanoparticles with a thickness of 0.94 nm and a diameter on the order of 200 nm (specific surface area $a_s \approx 800\,m^2/g$), whereas quartz grains are roughly spherical with a diameter of ~10 μm (specific surface area $a_s \approx 0.02\,m^2/g$). Because of their small crystal size, clay minerals can strongly influence the properties of porous media, for example, by forming surface coatings on larger minerals.

As tectonic forces continue to act on the rock formations over geological time, folding, faulting, and fracturing of the sediments may occur on large and small scales, giving the once-sub-horizontal sediments upward and downward-arching structures (anticlines and synclines,

(a)   (b)

Figure 9.2.4. Layers of clastic sediments

(a) Sandstone. *Figure from American Association of Petroleum Geologists.*

(b) Highly porous and permeable sandstone reservoir rock of the Farewell Formation, Whanganui Inlet, Golden Bay, New Zealand. *Photo by Lesli Wood, University of Texas at Austin.*

respectively). The dip angle (angle between the horizontal and the slope of the strata) or a qualitative description thereof (e.g., steeply dipping, gently dipping) is used to describe the degree to which the sub-horizontal sediments have been tilted downward. The term closure refers to the extent of the lowermost flexure of the caprock, an indicator of the closed region within which a buoyant fluid such as oil, natural gas, or CO_2 will be trapped. A structure with closure all around is referred to as a dome. If you picture an overturned saucer catching an updraft of smoke, smoke will spill out from under the saucer at the lowest point, called the spill point, if the smoke volume exceeds the closure volume of the saucer.

The commonly observed alternating sequence of reservoir rock (high porosity) and caprock (low porosity) in folded and faulted structures is well-exemplified by the thick sedimentary rocks of the Great Central Valley of California. Shown in **Figure 9.2.5** is a geological cross-section of a part of the Southern San Joaquin Valley, California [9.2]. Note that although the vertical scale of the cross-section is exaggerated, the vertical thickness of the sediments is still considerable (over 3,000 m). The alternating sandstone and shale formations are the result of continuous down-warping of the basin and infilling by sediments eroded, first from the North American continent and later from the rapidly rising Sierra

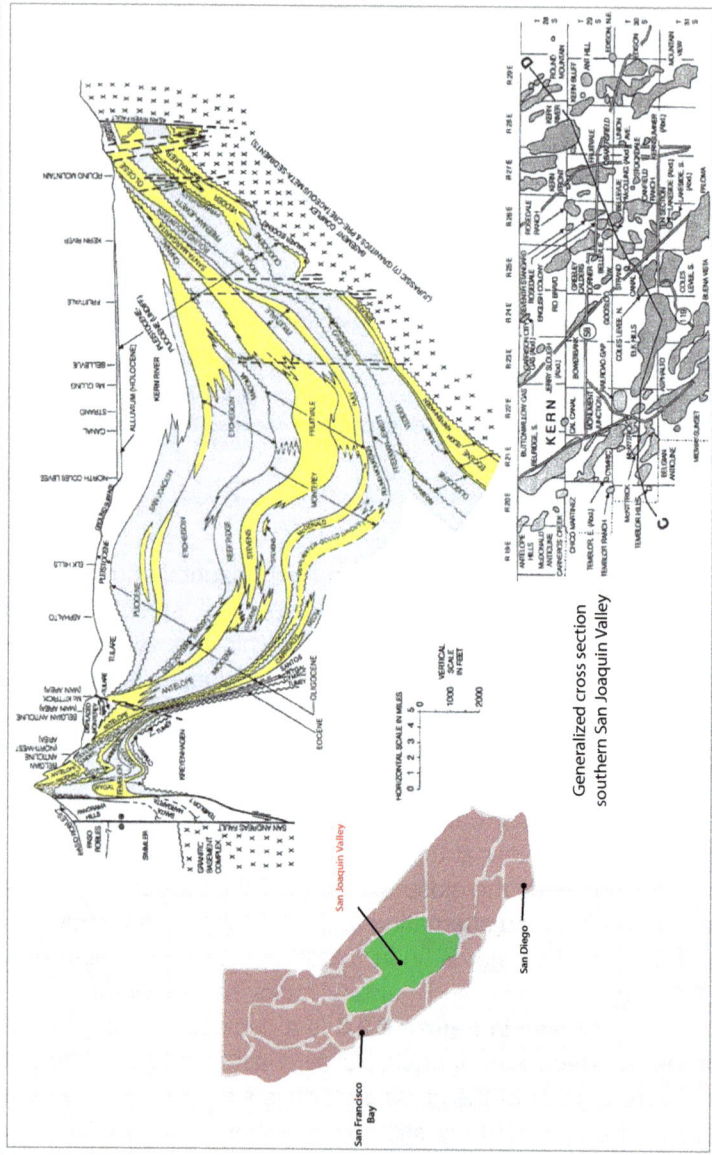

Generalized cross section
southern San Joaquin Valley

Figure 9.2.5 Southern San Joaquin Valley

Geological cross-section of the Southern San Joaquin Valley showing deep alternating layers of sandstones (yellow) and shales (gray). *Figure adapted from Myer et al. [9.2].*

Nevada. These sediments were deposited either in fluvial (sandstone) or shallow tidal (shale) environments depending on factors such as climate-related sea-level changes. These formations in the San Joaquin Valley contain significant amounts of oil and natural gas (i.e., hydrocarbon reservoirs). In the shallower parts of the structures, important water resources are also found (i.e., groundwater aquifers) that help support the large agricultural industry in the Central Valley. The presence of both hydrocarbons and valuable groundwater has led to an extensive study of the sedimentary rocks in California's Central Valley throughout their depth.

Alternative formations

Besides the clastic sedimentary rocks, chemical sedimentary rocks such as limestone and dolomite are also significant as potential CO_2 storage formations. Limestone and dolomite are particularly interesting because, unlike quartz-rich sandstone rocks that are relatively inert to aqueous CO_2 solutions, they are composed of carbonate minerals and are therefore potentially highly reactive with CO_2-rich fluids [9.3]. A natural example of such reactivity is the occurrence of large caves and karst structures that result from dissolution of limestone caused by naturally infiltrating groundwater that is weakly acidic from atmospheric CO_2. It is not difficult to imagine that with higher concentrations of dissolved CO_2 originating from an injection well, rock-water reactions could be extensive.

Igneous rocks such as granite and volcanic rocks such as basalt either lack the requisite porosity or are too fractured and lacking in cap-rock to contain fluid that is injected into them. However, some concepts for geological carbon sequestration exploit the large reactivity of igneous (volcanic and plutonic) rocks that contain abundant magnesium and iron in the form of highly reactive minerals [9.4]. Nevertheless, the majority of CO_2 storage capacity is believed to be in sedimentary rocks in basins both on-shore and off-shore [9.5]. These sedimentary basins will be the emphasis in this and the next chapter.

This field of geology entails considerably more complexity and subtlety than what was summarized above. We refer readers interested in additional information to a number of excellent textbooks or online resources [9.6, 9.7].

Porosity

The volume fraction of a rock that is occupied by pore space is known as the **porosity** (ϕ). The clastic sedimentary rocks provide abundant *intergranular* pore space, i.e., the void space located between solid grains. This pore space is highly irregular with very narrow channels, say on the order of 10–100 microns, between grains, referred to as pore throats, and larger openings or cavities, say on the order of hundreds of microns, referred to as pore bodies. The image in **Figure 9.2.6** is a 3D microtomographic image of the pore space of a sandstone sample from the Frio Formation (Texas Gulf Coast) showing the complex nature of pore throats and pore bodies. The right-hand image of **Figure 9.2.6** is the corresponding skeleton (solid material) of the rock [9.8].

Intergranular pores are commonly well-connected through pore throats. These connections allow fluids (gasses and/or liquids) to flow throughout the pore space as driven by gravitational or pressure forces. This permeability is necessary for sequestration in order to allow injection of CO_2 and widespread storage throughout the formation using a finite number of wells.

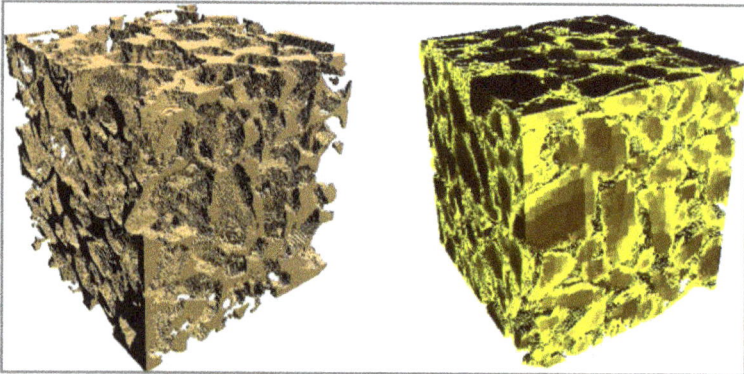

Figure 9.2.6   Sandstone pore space

Actual sandstone pore space shown by the solid gold color (left) and the corresponding skeleton (rock grains) (right) constructed by application of synchrotron microtomography methods at the Advanced Light Source [9.8]. Each side of the cube is 0.9 mm in length. *Image reproduced with permission from Springer Science + Business Media* [9.8, Figure 5].

Porosity in rocks can also be provided by fractures or faults in the rock; this is referred to as *fracture porosity,* defined as the void space in the apertures of the fractures and faults of the rock divided by the bulk rock volume.

Section 3

Fluids

Now that we have considered some of the salient properties of rocks, we turn to understanding the properties of fluids confined in these rocks. This is critical to geological CO_2 sequestration, because we ultimately seek to displace those fluids naturally present in rocks with our super-critical CO_2 from the capture process.

Deep groundwater environment

The vast majority of deep subsurface pore space in sedimentary basins is filled with saline groundwater. Some geological formations, of rare occurrence but great economic value, have pores that are filled predominantly with hydrocarbon fluids (oil and/or natural gas) and various other gasses (e.g., He or CO_2). Regardless of the composition, fluids at the depths relevant to geological carbon sequestration are at high pressure (6–25 MPa).

The **hydrostatic pressure** is the equilibrium pressure due to gravity alone, and most fluids in sedimentary basins are at this pressure. However, the degree of interconnectedness of the pores and changes in crustal loading can give rise to notable exceptions. For example, in sedimentary basins that were overlain by ice during the last ice age (e.g., in the upper midwestern USA), reservoirs isolated from surrounding aquifers by caprock may be underpressured, i.e., at less than hydrostatic pressure, because the rock has physically expanded due to the unloading of ice [9.9]. Similarly, in the Gulf of Mexico, there are deep overpressured reservoirs caused by the loading and compression of the overlying rock caused by deposition of sediment carried by the Mississippi River [9.10].

Table 9.3.1 Density and viscosity of water, brine and different gasses

		(Pressure/ Temperature)	Brine	Water	Air	CO_2	CH_4
Density (kg/m^3)	Ground surface	(0.1 MPa, 10°C)	1,190	990	1.2	1.9	0.68
	Deep system (2 km)	(20.1 MPa, 60°C)	1,200	1,000	200	725	130
Viscosity $(10^{-6}$ Pa s)	Ground surface	(0.1 MPa, 10°C)	1,800	1,300	18	14	11
	Deep system (2 km)	(20.1 MPa, 60°C)	940	470	24	60	18

Density of pure water and sodium chloride brine at two different representative conditions [9.11].

The prevailing temperature in deep sedimentary basins is controlled by the geothermal gradient, generally around 25°C/km. In addition to pressure and temperature, composition controls the properties of the fluids. Groundwater at depths greater than several hundred meters is typically salty. This is the result of long-term dissolution of minerals and the tendency for downward migration of saline water due to its greater density relative to fresh water. At depths of 1 km or more, groundwater is often very salty. If its salinity is greater than that of seawater (~35,000 mg/L (ppm)), we refer to this water as brine. The densities and viscosities of pure water and of brine at representative 2 km deep subsurface and shallow ground surface conditions are compared in **Table 9.3.1**. Due to its relative incompressibility, the density of groundwater does not vary much with depth. The viscosity of a fluid increases at higher density but decreases at higher temperature. **Table 9.3.1** shows that the temperature effect dominates.

Properties of CO_2

As we saw in Chapter 1, the critical temperature and pressure of CO_2 are 31°C and 7.4 MPa, respectively. In what seems like a quirk of nature, the hydrostatic pressure and geothermal gradients in the subsurface at around 800 m depth coincidentally bring CO_2 to its critical point. Under typical static conditions free phase CO_2 will be either gaseous, e.g., at shallow depths, or supercritical, e.g., in the subsurface at depths greater than approximately 800 m [9.11].

Table 9.3.1 compares the density and viscosity of CO_2 at surface and deep underground conditions with those of other gasses. Unlike CO_2, methane and air are not supercritical at the deep conditions and therefore have much lower (gas-like) densities. Because of the density increase for supercritical CO_2 the viscosity also increases, yet the viscosity of supercritical CO_2 is still about an order of magnitude lower compared to that of a liquid.

Figure 9.3.1 shows a modified phase diagram convenient for understanding CO_2 properties in the subsurface as a function of pressure, depth, and temperature. Superimposed on the diagram are the lines that connect the temperatures and pressures for which CO_2 has a constant density (isochores). In addition, the diagram shows two hypothetical paths; one in which the temperature increases 15°C per km and one for 30°C per km, which are the typical temperature gradients found in different geological formations. If CO_2 rises from the depths to the atmosphere, it will follow a path between these 15°C and 30°C boundaries, and we see that there is a narrow (green) triangular region where it may transition to a liquid. Furthermore, the upward migrating CO_2 will undergo a large expansion, during which the Joule-Thomson effect will cause the fluid to cool. Both effects may result in the formation of a liquid with very different properties from either gaseous or supercritical CO_2 [9.12, 9.13].

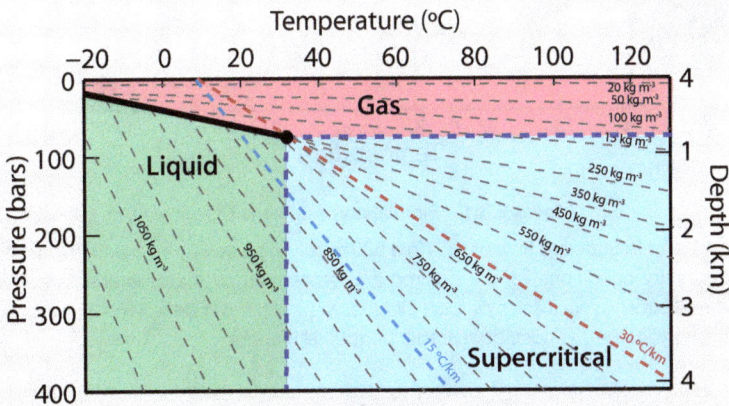

Figure 9.3.1 CO_2 phase diagram

Inverted portion of the CO_2 phase diagram relevant to sequestration with superimposed constant density lines (isochores) convenient for understanding phase conditions of CO_2 in the subsurface. *Figure reproduced with permission from Oldenburg [9.11].*

Oldenburg *et al.* have carried out simulations of rapid upward rise of CO_2 to observe the effects of decompression (see **Movies 9.3.1** and **9.3.2**). These simulations mimic a 500 m column above a plume of CO_2 (see **Figure 9.3.2**). These simulations confirm that decompression

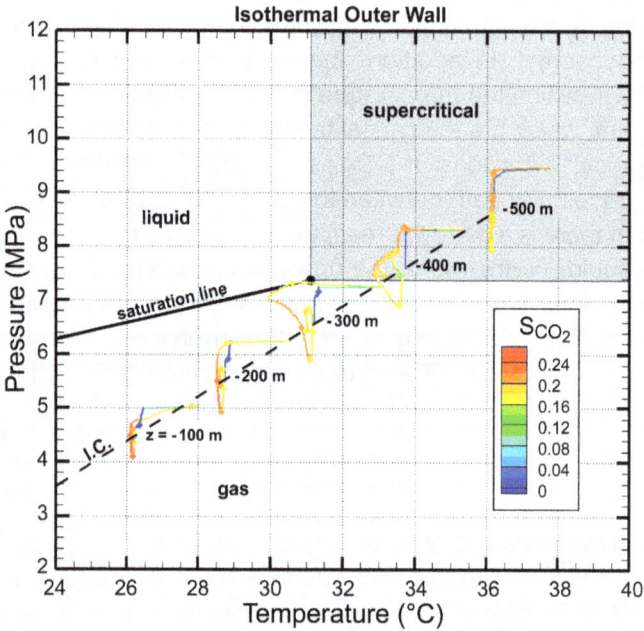

Movie 9.3.1   Simulation of upward CO_2 flow

Simulation of the upward flow of CO_2 in a 500 m column (see **Figure 9.3.2**). In this movie, the boundary conditions on the column are isothermal, so the heat effects are dissipated and play no role. *Movie by Curt Oldenburg* [9.13]. *It can be viewed at:* http://www.worldscientific.com/worldscibooks/10.1142/p911#t=suppl

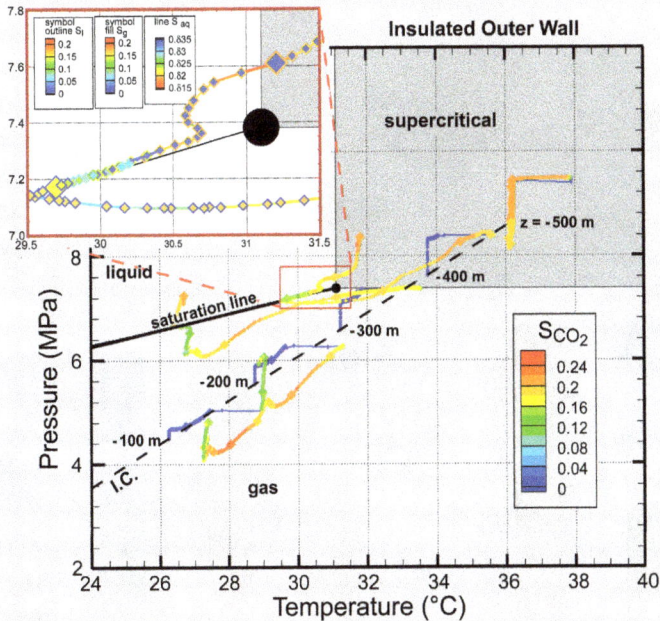

Movie 9.3.2   Simulation of the formation of liquid CO_2

Simulation of the upward flow of CO_2 in a 500 m column (see **Figure 9.3.2**). In this movie, the boundary conditions on the column are insulating, so the heat effects cause liquid CO_2 to form. *Movie by Curt Oldenburg* [9.13]. *It can be viewed at*: http://www.worldscientific.com/worldscibooks/10.1142/p911#t=suppl

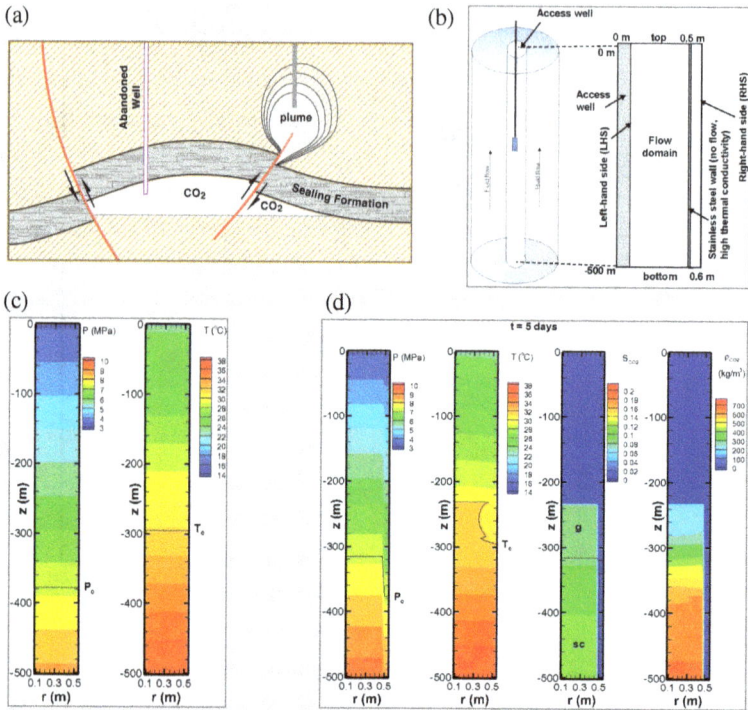

Figure 9.3.2 Stimulation of the upwelling of CO_2

(a) The caprock or seal is preventing the injected CO_2 to move upward. In some cases (see Chapter 10 for more details), faults or abandoned wells can be leakage pathways for CO_2 in these simulations, we are interested in simulating the flow in the rock formation as indicated by the grey cylinder, above the CO_2 plume that is escaping through a fault. In the movies, the effect of heat transfer on the behavior of CO_2 is studied: in **Movie 9.3.1**, heat that is generated during the expansion is dissipated, while in **Movie 9.3.2** the walls of the cylinder are insulated.

(b) Sketch of the 500 m deep and 1 m wide column. In the simulations, the column is assumed to be filled with sand.

(c) Initial conditions for the system are fully brine-saturated (no CO_2 present). The figure shows with a color coding the hydrostatic P (left) and geothermal-gradient T (right) profiles in the column.

(Continued)

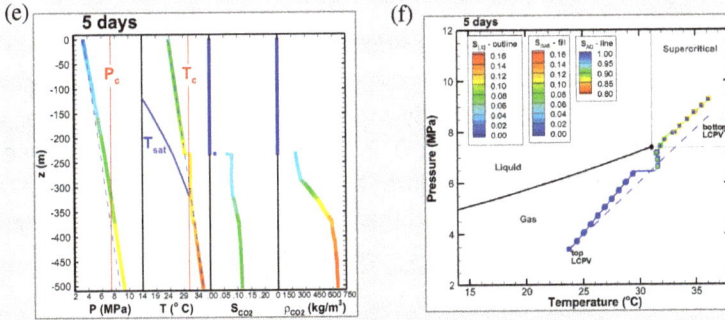

Figure 9.3.2 (*Continued*)

(d) Simulation results for the constant-CO_2 injection case after 3 days. The figures show in addition to pressure and temperature the CO_2 saturation (S_{CO_2}) and density (ρ_{CO_2}) for the case of the fixed geothermal gradient boundary condition. The critical pressure (P_c) and temperature (T_c) are indicated by the light solid curves on the P and T plots.

(e) In our column, the properties are nearly one-dimensional, which allows us to plot the average properties as a function of the depth. The data are after 5 days for the constant-injection case with a fixed geothermal gradient boundary condition. The color of the lines represents the value of the variable being plotted against depth, for visual emphasis. Initial conditions for P and T are shown by the dashed lines. The liquid-gas phase boundary is shown in the temperature frame by the line labeled T_{sat}.

(f) To see whether during the upwelling the phase behavior of CO_2 changes, the results of the previous figure are plotted in the CO_2 phase diagram. The results are again after 5 days for the constant-injection case with a fixed geothermal gradient boundary condition. The dashed line shows the initial conditions. The color of the outline of the symbols represents liquid or supercritical CO_2 saturation, the fill color represents gaseous CO_2 saturation, and the line color represents aqueous phase saturation with value ranges shown by the three legends. Note that the abrupt color change that occurs as the profile crosses the border between the supercritical and gas regions does not correspond to an abrupt change in properties, merely a change in nomenclature.

Figures reproduced from Oldenburg et al. [9.13].

cooling can lead to the formation of liquid CO_2, depending on the ability of the system to provide heat from the surrounding formation. The formation of liquid CO_2 during upflow could inhibit the upward flow, providing a negative feedback for fast upward leakage as described in **Movies 9.3.1** and **9.3.2**.

Solubility of CO_2 in groundwater

The solubility of CO_2 in water is another key physical property that has important consequences for controlling solubility trapping (see Chapter 8). **Figure 9.3.3** shows the solubility profiles for CO_2 in groundwater as a function of depth or pressure for 0, 2, and 6 molar NaCl brine at two different *P-T* conditions. The figure shows that the solubility of CO_2 increases rapidly with depth, and then declines slightly as both pressure and temperature increase and CO_2 becomes supercritical.

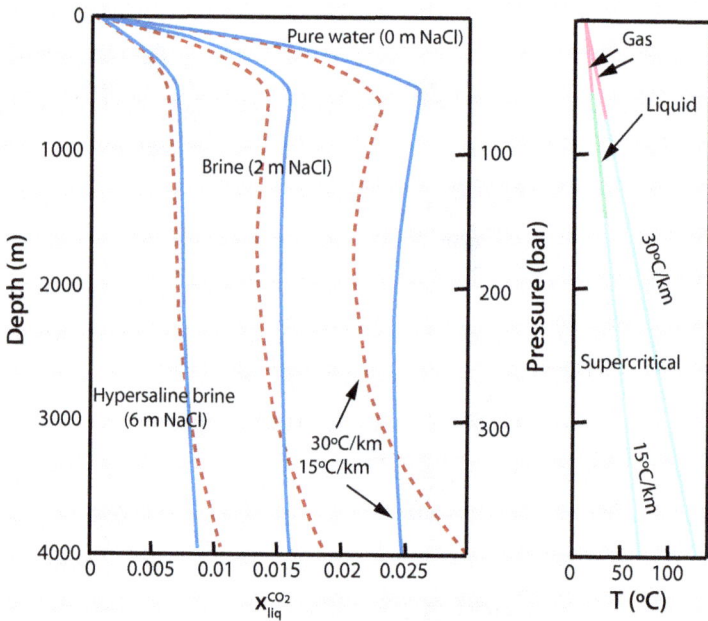

Figure 9.3.3 Solubility of CO_2 in fresh and saline water as a function of depth

Solubility of CO_2 in fresh and saline water as a function of depth or pressure at varying *P-T* conditions (left) and temperature profiles that highlight the various phases that CO_2 assumes (right). The red dashed lines are for a temperature profile of 30°C/km and the blue solid lines for a profile of 15°C/km. *Figure adapted from Oldenburg* [9.11].

Section 4

Wetting and capillary effects

During the initial stage of the injection, a plume of supercritical CO_2 will form and this plume will only very slowly dissolve into the brine. Therefore, the supercritical CO_2 and brine will for a very long time behave as two separate phases. The fact that we have two phases will have profound consequences on both the static and dynamic behavior of these fluids in the pores of a rock.

Wetting

Wetting is the ability of a fluid to wet a substrate. If a fluid wets a substrate it will spread as a thin film, while a non-wetting fluid will form a droplet (see **Figure 9.4.1**). Fluids can also partially wet a substrate, and this wetting behavior is quantified by the contact angle.

Suppose we know the liquid-gas (γ_{LG}), gas-solid (γ_{SG}), and liquid-solid (γ_{LS}) surface tensions and our surface is perfectly smooth. In equilibrium there should be no net force on our droplet, which implies that at the point where the solid, gas, and liquid meet, the net force per unit length should be zero (**Figure 9.4.2**), or:

$$\gamma_{SG} = \gamma_{SL} + \gamma_{LG} \cos\theta,$$

where θ is the contact angle. This is the well-known Young-Laplace equation.

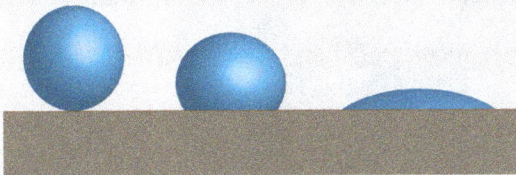

Figure 9.4.1 Wetting a substrate

Differences in wetting behavior: non-wetting (left), partial wetting (middle), and nearly complete wetting (right).

We see that the contact angle, and hence the wetting behavior, not only depends on the interactions of the fluid with the substrate (i.e., γ_{LS}) but also on the liquid-gas surface tension. Let us now look at two model pores. Our models are cylindrical pores, one with a large diameter and one with a small one. **Figure 9.4.3** shows one pore (left) in contact with a gas that wets the substrate and one (right) with a gas that does not wet the walls of the pores. We see that in the first case a wetting layer covers

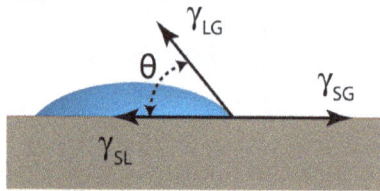

Figure 9.4.2 Partial wetting

Partial wetting of a liquid droplet on a substrate. θ is the contact angle and γ the interfacial tension between the solid (S), liquid (L), and gas (G) phase.

Figure 9.4.3 Capillary condensation

Porous media in contact with a wetting fluid (left) and a non-wetting fluid (right). The top figures represent a large pore and the bottom ones a narrow pore.

the substrate. If the pores are below a certain thickness we see that this wetting layer extends through the entire pore, and we observe capillary condensation. A liquid is formed inside the pores at conditions where the outside fluid is still in the gas phase.

If we look in detail at the gas-liquid interface in the narrow pore (see **Figure 9.4.3**), we see that because of the wetting of the fluid with the walls our meniscus is curved. Such a curved interface causes a pressure difference between the liquid and gas phase, called the **capillary pressure**. Consider a balloon: the surface tension of the skin of the balloon is compensating for the fact that the pressure inside the balloon is higher than the pressure outside. For our fluid in a pore, we can relate this pressure difference to the liquid-gas surface tension using the **Young-Laplace equation** (see **Box 9.4.1**), which reads for a cylindrical pore and a fully wetting fluid (zero contact angle):

$$p_{gas} - p_{liquid} = \frac{2}{R}\gamma_{LG},$$

where R is the radius of the pore. If we observe a non-zero contact angle, the surface has a slightly different shape and the Young-Laplace equation becomes:

$$p_C = p_{gas} - p_{liquid} = \frac{2\gamma_{LG}R}{\cos\theta},$$

where θ is the contact angle and $R/\cos\theta$ is now the radius of curvature of the film. This difference in pressure is called the capillary pressure, p_C. See **Question 9.4.1** for a biological application of the Young-Laplace equation.

Box 9.4.1 Young-Laplace equation

Let us assume that our pore, gas, and liquid are at constant temperature. In addition, the total volume and the total number of particles are constant. From thermodynamics, we know that at these conditions the Helmholtz free energy (A) takes its minimum value:

$$dA = -SdT - pdV + \gamma dA,$$

where we see two work terms: one if we change the volume V and one if we change the area A.

(Continued)

Box 9.4.1 (Continued)

Let us now change the radius of our meniscus, R, by an infinitesimal amount. As we do not change the temperature, the change in the Helmholtz free energy is given by:

$$dA = \left(p_{gas} - p_{liquid}\right)dV + \gamma_{LG}dA,$$

with:

$$V = \frac{1}{2}\left(\frac{4}{3}\pi R^3\right) + V_{rest} \quad \text{and} \quad A = \frac{1}{2}4\pi R^2,$$

where V_{rest} is the part of the volume that does not depend on R. This gives:

$$dV = 2\pi R^2 dR \quad \text{and} \quad dA = 4\pi R dR$$

At equilibrium, the Helmholtz free energy takes a minimum value:

$$\frac{dA}{dR} = -2\pi R^2\left(p_{gas} - p_{liquid}\right) + 4\pi R\gamma_{LG} = 0,$$

or:

$$p_{gas} - p_{liquid} = \frac{2}{R}\gamma_{LG},$$

which is the Young-Laplace equation. In the derivation we have assumed a zero contact angle, which only holds for a fluid that completely wets the substrate. If we observe a non-zero contact angle, the surface has a slightly different shape and the Young-Laplace equation becomes:

$$p_{gas} - p_{liquid} = \frac{2}{R}\gamma_{LG}\cos\theta,$$

where θ is the contact angle. Note that $R' = R/\cos\theta$ is now the radius of curvature of the film.

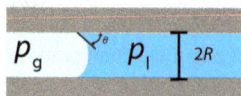

Let us now look at a mixture of brine and supercritical CO_2. In this particular case we have a two-phase system of a supercritical fluid and a liquid. An important question is which phase will wet the rocks? The answer to this question is illustrated at the molecular level by a molecular dynamics simulation shown in **Movie 9.4.1.** This simulation nicely illustrates that if we start with both fluids in contact with the clay surface, the water phase will slowly wet the walls. Indeed, most of the minerals that are in the aquifers are hydrophilic and hence will be wetted by the water phase.

Question 9.4.1 Trees

Calculate, using the Young-Laplace equation the maximal height of a tree if only capillary forces are allowed to transport water to the leaves.

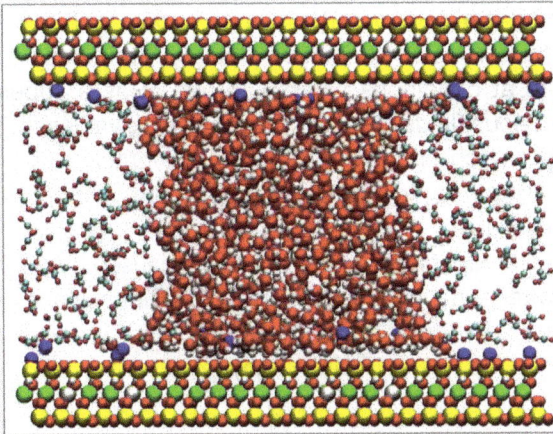

Movie 9.4.1   Molecular dynamics simulation of clay wetting

Movie of a molecular dynamics simulation of water and CO_2 in a 4-nm-wide nanopore between parallel smectite clay surfaces. The clay mineral has a negative structural charge that is balanced by adsorbed exchangeable sodium ions (blue spheres). At the beginning of the simulation, the water phase (O and H atoms in red and white) and the CO_2 phase (O and C atoms in red and light blue) are placed such that the CO_2-water-clay wetting angle is 90°. During the simulation, the wetting angle decreases as water spreads to coat the hydrophilic clay surface. *Movie produced by Ian Bourg. It can be viewed at*: http://www.worldscientific.com/worldscibooks/10.1142/p911#t=suppl

As for the gas-liquid system, we can derive the Young-Laplace equation for the capillary pressure between $CO_{2(g)}$, and brine (w):

$$p_g - p_w = \frac{2}{R}\gamma_{gw}\cos\theta,$$

where we know the surface tension between the CO_2 and brine phase. Because of capillary forces, the wetting phase will be preferentially drawn into the small pores.

Brine-CO₂ interfacial tension

A key parameter in the expression for the capillary pressure is the brine-CO_2 interfacial tension (γ_{gw}). In many practical applications, the values of the capillary pressure are estimated from data on other gasses (e.g., N_2, CH_4). To see whether these estimates are valid, Nielsen et al. [9.14] carried out molecular level calculations to study the effect of pressure on the brine-CO_2 interfacial tension. The results shown in **Figure 9.4.4** show that the interfacial tension decreases strongly with pressure to a value of γ_{gw} ~25 mN/m at conditions where CO_2 exists as a supercritical fluid. Other non-aqueous fluids (e.g., N_2, CH_4) have γ_{gw} values that are roughly twice as large, which brings into question the practice of using data on other non-wetting fluids for those systems where data on the behavior of CO_2 are not available.

The pressure dependence of γ_{gw} values is caused by the adsorption of CO_2 on the water surface as described by the Gibbs adsorption equation:

$$\frac{d\gamma_{gw}}{d\ln f} = -RT\Gamma_g^w,$$

where f is the fugacity of CO_2 (a function of p_g) and Γ_g^w is the Gibbs surface excess of CO_2 relative to H_2O, i.e., the quantity of CO_2 adsorbed on the water surface (**Figure 9.4.4**). This equation, which should be generally applicable to any species in the brine-CO_2 system, shows that the addition of species that adsorb at the brine-CO_2 interface tends to lower the interfacial tension (and, conversely, species that are excluded from the interface tend to increase the interfacial tension). For example, it predicts that if NaCl is excluded from the vicinity of the brine-CO_2 interface, γ_{gw} should increase with brine salinity, as observed experimentally.

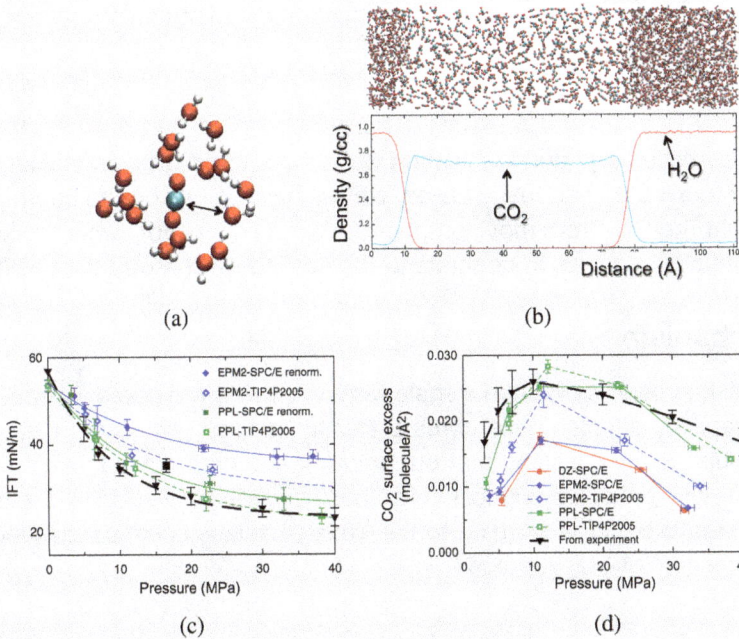

(a)

(b)

(c)

(d)

Figure 9.4.4 Molecular-scale predictions of the CO_2-water interfacial tension

(a) Molecular dynamics (MD) simulation snapshot of a $CO_{2(aq)}$ molecule and the closest water molecules (first solvation shell).

(b) Snapshot of a simulation cell in which a region of supercritical CO_2 is in contact with a region of liquid water. Average density profiles of water and CO_2 in the direction normal to the CO_2-water interfaces are shown below the snapshot.

(c) Interfacial tension γ_{gw} as a function of pressure at 373 K; the black triangles are experimental values; colored symbols are MD simulation predictions obtained with different CO_2-water models.

(d) Density of CO_2 adsorbed on the water surface as a function of pressure at 373 K; the colored symbols are MD simulation predictions; black triangles were deduced from the experimental data in (c) using the Gibbs adsorption equation.

Figures reproduced from Nielsen et al. [9.14].

Section 5

Caprock

We have seen that during injection we form a plume of supercritical CO_2 (see **Figure 9.5.1**). In many cases, we will have to rely on the caprock to prevent the CO_2 from moving upward. We will now apply what we know about fluids in pores to understand the sealing mechanism of caprock.

Mechanism

Recall that caprock is often a shale with relatively small pores. Because of the capillary effects, these pores prefer to be filled with brine relative to CO_2. Let us now assume that our caprock can be modeled as a "membrane" with parallel cylindrical holes that are perfectly wetted by the brine (see **Figure 9.5.2**). According to the Young-Laplace equation, as long as the pressure of CO_2 is below:

$$p_g = p_w + \frac{2}{R} \gamma_{gw},$$

the capillary force will prevent the CO_2 from moving upward. We also see that the smaller the pores, the larger the pressure we can sustain below our caprock (see also **Question 9.5.1**).

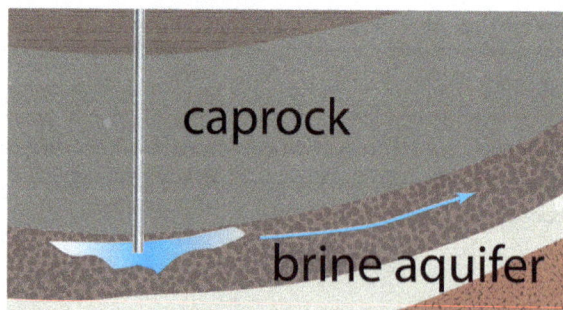

Figure 9.5.1 Caprock

Because of the caprock, the injected CO_2 does not move upward.

Figure 9.5.2 Caprock with model pores

Question 9.5.1 CO_2 versus methane

By the same mechanism that prevents the CO_2 plume from moving upward, caprock can act as a seal for natural gas. Let us now replace the natural gas in this reservoir by CO_2. Argue whether we can put more, less, or an equal amount of CO_2 in this reservoir.

Figure 9.5.2 is, of course, a far too simplified picture of a real caprock. The pores in a caprock have a distribution of shapes and diameters. The sealing capacity of the caprock is determined by the largest pores, which determine the *capillary breakthrough pressure* $p_{c,b}$ of the seal formations, defined as the lowest p_C value at which CO_2 finds a connected pathway through the rock sample.

Core scale measurements of $p_{c,b}$

Experimental data on the capillary breakthrough pressure ($p_{c,b}$) values of seal formations are available only for a few rock samples. Measurements require retrieving an intact water-saturated sample of rock, placing a CO_2-rich phase in contact with one side of the sample, and slowly increasing the pressure of the CO_2-rich phase until CO_2 flows through the rock sample. Experiments with small core samples, which typically require about one month, show that core samples of caprocks act as strong barriers to the flow of inert gasses (N_2, CH_4) and as somewhat weaker barriers to the flow of CO_2. **Tables 9.5.1** and **9.5.2** provide some of these experimental data. If we assume that the ideal Young-Laplace equation holds for these samples, the capillary breakthrough pressure and interfacial tensions listed in **Table 9.5.2** suggest that the critical pore throat radii of seals are on the order of 4 to 200 nm (see also **Question 9.5.2**).

Table 9.5.1 Capillary breakthrough pressure of CO_2 in reservoir and seal rocks

$p_{c,b}$ (MPa)	Porous medium
	Reservoir rocks
0.01 to 0.1	generic sandstone used in field scale models
	Seals
0.21 to 1.85	mudstone
0.64	marlstone
0.74	limestone
3.5 to 4.3	powdered claystone
5.0 to 11.2	evaporite
0.8 to 12.0	generic seals used in field scale models

Capillary breakthrough pressure ($p_{c,b}$) of CO_2 measured in core samples of reservoir and seal rocks. *Data from Hildenbrand et al.* [9.15]; *Li et al.* [9.16]; *Wollenweber et al.* [9.17]; *Skurtveit et al.* [9.18].

Table 9.5.2 Capillary breakthrough pressures of several caprock samples

System	$p_{c,b}$ (MPa)	γ_{gw} (mN m^{-1})
	Sample 1	
CO_2/brine	9.2	21
N_2/brine	27.9	57
	Sample 2	
CO_2/brine	11.2	20
N_2/brine	29.7	56.4
	Sample 3	
CO_2/brine	5.0	25.1
CH_4/brine	12.8	57.4

Capillary breakthrough pressures of caprock samples, with corresponding values of the fluid-fluid interfacial tension measured with different non-wetting fluids (CO_2, N_2, or CH_4). *Data on core samples obtained by Li et al.* [9.16].

Question 9.5.2   Wetting angles, critical pore throat radii, and maximum amount of CO_2 that can be immobilized under a seal

Calculate the CO_2-water-mineral wetting angle (θ) and the critical pore throat radii of the caprock samples in **Table 9.5.2**, assuming that $\theta = 0$ in experiments where the invading fluid was CH_4 or N_2. Then, calculate the range of critical pore throat radii of reservoir rocks and caprocks from the $p_{c,b}$ values in **Table 9.5.1** using γ_{gw} ~22 mN/m and the θ value calculated in the first part of this question. What is the maximum amount of CO_2 that can be immobilized under a seal that has $p_{c,b} = 1$ MPa? Express the results as: (a) the thickness of the CO_2 plume, (b) the mass of CO_2 per area, or (c) the area necessary to store the CO_2 captured annually from a 1 GW gas-fired power plant. Assume that $\phi = 0.25$, $\rho_{H_2O} = 1{,}000$ kg/m^3, $\rho_{CO_2} = 600$ kg/m^3. A modern gas-fired power station with 1 GW net electrical capacity and 90% capture produces about 2.5 Mt CO_2/year for storage.

Fractures and faults

In the previous section, we have analyzed a small sample of a caprock, which we hope is representative of the entire caprock. However, on a distance of several kilometers one could expect the caprock to have fractures and faults, which could lead to potential leakage of CO_2.

Large faults are likely to be detected through remote sensing studies and accounted for in the planning of a sequestration operation. Small faults and fractures, however, may be ubiquitous and essentially impossible to detect. These potential preferential flow paths include healed fractures in the seal rock that could re-open as a consequence of the pressure changes caused by CO_2 injection and ~10 μm wide "micro-annuli" that may exist in wells at the boundaries between the well casings, the well cement, and the caprock formations (**Figure 9.5.3**).

Small cracks and fractures in caprocks are poorly understood. They may become less important with depth if lithostatic pressure, the pressure exerted by the other rocks, helps to re-seal fractures (see **Figure 9.5.4**) [9.19]. Their properties may depend on the mechanical properties of the rock. For example, shale gas production data suggest that wells drilled in shale formations that have high clay and/or organic contents have shorter gas production lifetimes, perhaps because these

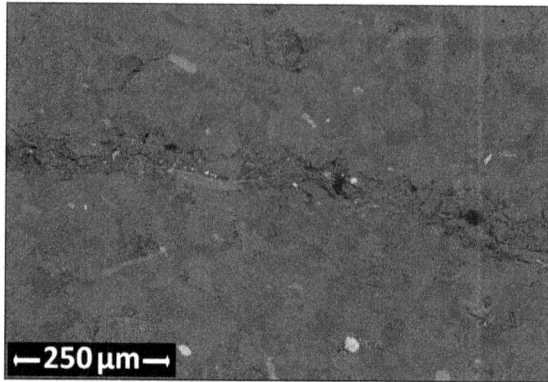

Figure 9.5.3 SEM image of a healed fracture

Scanning electron microscopy (SEM) image of a phyllosilicate-filled fracture (that could, potentially, be reactivated as a preferential flow path) in the Eau Claire caprock (depth 2846 ft.), southwest Ohio. *Image courtesy of Alexander Swift and Julia Sheets (Ohio State University).*

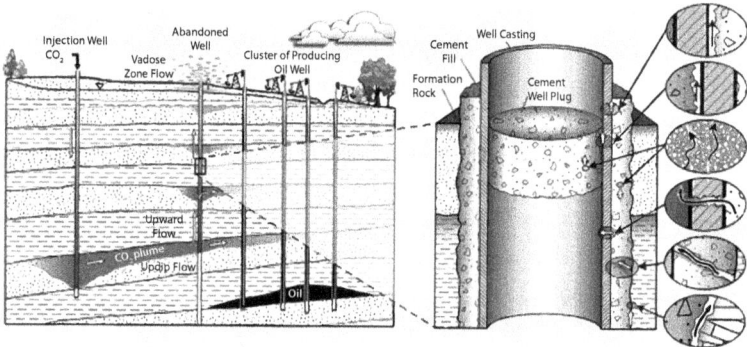

Figure 9.5.4 Potential defects in cemented wells

Schematic figure showing potential defects that could cause wells to act as preferential conduits for fluid flow through seals. *Figure redrawn from Nordbotten et al.* [9.19].

attributes enhance the ability of rock fractures to heal themselves. This tendency, though detrimental to the efficiency of shale gas retrieval, would be beneficial to the security of the CO_2-sequestration process.

Influence of chemical reactions on fracture permeability

If fracture flow is important, seal k_v and $p_{c,b}$ values may vary relatively rapidly because acidic fluids flowing through a fracture will react with the fracture surfaces, potentially modifying the fracture aperture. A key question in the long-term security of geological sequestration, then, is whether fluid (brine or CO_2) flow in seal fractures is self-enhancing or self-limiting. The answer to this question may depend on the composition of the fluid that flows in the fracture and on the rates of geochemical and geomechanical interactions between the fracture fluid and the seal rock matrix (**Figure 9.5.5**) [9.20].

At the base of a seal formation, the fluid flowing through fractures may be either pristine reservoir water that has been displaced by the CO_2 injection or, close to the CO_2 plume, acidic fluids: CO_2-enriched water or

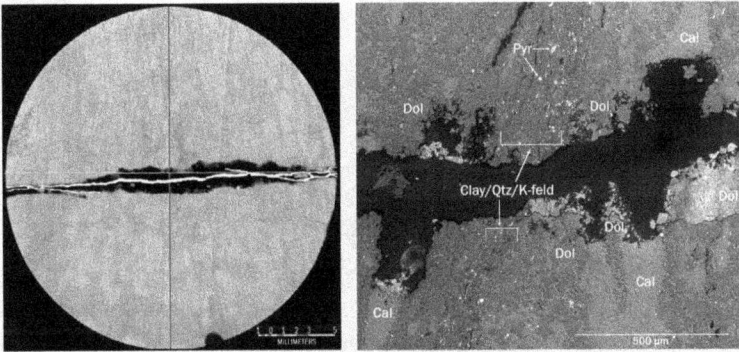

Figure 9.5.5 Flow and geochemistry in an artificially fractured caprock

Experimental results on the evolution of an artificially fractured caprock (the top seal of the Amhersburg formation, a carbon sequestration pilot site in Northern Michigan). The left side of the figure shows a cross-section of the core sample at the beginning of the experiment (the initial state of the fracture is shown in white) and after one week of exposure to acidic brine (the final state of the fracture is shown in black) as reconstructed from micro-CT scans. The right side of the figure (a back-scattered electron scanning electron microscope (BSE-SEM) image of a portion of the fracture after exposure to acidic brine) shows that the dissolution of fracture surfaces occurs preferentially in regions occupied by calcite grains ("Cal") rather than in regions occupied by dolomite ("Dol") or silicate minerals. *Images reproduced from Ellis et al.* [9.20], *with permission from Elsevier.*

a two-phase mixture of water and CO_2, depending on whether CO_2 can overcome the capillary entry pressure of the fracture. Acidic fluids would tend to dissolve the fracture walls and increase the fracture aperture, as shown by experimental studies with artificially fractured seal rocks (**Figure 9.5.5**). As the acidic fluid or fluid mixture travels upward through the fracture, weathering reactions may promote the precipitation of carbonate minerals farther up in the fracture. The overall geochemical process is likely to be analogous with the mineral trapping reactions that occur within the reservoir (which will be discussed in Section 9.8), but the dissolution and precipitation reactions will occur at different locations along the fracture. The hypothesis that carbonate mineral precipitation can seal fractures is supported by studies of natural CO_2 seeps [9.20].

Section 6

Permeability

In contrast to caprock, a good reservoir sandstone will have, in addition to sufficient porosity (i.e., storage capacity), a high permeability (i.e., ease with which fluids can be transported through the reservoir). In these reservoirs, the permeability is determined by the degree of connectedness of the pores.

The permeability is measured in units of m^2 or equivalently, darcys (d), where 1 darcy equals 10^{-12} m^2 and 1 md = 10^{-15} m^2 (see **Box 9.6.1** for the difference between this definition and the permeability of a membrane, which we introduced in Chapter 7). A good reservoir rock will have permeability on the order of 100 md or higher. For comparison, a good caprock has very low permeability of 0.1 md or less.

Permeability is a property that varies over many orders of magnitude for common rocks. Porosity and permeability are generally correlated, but the relation between porosity and permeability is complex and no universally applicable correlation exists. Even in the same geological formation, the relation between porosity and permeability is non-unique and dependent on the characteristics pertaining to the interconnectedness of the pores. **Figure 9.6.1** illustrates the permeability as a function of porosity for sandstones in the St. Peter sandstone of the Illinois Basin [9.21]. A graphical representation of the variability in permeability is shown in **Figure 9.6.2** [9.22]. Generally, as grain size increases in rocks, the permeability increases by over eight orders of magnitude.

Fracture permeability

The data in **Figure 9.6.2** also show the effect of fractures on the permeability of some rocks. Fractures can impart very high permeability to otherwise low-permeability rocks. In fact, even a single fracture with a small aperture can impart a fairly large overall permeability to a rock. It has been shown that fractures provide permeability k (in m²) according to the so-called Cubic Law, which follows from the flow of a fluid between two parallel plates:

$$k = \frac{Nb^3}{12},$$

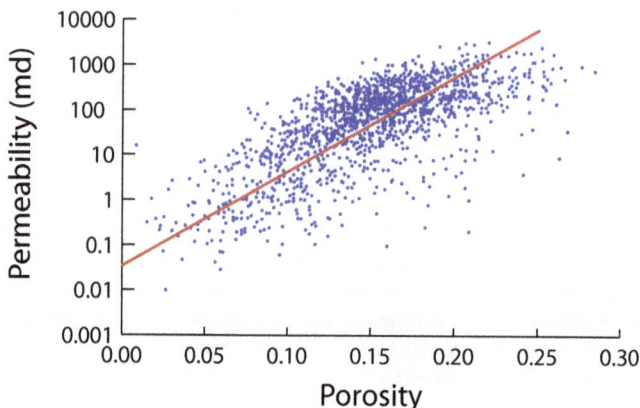

Figure 9.6.1 Relation between permeability and porosity
Permeability (md) as a function of porosity (%) for thousands of core samples from the St. Peter sandstone in Illinois. *Figure based on Leetaru et al.* [9.21].

where N is the number of fractures per meter, and b is the fracture aperture. For example, one fracture with thickness equal to that of a sheet of paper across 10 m is equivalent to a permeability of $k = 0.1$ m^{-1} × $(1 \times 10^{-4}$ m$)^3/12 = 8 \times 10^{-15}$ m$^2 = 8$ md, which is is approximately equivalent to the overall permeability of silt.

Fracturing can be caused by pressure from the CO_2 injection. **Figure 9.6.3** shows various pressure profiles in the subsurface relevant to fluid injection. The hydrostatic pressure *profile* is determined by gravity acting on fluid in the connected pore space of the geological system. Assuming continuous pore connectivity and equilibrium conditions, pressure increases with depth as a function of fluid density. At these conditions, the pressure inside the pores is typically very nearly hydrostatic. In order to inject fluid into a subsurface formation, the injection pressure must be larger than the local pore pressure. The curve on the far right-hand side of **Figure 9.6.3** is the *lithostatic gradient*. This pressure arises from the weight of the rock itself, and is a function of total rock density and depth. If the fluid injection pressure exceeds the lithostatic pressure, the rock will open up along horizontal planes and lift the rock column to form a cavity or void space (see also **Question 9.6.1**).

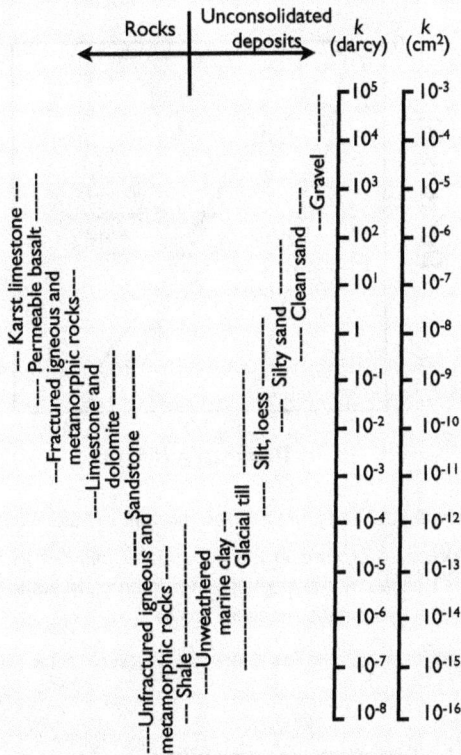

Figure 9.6.2 Permeability of different kinds of rocks

General ranges of permeability (k) in two different units for different kinds of rocks. *Data from Freeze and Cherry* [9.22].

The intermediate curve between hydrostatic and lithostatic profiles is the more interesting one, and is referred to as the *fracture gradient*, or frac gradient. At pressures greater than the frac gradient pressure at any depth, rock can fracture, creating high-permeability vertical pathways. The frac gradient can be determined in a well by a well-testing procedure known as a leak-off test, whereby the bottom of the well is insulated from the rest of the well and high fluid pressure is applied through tubing. The frac pressure is the pressure at the point when a sharp increase in flow rate occurs as fluid pressure increases, at this point the formation begins to fracture. When this test is conducted at multiple depths during the drilling of the well, we obtain an average fracture gradient.

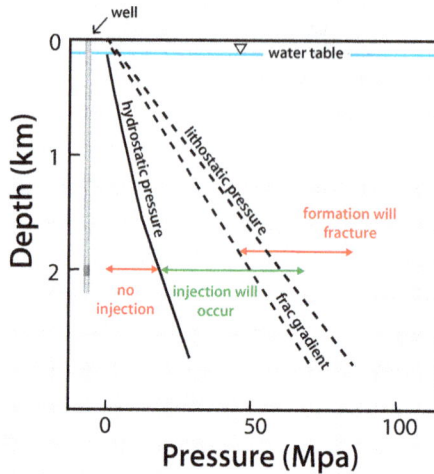

Figure 9.6.3 Pressure profiles relevant to fluid injection

Sketch of various pressure profiles in the subsurface relevant to fluid injection. The water table, which is defined as the level of water in an open static well at equilibrium, is the starting point of a hydrostatic pressure profile.

Question 9.6.1 Fracture permeability

Examples of fractures in a cave carved into soft chalk (calcium carbonate) are shown in the figure. These fractures were probably caused by near-surface subsidence or stress changes related to the excavation of the cave. If there is on average one fracture every 2 m with aperture 0.5 mm running through this otherwise low-permeability rock, what is the effective permeability of the rock? Hint: The correct answer gives a permeability similar to that of clean sandstone.

Fractures exposed in a cave impart a large overall permeability to an otherwise low-permeability rock. *Photo by Curt Oldenburg, 2011.*

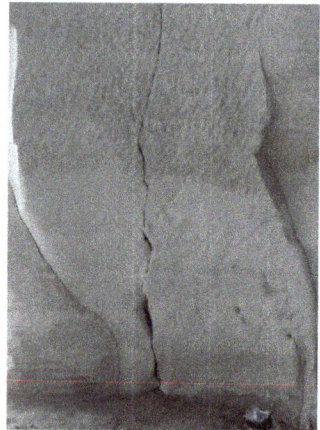

Hydraulic fracturing or **fracking** is permitted by regulatory agencies for producing natural gas and oil from shale (see **Box 9.6.2**). Hydraulic fracturing is, however, prohibited for most other injection wells. Because hydraulic fractures can be hard to control and can alter the properties of the formations, fracking may allow potential migration of the disposed fluid into protected groundwater resources. To protect these groundwater resources, regulatory agencies require fluid pressure to remain below the so-called frac gradient pressure. We will be discussing this aspect further in the section on potential impacts of geological sequestration.

Scale-dependence of permeability

The permeability of rocks not only varies over many orders of magnitude, but also depends on the scale of the rock sample. This scale-dependence of permeability originates from heterogeneity in rocks and the way permeability is measured. The permeability is obtained from the flow rate for a given pressure gradient. Flow paths of different length scale will be accessible to fluids depending on the length scale of measurement. If the pressure gradient is placed over a long distance, more flow pathways will be accessed, and the likelihood of intersecting a long fast-flow path is higher. Because the flow rate will reflect the highest permeability pathways sampled, analogous to an electric current with various resistors positioned in parallel rather than in series, the measured permeability increases when the length scale of the measurement increases.

Box 9.6.2   Hydraulic fracturing (or "fracking")

The ability to create high-permeability vertical fractures is the basis for the relatively new approach to extracting hydrocarbons called hydraulic fracturing (or "fracking"). Hydraulic fracturing is allowing the production of large amounts of natural gas and oil from otherwise very low-permeability shale rocks. In hydraulic fracturing, long horizontal wells are drilled through the shale formations that contain either natural gas (primarily methane) or oil in economically useful quantities. Fluid at very high pressure is then injected into the horizontal wells to create sub-vertical fractures in the shale. Proppants (e.g., sand) and other chemical agents in the injected fluid flow into these fractures and serve to hold them open even after the injection over-pressure is removed. These open fractures provide the permeability needed to economically extract hydrocarbons from the rock for fuel.

Figure 9.6.4 illustrates this scale-dependence of the permeability for measurements from laboratory to borehole scale [9.23].

The implications of the scale dependence of permeability for carbon sequestration are profound. First, because injection wells are relatively small-scale access points to the formation, it is most likely that they will not intersect rare high-permeability features. But if they do intersect fractures or faults, or once the injected CO_2 encounters a nearby permeable fracture or fault, these features will take up most of the injection mass, thereby bypassing nearby pore space. This tendency continues throughout the injection process. Under these conditions, CO_2 does not fill the pore space efficiently. Second, at long times following the injection period, a similar effect occurs as CO_2 buoyancy provides a constant upward driving force and the injected CO_2 plume underlies a large area of caprock. In this configuration, any weakness in the sealing integrity of the caprock over the large footprint of the CO_2 plume will likely result in leakage into or through the caprock. The analogy with rainfall and an old roof on a house comes to mind; if there is a way for water to leak in, it will.

We will return to some of these ideas related to how permeability and its variation affect injection and migration of CO_2 in the section on Capacity Assessment (Chapter 10).

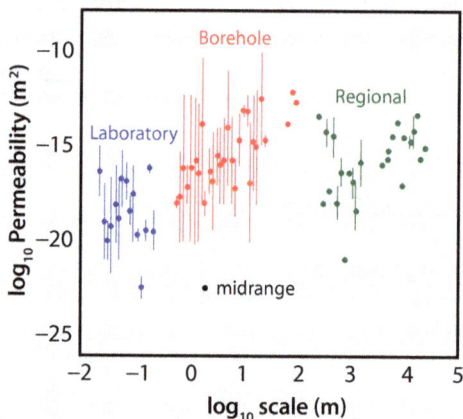

Figure 9.6.4   Scale effect of permeability

Scale effect of permeability as measured for crystalline rocks. Sedimentary rocks are believed to show the same general trends in permeability as a function of measurement scale. *Data from from Clauser* [9.23].

Section 7

Two-phase flow

In the previous section, we looked at the permeation of a single phase through a geological formation. After injection, most of the CO_2 will remain in its supercritical form as a free phase fluid that is mostly immiscible with native formation brine for decades to hundreds of years. Hence, the transport of CO_2 through an aquifer is a two-phase flow process.

Relative permeability

In the previous section (**Box 9.6.1**), we defined the single phase permeability using **Darcy's law**:

$$j = -\frac{k}{\mu}\nabla p,$$

where ∇p is the pressure gradient, μ the viscosity of the fluid, and k is the single phase permeability of the rock. To extend Darcy's law to two-phase flow, we can define an *effective* permeability of phase i:

$$j_i = -\frac{k_{ij}}{\mu_i}\nabla p_i = -\frac{k_{r,i}k}{\mu_i}\nabla p_i,$$

where k_{ij} is the effective permeability of phase i in the presence of phase j. In the second equation, we have introduced the **relative permeability** $k_{r,i}$ of phase i in the presence of phase j, which is defined as the effective permeability divided by the single phase permeability (**Box 9.7.1**).

To describe the permeation of a two-phase fluid through our rock, we need to know how the relative permeability depends on the volume fraction of the pores that is occupied by each of the phases. The fraction of the pure volume that is occupied by phase i is called the **phase saturation** (e.g., brine saturation or CO_2 saturation):

$$S_i = \frac{V_i}{V_p} = \frac{V_i}{\phi V_T},$$

where V_i is the total volume of phase i, V_p is the pore volume. In the second equation, ϕ is the porosity and V_T is the total volume of the rock.

The simplest model for two phase flow is to assume that brine and CO_2 flow completely independently. Suppose in our rock we have $S_g = 0.1$, i.e., 10% of the pore volume is occupied by CO_2 and 90% by brine. If the flow can be seen as two independent single flows, we have 10% of the rock in which CO_2 permeates as if it were a single phase and in 90% of the rock, single phase flow of brine. Or, in more general terms, if S_i is the saturation of phase i, we have a relative permeability, simply given by:

$$k_{r,i} = S_i$$

For most fluids $k_{r,i}$ is smaller than 1. Let us now see how differences in wettability of the two phases change this simple picture of two-phase flow.

We consider the two phase flow of brine and CO_2. As we have seen before, brine is the wetting phase. **Figure 9.7.1** presents two scenarios for a high (a) and low (b) brine saturation. In **Figure 9.7.1 (a)**, the brine phase occupies most of the pore space, and thereby forms a connected

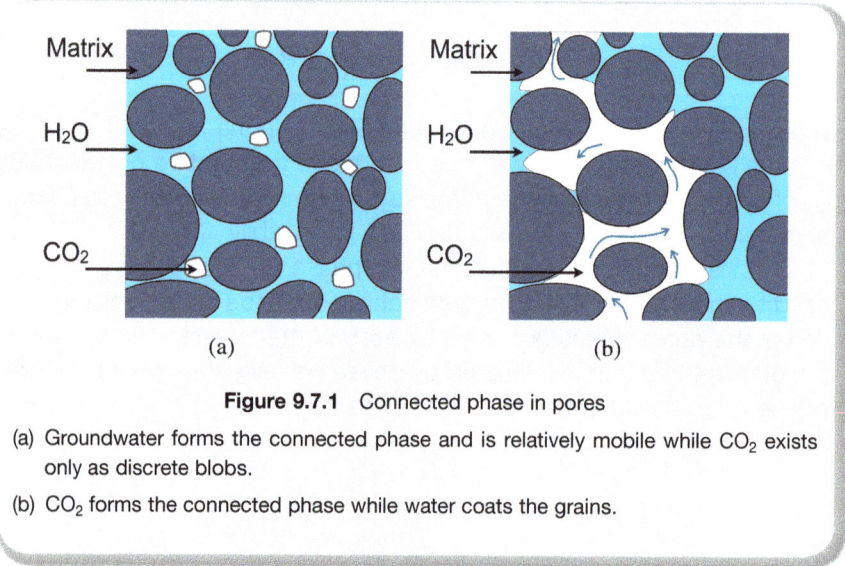

Figure 9.7.1 Connected phase in pores

(a) Groundwater forms the connected phase and is relatively mobile while CO_2 exists only as discrete blobs.

(b) CO_2 forms the connected phase while water coats the grains.

phase that can move relatively freely in response to the pressure gradient. In contrast, CO_2, which is at low saturation, exists only as discrete non-connected blobs. These blobs have a relative permeability of zero, because we cannot apply a pressure gradient across the discontinuous CO_2 phase. We see that in this scenario the actual mobility of CO_2 is much lower compared to what one would expect on the basis of the ideal relative permeability.

Figure 9.7.1 (b) shows the opposite situation in which the CO_2 is the connected phase and now the brine phase is relatively immobile. As the brine is wetting the grains, we see a wetting layer around the grains and, because of the capillary forces, brine is filling the narrow pores. We see that this is a qualitatively different picture compared to the one with the CO_2 blobs. But also in this case, the brine phase will become immobile below a certain saturation.

In general, we see that the phase with greater saturation is more mobile than the phase with lower saturation. In some situations, the mobility of a phase becomes zero before its saturation reaches zero.

We can also look at **Figure 9.7.1** in the context of the capillary pressure which we have defined as the non-wetting-phase pressure minus the wetting-phase pressure. In a two-phase system, the capillary pressure is zero if we have a pore which is saturated with the wetting (brine) phase. If we inject CO_2, we need to apply a pressure above the hydrostatic pressure in order to induce the flow. If the pressure above the hydrostatic pressure is small, the capillary pressure is small and we are only able to displace brine from the large pores. To further decrease the brine saturations, we need to increase our pressure on the CO_2 to displace the water in the smaller pores. We can continue to increase the pressure until the non-wetting phase dominates the pore space. At this point we have a disconnected, immobile, brine phase. As the CO_2 can bypass these pockets of brine, a further increase of the CO_2 pressure does not displace any brine. The relation between the capillary pressure and the brine saturation is illustrated in **Figure 9.7.2.** The capillary pressure increases when the volume fraction of the non-wetting phase in the pores increases. The maximum capillary pressure is reached when the wetting phase is at its lowest point of saturation, the point of **residual saturation** or **irreducible saturation**.

Models have been developed to describe the relative permeability and capillary pressure as a function of the saturation. The functions $k_{r,g}(S_g)$, $k_{r,w}(S_g)$, and $p_c(S_g)$ are referred to as the *characteristic curves* of the porous medium. Several semi-empirical relations have been proposed for modeling these curves, for example:

$$k_{r,w} = \left(S^*\right)^{0.5}\left\{1-\left(1-\left[S^*\right]^{1/m}\right)^m\right\} \tag{I}$$

$$p_c = p_0\left(\left[S^*\right]^{-1/m} - 1\right)^{1-m} \tag{II}$$

$$k_{r,g} = \left(1-S'\right)^2\left(1-S'^2\right) \tag{III}$$

where p_0 (MPa) and m are fitting parameters and S^* and S' are rescaled brine saturations:

$$S^* = \frac{S_w - S_{w,r}}{1 - S_{w,r}} \quad \text{and} \quad S' = \frac{S_w - S_{w,r}}{S_w - S_{w,r} - S_{g,r}}$$

The rescaled brine saturations take into account the residual brine saturation of the porous medium $S_{w,r}$ and the residual CO_2 saturation $S_{g,r}$.

The proposed equations (I–III) suitably describe expected trends: p_c is greater than zero and increases with S_g, as pressure must be applied

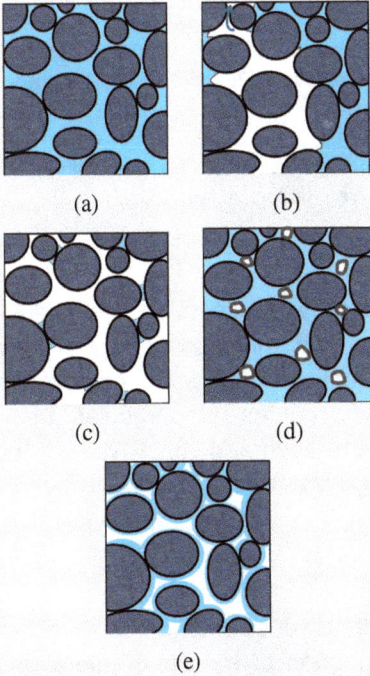

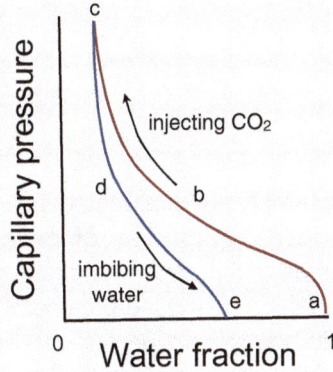

Figure 9.7.2 Capillary hysteresis

The graph gives the brine saturation as a function of the capillary pressure. We start with a formation with only brine (a). If we increase the pressure of CO_2 above the hydrostatic pressure, CO_2 will flow into the largest pores (b). The amount of brine will decrease because of the increase in capillary pressure, we displace brine from increasingly smaller pores. The brine saturation will decrease until we reach the minimum brine saturation for which the brine mobility is zero because the brine is trapped in disconnected capillaries (c). If we now decrease the CO_2 pressure, we see that brine is wetting the grains (d) and filling the small pores first, thus isolating the CO_2 phase. As a result, the wetting curve has for the same capillary pressure a lower brine saturation. If we again reach the hydrostatic pressure, droplets of CO_2 remain trapped in the formation (e).

in order to force CO_2 to invade the porous medium, and $k_{r,i}$ increases with S_i, as each fluid flows more readily if it occupies a larger fraction of the pore space. Characteristic curves have been extensively investigated in the case where the wetting fluid is water and the non-wetting fluid is air (for example, in unsaturated soils), but they remain much less extensively examined in the case of brine-CO_2 fluid mixtures at high T and P. Because of their relative simplicity, Equations (I–III) do not capture the full complexity of measured characteristic curves. In particular, they are poorly adapted to porous media where several distinct types of pores with very different sizes contribute significantly to the porosity.

Hysteresis

The reader may have noticed that in our discussion of the relative permeability and capillary pressure as a function of the brine saturation, we conveniently started with a sample that was fully saturated with brine.

Let us now do the following experiment (see **Figure 9.7.2**). We start with our sample for which the pores are fully brine-saturated ($S_w = 1$). We have seen that we can inject CO_2 if we apply a pressure higher than the local hydrostatic pressure of the fluid (see **Figure 9.6.3**). This will displace the brine (the drainage process) in the largest pores, and by increasing the CO_2 pressure we can displace brine from increasingly smaller pores until CO_2 forms a continuous path and we are left with our residual brine saturation.

The next step in our experiment is to reverse the invasion of CO_2. We stop the injection and the pressure of the CO_2 phase will slowly decrease and brine will flow back (the imbibition process). As brine is wetting the grains, the capillary forces cause it to wet the pore throats first. This process traps bubbles of CO_2. At this point, it is important to note that imbibition and drainage present qualitatively different situations: because of the trapped bubbles of CO_2, at the same capillary pressure the water saturation is lower during imbibition. Hence, imbibition and drainage do not follow the same capillary pressure curve. The **imbibition** process will continue until brine is in the continuous phase. The remaining bubbles of CO_2 are trapped and the mobility of the CO_2 has now dropped to zero. Once the permeability of the CO_2 phase has dropped to zero, we cannot remove any additional CO_2. After the experiment, we have again reached the hydrostatic pressure, but we do not have the fully saturated brine phase with which we started our experiment! This phenomenon by which the relative permeability of a phase (brine during drainage, CO_2 during imbibition) decreases to zero while a significant quantity of that phase (the residual phase saturation) remains present in the porous medium is known as **residual phase trapping**.

This phenomenon of following different paths during drainage and imbibition is called hysteresis and the resulting residual CO_2 saturation is the basis of the capillary or residual trapping process that was introduced in Chapter 8 (**Figure 8.2.3** and **Movie 8.2.1**).

Significance of residual-phase trapping

Residual phase trapping has both a positive and negative effect on geological CO_2 sequestration.

The negative effect is that if we inject CO_2 in an aquifer, the residual saturation of the water will reduce the effective capacity of a particular geological formation.

The positive effect is illustrated in **Figure 9.7.3**, which shows a post-injection situation in which long-term CO_2 injection has produced a large CO_2 plume that has migrated upward and is trapped against a gently dipping caprock. At the leading edge of the CO_2 plume, water will be displaced by CO_2 and the system will follow a drainage curve with respect to relative permeability and capillary pressure. Along the drainage curve, relative permeability of the rock to CO_2 is high and migration is facilitated. At the trailing edge of the plume, the system will follow a wetting curve as CO_2 migrates up-dip and groundwater imbibes the pore space behind the plume. Along the wetting curve, relative permeability of CO_2 in the rock is lower than along the drainage curve and, most significantly, at the residual saturation of CO_2, the relative permeability of CO_2 becomes zero. This residual CO_2 is effectively trapped. By this process, an upward-migrating CO_2 plume eventually will lose all of its mobile fraction to residual gas saturation. What this thought experiment illustrates is that upward-migrating CO_2 plumes eventually become completely

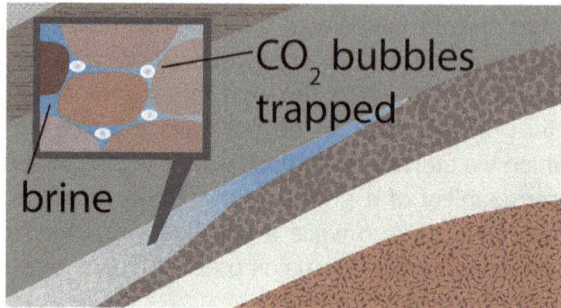

Figure 9.7.3 Residual trapping

Schematic figure showing a CO_2 plume immobilized by residual gas trapping hundreds of years after the end of CO_2 injection. The figure neglects any stratigraphic, solubility, or mineral trapping.

trapped by residual gas saturation, even in the absence of a closed structure (anticline or dome) and associated caprock. Residual CO_2 trapping, then, is one of the key mechanisms of carbon sequestration (**Question 9.7.1**). Higher residual trapping yields smaller plume migration distances, greater storage security, and higher storage capacities.

Residual gas trapping is measured using *core flooding* experiments, where core samples of reservoir rocks, initially brine-saturated, are successively invaded with CO_2, then subject to brine imbibition at controlled pressure and temperature. The value of S_g at the beginning of brine imbibition is known as the initial CO_2 saturation $S_{g,i}$. The steady-state value of S_g reached during brine imbibition is the residual CO_2 saturation $S_{g,r}$ (**Figure 9.7.4**). In the final state, any CO_2 that remains in the column consists of immobile, disconnected saturation (bubbles). At present, relatively few high-quality data are available on the $S_{g,r}$ values of CO_2 in reservoir rocks. Experimental results have been obtained only for small core samples (with lengths ~5 cm) and their applicability at much larger scales (meter to kilometer scales of reservoir models) is unclear. Values used in field scale models and storage capacity estimates range from $S_{g,r} = 0.05$ to 0.4 (see **Table 9.7.1**).

Given the importance of residual trapping, current research is focused on obtaining a better understanding of how residual trapping is related to the properties of the rocks. Such understanding is important in

Question 9.7.2 Brine imbibition in a capillary network

The two branches shown in the figure are capillary tubes with the same length and different diameters. Does brine (blue) imbibe more rapidly in the narrower tube, in the broader tube, or at the same rate in both tubes?

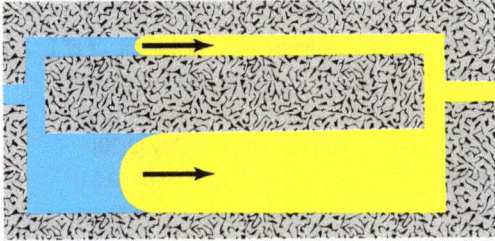

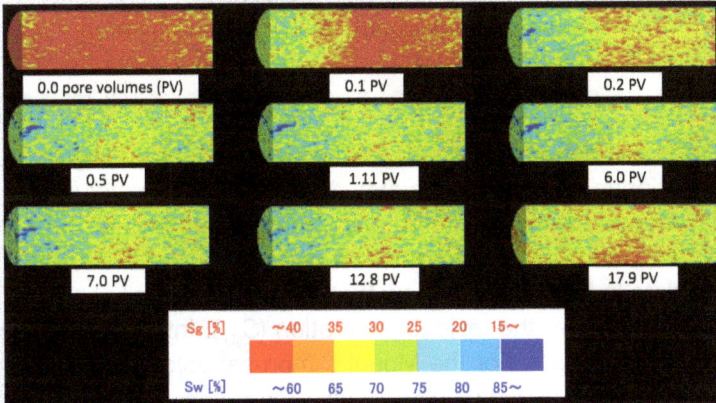

Figure 9.7.4 Brine imbibition in a sandstone core sample

X-ray CT images of CO_2 saturation during brine imbibition in a core sample (14.5 cm long, 3.68 cm in diameter) of Tako sandstone at 40°C and 10 MPa. The figure in the upper left corner shows the initial state of the sample after invasion by CO_2. The other figures show CO_2 saturation in the core sample after flushing by up to 17.9 pore volumes of brine. After about one pore volume of brine has been flushed through the sample, CO_2 saturation stabilizes at $S_{g,r}$ ~0.3. *Image reproduced from Shi et al.* [9.24], *with permission from Elsevier.*

Table 9.7.1 $S_{g,r}$ values of reservoir sandstones used in field-scale carbon sequestration models or measured in core-scale experiments

$S_{g,r}$	Source
Values used in field-scale models	
0.05	André *et al.* [9.26]; Liu *et al.* [9.27]
0.18	Alkan *et al.* [9.28]
0.20 to 0.25	Zhou *et al.* [9.29]
0.27*	Doughty [9.30]
0.30	Okwen *et al.* [9.31]
0.4*	Juanes *et al.* 9.32]
Values obtained from core-scale measurements	
0.10 to 0.30	Bachu and Bennion [9.33]
0.28	Shi *et al.* [9.24]
0.21 to 0.33*	Krevor *et al.* [9.25]

* Maximum possible values of $S_{g,r}$ (in studies where $S_{g,r}$ was modeled or measured as a function of the maximum historical CO_2 saturation $S_{g,i}$).

order to make better predictions of the potential for geological carbon sequestration in an aquifer. The experimental data on residual gas trapping in otherwise water-saturated porous media suggest the existence of correlations between $S_{g,r}$ and other properties.

The data in **Figure 9.7.5** show that the higher the initial saturation of CO_2 ($S_{g,i}$), the larger the residual saturation ($S_{g,r}$). Interestingly, this correlation is not perfect. For the Mount Simon sandstone we see that the experimental data show an optimal initial saturation. A possible explanation is the wetting behavior of this sandstone. The Mount Simon sandstone data may be characteristic of porous media that are partially wet rather than completely water-wet.

Existing studies also suggest that $S_{g,r}$ values are highest at conditions where brine imbibition is slow and dominated by capillary forces and are lower if brine imbibition is rapid and dominated by viscous forces (as is expected to be the case during geological sequestration, but may not be the case in laboratory experiments). However, even in studies that used high initial gas saturations and low brine flow rates, reported values of $S_{g,r}$ ranged broadly.

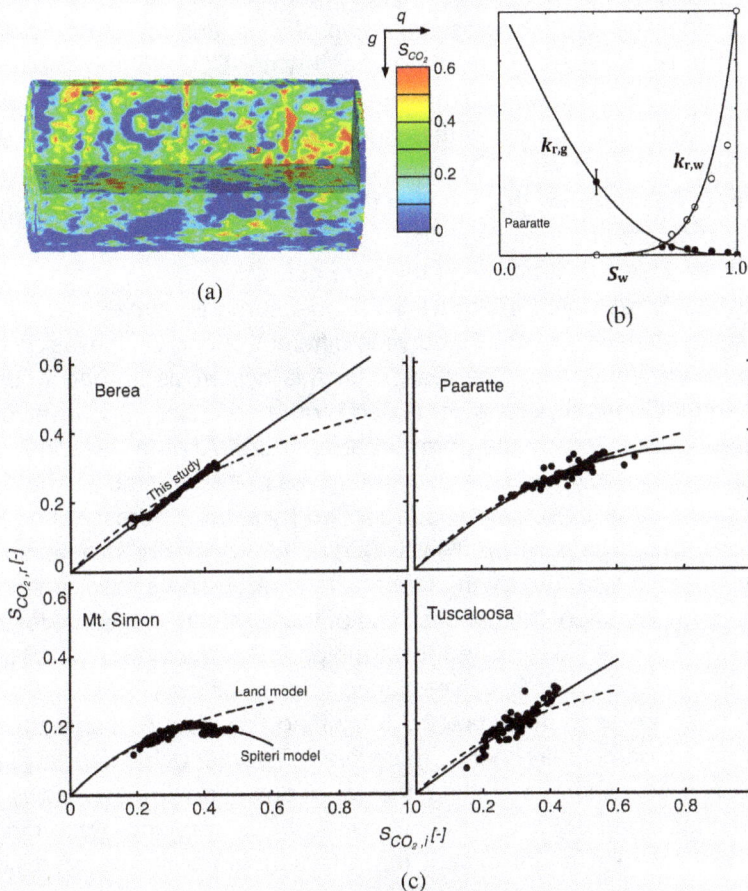

(a)

(b)

(c)

Figure 9.7.5 Core-scale data on residual CO_2 trapping

Core-scale data on residual CO_2 trapping in four sandstones (Berea, Paaratte, Mount Simon, Tuscaloosa) at 50°C and 9 MPa.

(a) X-ray CT reconstruction of CO_2 saturation in a sample of Paaratte sandstone when the inlet fluid mixture is 90% CO_2 (flow is from left to right; gravity from top to bottom).

(b) Relative permeability curves measured during CO_2 invasion in the Paaratte sandstone.

(c) Residual CO_2 saturation in the four sandstones, plotted against the maximum historical CO_2 saturation. The data are fitted with two different mathematical models of the relationship between $S_{g,r}$ and $S_{g,i}$. *Images reproduced from Krevor et al.* [9.25], *with permission from John Wiley and Sons.*

To understand why $S_{g,r}$ values are so challenging to characterize, let us consider the capillary pressure curve of a generic rock sample. If we knew the pore radius R and the mineral-water-CO_2 wetting angle θ of each individual pore, we could predict the capillary pressure curve of our sample in the ideal case where each pore is filled with brine if $P_c < (2/R)$ $\gamma_{gw} \cos\theta$ and filled with CO_2 otherwise. This "ideal" P_c curve (shown in **Figure 9.7.6**) assumes that the fluid filling each pore depends only on the size of the pore and the wettability of the pore walls and does not depend on the larger-scale structure of the pore network.

In reality, as noted in **Figure 9.7.2**, measured P_c curves have a significant hysteresis, i.e., the P_c curve is different during CO_2 invasion than during brine imbition. One reason for this hysteresis is that, whereas brine imbition in a pore is controlled by the pore size, CO_2 invasion in a pore is controlled by the pore throat aperture. Another reason is that the finite connectivity of pore networks gives rise to situations where, for example, a large pore may be surrounded by smaller pores: during CO_2 invasion, this large pore will fill with CO_2 at a P_c value that is determined, not by its own size, but by the size of the surrounding smaller pores. As a first approximation, the net outcome of these effects is that the ideal P_c curve in **Figure 9.7.6** is shifted toward higher P_c values during CO_2 invasion and toward lower P_c values during brine imbition.

As can be seen in **Figure 9.7.6**, shifting the ideal P_c curve upward does not fundamentally change the overall character of the capillary invasion process, whereas shifting the ideal curve downward allows the emergence of a new phenomenon: residual CO_2 trapping. According to **Figure 9.7.6**, one might expect that the conditions that favor residual trapping include an almost flat P_c curve (i.e., the porous medium must contain many large pores) and a large hysteresis (i.e., the large pores must be surrounded by narrow pore throats or smaller pores). The degree by which the ideal P_c curve is shifted downward during brine imbition (which has a very large impact on $S_{g,r}$) cannot be predicted solely from the properties of individual pores: it requires an understanding of the relative size of the pores and pore throats, the structure of the pore network, and the fluid dynamics of brine imbition in the porous medium.

This fundamental difference between CO_2 invasion and brine imbition is illustrated in **Figures 9.7.7** and **9.7.8**. The first figure shows (on the right side) an X-ray micro-CT (computed tomography) image of CO_2 distribution during CO_2 invasion in a sandstone sample. The finely

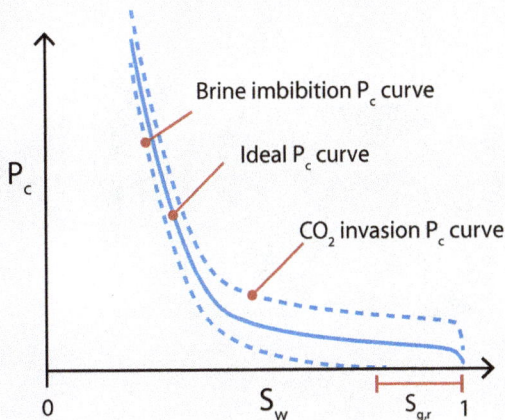

Figure 9.7.6 Ideal capillary pressure curve

Illustration of the hysteresis of the capillary pressure curve of a rock sample. The solid line shows a schematic view of the ideal P_c curve that would be calculated from the distribution of capillary entry pressures of all pores in the porous medium. The dashed lines illustrate how the ideal P_c curve is shifted to higher or lower P_c values during CO_2 invasion and brine imbibition, respectively. Shifting the curve upward does not fundamentally change the character of the multiphase flow process; shifting the curve downward gives rise to a new phenomenon: residual CO_2 trapping.

structured distribution of the invading fluid suggests that the invasion process is very complicated. Upon closer inspection, however, the authors of this study discovered that the CO_2 distribution is almost identical to the distribution that would be predicted from the Young-Laplace equation: the larger pores are filled with CO_2 while the smaller pores remain filled with water. The CO_2 distribution looks complicated because of the heterogeneous distribution of pore sizes in the sample (for example, the region highlighted on the left side of **Figure 9.7.7** has somewhat smaller pores than the rest of the sample, therefore it is almost entirely bypassed by the invading CO_2) but the basic processes that determine the CO_2 invasion pattern are those that are embodied in the ideal P_c curve in **Figure 9.7.6**.

The second figure (**Figure 9.7.8**) shows a photograph of brine and residually trapped CO_2 bubbles in a glass micromodel. The experimental

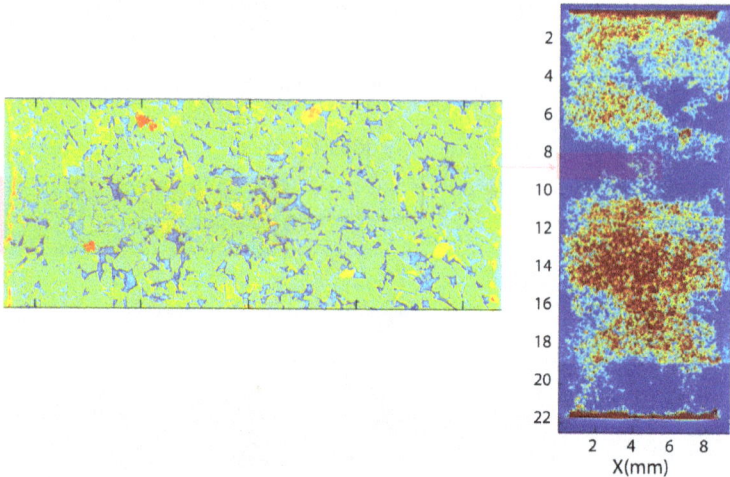

Figure 9.7.7 CO_2 invasion in sandstone at the pore network scale

Illustration of the influence of pore size heterogeneity on the pattern of CO_2 invasion in brine-filled sandstone. The X-ray radiography image on the right side shows CO_2 distribution (CO_2 in red, brine in blue) in a small core (about 2 cm long and 1 cm in diameter) of Domengine sandstone, a target sequestration formation in the Sacramento basin, CA, after injection of CO_2 at 50°C and 8.6 MPa. The CO_2 was injected from the top of the core at a low flow rate. The figure on the left side (a cross-section of a micro-CT reconstruction of the region in the center of the core sample that is largely bypassed by CO_2) shows that small variations in pore size dominate CO_2 flow patterns at the pore network scale. *Figures courtesy of Jonathan Ajo-Franklin (Lawrence Berkeley National Laboratory).*

setup consists of two silica plates that were etched with a network of channels and fused to each other to form a two-dimensional pore network [9.34]; the fabrication technique can generate pore networks with a minimum channel size of about 50 mm. This allows the direct observation of multiphase flow in two-dimensional porous media that mimic certain properties of silica-rich sandstone, such as pore size and the surface chemistry of the pore walls. The experiment shown in **Figure 9.7.8** reveals that $S_{g,r}$ can be quite high (in this case, $S_{g,r} = 0.19$) if the pore throats are much larger than the pore bodies, even in the absence of any pore network heterogeneity. The inset shows a closer view of several disconnected CO_2 bubbles that became trapped in the pore bodies when the imbibing brine filled the pore throats. A simple calculation based on

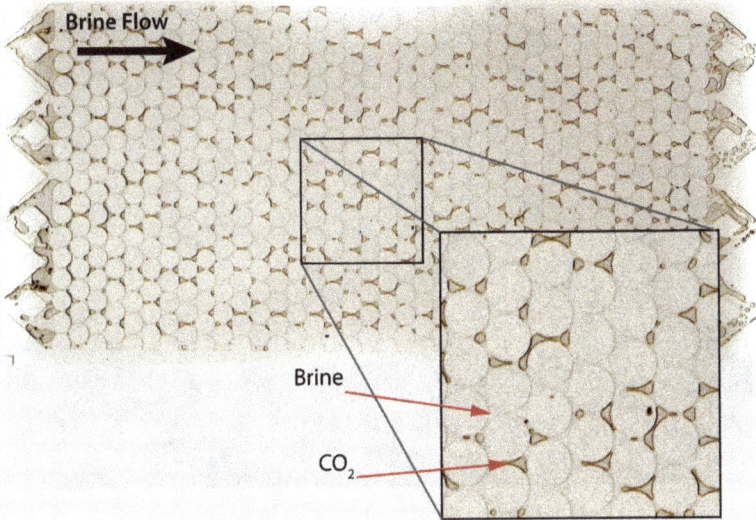

Figure 9.7.8 Residual trapping in a glass micromodel

Photograph showing residually trapped CO_2 bubbles in a glass micromodel (2 x 1 cm) at 8.5 MPa and 45°C. The pore space in the micromodel was initially filled with a 5 M NaCl brine; CO_2 was then injected at constant flow rate until CO_2 saturation reached a stable value ($S_{g,i}$ = 0.91); finally, brine was injected at a constant flow rate until CO_2 saturation reached a new stable value ($S_{g,r}$ = 0.19). *Figure courtesy of Jiamin Wan (Lawrence Berkeley National Laboratory).*

the Young-Laplace equation would predict that each pore should contain a trapped CO_2 bubble. Clearly this is not the case, so the fluid dynamics of the imbibition process must influence $S_{g,r}$ (see also **Question 9.7.2**).

We note in passing that CO_2 residual trapping does not occur only during brine imbibition: CO_2 exsolution from brine in porous rocks also results in a distribution of disconnected CO_2 bubbles (**Movie 9.7.1**). One implication of this is that if a CO_2-saturated brine becomes depressurized (for example, during leakage of CO_2-saturated brine), the exsolved CO_2 bubbles could have a very low mobility because of residual CO_2 trapping.

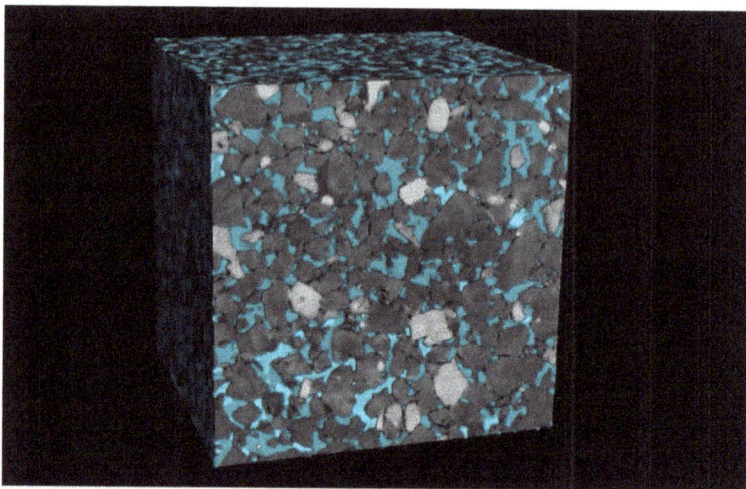

Movie 9.7.1 CO_2 exsolution in water-filled sandstone

Movie created from an X-ray CT experiment studying the exsolution of CO_2, i.e., the degassing of CO_2 from the aqueous phase, caused here by a pressure decrease, in a water-filled sandstone. A sample of Domengine sandstone (the same formation as in **Figure 9.7.7**) was flooded with CO_2-saturated water (with 1 M KI used as a contrast agent) at 70 psi, then pressure was decreased such that CO_2 exsolved from the water phase. The cube of rock shown in the movie is about 4.5 mm wide. The movie shows the structure of the sandstone grains in gray, the aqueous brine in blue, and the exsolved CO_2 in yellow. At the end of the movie, the sandstone grains are made invisible to highlight the distribution of the two fluid phases. *Movie courtesy of Marco Voltolini and Jonathan Ajo-Franklin (Lawrence Berkeley National Laboratory). It can be viewed at:* http://www.worldscientific.com/worldscibooks/10.1142/p911#t=suppl

Section 8

Chemical reactions

Figure 9.8.1 shows the final two steps of geological sequestration: CO_2 dissolving in the brine (solubility trapping) and its chemical conversion into minerals (mineral trapping). We have seen that these reactions are

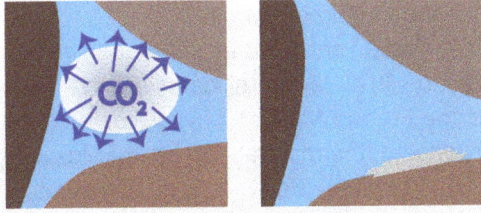

Figure 9.8.1 The final steps in geological sequestration

The figure on the left illustrates the dissolution of CO_2 in the brine and the figure on the right the conversion into carbonates.

very slow; the expected time scale for converting CO_2 into minerals is (very roughly) on the order of 10,000 years.

Mineral weathering

The rate-limiting step in the sequestration of CO_2 as carbonate minerals is the rock weathering rate, in particular the *in-situ* dissolution rate of silicate minerals.

Two key classes of mineral weathering reactions that occur in CO_2 storage formations are the dissolution of calcite, which buffers the pH of water near the plume to ~4.9:

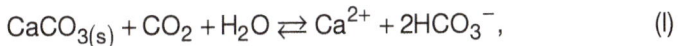

$$CaCO_{3(s)} + CO_2 + H_2O \rightleftharpoons Ca^{2+} + 2HCO_3^-, \tag{I}$$

and the dissolution of silicate minerals, which leads to CO_2 trapping in carbonate minerals (MCO_3, where M = Ca, Mg, or Fe):

$$M_{rich-silicates} + CO_2 + H_2O \rightleftharpoons MCO_{3(s)} + M_{poor-silicates}. \tag{II}$$

Reaction (I) is relatively rapid and often treated as an equilibrium phenomenon. Reaction (II), on the other hand, occurs on time scales of hundreds of years. Its inclusion in sequestration models requires knowledge of the rates of silicate dissolution and precipitation reactions at geological conditions: high temperature, pressure, and salinity, high solid-water ratios, and very low flow rates.

In geochemical models and studies of natural analogs, the key minerals involved in reaction (II) as M-rich-silicates are the Ca-bearing feldspars (plagioclase) and the Fe-and Mg-rich phyllosilicates (chlorite, glauconite, smectite); as carbonates, the minerals dolomite ($Mg_{0.5}Ca_{0.5}CO_3$), ankerite [$(Mg,Ca,Fe,Mn)CO_3$], siderite ($FeCO_3$), and dawsonite [$NaAlCO_3(OH)_3$]; and as M-poor silicates, quartz, kaolinite, and alkali feldspars (see **Figure 9.2.3**).

Weathering rate models and data

Mineral dissolution and precipitation rates are often modeled with the following semi-empirical relation:

$$r = ka_r \left[1 - \left(\frac{Q}{K_s} \right)^n \right]^m,$$

where a_r is the specific reactive surface area of the mineral of interest (m^2/g), k is the reaction rate constant (mol/m^2 s), n and m are power terms (often assumed equal to one), K_s is the thermodynamic equilibrium constant of the dissolution reaction, and Q is the ion activity product. The ratio Q/K_s is related to the Gibbs free energy of the dissolution reaction [$\Delta G_r = RT \ln(Q/K_s)$]. The reaction rate constant k is described with the expression:

$$k = k_n \exp\left[-\frac{E_n}{R}\left(\frac{1}{T} - \frac{1}{T_0} \right) \right] + k_H \exp\left[-\frac{E_H}{R}\left(\frac{1}{T} - \frac{1}{T_0} \right) \right] a_H^{n_H}$$

$$+ k_{OH} \exp\left[-\frac{E_{OH}}{R}\left(\frac{1}{T} - \frac{1}{T_0} \right) \right] a_{OH}^{n_{OH}},$$

where $i = n$, H, and OH for neutral, proton-promoted, and hydroxyl-promoted dissolution mechanisms, a_H and a_{OH} are the proton and hydroxyl activity, n_H and n_{OH} are power terms, and k_i and E_i are the rate constant and the activation energy associated with each reaction mechanism (at $T_0 = 298$ K). The two equations take into account that mineral dissolution and precipitation rates increase with temperature and equal zero if

$Q = K_{sp}$ (thermodynamic equilibrium) and are consistent with the observations that the rates exhibit minima at intermediate pH values (**Figure 9.8.2**) [9.35]. These minima exist because in most cases dissolved hydronium or hydroxyl ions can catalyze the rate-limiting step of the dissolution reaction. (In the case of silicate minerals, this rate-limiting step is often associated with the hydrolysis of a >Si-O-Si< bridge). Most experimental data on the dissolution of well-sorted, pure mineral grains can be described with these two equations.

Dissolution-precipitation model parameters of several mineral phases are listed in **Table 9.8.1**. Dissolution rates at 50°C calculated from the data in **Table 9.8.1** are plotted vs. pH in **Figure 9.8.3**. The dissolution rates range over ten orders of magnitude. According to **Figure 9.8.3**, clays and feldspars dissolve on time scales of tens of years at pH 5 (eventually leading to mineral sequestration) whereas quartz grains are essentially inert on this time scale. As shown in **Figure 8.2.3**, mineral trapping is expected to occur very slowly (on time scales of thousands to tens of thousands of years), hundreds or thousands of times more slowly than the weathering rates of M-rich clays and feldspars measured in laboratory experiments. The reason for this difference is the topic of the remainder of this section.

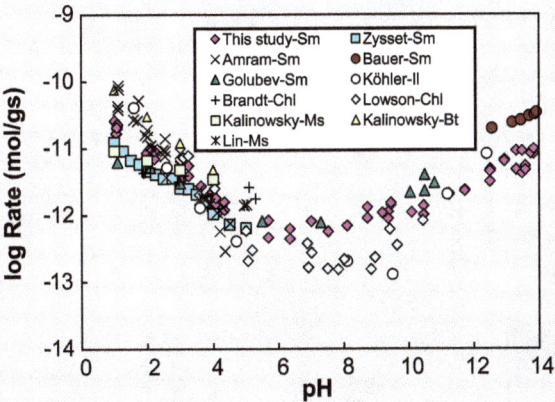

Figure 9.8.2 Dissolution rate constants of smectite and several other 2:1 structure phyllosilicates

Compilation of experimental data on the logarithm of the dissolution rate constant (more precisely, the product $a_r k$, in mol/g s) of smectite and several other 2:1 structure phyllosilicates as a function of pH at 25°C. *Figure reproduced from Rozalén et al. [9.35], with permission from Elsevier.*

Table 9.8.1 Dissolution rate law parameters of selected minerals

Mineral	Neutral mechanism		Proton-promoted mechanism			Chemical composition
	$\log k_n$ (mol m^{-2} s^{-1})	E_n (kJ mol^{-1})	$\log k_H$ (mol m^{-2} s^{-1})	E_H (kJ mol^{-1})	n_H	
$M^{(II)}$-rich silicates						
Oligoclase (a type of plagioclase feldspar)	−11.84	69.8	−9.67	65.0	0.457	$(Na,Ca)(Al,Si)_4O_8$ where $Ca/(Ca+Na)$ = 10 to 30%
Smectite (a M-rich clay)	−12.78	35.0	−10.98	23.6	0.340	$Ca_{0.52}(Al_{2.8}Fe_{0.5}Mg_{0.7})$ $(Si_{7.65}Al_{0.35})O_{20}(OH)_4$
$M^{(II)}$-poor silicates						
Quartz	−13.99	87.6				SiO_2
Albite (Na-feldspar)	−12.56	69.8	−10.16	65.0	0.457	$NaAlSi_3O_8$
Kaolinite (a M-poor clay)	−13.18	22.2	−11.31	65.9	0.777	$Al_2Si_2O_5(OH)_4$
Carbonate minerals						
Calcite	−5.81	23.5	−0.30	14.4	1.000	$CaCO_3$
Magnesite	−9.34	23.5	−6.38	14.4	1.000	$MgCO_3$

Dissolution rate law parameters of selected carbon sequestration-relevant minerals as compiled by Palandri and Kharaka [9.36]. Rate law parameters for the hydroxyl-promoted dissolution mechanism are not included in the table.

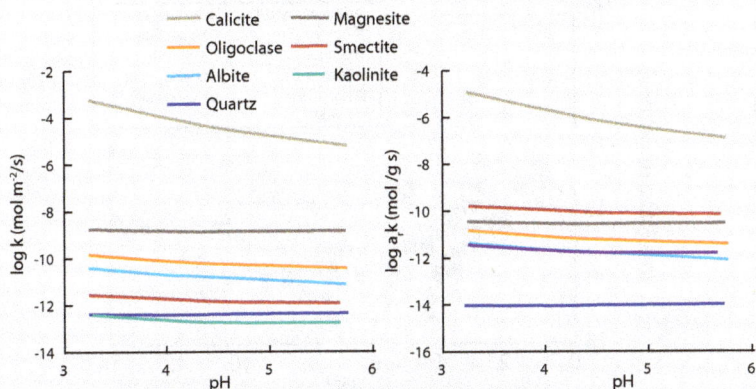

Figure 9.8.3 Dissolution rates of selected minerals at 50°C

Logarithm of (left side) the dissolution rate constant (log k, mol/m^2 s), and (right side) the product of the dissolution rate constant and the reactive surface area (log $a_r k$, mol/g s) of selected minerals relevant to carbon sequestration at 50°C as a function of pH. The curves were calculated with the parameters in **Table 9.8.1**, with a_r values in the mid-range of those reported in mineral dissolution studies ($a_r = 0.02$ m^2/g for quartz, calcite, and magnesite, 0.1 m^2/g for feldspars, 8 m^2/g for kaolinite, and 50 m^2/g for smectite). M$^{(II)}$-rich silicates are shown in orange, M$^{(II)}$-poor silicates in blue, and carbonate minerals in brown.

Accuracy of data on the weathering rates

An important focus of fundamental research in sequestration is to understand the weathering rates of minerals under *in-situ* conditions. A first step toward understanding these rates consists of looking at individual mineral grains and understanding that different crystallographic surfaces of the same grains may have very different reactivities. This realization, which may warrant a re-examination of the meaning of the reactive surface area a_r, is illustrated by the data shown in **Figure 9.8.4** for the case of smectite and other 2:1 structure phyllosilicates.

As noted in Section 9.2, phyllosilicates are composed of flake-shaped lamellae with thicknesses of ~1 nm (in the case of the 2:1 structure) and diameters of several hundreds of nm or more. Each individual lamellum is a natural nanoparticle with a specific surface area of about 800 m^2/g. In most conditions, the lamellae form ordered stacks where

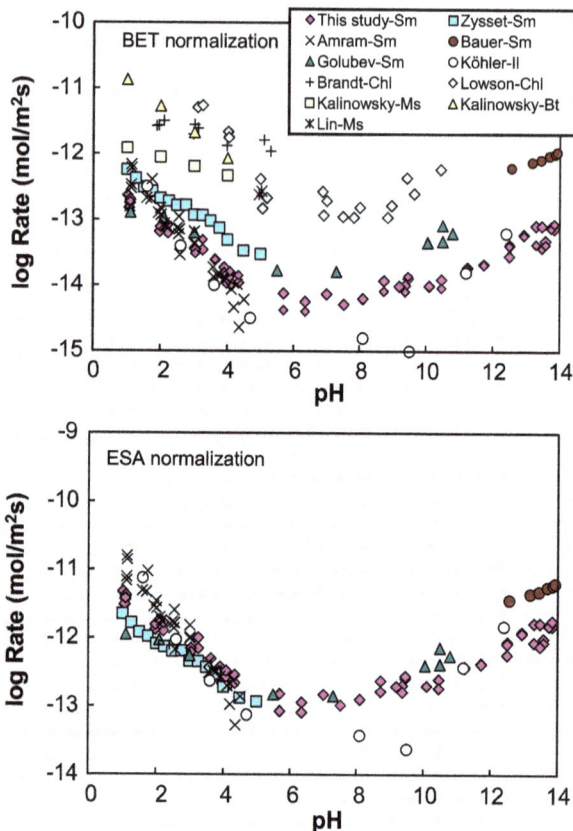

Figure 9.8.4 Dissolution rate constant of smectite and several other 2:1 structure phyllosilicates

Same data as in **Figure 9.8.2**, but the measured rates (in mol/g s) are normalized to the specific surface area. The upper figure shows dissolution rates normalized to the N$_2$-BET surface area. The lower figure shows the same rates normalized to the edge surface area (ESA) of the clay lamellae (either measured or estimated as ~6.5 m^2/g). *Figures reproduced from Rozalén et al. [9.35], with permission from Elsevier.*

the interlayer space is either accessible (in the case of smectite; see **Figure 9.2.3**) or inaccessible to water (in the case of other 2:1 structure phyllosilicates). The external surface area of these stacks is measured by adsorption of nitrogen gas (with the so-called N$_2$-BET method of Brunauer, Emmett and Teller) [9.37]. **Figure 9.8.4** shows that the phyllosilicate dissolution rates normalized to the N_2-BET surface area of dry powders exhibit a large scatter, suggesting that this surface area is a poor measure of reactive surface area. Dissolution rates show much less scatter if they are normalized to the edge surface area of the clay particles, i.e., the highly-reactive surface exposed at the edges of the clay flakes, measured by atomic force microscopy or by high-resolution low-pressure gas adsorption. This is consistent with direct observations of the dissolution of individual smectite clay particles, where the reaction occurs primarily on the edge surfaces (**Figure 9.8.5**) [9.38].

A second key concept is that minerals do not grow only by a simple process of solute attachment and detachment at growth sites; in reality, their growth is much more complex. For example, mineral precipitation may involve the formation of metastable precursors such as amorphous silica or amorphous calcium carbonates that later slowly transform into

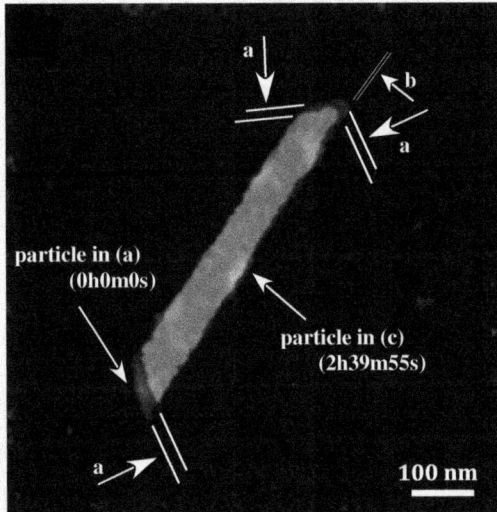

Figure 9.8.5 Dissolution of a single smectite lamellum

Superposition of two AFM (atomic force microscopy) images of a single smectite lamellum observed before and after 160 minutes of dissolution in a 0.01 M NaOH solution. The basal plane of the particle lies within the plane of the image; the thickness of the particle in the direction normal to the plane of the image is 1 nm. The figure shows that dissolution occurs on the edge surfaces and different edge faces have different dissolution rates. *Figure reproduced from Kuwahara [9.38], with permission from the Mineralogical Society of America.*

crystalline phases. The nucleation of solid phases or growth sites, rather than the binding of solutes at existing growth sites, is increasingly recognized as a key factor that controls mineral precipitation rates. This is illustrated in **Figure 9.8.6**, which shows that kaolinite dissolution at pH 4 and 22°C is well described by the rate equations, but kaolinite growth at the same conditions is much better described by a model that accounts for the rate of nucleation of growth sites on the kaolinite surface [9.39].

Dissolution at *in-situ* conditions

The long-term weathering rates of minerals at geological conditions can be reconstructed from a range of sources. For example, silicate glass weathering experiments with durations of tens of years have been carried

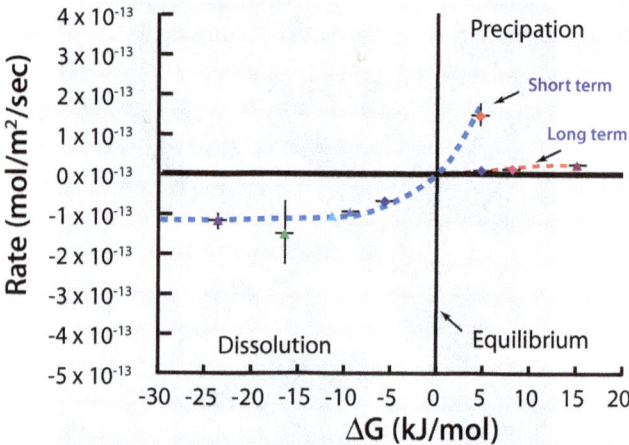

Figure 9.8.6 Dissolution and precipitation rates of kaolinite

Dissolution and precipitation rates of kaolinite as a function of the Gibbs free energy of reaction at pH 4 and 22°C. The symbols represent measured values of the steady-state reaction rate. The dashed blue curve was obtained by fitting the rate equation described in this section (with $n = 0.5$ and $m = 1$) to the dissolution data. The dashed red curve was obtained by fitting a two-dimensional nucleation model to the precipitation data. At $\Delta G_r \approx 5$ kJ/mol two experimental values are reported on the graph: the faster growth rate was observed on a time scale of hours for a kaolinite that had previously been subjected to steady-state dissolution for five months and the slower growth rate was observed on longer time scales. Figure redrawn from Yang and Steefel [9.39].

out for the purpose of evaluating the corrosion rates of these glasses in high-level radioactive waste repositories. Ancient glass artifacts buried in soils or immersed in seawater for thousands of years provide a longer-term view of weathering processes. Geological media with well-known ages enable reconstruction of mineral weathering rates on time scales of millions of years (**Figure 9.8.7**) [9.40]. In all cases, long-term weathering rates are orders of magnitude smaller than the rates observed in short-term laboratory dissolution experiments. This is the main reason why geochemists expect that mineral trapping will take tens of thousands of years (**Figure 8.2.3**), whereas laboratory experiments suggest that it could occur hundreds of times more rapidly (**Figure 9.8.3**).

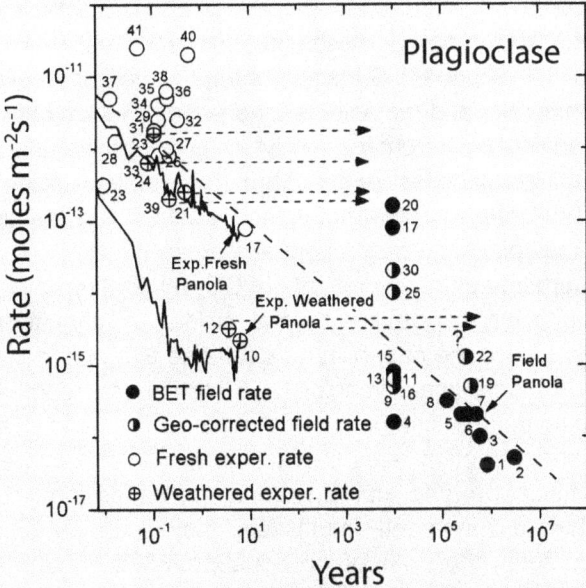

Figure 9.8.7 Dissolution rate of plagioclase feldspar as a function of observation time scale

Compilation of plagioclase feldspar dissolution rates as a function of the time scale of observation. Measurements of the short-term dissolution rates of fresh samples of pure feldspar yield relatively high rates ($\sim 10^{-12}$ mol/m^2 s). Geochemical reconstructions of the long-term dissolution rates of feldspar samples recovered from natural formations yield much lower rates ($\sim 10^{-16}$ mol/m^2 s). *Figure reproduced from White and Brantley [9.40], with permission from Elsevier.*

The origin of the discrepancy between the experimental field-scale weathering rates and the prediction of mathematical models describing these geochemical processes are caused by the approximations that are made in these models. First, as noted above, the models assume that the same rate equations describe both dissolution and precipitation, whereas experiments show that mineral growth can follow a different rate law than mineral dissolution (**Figure 9.8.6**). Since dissolution and precipitation reactions are coupled, slow growth of M-poor silicates could limit the overall rate of carbonate mineralization. Second, the reactive surface area of complex mineral assemblages may differ significantly from the geometric surface areas of its mineral constituents, because a portion of the grain surfaces may be occluded by other minerals (**Figure 9.8.8**) [9.41] and because mineral reactivity may be inhibited by slow fluid flow (**Figure 9.8.9**) [9.42]. Third, surface coatings and metastable amorphous phases often form during dissolution in near-equilibrium conditions (**Figure 9.8.10**) [9.43, 9.44]. Surface coatings can modify the reactivity of dissolving minerals, and metastable amorphous phases often are poorly characterized or absent from thermodynamic databases. The influence of surface coatings, inaccessible (or poorly-accessible) surface areas, and other effects is often approximately accounted for in the geochemical models by multiplying all mineral surface areas by $\sim 10^{-2}$. This suggests that the uncertainty of the overall rate of CO_2 mineral trapping in sequestration is quite large (possibly several orders of magnitude). Existing research tools have the capability to decrease this uncertainty by combining nano-to core-scale imaging and multiphase reactive transport modeling to reveal the distribution of minerals, brine, and CO_2 as well as their weathering reaction rates in reservoir rocks. The challenge is that the characterization and computational demands are considerable because rocks are inherently multiscale, their scale ranging from less than a micrometer (the scale of individual pores) to hundreds of meters (the thickness of geological formations).

Outlook

In this chapter, we have reviewed the physico-chemical basis for the sequestration of CO_2 in rocks. We see that we are faced with many research challenges and opportunities that are needed to further understand and improve our predictions about the transport, equilibria and

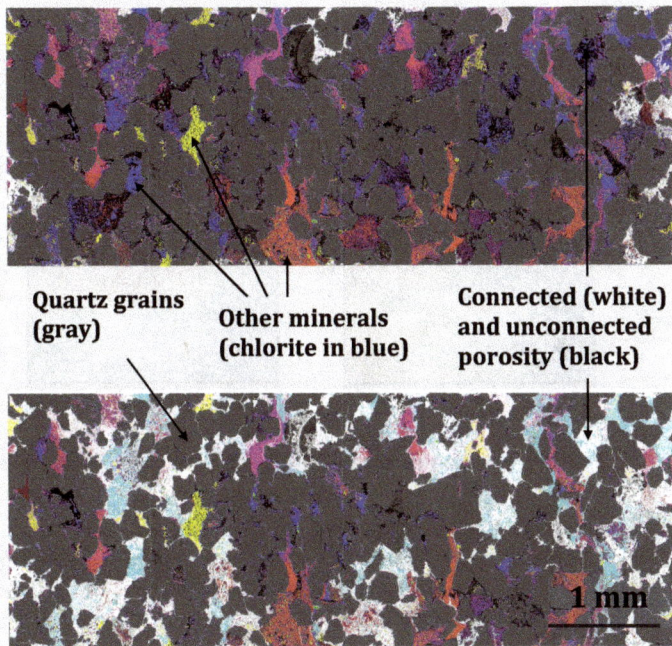

Figure 9.8.8 SEM reconstruction of the mineralogy and connected porosity of a sandstone sample

Scanning electron microscopy (SEM) images of a sandstone sample from the Cranfield CO_2 sequestration pilot site in Mississippi (6 by 2.4 mm area) with pore space and mineralogy identified by electron microscopy and X-ray spectroscopy techniques. In the upper image, where mineralogy is mapped with a resolution of 1 µm, most of the pore space appears to be unconnected (in black). In the lower image, where mineralogy is mapped with a finer resolution (0.33 µm) and the chlorite mineral fraction (in blue) is treated as a phase with well-connected internal nanoporosity (as revealed by nanotomography experiments), most of the pore space is seen to be connected (in white). The figure shows that the more finely one looks at a rock sample, the more pores appear to be connected. This type of very fine-scale characterizations becomes extremely challenging if one wishes to apply it to a large enough region of the rock sample to capture the heterogeneity of the rock on scales of centimeters or more. *Images reproduced from Landrot et al.* [9.41], *with permission from Elsevier.*

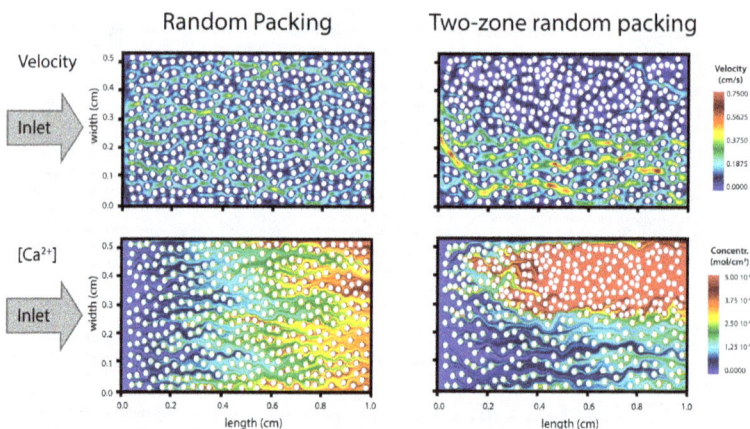

Figure 9.8.9 Reactive transport simulation of CO_2-acidified water flowing through a bed of calcite beads

Two-dimensional adaptive mesh refinement simulation of CO_2-rich water flowing through a bed of calcite beads. The figures on the left side show the case where the beads are randomly distributed; the right side shows the case where the beads are more densely distributed in one half of the simulated system. Upper and lower figures show the velocity field and Ca^{2+} concentrations, respectively. The inlet water is at pH 5 with $p_{CO_2} = 3.15 \times 10^{-4}$ bar, 0.01 M NaCl, and 0.1 cm s^{-1} velocity. The calculations predict that the average (upscaled) dissolution rate of calcite in the heterogeneously-packed system (right) is lower than in the homogeneously-packed system (left), i.e., heterogeneity can be viewed as reducing the "effective" reactive surface area. *Figure reproduced from Molins et al. [9.42], with permission from John Wiley and Sons.*

reaction in porous media. The demonstration projects on sequestration (see Chapter 8), however, have been successful both as pilot models for large-scale storage and as valuable sources of data that can be used to test models. The success of the existing demonstration projects shows that geological storage of CO_2 is a technology that is already applicable at the industrial scale. Opportunities exist for improving existing models and for significantly optimizing the efficiency of CO_2 storage operations.

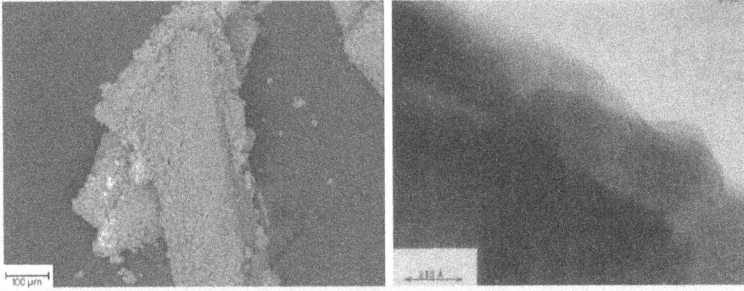

Figure 9.8.10 Surface coatings and amorphous phases

Illustration of the formation of surface coatings and amorphous phases during silicate weathering. The figure on the left side shows a calcite coating formed on the surface of a fast-dissolving silicate (wollastonite, $CaSiO_3$) upon exposure to CO_2-rich water at pH 6 and 90°C. The figure on the right side shows an amorphous gel layer formed on a feldspar surface during dissolution at low pH. Geochemical models generally neglect the existence of these surface coatings and amorphous solid phases. *Images reproduced from Daval et al. [9.43] and Casey et al. [9.44], both with permission from Elsevier.*

Section 9

Review

9.1. Reading self-test

1. What is the typical grain size of shale?
   a. > 2 mm
   b. 0.06–2 mm
   c. 0.008–0.06 mm
   d. < 0.008 mm

2. What is a typical ratio of the porosity of sandstone and shale?
   a. 10
   b. 2
   c. 0.5
   d. 1

3. What are the typical temperature gradients found in geological formations?
   a. Between 5°C and 20°C per km
   b. Between 10°C and 25°C per km
   c. Between 15°C and 30°C per km
   d. Between 25°C and 35°C per km

4. What is a typical ratio of the breakthrough pressure of CH_4 and CO_2 for caprock?
   a. 0.05
   b. 0.5
   c. 2
   d. 20

5. Which statement about capillary hysteresis is not correct?
   a. Capillary hysteresis cannot occur in formations in which the rocks are CO_2 wetting
   b. The mobility of the minority phase drops to zero if the majority phase is continuous

c. CO_2 is trapped as isolated bubbles

d. Brine is trapped in the smaller capillaries

9.2. Mineral dissolution

1. What properties influence the characteristic time scales of mineral dissolution? (Choose all that apply)

   a. Capillary pressure

   b. Surface coatings

   c. Mineral type (for example, clays dissolve more rapidly than calcite)

   d. The time scale of observation

9.3. Capillary breakthrough pressure

1. What parameters determine the capillary breakthrough pressure of a rock sample? (Choose all that apply)

   a. The size of the largest pores

   b. The size of the smallest pores

   c. The size of the critical pore

   d. The CO_2-water interfacial tension

   e. The CO_2-water-mineral wetting angle

9.4. Residual CO_2 saturation

1. What is the expected range of residual CO_2 saturation in sandstone?

   a. 0.10 to 0.40 according to most studies

   b. About 0.75

   c. 0.30 to 0.50

   d. It depends on the initial CO_2 saturation

Section 10

References

1. Xu, T., J.A. Apps and K. Pruess, 2005. Mineral sequestration of carbon dioxide in a sandstone-shale system. Chemical Geology, **217** (3–4), 295. http://dx.doi.org/10.1016/j.chemgeo.2004.12.015

2. Myer, C.L., C. Downey, J. Clinkenbeard, *et al.*, 2005. Preliminary geologic characterization of West Coast States for geologic sequestration, WESTCARB Topical Report. http://dx.doi.org/10.2172/907916

3. Andre, C.L., P. Audigane, M. Azaroual, and A. Menjoz, 2007. Numerical modeling of fluid-rock chemical interactions at the supercritical CO_2-liquid interface during CO_2 injection into a carbonate reservoir, the Dogger aquifer (Paris Basin, France). Energ. Convers. Manage., **48** (6), 1782. http://dx.doi.org/10.1016/J.Enconman.2007.01.006

4. Matter, J.M., T. Takahashi, and D. Goldberg, 2007. Experimental evaluation of in situ CO_2-water-rock reactions during CO_2 injection in basaltic rocks: Implications for geological CO_2 sequestration. Geochem. Geophy. Geosys., **8** (2). http://dx.doi.org/10.1029/2006gc001427

5. Metz, B., O. Davidson, H. deConinck, M. Loos, and L. Meyer, 2005. IPCC Special Report on Carbon Dioxide Capture and Storage. http://www.ipcc.ch/pdf/special-reports/srccs/srccs_wholereport.pdf

6. Grotzinger, J.H.J., F. Press, and R. Siever, 2006. Understanding Earth. USA: WH Freeman.

7. Prothero, D.R., and F. Schwab, 2003. Sedimentary Geology. USA: WH Freeman.

8. Silin, D., L. Tomutsa, S.M. Benson, and T.W. Patzek, 2011. Microtomography and pore-scale modeling of two-phase fluid distribution. Trans. Por. Med., **86** (2), 495–525. http://dx.doi.org/10.1007/S11242-010-9636-2

9. Neuzil, C.E., and D.W. Pollock, 1983. Erosional unloading and fluid pressures in hydraulically tight rocks. J. Geol., **91** (2), 179.

10. Burrus, J., 1998. "Overpressure models for clastic rocks, their relation to hydrocarbon expulsion: a critical reevaluation" in Abnormal Pressures in Hydrocarbon Environments: An Outgrowth of the AAPG Hedberg Research Conference, Golden, Colorado, June 8–10, 1994, edited by B.E. Law, G.F. Ulmishek, and V.I. Slavin. USA: Amer. Assn. of Petroleum Geologists.

11. Oldenburg, C.M., 2007. "Migration mechanisms and potential impact of CO_2 leakage and seepage" in Carbon Capture and Sequestration: Integrating Technology, Monitoring and Regulation, edited by E.J. Wilson and D. Gerard. Iowa: Blackwell.

12. Pruess, K. 2005. "Numerical simulations show potential for strong nonisothermal effects during fluid leakage from a geologic disposal reservoir for CO_2" in Dynamics of Fluids and Transport in Fractured Rock, Vol 162, edited by B. Faybishenko, P.A. Witherspoon, and J. Gate. Washington, DC: AGU. pp. 81.

13. Oldenburg, C.M., C. Doughty, C.A. Peters, and P.F. Dobson, 2012. Simulations of long-column flow experiments related to geologic carbon sequestration: effects of outer wall boundary condition on upward flow and formation of liquid CO_2. Greenhouse Gases, 2 (4), 279. http://dx.doi.org/ 10.1002/Ghg.1294

14. Nielsen, L.C., I.C. Bourg, and G. Sposito, 2012. Predicting CO_2-water interfacial tension under pressure and temperature conditions of geologic CO_2 storage. Geochimica et Cosmochimica Acta, 81, 28. http://dx.doi.org/ 10.1016/j.gca.2011.12.018

15. Hildenbrand, A., S. Schlömer, B.M. Krooss, and R. Littke, 2004. Gas breakthrough experiments on pelitic rocks: comparative study with N_2, CO_2, and CH_4. Geofluids, 4 (1), 61. http://dx.doi.org/10.1111/j.1468-8123.2004.00073.x

16. Li, S., M. Dong, Z. Li, S. Huang, H. Qing, and E. Nickel, 2005. Gas breakthrough pressure for hydrocarbon reservoir seal rocks: implications for the security of long-term CO_2 storage in the Weyburn field. Geofluids, 5 (4), 326. http://dx.doi.org/10.1111/j.1468-8123.2005.00125.x

17. Wollenweber, J., S. Alles, A. Busch, B.M. Krooss, H. Stanjek, and R. Littke, 2010. Experimental investigation of the CO_2 sealing efficiency of caprocks. International Journal of Greenhouse Gas Control, 4 (2), 231. http://dx.doi. org/10.1016/j.ijggc.2010.01.003

18. Skurtveit, E., E. Aker, M. Soldal, M. Angeli, and Z. Wang, 2012. Experimental investigation of CO_2 breakthrough and flow mechanisms in shale. Petroleum Geoscience, 18 (1), 3. http://dx.doi.org/10.1144/1354-079311-016

19. Nordbotten, J.M., D. Kavetski, M.A. Celia, and S. Bachu, 2009. Model for CO_2 leakage including multiple geological layers and multiple leaky wells. Environ. Sci. Technol., 43 (3), 743. http://dx.doi.org/10.1021/es801135v

20. Ellis, B.R., G.S. Bromhal, D.L. McIntyre, and C.A. Peters, 2010. Changes in caprock integrity due to vertical migration of CO_2-enriched brine. Energy Procedia, 4, 5327. http:// dx.doi.org/10.1016/j.egypro.2011.02.514

21. Leetaru, H., D. Harris, J. Rupp, D. Barnes, J. McBride, and J. Medler, 2010. Reservoir Properties of the St. Peter Sandstone in Illinois. http://knoxstp. com/reservoir.htm

22. Freeze, R.A., and J.A. Cherry, 1979. Groundwater. N.J.: Prentice-Hall.

23. Clauser, C., 1992. Permeability of crystalline rocks. Eos, Trans. AGU, 73 (21), 233. http://dx.doi.org/10.1029/91eo00190

24. Shi, J.-Q., Z. Xue, and S. Durucan, 2011. Supercritical CO_2 core flooding and imbibition in Tako sandstone — Influence of sub-core scale heterogeneity.

International Journal of Greenhouse Gas Control, **5** (1), 75. http://dx.doi. org/10.1016/j.ijggc.2010.07.003

25. Krevor, S.C.M., R. Pini, L. Zuo, and S.M. Benson, 2012. Relative permeability and trapping of CO_2 and water in sandstone rocks at reservoir conditions. Water Resources Research, **48** (2), W02532. http://dx.doi.org/10.1029/ 2011WR010859

26. André, L., P. Audigane, M. Azaroual, and A. Menjoz, 2007. Numerical modeling of fluid-rock chemical interactions at the supercritical CO_2-liquid interface during CO_2 injection into a carbonate reservoir, the Dogger aquifer (Paris Basin, France). Energy Conversion and Management, **48** (6), 1782. http://dx.doi.org/10.1016/j.enconman.2007.01.006

27. Liu, F., P. Lu, C. Zhu, and Y. Xiao, 2011. Coupled reactive flow and transport modeling of CO_2 sequestration in the Mt. Simon sandstone formation, Midwest U.S.A. International Journal of Greenhouse Gas Control, **5** (2), 294. http://dx.doi.org/10.1016/j.ijggc.2010.08.008

28. Alkan, H., Y. Cinar, and E.B. Ülker, 2010. Impact of capillary pressure, salinity and in situ conditions on CO_2 injection into saline aquifers. Transport in Porous Media, **84** (3), 799. http://dx.doi.org/10.1007/s11242-010-9541-8

29. Zhou, Q., J.T. Birkholzer, E. Mehnert, Y.-F. Lin, and K. Zhang, 2010. Modeling basin- and plume-scale processes of CO_2 storage for full-scale deployment. Groundwater, **48** (4), 494. http://dx.doi.org/10.1111/j.1745-6584.2009.00657.x

30. Doughty, C., 2010. Investigation of CO_2 plume behavior for a large-scale pilot test of geologic carbon storage in a saline formation. Transport in Porous Media, **82** (1), 49. http://dx.doi.org/10.1007/s11242-009-9396-z

31. Okwen, R.T., M.T. Stewart, and J.A. Cunningham, 2010. Analytical solution for estimating storage efficiency of geologic sequestration of CO_2. International Journal of Greenhouse Gas Control, **4** (1), 102. http://dx.doi. org/10.1016/j.ijggc.2009.11.002

32. Juanes, R., E.J. Spiteri, F.M. Orr Jr., and M.J. Blunt, 2006. Impact of relative permeability hysteresis on geological CO_2 storage. Water Resources Research, **42** (12), W12418. http://dx.doi.org/10.1029/2005WR004806

33. Bachu, S., and B. Bennion, 2008. Effects of in-situ conditions on relative permeability characteristics of CO_2-brine systems. Environmental Geology, **54** (8), 1707. http://dx.doi.org/10.1007/s00254-007-0946-9

34. Kim, Y., J. Wan, T.J. Kneafsey, and T.K. Tokunaga, 2012. Dewetting of silica surfaces upon reactions with supercritical CO_2 and brine: Pore-scale studies in micromodels. Environ. Sci, Technol., **46** (7), 4228. http://dx.doi.org/10.1021/ es204096w

35. Rozalén, M.L., F.J. Huertas, P.V. Brady, J. Cama, S. García-Palma, and J. Linaresa, 2008. Experimental study of the effect of pH on the kinetics of montmorillonite dissolution at 25°C. Geochimica et Cosmochimica Acta, **72** (17), 4224–4253.

36. Palandri J.L., and Y.K. Kharaka, 2004. A Compilation of Rate Parameters of Water-Mineral Interaction Kinetics for Application to Geochemical Modeling, US Geological Survey Open File Report 2004-1068. http://pubs.usgs.gov/of/2004/1068

37. Brunauer, S., P.H. Emmett, and E. Teller, 1938. Adsorption of gases in multimolecular layers. J. Am. Chem. Soc., **60** (2), 309. http://dx.doi.org/10.1021/ja01269a023

38. Kuwahara, Y., 2006. In-situ AFM study of smectite dissolution under alkaline conditions at room temperature. American Mineralogist, **91** (7), 1142. http://dx.doi.org/10.2138/am.2006.2078

39. Yang, I., and C.I. Steefel, 2008. Kaolinite dissolution and precipitation kinetics at 22°C and pH 4. Geochimica et Cosmochimica Acta, **72** (1), 99. http://dx.doi.org/10.1016/j.gca.2007.10.011

40. White, A.F., and S.L. Brantley, 2003. The effect of time on the weathering of silicate minerals: why do weathering rates differ in the laboratory and field? Chemical Geology, **202** (3–4), 479. http://dx.doi.org/10.1016/j.chemgeo.2003.03.001

41. Landrot, G., J.B. Ajo-Franklin, L. Yang, S. Cabrini, and C.I. Steefel, 2012. Measurement of accessible reactive surface area in a sandstone, with application to CO_2 mineralization. Chemical Geology, **318–319**, 113. http://dx.doi.org/10.1016/j.chemgeo.2012.05.010

42. Molins, S., D. Trebotich, C.I. Steefel, and C. Shen, 2012. An investigation of the effect of pore scale flow on average geochemical reaction rates using direct numerical simulation. Water Resources Research, **48** (3), W03527. http://dx.doi.org/10.1029/2011WR011404

43. Daval, D., I. Martinez, J. Corvisier, N. Findling, B. Goffé, and F. Guyot, 2009. Carbonation of Ca-bearing silicates, the case of wollastonite: Experimental investigations and kinetic modeling. Chemical Geology, **265** (1–2), 63. http://dx.doi.org/10.1016/j.chemgeo.2009.01.022

44. Casey, W.H., H.R. Westrich, G.W. Arnold, and J.F. Banfield, 1989. The surface chemistry of dissolving labradorite feldspar. Geochimica et Cosmochimica Acta, **53** (4), 821. http://dx.doi.org/10.1016/0016-7037(89)90028-8

Chapter 10

Large-Scale Geological Carbon Sequestration

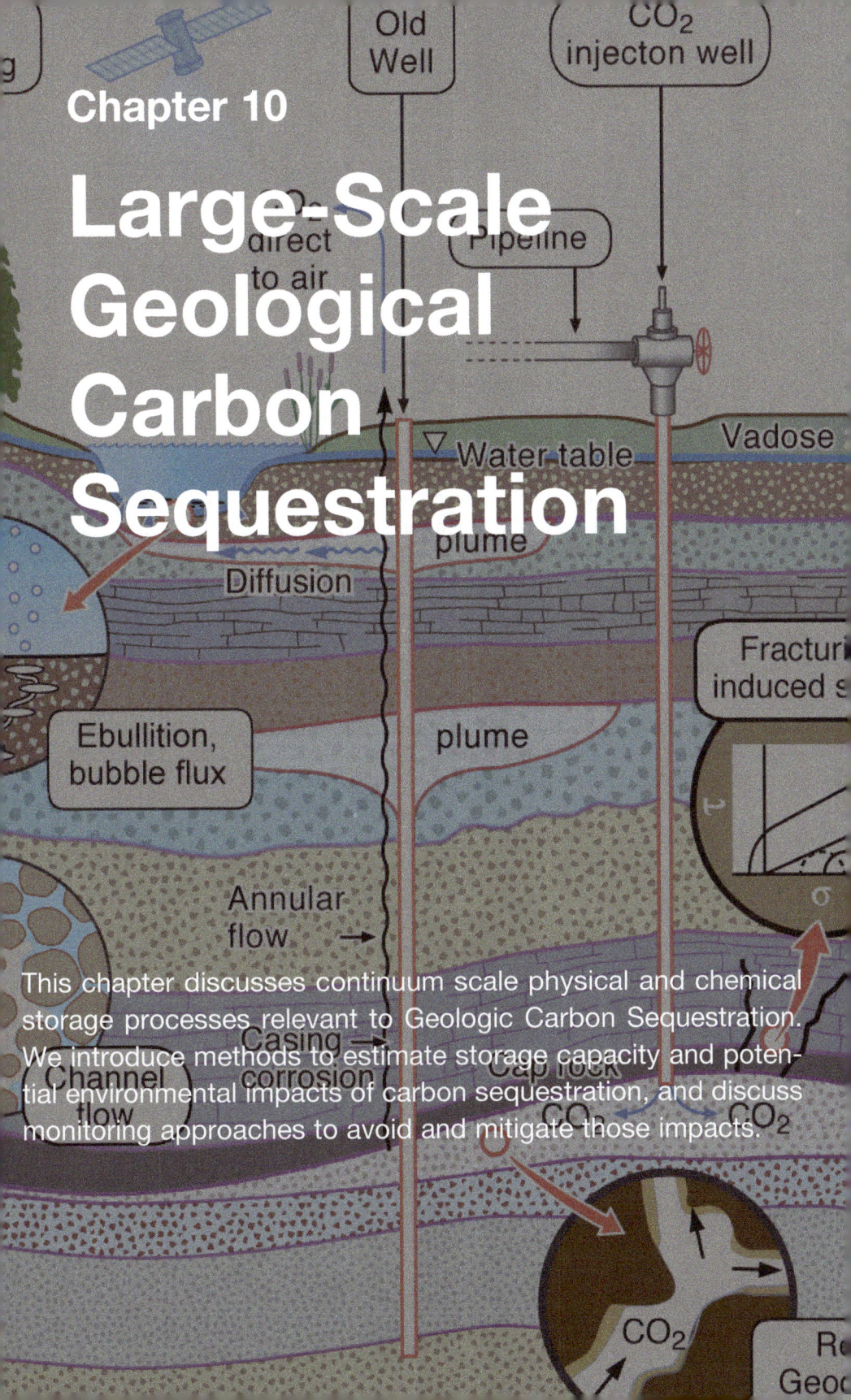

This chapter discusses continuum scale physical and chemical storage processes relevant to Geologic Carbon Sequestration. We introduce methods to estimate storage capacity and potential environmental impacts of carbon sequestration, and discuss monitoring approaches to avoid and mitigate those impacts.

Section 1

Introduction

The implicit conclusion in Chapter 8 is that geological carbon sequestration will be a safe and effective approach for long-term storage of CO_2 and that CO_2 can be isolated indefinitely from the atmosphere. In this chapter, we focus on the following large-scale questions:

- Given the geological system and the properties of CO_2, what is the capacity for storing CO_2, and on what factors does it depend?
- Are there negative environmental impacts related to carbon sequestration and if so how can they be mitigated?
- What methods will be used to monitor and account for CO_2 that is sequestered deep underground?

We will address these questions from a continuum scale perspective. This scale is loosely defined as being much larger than the pore scale in a sedimentary rock, typically on the order of 100 microns (μm). We refer to analyses and characterization as continuum scale because smaller scale features and processes (for example, the shape of individual pores and the connectivity of the pore network) are spatially averaged and lumped into a homogeneous macroscopic continuum.

In Chapter 8, we introduced the time scales of the different processes that are relevant for geological sequestration (see **Figure 8.2.3**). This figure summarizes a wealth of understanding of the processes related to carbon sequestration gleaned over more than a century's worth of research in the earth sciences. In its abstraction, however, this figure hides the rationale and evidence for its validity. Here we try to peel away the multiple layers of abstraction to reveal a fuller picture of the evidence and experience that indicates carbon sequestration is a highly promising approach to mitigating CO_2 emissions.

Section 2

Field-scale models

Geological carbon sequestration (GCS) operations rely on field-scale models to predict the behavior of injected CO_2 on length scales of tens of kilometers and time scales of thousands of years. These models are essential for optimizing CO_2 injection, predicting the economic cost of carbon sequestration and the storage capacity of target formations, purchasing CO_2 storage rights, planning potential remedial measures, and securing public support and regulatory approval for carbon sequestration. Key questions that the field-scale models must accurately answer include: How much CO_2 can be trapped in target formations? How rapidly can CO_2 be injected? What will be the eventual extent of the CO_2 plume? How will CO_2 injection impact regional hydrogeology? How long will monitoring operations need to continue beyond the time of CO_2 injection?

Field-scale simulations also provide useful conceptual insights into the influence of different phenomena on the performance of carbon sequestration operations. These operations rely on predictions using field-scale models of multiphase flow and geochemistry in rock formations. In this section, we briefly examine these field-scale models. We focus on two questions: how accurately do the field-scale models predict the fate of CO_2 in porous rocks? And how could these models be improved through fundamental science?

Accuracy

The accuracy of field-scale models can be evaluated in two ways. The first type of test compares the blind predictions of a geological carbon sequestration model with existing field-scale data (**Figure 10.2.1**) [10.1]. This exercise is essential for building confidence in the predictive ability of models, but it is seldom carried out because of the limited availability of appropriate experimental data. The second type of test probes the sensitivity of model predictions to known uncertainties in model design and input parameters (**Figure 10.2.2**) and provides insight into the relative importance of different fundamental properties and processes [10.2, 10.3]. As illustrated in **Figures 10.2.1.** and **10.2.2**, both approaches show

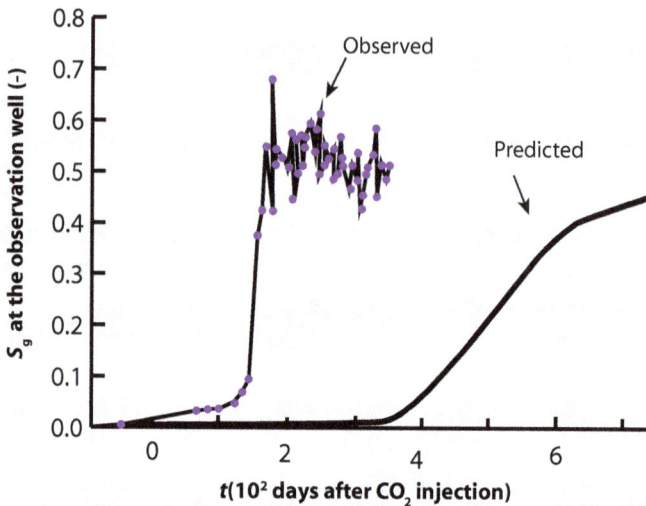

Figure 10.2.1   Blind prediction of data from a pilot site

Blind prediction of field-scale sequestration data from the CO2CRC Otway pilot site in Australia, where 65 kilotonne of CO_2 was injected into a depleted gas reservoir. The graph shows predicted and measured values of CO_2 saturation, S_g, as a function of time at an observation well along the path of the CO_2 plume. The increase in CO_2 saturation indicates the arrival of the CO_2 plume at the observation well. *Figure redrawn from Underschultz et al.* [10.1].

that existing field-scale carbon sequestration models yield useful qualitative information on the behavior of CO_2 in geological formations but large uncertainties are manifested in their quantitative predictions of basic properties, such as the velocity of the CO_2 plume or the rates of various CO_2 trapping mechanisms.

The quality of field-scale model predictions depends on two features: the design of the model, which must account for all relevant physical phenomena, and the input parameters of the model, which must accurately describe fundamental properties such as the CO_2-brine equation of state and specific properties of the rock formations, such as their permeability. A model may fail to predict the fate of CO_2 because of inaccuracies in its design or because of a poor choice of input parameters. In the case of carbon sequestration, both potential sources of inaccuracy present significant challenges: geological formations are complex sites

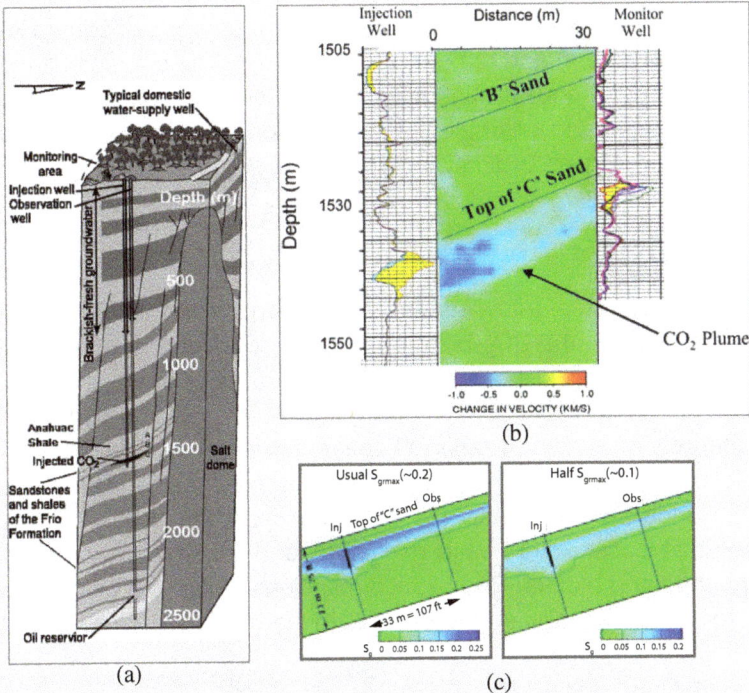

Figure 10.2.2 Sensitivity of field-scale model predictions to residual gas saturation

Illustration of the sensitivity of field-scale model predictions to model input parameters. (a) Schematic description of the Frio I sequestration pilot site (a sandstone formation near Houston, Texas, where about 1.6 kilotonne of CO_2 was injected at a depth of 1,530 m).

(b) Experimental data on CO_2 saturation (S_g) during the CO_2 injection test; the left and right sides of the figure show S_g values (more precisely, changes in resistivity caused by the presence of CO_2) measured as a function of depth at the injection (left-hand side) and observation wells (right-hand side); the central part of the figure shows a reconstruction of CO_2 saturation in the region between the wells from cross-well seismic measurements (**Figure 10.5.5**).

(c) Model predictions of S_g for two different values of the maximum residual gas saturation ($S_{g,r,max}$ = 0.2 or 0.1 on the left-hand and right-hand sides), a model parameter discussed in more detail in Section 9.7. The model correctly predicts the overall shape of the CO_2 plume, but not its fine structure; model predictions are highly sensitive to $S_{g,r,max}$, a poorly-constrained input parameter (**Table 9.7.1**).

(a) and (c): *Reproduced from Doughty et al.* [10.3]. (b): *Reproduced from Daley et al.* [10.2]. *All figures reproduced with kind permission from Springer Science + Business Media.*

where many processes occur, some of which are very poorly understood; in addition, the formations of interest are vast (hundreds of square kilometers in areal extent) and remote (> 800 m deep). The characterization of potential carbon sequestration sites is expensive (the drilling of characterization wells can constitute a major portion of carbon sequestration project costs) and accordingly incomplete.

Simulation grid

With the exception of a few studies that rely on models that employ analytical solutions to the CO_2-brine multiphase flow equations in simplified geometries, most field-scale carbon sequestration models and all models that couple multiphase flow and geochemistry use numerical methods that rely on a discretization of space into an ensemble of finite size simulation grid blocks, each treated as a well-mixed reactor (**Figure 10.2.3**) [10.4]. During each time step, the model successively calculates the processes that occur within each grid block and the fluid fluxes between neighboring grid blocks. Grid block dimensions typically are on the order of 1 to 10 m vertically and 10 to 100 m horizontally. Each grid block

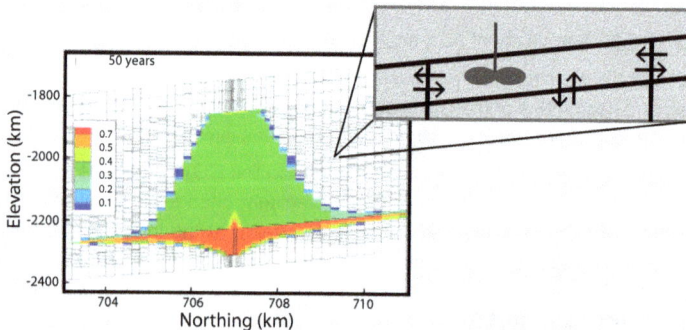

Figure 10.2.3 Prediction of CO_2 saturation near an injection well

Model prediction of CO_2 saturation near an injection well in the Illinois Basin after 50 years of injection at the rate of 5 megatonne CO_2 per year. The simulation grid is shown by the thin gray lines. Each grid block is 20 to 1,000 m wide (increasing with distance from the well) and up to 10 m high. The inset shows a schematic view of a grid block as a well-mixed reactor (shown by the mixing paddle) that exchanges fluids with the surrounding grid blocks (shown by the arrows). *Figure reproduced from Zhou et al, with permission from John Wiley and Sons* [10.4].

contains a mixture of mineral phases and pore space, the pore space being filled either by brine, CO_2, or a two-phase mixture of brine and CO_2, i.e., for each grid block our model must keep track of the porosity Φ and the water and CO_2 saturations S_w and S_g.

The discretization of space into an ensemble of simulation grid blocks implies several approximations. First, fluid and rock properties are not homogeneous at the grid-block scale. This means that the effective grid-block scale properties of the porous media of interest cannot necessarily be determined from measurements made on a scale smaller than the grid blocks (for example, from experiments on 5–10 cm core samples) because many properties are influenced by the heterogeneity of matter within the scale of the grid blocks. This is illustrated by the permeability anisotropy of natural formations: at the grid-block scale, horizontal permeability tends to be higher than vertical permeability, because discrete regions of higher or lower permeability within grid blocks are oriented in the direction of the bedding (**Figure 10.2.4**). A single core sample may not well represent a region of the formation large enough to capture this effect. This effect is often modeled by assuming that permeability in the direction parallel to the bedding is greater (for example, by a factor of 10) than in the direction normal to the bedding.

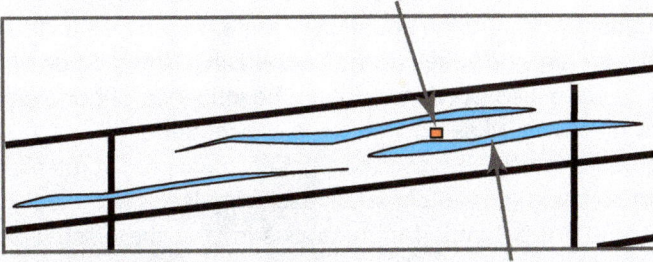

Rock properties are often detemined from core samples and well logs (on a ~10^{-1} m length scale)

Significant heterogeneity (e.g., lower permeability zones) may exist within grid blocks.

Figure 10.2.4 Sub-grid-block scale permeability features

Schematic figure illustrating the tendency of high- or low-permeability zones on a sub-grid-block scale (such as sand or shale lenses) to be oriented in the direction of the bedding.

A second approximation inherent in carbon sequestration models arises from the need to estimate rock properties in the entire region of a CO_2 storage site, over hundreds of square kilometers in areal extent and hundreds of meters in depth, from a relatively small number of local measurements and remote-sensing studies. One method for guessing the rock properties in the un-sampled regions between wells consists of representing rock formations as a set of discrete geological strata (**Figure 10.2.5**) and assuming that these strata extend continuously in the un-sampled regions between wells. A more elaborate geostatistical method consists of using the properties measured at characterization wells to generate a large number of possible rock property distributions that are then used to predict the range of possible outcomes of CO_2 injection.

Thermodynamics and reaction kinetics within a simulation grid block

Field-scale carbon sequestration models treat each simulation grid block as a well-mixed reactor containing water, several mineral phases, and eventually a CO_2-rich phase. Key geochemical phenomena that occur within grid blocks are summarized in **Figure 10.2.6**. Parameters required to describe these phenomena include variables that characterize the state of the grid block such as temperature, pressure, CO_2 saturation, and the mineralogy and porosity of the rock formation. Other required parameters are those that describe the geochemistry of mineral-brine-CO_2 systems: the CO_2-brine equation of state, which tells us the relationship between pressure, temperature, density, and composition. This equation of state also needs to include the influence of brine salinity and the behavior of impurities co-injected along with the CO_2.

Field-scale carbon sequestration models must predict rates of dissolution or precipitation of solid phases in each simulation grid block. On short time scales, the most important reactions involve the dissolution of carbonate minerals (primarily calcite) that occurs rapidly in response to the acidification of pore water caused by the dissolution of $CO_{2(sc)}$ or supercritical CO_2. These reactions were discussed in the previous section.

The derivation of self-consistent thermodynamic databases is an arduous task that is further complicated in the case of carbon sequestration by the coexistence of two fluid phases and the high temperature, pressure, and salinity in the formations of interest.

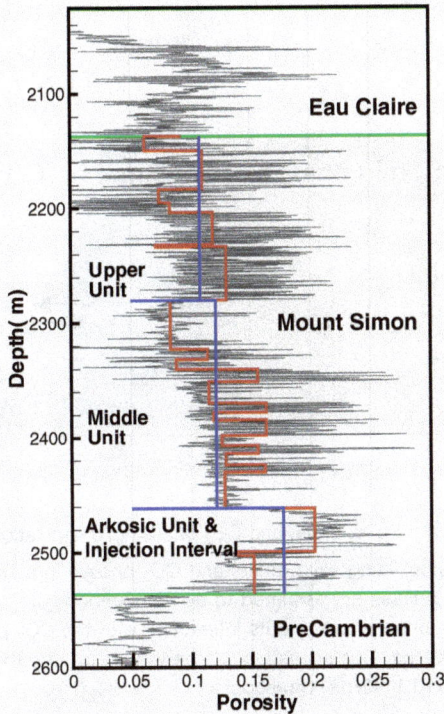

Figure 10.2.5 Well log measurements of porosity

Well log measurements of porosity (thin gray lines) vs. depth in the Mount Simon sandstone, a target sequestration formation in Illinois. The figure also shows measurements from a portion of the overlying Eau Claire shale (the caprock) and from the underlying PreCambrian granite. The red lines show the simplified model of sandstone strata used in the field-scale sequestration model of Zhou *et al.* [10.4]. The blue lines show an even more simplified model in which the sandstone formation is described as only three units. *Figure reproduced from Zhou et al. [10.4], with permission from John Wiley and Sons.*

For combinations of species other than water, CO_2, and simple salts such as NaCl at conditions relevant to sequestration, thermodynamic databases are incomplete. For example, experimental data are lacking on the behavior of CO_2-H_2S or CO_2-SO_2 mixtures in equilibrium with water. This is an important knowledge gap to fill; the co-injection of H_2S or SO_2 with CO_2 could significantly improve the economic viability of sequestration when applied to the combustion of sulfur-rich coal and

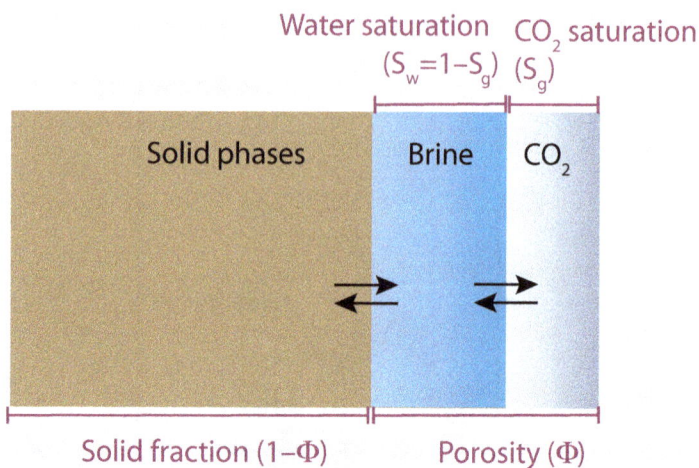

Figure 10.2.6 Solid, brine, and CO_2 phases in a simulation grid block

Schematic view of coexisting solid, brine, and CO_2 phases in a simulation grid block. The brine and CO_2 phases are assumed to be homogeneous in each grid block. The geochemistry of the brine phase and its interaction with the CO_2 phase are described with equilibrium relations. *This animation can be viewed at*: http://www.worldscientific.com/worldscibooks/10.1142/p911#t=suppl

natural gas because H_2S and SO_2 are products of coal gasification and combustion, respectively. Experimental data also are lacking on the thermodynamic activity of many aqueous species (such as $Al^{3+}_{(aq)}$, an important constituent of many silicate minerals) in concentrated brines.

Mass fluxes between simulation grid blocks

During each simulation time step, field-scale sequestration models calculate the mass fluxes of the aqueous and CO_2-rich phases between adjacent grid blocks. These fluxes are driven by buoyancy (the CO_2-rich phase is lighter than the aqueous phase) and capillary pressure. Water has more affinity for the solid grains of the matrix bordering the pore space than CO_2 does in hydrophilic rocks, and therefore is held more strongly in the rock than CO_2.

Calculation of the advective fluxes between neighboring simulation grid blocks requires solving Darcy's law in each fluid phase. Darcy's law was introduced in **Box 9.6.1** for flow in the horizontal directions. In the vertical direction one has to account for the effect of gravity, which gives:

$$u_j = -\frac{kk_{r,i}\left(S_g\right)}{\mu_i}\left(\frac{\partial p_i}{\partial z} - \rho_i g\right) \tag{I}$$

where i = w or g for the water- and CO_2-rich phases, p_i is the density of the phase of interest and g is the gravitational acceleration. Values of ρ_g and ρ_w differ by the capillary pressure p_c, a function of CO_2 saturation: $p_g - p_w = p_c(S_g)$. The characteristic curves relate these properties to the saturation (see Section 9.7).

Field-scale model assumptions

Here we discuss several phenomena for which we do not yet have the fundamental scientific understanding required to optimally include them in our field-scale sequestration models: mixing, the impact of weathering on permeability, organic and biological effects, confinement effects, and reactions in adsorbed water films. The assumptions incorporated in our models simplify or neglect these effects. A better understanding of these phenomena would significantly enhance the reliability of our model predictions.

Mixing

An important assumption in field-scale simulations, inherent in their grid-based discretization of space, is that the system is well-mixed within

---

Question 10.2.1 Characteristic time scale of CO_2 dissolution in brine

Field-scale sequestration models assume that $CO_{2(aq)}$ concentration is uniform in the brine-filled pore space between the CO_2 fingers. In reality, on what time scale does $CO_{2(aq)}$ diffuse into the space between CO_2 fingers? The characteristic time scale τ of diffusion (with diffusion coefficient D) over a distance d is given by the relation $\tau \approx d^2/2D$. The diffusion coefficient of CO_2 in water at 50°C is $D \approx 4 \times 10^{-9}\,m^2\,s^{-1}$. Calculate τ if the distance between CO_2 fingers is either 2 mm or 2 m.

---

each simulation grid block. An important consequence of this assumption is that if a simulation grid block contains even the smallest amount of CO_2-rich phase, the entire aqueous phase in the grid block is assumed to be in equilibrium with the CO_2-rich phase. In reality, CO_2, water, and minerals may not be instantaneously well-mixed at the grid-block scale (see for e.g., **Question 10.2.1**). Because of effects like capillary and viscous fingering caused by heterogeneous mineral distribution (**Figure 10.2.7**) or heterogeneity in the flow properties of geological formations (**Figure 10.2.8**) [10.6], the durations required to equilibrate the CO_2, brine, and mineral phases in a simulation grid block can be substantial (**Figure 10.2.7**), as in **Question 10.2.1**. One implication of slow mixing is that the

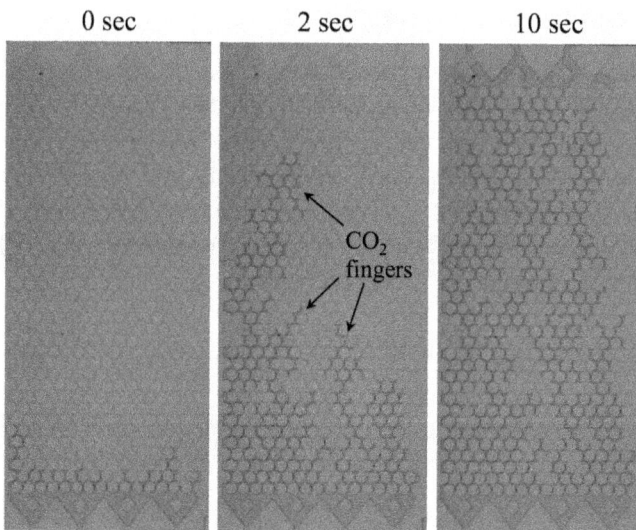

Figure 10.2.7 CO_2 invasion in water-filled porous media

Upon invasion in a porous medium, CO_2 forms capillary or viscous "fingers." The figure shows a photograph of a CO_2 invasion pattern (in darker gray) in an initially brine-filled micromodel (an idealized porous medium made of fused etched silica plates). It shows that finger formation can occur even if the porous medium is essentially homogeneous. *Figures reproduced with permission from Kim et al. [10.5]. Copyright (2012) American Chemical Society.*

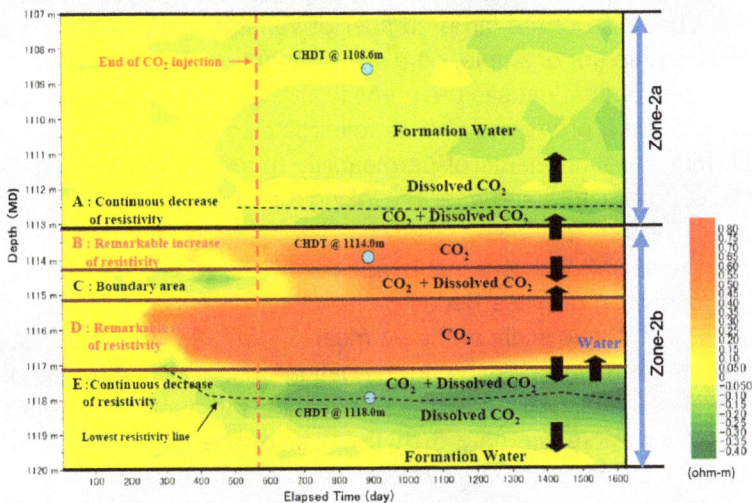

Figure 10.2.8 CO_2 flow and diffusion in water-filled porous media

The figure shows measured electrical resistivity changes as a function of depth (vertical axis) and time (horizontal axis) at an observation well at the Nagaoka sequestration pilot site in Japan; red shows resistivity increases caused by the arrival of supercritical CO_2 at the observation well; green shows resistivity decreases caused by CO_2 dissolution in water. The regions of CO_2-rich water that form above and below the CO_2 plume (dark green) grow very slowly — on time scales of hundreds of days. *Figure reproduced from Sato et al.* [10.6], *with permission from Elsevier.*

weathering reactions involving minerals, water, and CO_2 may be rate-limited by the slow diffusive mixing of these three components.

Impact of weathering reactions on permeability

In many sequestration simulations, the influence of weathering reactions on flow parameters such as permeability and capillary pressure is neglected. This is a significant approximation, because weathering reactions could cause large porosity changes (up to ~50% porosity increase or decrease in different regions of the storage formation if the injected CO_2 contains sulfur impurities [10.7]). Porosity changes are well known to influence the permeability of porous media, but the various models that have been proposed to describe this phenomenon predict very different relationships between ϕ and k.

Characterization of the full relationship between ϕ and k is an arduous task because of the large number of variables that might influence this relationship (for example, pore size distribution, flow regime, and the type of reaction that causes the porosity change). Modern imaging techniques provide a promising avenue towards elucidating this relationship by allowing measurements of permeability, porosity, and pore network structure, both before and after provoking a porosity change in a porous medium. This approach is illustrated in **Figure 10.2.9** and **Movie 10.2.1** [10.8]. **Figure 10.2.9** shows that biologically induced precipitation of calcite in a network of glass beads causes a large permeability decrease (note the logarithmic scale of the permeability axis). **Movie 10.2.1** shows that calcite forms a roughly uniform coating on the entire surface of the silica grains. A uniform coating will influence the aperture of the pore throats much more strongly than the size of the pore bodies, which explains the strong permeability decrease.

Another illustration of the strong influence of porosity on permeability is the relationship between fracture permeability and fracture aperture. As discussed in Section 9.6, for fluid flow in a planar fracture, hydrodynamic theory predicts that the fracture permeability k has a cubic-law dependence on fracture aperture b (i.e., $k \propto b^3$). Several studies have applied the imaging approach described above to determine fracture aperture and permeability both before and after reaction with a CO_2-acidified brine [10.9, 10.10]. What they found is that calcite dissolution causes a rapid increase in the average fracture aperture if the invading brine is undersaturated with respect to calcite (**Figure 9.5.5**). However, because of the heterogeneity of the rock, the preferential dissolution of calcite relative to other minerals can also cause a significant increase in the roughness of the fracture surface, which tends to moderate the influence of fracture aperture on permeability [10.9].

In short, in order to predict the relationship between porosity (or aperture, in the case of a fracture) and permeability, we need to understand where precipitation and dissolution occur in the pore space, because the constrictions of major flow pathways have a much greater influence on permeability than other parts of the pore network. This requires understanding the feedbacks between hydrodynamics and geochemistry (in carbonate-rich rocks, these feedbacks can result in the formation of high-permeability channels or "wormholes" [10.11]),

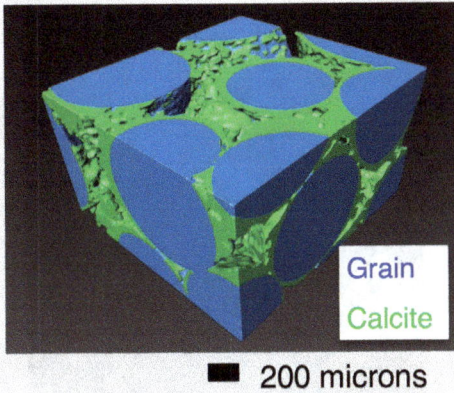

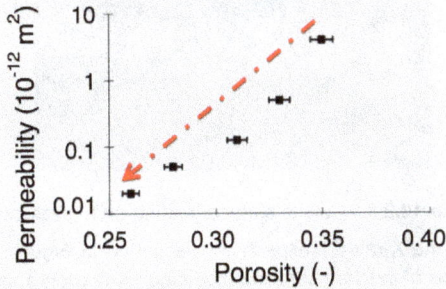

Figure 10.2.9 Calcite precipitation in a network of glass beads

Experimental results showing the correlation between porosity Φ and permeability k during calcite precipitation in a network of glass beads. The top figure shows an X-ray computed tomography image of a portion of the porous medium. The lower figure shows that a porosity decrease of ~30% causes a permeability decrease of about two orders of magnitude. *Figures reproduced from Armstrong and Ajo-Franklin [10.8], with permission from John Wiley and Sons.*

the heterogeneity of rocks (for example, the layered structure of many sedimentary rocks would tend to attenuate the effect of weathering on fracture permeability in the direction normal to the bedding [10.9]), and other phenomena such as nanoparticle transport and aggregation (clay minerals, in particular, can disproportionately affect permeability, probably because of their nanoparticulate nature and high surface area [10.12]).

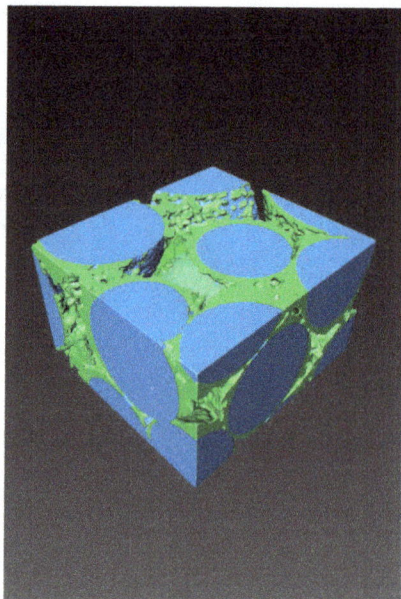

Movie 10.2.1   Calcite surface coating on a glass bead

Movie created from the X-ray CT experiment described in **Figure 10.2.9**. The movie shows the distribution of calcium carbonate coating formed on a single glass bead in the porous medium. *Movie courtesy of Jonathan Ajo-Franklin (Lawrence Berkeley National Laboratory). It can be viewed at*: http://www.worldscientific.com/worldsci-books/10.1142/p911#t=suppl

Organic and biological effects

Most field-scale models assume that organic and biological effects have a negligible influence on the fate of CO_2 in geological formations. In reality, organic matter and microorganisms are ubiquitous in the subsurface and they can strongly influence (and be influenced by) the injection of super-critical CO_2. This is obviously the case in depleted hydrocarbon reservoirs, where the well-established practice of CO_2-enhanced oil recovery (CO_2-EOR) relies on using CO_2-hydrocarbon interactions to increase the fraction of the oil initially in place (OIIP) that is recovered. Evidence of significant interaction between CO_2 and organic matter also has been detected at GCS sites that do not contain exploitable hydrocarbon resources. For example, at the Frio I carbon sequestration pilot site

(**Figure 10.2.2**) a hundred-fold increase in the concentration of dissolved organic molecules (such as acetate, $CH_3CO_2^-$) was detected in the vicinity of the CO_2 plume [10.12]. These organic molecules may have been solubilized by the geochemical changes associated with the injection of CO_2.

The presence of organic molecules in geological formations may influence the fate of injected CO_2 in several ways. In the short term, dissolved organics could adsorb at the CO_2-brine interface, decreasing the CO_2-brine interfacial tension. Less soluble organic molecules could form hydrophobic coatings on mineral surfaces, altering the wetting properties of rock formations [10.13, 10,14]. As discussed in Section 9.4, the capillary pressure P_c associated with CO_2 invasion in porous rocks is strongly dependent on the CO_2-brine interfacial tension and the wettability of the pore walls by brine vs. CO_2. In short, organic molecules could modify the mineral-brine-CO_2 multiphase flow properties in ways that could either enhance or decrease CO_2 storage efficiency. In the longer term, dissolved organic molecules are known to either catalyze or inhibit mineral weathering reactions such as those that were described in Section 9.8 [10.15]. This will either accelerate or slow down the permanent trapping of CO_2 as mineral carbonates.

In the case of shale formations, the potential role of organic matter is evidenced by the existence of significant quantities of organic grains (roughly 0.5 to 13% vol.) in these formations. A detailed view of one of these organic grains is shown in the inset in **Figure 10.2.10** (another organic grain is shown in purple in **Figure 10.2.11**). The electron microscopy study highlighted in **Figure 10.2.10** revealed that the largest pores in the clayshale samples under investigation were associated with organic grains. This finding indicates that organic matter could disproportionately affect fluid flow properties in this type of sample, because the largest pores are also the ones where fluid flow is most likely to take place.

Finally, microbiological effects are thought to play several potential roles in GCS. Few microorganisms are able to survive in the presence of supercritical CO_2, but at least one microorganism has been found to survive in the extremely harsh conditions (high P, T, salinity, and CO_2 concentration) of GCS sites [10.16]. Other such microorganisms are likely to be discovered in the future. At the carbon sequestration pilot site at Ketzin, Germany, microorganism numbers and activity were found to rebound relatively rapidly (within a few months) after the passage of the CO_2 plume at an observation well [10.17]. One potential impact of

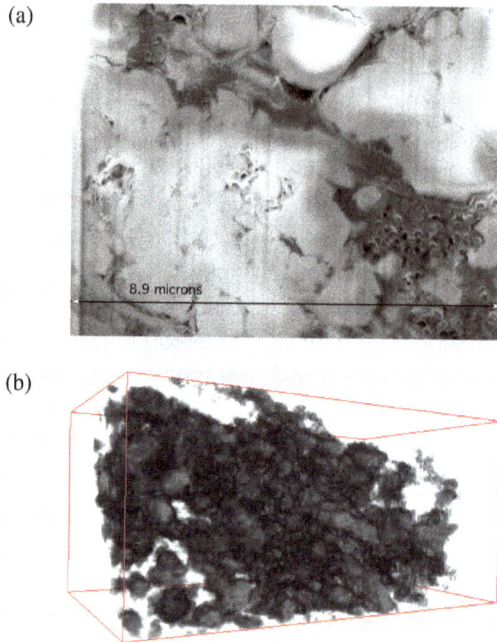

Figure 10.2.10   Organic matter in clayshale

Pore structure of a clayshale mapped by scanning electron microscopy (SEM) with a 5 nm resolution. The lower part of the figure shows a three-dimensional reconstruction of the structure of an organic grain generated from SEM images of a large number of very thin slices of the sample. A significant fraction of the largest (~100 nm wide) pores in this sample are associated with organic matter. *Image courtesy of Tim Kneafsey, Lawrence Berkeley National Laboratory. The animated version of* **Figure 10.2.10 (b)** *can be viewed at:* http://www.worldscientific.com/worldscibooks/10.1142/p911#t=suppl

microorganisms on the fate of CO_2 is their ability to accelerate the rates of dissolution and precipitation reactions that will eventually sequester CO_2 as mineral carbonates, for example, by catalyzing mineral nucleation reactions [10.18]. Studies of natural CO_2 seeps suggest that microorganisms may play exactly this role in the natural environment [10.19].

Confinement effects

Another assumption of field-scale models is that the aquatic geochemistry of pore water located in geological formations is identical to that of

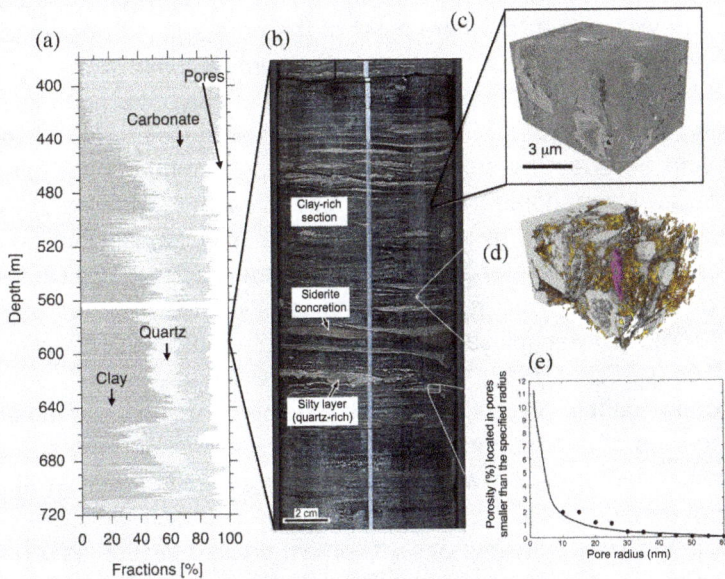

Figure 10.2.11 Microstructure of Opalinus clayshale

The Opalinus clay formation has been extensively studied as a potential host formation for a Swiss high-level radioactive waste repository. (a) On length scales of tens of meters, the average mineralogy of the shale is relatively homogeneous (~60% clay minerals, 20% quartz, 20% carbonates). (b) On length scales of centimeters, the mineralogy is heterogeneous, with alternating clay-rich beds, silt lenses, and carbonate concretions. (c) On length scales of micrometers, focused ion beam scanning electron microscopy images show that the non-clay grains (quartz, carbonates, organics) are distributed in the clay matrix. Figure (d) shows a reconstruction of the mineralogy of the sample shown in (c), where gray regions are quartz or carbonate grains, the purple region is organic matter, yellow regions are pores larger than 10 nm (detected by SEM), and the transparent region is the nanoporous clay matrix. Figure (e) shows the cumulative pore-size distribution of the sample shown in (c): the data points (circles) are from the pore space probed by FIB-SEM; the measured porosity is 2%, but the technique cannot detect pores smaller than 10 nm. The line shows the pore space probed by N_2 adsorption (the measured porosity is 11%, but the technique cannot probe pores narrower than about 2 nm). The total water-accessible porosity of Opalinus clay is between 12 and 16% (this includes pores located within clay grains, such as smectite interlayer nanopores shown in **Figure 9.2.3**). (a): *Reproduced from Wenk et al.* [10.20], *2009, Figure 2, with kind permission of The Clay Minerals Society, publisher of Clays and Clay Minerals.* (b): *Reproduced from Marschall et al.* [10.21], *with permission from Revue IFP.* (c) and (d): *Reproduced from Keller et al.* [10.22], *with permission from Elsevier.* (e): *Reproduced from Keller et al.* [10.23], *with permission from Pergamon.*

bulk liquid water. This assumption is likely to be incorrect in the case of seals and well cements, where most of the pore space consists of very small pores. **Figure 10.2.11** highlights several studies of the Opalinus clay in Switzerland that illustrate the difficulty of characterizing this nano-porosity, both because of the multiscale heterogeneity of sedimentary rocks and because the smallest pores in clay formations (for example, the interlayer nanopores of smectite clay minerals described in **Figure 9.2.3**) are not detected by most experimental techniques. Detailed char-acterization of micrometer-scale samples of the Opalinus clay shows that most of the pore space in these samples is located in pores narrower than 10 nm, and a significant fraction is located in pores narrower than 2 nm [10.20, 10.21, 10.22, 10.23].

Confinement in porous media with pore sizes on the order of nanom-eters to tens of nanometers (such as clays, zeolites, nanoporous silica, or carbon nanotubes) is well known to influence fluid properties such as freezing temperatures, dielectric constants, and self-diffusion coeffi-cients. In fact, this influence of confinement on fluid properties is used in the design of nanofluidic devices [10.24]. The potential implications of this "confinement effect" for carbon sequestration are suggested by sev-eral studies. For example, **Figure 10.2.12** shows that CO_2 may have a much greater solubility in clay interlayer nanopores than in bulk liquid water [10.25]. This enhanced solubility could be viewed as an adsorption phenomenon driven by CO_2-clay interactions or as an absorption phe-nomenon driven by the confinement-induced changes in the properties of water. One implication of this finding is that shale formations may act as barriers to CO_2 leakage, not only because of their low permeability and high capillary entry pressure, but also because of their ability to absorb (or adsorb) CO_2.

The nanoporosity of rocks may also influence mineral weathering reactions, as revealed by the synchrotron X-ray experiments described in **Figure 10.2.13**. This study showed that the precipitation of calcium car-bonate is inhibited in the 7.5 nm diameter nanopores of a nanoporous glass. This result is unexpected, because silica surfaces are known to promote the nucleation of solid carbonates [10.26]. Geochemical models generally predict that, when mineral precipitation occurs in a porous medium, the smallest pores become filled by the precipitating solid more rapidly than the larger ones, because they have a higher ratio of surface area to pore volume. **Figure 10.2.13** clearly shows the opposite effect.

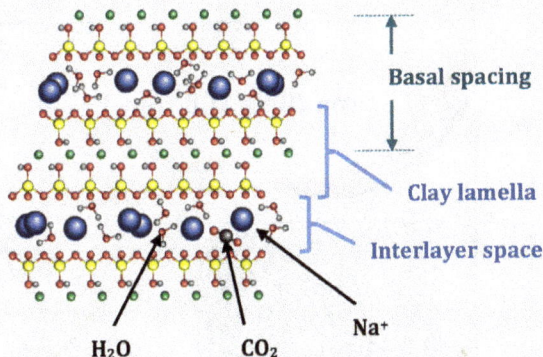

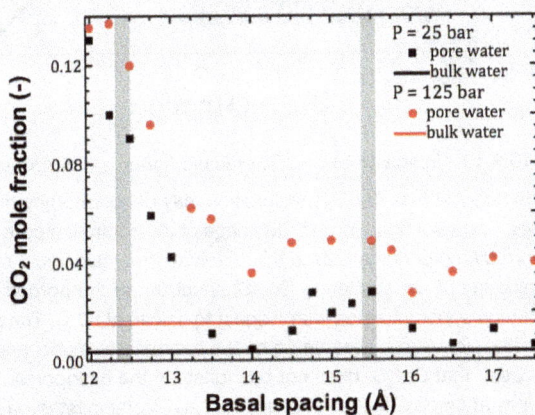

Figure 10.2.12 CO_2 solubility in smectite clay interlayer nanopores

Molecular simulations suggest that CO_2 solubility is higher in smectite interlayer nanopores than in bulk water. The upper part of the figure is an illustration of the structure of hydrated smectite clay. The lower part of the figure shows predictions of the solubility of CO_2 in the nanopore water (squares) and in bulk liquid water (horizontal lines) as a function of basal spacing (the width of one lamella plus one interlayer nanopore) at $P = 2.5$ MPa (black) and 12.5 MPa (red). The vertical shaded bars indicate stable swelling states where the nanopores contain one or two statistical water monolayers. *Figures reproduced with permission from Botan et al. [10.25]. Copyright (2010), American Chemical Society.*

One implication for GCS is that if solid carbonates precipitate in a clay-shale caprock, as predicted under certain conditions [10.27], this precipitation may preferentially clog the largest pores in the rock and enhance its sealing properties.

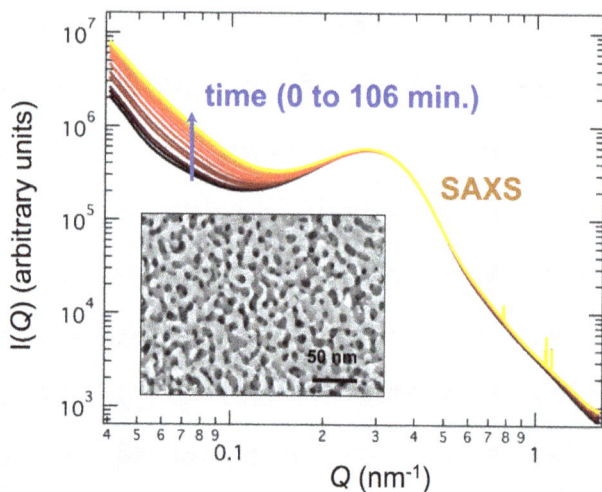

Figure 10.2.13 Inhibition of CaCO₃ precipitation in nanoporous silica

Experimental data suggest that CaCO₃ precipitation may be thermodynamically inhibited in silica nanopores. The inset figure is an SEM image of the controlled pore glass CPG-75 showing the 7.5 nm diameter nanopores in this medium. The main figure shows the evolution of the small angle X-ray scattering (SAXS) spectrum of the porous medium upon exposure to a solution supersaturated with respect to calcite at 90°C. The peak at $3 \, nm^{-1}$ in the SAXS spectrum (caused by scattering by the nanopores) is unchanged during the experiment, indicating that CaCO₃ does not precipitate in the nanopores. The change in the SAXS spectrum at small Q values shows that CaCO₃ precipitates on the outside of the CPG-75 grains. *Images courtesy of Andrew Stack, Oak Ridge National Laboratory.*

Reactions in adsorbed water films

Finally, field-scale models assume that the CO_2-rich phase does not interact directly with solid surfaces: i.e., that CO_2-mineral interactions are always mediated by the aqueous phase. This assumption is consistent with the expectation that most solid surfaces in carbon sequestration sites are hydrophilic, hence at low capillary pressures they should be coated with a film of adsorbed or capillary water. However, in certain regions of a CO_2 storage site (near the injection well, where dry CO_2 is continuously injected, and near the top of a thick CO_2 plume, where capillary pressure is high) these water films may be extremely thin. Indeed, studies of mineral weathering reactions suggest that the aquatic geochemistry of very thin adsorbed water films may differ from that of bulk liquid water. One of the few existing studies of the geochemical impact of these films showed that the weathering reaction of the mineral forsterite (Mg_2SiO_4) by supercritical CO_2 in the presence of an adsorbed water film resulted in CO_2 mineral trapping as magnesite ($MgCO_3$) on a time scale of weeks [10.28]. This result is remarkable, because the precipitation of magnesite in bulk liquid water in the same conditions is too slow to detect in the laboratory. This difference suggests that the very slow desolvation of Mg^{2+} (thought to be the rate-limiting step of magnesite precipitation) occurs more rapidly in an adsorbed water film than in bulk liquid water.

Section 3

Capacity assessment

Having discussed above the continuum scale aspects of the rock formations, their structures, pore-space properties, and two-phase flow processes, we are in a position to consider the larger-scale question of how much CO_2 can be reasonably expected to be injected and stored in suitable geological formations. This question turns out to be quite complicated, and as there is a lack of experimental data from actual large-scale injections, it is the subject of considerable speculation (e.g., [10.29]). The unsettled nature of capacity assessment is shown in **Figure 10.3.1** by the large variability in capacity assessments done by independent groups around the world.

To cope with these uncertainties, researchers can return to existing industries and build on the knowledge and experience base from oil and gas exploration and production to develop sound concepts for CO_2 storage capacity. In the oil and gas extraction industry, the amount of recoverable hydrocarbon is the fundamental measure of the economic value of a given oil and gas reservoir prior to production. This idea of recoverable hydrocarbon is analogous to CO_2 storage capacity as discussed further in this section.

Resource capacity and reserve capacity

The concepts of resource and reserve have been borrowed from the hydrocarbon industry to refer to the total CO_2 potential storage capacity provided by a formation, and to the practical storage capacity, respectively. Many of the same limiting factors that make hydrocarbon reserves a fraction of hydrocarbon resources also make CO_2 storage reserves a small fraction of the total CO_2 storage resources. For example, the economics of the system are critical; today, without any significant policy or economic incentive for capturing and storing CO_2, there is effectively no storage reserve anywhere. But this narrow economic criterion for defining reserve does not reflect the potential future value of geological carbon

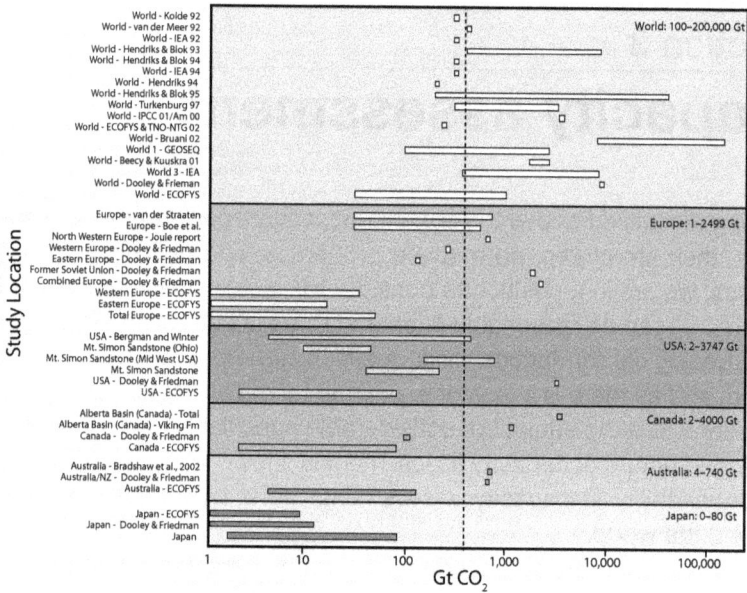

Figure 10.3.1 Storage capacity

Capacity assessments have been made all around the world using various different approaches and assumptions, the result of which is large variability and uncertainty in capacity assessment. The dotted line gives the estimated worldwide capacity. *Figure reproduced from Bradshaw et al. [10.29], with permission from Elsevier.*

sequestration sites that is recognized widely today. Some industries are building considerable expertise and knowledge about how to carry out sequestration effectively on the assumption that someday economic incentives and/or legal imperatives will be in place. Regardless, regulations on injection, monitoring, and verification of storage, legal aspects of ownership and liability related to the injected CO_2, and land access comprise some of the other non-technical factors that control reserve capacity. As for the technical aspects, most of the issues discussed in the earlier sections of this chapter relate to aspects of sequestration processes that reduce the total resource storage capacity by a large amount to create the reserve capacity. Here we focus on these process-oriented or technical aspects of CO_2 storage capacity.

Storage capacity

The injection of CO_2 into a brine-filled porous reservoir is a complex process that cannot be tightly controlled or engineered once the CO_2 leaves the wellbore. Instead, the vagaries of the natural system largely control how the CO_2 fluid invades and occupies the pore space. An understanding of the processes involved has been developed through theory, experience with analogous systems, and simulation, but controlling these processes remains elusive. **Figure 10.3.2** illustrates the different factors that control the filling of the pore space by injected CO_2. Specifically, **Figure 10.3.2 (a)** illustrates multiphase flow effects which, as discussed in the last chapter, tend to place CO_2 (the non-wetting phase) in the centers of pores while native groundwater wets the solid grains of the matrix. **Figure 10.3.2 (b)** illustrates gravity effects which provide a strong upward driving force that causes CO_2 to accumulate in the upper regions of the storage formation because of the density contrast between supercritical CO_2 and brine. Shown in **Figure 10.3.2 (c)** are the effects resulting from heterogeneity in porosity and permeability, which also strongly control where the CO_2 invades and how it is ultimately trapped. Finally, **Figure 10.3.2 (d)** shows structural effects, specifically the role of dipping strata in rocks that will encourage long lateral migrations. Doughty et al. [10.30] defined an overall capacity factor C as the product of the porosity and four capacity factors that represent each of these main effects. The idea here is that porosity alone does not provide a realistic estimate of capacity in a porous natural rock. Instead, capacity is potentially reduced by

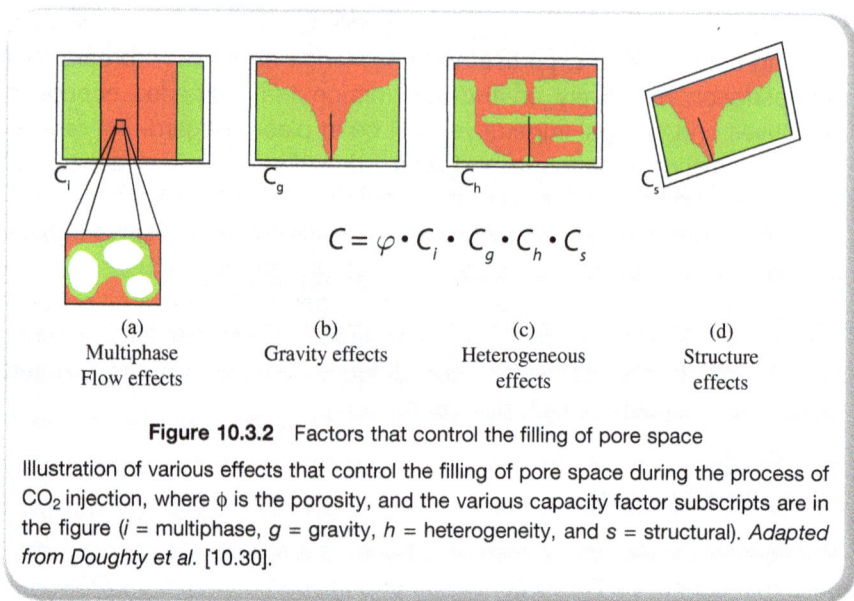

$$C = \varphi \cdot C_i \cdot C_g \cdot C_h \cdot C_s$$

(a) Multiphase Flow effects

(b) Gravity effects

(c) Heterogeneous effects

(d) Structure effects

Figure 10.3.2 Factors that control the filling of pore space

Illustration of various effects that control the filling of pore space during the process of CO_2 injection, where ϕ is the porosity, and the various capacity factor subscripts are in the figure (i = multiphase, g = gravity, h = heterogeneity, and s = structural). *Adapted from Doughty et al.* [10.30].

these four other effects. The reserve capacity as defined using this conceptual framework would be a small fraction of the total resource capacity that might be represented approximately by the porosity multiplied by the formation volume.

Capacity assessment

Different agencies and groups have developed approaches for estimating CO_2 storage capacity. Here we compare the different approaches used by the US Department of Energy (e.g., [10.31]), the Carbon Sequestration Leadership Forum (CSLF), and the United States Geological Survey (USGS). **Table 10.3.1** gives an overview of these three approaches. All three approaches require information about the areal extent of the subsurface formation and the depth, thickness, porosity, and structure(s) present (e.g., anticlines, synclines, fault traps). The DOE and CSLF approaches combine all of the complex processes (e.g., those discussed in **Figure 10.3.2**) into an overall efficiency factor, E or C_c. The difference between the E and C_c efficiency factors is that the E factor includes the effect of residual saturation whereas C_c does not; the CSLF approach considers it separately. The bigger difference between the DOE and CSLF approaches is that the CSLF approach makes a conservative assumption that only the volume of the formation with structurally closed traps is available for storage. In contrast, the DOE approach includes the entire formation thickness as potential storage volume in acknowledgment of the potential importance of residual-phase trapping. This difference can be large, as sketched in **Figure 10.3.3**, and this points out an unresolved source of uncertainty and variability that must be considered when evaluating various published capacity estimates.

Also listed in **Table 10.3.1** for completeness is the USGS approach which is based on principles used in the National Oil and Gas Assessment [10.32]. Briefly, the USGS approach considers the uncertainty in properties of the storage formations and performs Monte Carlo analyses to arrive at probabilistic estimates of capacity [10.33,10.34]. Although a

Table 10.3.1 Approaches used for estimating reserve capacity

Group	Approach	Notes
US DOE	$C = A \, h_g \, \phi_{tot} \, \rho \, E$	Applies to entire area being assessed, and includes entire gross thickness of formation. Efficiency factor accounts for all of the effects that limit CO_2 pore-filling.
CSLF	$C = A_{trap} \, h_{trap} \, \phi_{trap} \, \rho \, (1 - S_{lr}) \, C_c$	Applies only to the closed structure parts of the reservoir (trap). Separates out effect of residual aqueous phase (S_w) from the efficiency factor (C_c).
USGS	Probabilistic assessment carried out by categorizing trapping as either primarily structural or primarily residual-phase. Based on geological model uncertainty, uses Monte Carlo simulation to determine likely storage capacity.	Requires expert judgment about trapping style, knowledge of structure, and other critical factors. Well-established precedent in the National Oil and Gas Assessment.

Definitions: A = area of region being assessed for CO_2 storage capacity; h_g = gross thickness of formation being assessed; ϕ_{tot} = average porosity over thickness h_g; ρ = density of CO_2 averaged over h_g; E = efficiency factor reflecting fraction of total pore volume filled by CO_2.

relatively large effort has gone into capacity assessment, one must treat these estimates with caution; as with many calculations involving the subsurface there is great uncertainty in the numbers.

Pyramid representations

Despite this large uncertainty inherent in large-scale prediction of capacity, more accurate assessments can be made on smaller scales when more site-specific information is available. By confining the assessment to that subset of the region where storage is economical and conforms to regulations and requirements, the derived capacity is naturally smaller. These simple concepts are illustrated in **Figure 10.3.4** by a pyramid in which the large base represents both large uncertainty and large storage volume associated with large-scale capacity assessment. At the top of the pyramid, we have a particular site where characterization data have been collected and evaluated. As smaller regions are examined in more detail, both the uncertainty and the storage volume estimates decrease.

In summary, the assessment of CO_2 storage capacity is an uncertain endeavor, but broad agreement exists within the geological carbon sequestration community that the large sedimentary basins of the world contain enormous capacity for future CO_2 storage [10.35].

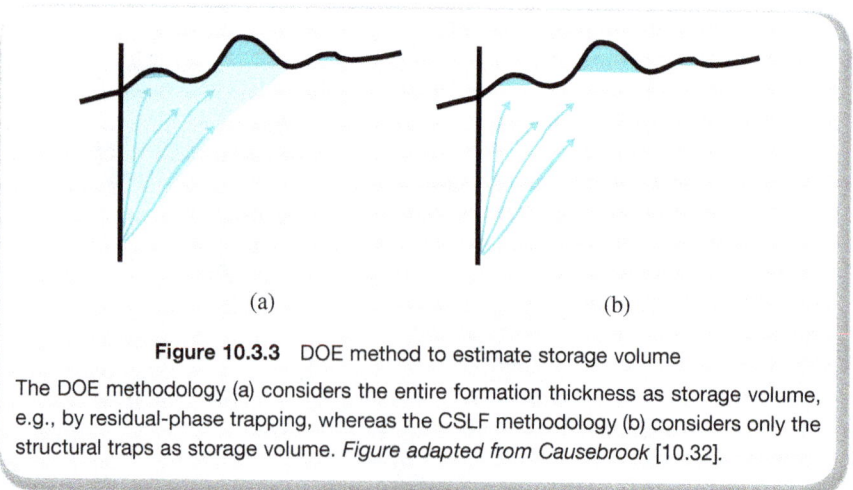

(a)   (b)

Figure 10.3.3 DOE method to estimate storage volume

The DOE methodology (a) considers the entire formation thickness as storage volume, e.g., by residual-phase trapping, whereas the CSLF methodology (b) considers only the structural traps as storage volume. *Figure adapted from Causebrook* [10.32].

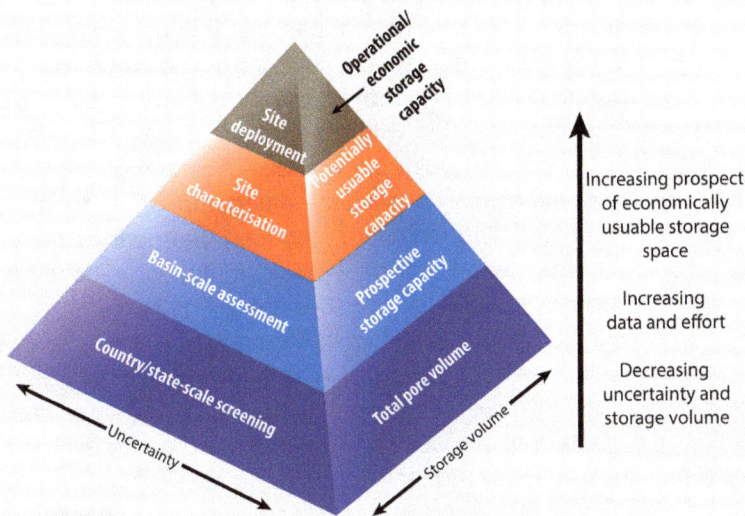

Figure 10.3.4 Storage capacity pyramid

Two dimensions of capacity assessment, uncertainty and storage volume, decrease with increasing focus on a particular site. *Data from CO2CRC.*

Section 4

Health, safety, and environmental impacts

The scale over which geological carbon sequestration must be carried out to reduce CO_2 emissions means that sequestration itself will have its own set of potential health, safety, and environmental (HSE) impacts, as well as present issues related to sustainability (e.g., [10.36]). **Figure 10.4.1** illustrates potential HSE impacts of sequestration in terms of their qualitative potential severity and the depth at which they might occur.

As shown in **Figure 10.4.1**, the impacts that occur at the greatest depths are not classified as having large HSE impact [10.37]. For example, the unplanned intrusion of CO_2 into hydrocarbon reservoirs due to leakage from the intended target formation may cause an economic impact by mixing with the oil or natural gas, but it is not considered an HSE impact. Similarly, brine displacement from one deep aquifer to another may change the salinity of the brine formations, but as long as all of the deep aquifers are filled with non-potable water, such changes do not impact HSE. We also point out that displacement of brine due to CO_2 injection is not necessarily a leakage-related impact. As for induced seismicity, the overwhelming majority of induced deep earthquakes will be so small that they will have minimal HSE impact. The situation becomes more controversial if earthquakes are felt by humans.

The shallower HSE impacts shown in **Figure 10.4.1** are all related to leakage of CO_2 out of its intended storage formation. Such leakage could occur through wells or through certain kinds of faults and fractures (see Section 9.6). A great deal of research is being carried out to evaluate the potential for faults and wells to leak. In HSE risk assessment, we define risk as the likelihood of a leakage event leading to an impact on HSE

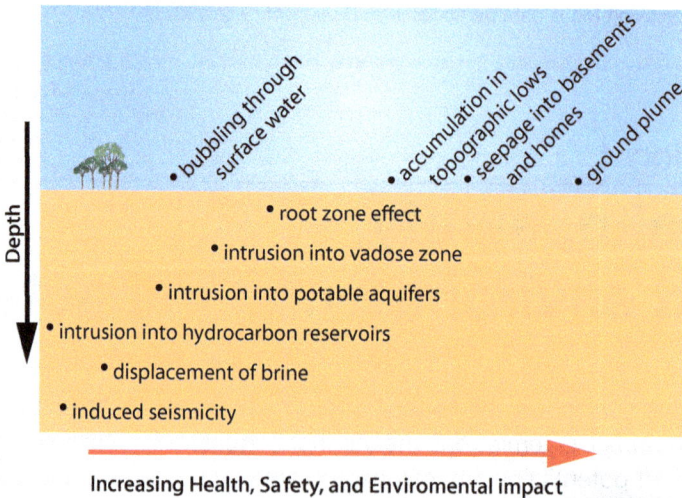

Figure 10.4.1 Potential environmental impacts of geological carbon sequestration

Qualitative categorization of potential environmental impacts in terms of depth (vertical axis) and HSE impact (horizontal axis). The vertical axis is not to scale, but the outlined area highlights the impacts that potentially might occur at shallow depths. *Figure adapted from Oldenburg* [10.37].

multiplied by the severity of that impact. By this definition, reducing either the likelihood of the event or the impact of the event can reduce HSE risk. For example, locating a sequestration site in an area with few existing wells decreases the likelihood of a leaking well. On the other hand, if a site (e.g., an abandoned oil field) contains many wells but no potable groundwater or human population present, the potential impact of leakage is low. In both cases, the HSE risk may prove to be acceptable even if likelihood and/or consequences of potential leakage are non-zero. In the context of risk assessment, there is never truly zero risk. In risk assessment, we use the term *de minimis* to indicate inconsequential risk, which implies the risk is too small to evaluate and can for all practical purposes be considered to be zero.

Groundwater quality

"Intrusion into potable aquifers" reflects the degradation of groundwater that can occur when CO_2 leaks into potable groundwater. Although nothing is intrinsically unhealthy about carbonated water, it does form carbonic acid and the associated drop in pH can change the hydrogeochemical equilibrium that was established over millennia in the aquifer. The main concern is that heavy metals such as arsenic (As) and lead (Pb) could leach out of minerals in the solid grains of the rock matrix to an extent that would measurably degrade groundwater quality. Because heavy metals are linked to a variety of negative human health impacts, this potential consequence of the leakage of CO_2 from a sequestration site warrants considerable study and risk assessment. In addition, brine from depth could leak upward from the CO_2 injection formation due to increased pressure related to injection. Deep brines could degrade groundwater quality through salinization as well as through introducing heavy metals. We present in **Figure 10.4.2** a sketch of some of the processes that are of concern for groundwater degradation related to geological carbon sequestration [10.38].

Induced seismicity

Although induced seismicity appears in **Figure 10.4.1** on the left-hand side indicating that the experts expect relatively small HSE impact, the subject has emerged recently in the technical literature [10.39].

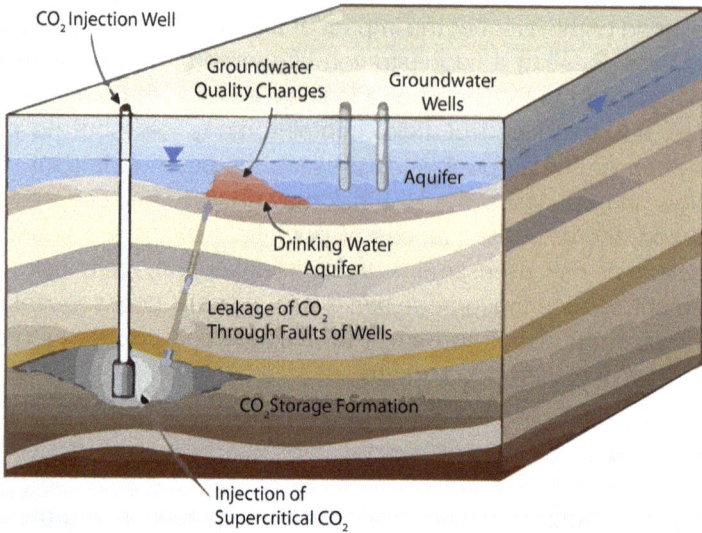

Figure 10.4.2   Potential CO_2 leakage

Sketch of potential leakage process for CO_2 or deep brine to impact potable groundwater. *Figure redrawn from Apps et al.* [10.38].

Earthquakes occur across an enormous range of magnitudes as indicated by the logarithmic scale used to describe them (the Richter scale). In addition, there is a long-recognized inverse logarithmic relationship between earthquake magnitude and frequency known as the Gutenberg-Richter Law. In short, it is very likely that a very small earthquake will occur and very unlikely that a large earthquake will occur, regardless of cause. Small earthquakes at depths greater than 2 km, e.g., those with Richter-scale magnitudes up to 2 or so, cannot be felt at the ground surface.

Earthquakes occur naturally along faults due to tectonic forces. Briefly, tectonic stress builds in the rock which resists deformation (strain) due to its inherent strength. Eventually the strength of the rock as controlled by its weakest elements, faults and fault zones, is overcome and slip occurs to relieve the stress. The slip event can produce seismic waves that travel great distances and change in amplitude depending on the type of rock through which they are traveling: this is the earthquake.

This seismic energy results in ground accelerations that can damage structures. When most people hear the word "earthquake," they associate it with danger and structural damage. Most of the earthquakes of which we speak as being caused by injection are too small to be felt. For this reason, the term *induced seismicity* is preferred in the geological carbon sequestration context.

Induced seismicity is a well-known by-product of changing the loading or pore pressure in rock. Induced seismic events have been recorded as consequences of the filling of large surface-water reservoirs [10.40] and geothermal energy production. Here we focus the discussion on changes in pore pressure as the impetus for induced seismicity.

In the earth, rocks are in a state of stress characterized by three mutually orthogonal principal components (see **Box 10.4.1**). The largest stress is σ_1, followed by σ_2, and σ_3. The orientations of σ_1, σ_2, and σ_3 depend on the depth and tectonic forces active in any given block of rock. In general, the deeper the rock of interest, the larger the vertical component of stress becomes relative to the horizontal stresses as the weight of the rock above increases. Negative normal stresses correspond to tensional stresses.

Shown in **Figure 10.4.3** is the classic Mohr-Coulomb plot of shear stress (vertical axis) vs. normal stress (horizontal axis). Recall **Figure 9.6.3** which shows the tendency to fracture the rock as a function of pressure and depth. This tendency to fracture corresponds to the values of shear and normal stress above which the intact (unfaulted) rock will break, the so-called *intact rock failure envelope* (see **Figure 10.4.3**). The other envelope in **Figure 10.4.3** describes the region above which a fault has a tendency to slip, the *fault slip envelope*. As shown in the figure, if the normal stress across the fault is very large, it will take a larger shear stress to re-activate the fault.

The two semi-circles in the figure represent σ_1 and σ_3 of a formation in which we inject CO_2. Because of the injection, the local fluid pressure increases. As a consequence, all of the compressive normal stresses decrease because the fluid pressure acts against compressive normal stresses and tends to dilate the rock. This is represented in the figure by the semi-circles moving to the left. Therefore the mechanism of induced seismicity is that pore pressure arising from fluid injection reduces the effective normal stress as shown in **Figure 10.4.3** to the point that existing shear stresses can re-activate the fault. What this means is that

Box 10.4.1 Mohr diagrams

Mohr circles provide a convenient and illustrative way to understand stress and brittle failure in rocks. The stress state of rock in the subsurface can be resolved into three principal mutually perpendicular vectors: σ_1, σ_2, and σ_3. These vectors are referred to as the maximum principal stress, intermediate stress, and minimum principal stress directions, respectively ($\sigma_1 \geq \sigma_2 \geq \sigma_3$). If we consider any plane within the earth whose strike (intersection of horizontal plane with this plane of interest) is parallel to σ_2, then (a) shows that the two-dimensional plane stress in the plane containing σ_1 and σ_3 can be resolved using the appropriate tensor operations into a shear stress (σ_s, often denoted by τ) and normal stress (σ_n).

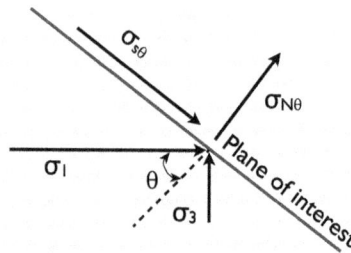

(a)

The shear and normal stresses on any plane of interest with strike parallel to σ_2 can be found using a Mohr diagram. Figure (b) shows a Mohr diagram, in which the shear stress (σ_s) is plotted on the y-axis and normal stress (σ_n) on the x-axis. The Mohr circle has diameter $\sigma_1 - \sigma_3 = \sigma_d$ (differential stress), and is centered at $\frac{1}{2}(\sigma_1 + \sigma_3)$ on the normal-stress axis.

Let us now determine the shear and normal stress of our plane in (a). Mathematically, these stresses are given by:

$$\sigma_s = \frac{1}{2}(\sigma_1 - \sigma_3)\sin(2\theta)$$

$$\sigma_n = \frac{1}{2}(\sigma_1 + \sigma_3) + \frac{1}{2}(\sigma_1 - \sigma_3)\cos(2\theta)$$

Figure (b) shows that these components can be obtained from a simple graphical construction of the Mohr diagram.

(Continued)

starting from a stress state in which a fault is stuck because shear stress is not large enough to overcome the normal stress on it, pore pressure increases can reduce the effective normal stress to the point that the fault will fail with no increase in shear stress.

Box 10.4.1 (*Continued*)

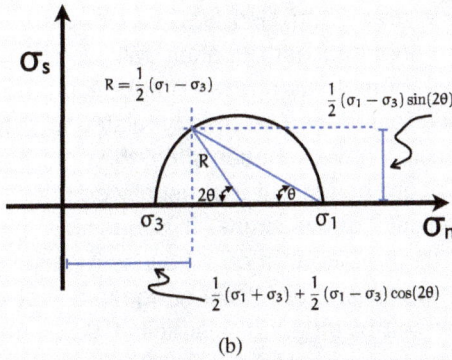

$$R = \frac{1}{2}(\sigma_1 - \sigma_3)$$

$$\frac{1}{2}(\sigma_1 - \sigma_3)\sin(2\theta)$$

$$\frac{1}{2}(\sigma_1 + \sigma_3) + \frac{1}{2}(\sigma_1 - \sigma_3)\cos(2\theta)$$

(b)

We see that for an angle $\theta = 0$ the shear stress is zero and the normal stress $\sigma_n = \sigma_3$. For an angle $\theta = 90°$ we obtain $\sigma_s = 0$ and $\sigma_n = \sigma_1$. The maximum shear stress is obtained for an angle of 45°.

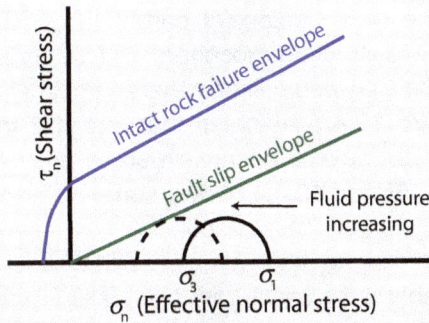

Figure 10.4.3   Shear stress vs. effective normal stress

Shear stress vs. effective normal stress for rocks undergoing changing pressurization due to injection.

Many factors complicate the above explanation of induced seismicity. One of these is the existence of faults that are very near to failure naturally, i.e., a natural earthquake may be only a few years away. The natural increase in tectonic stress is augmented by an injection process which effectively triggers the earthquake. Such an earthquake may be quite large and not related directly to the amount of fluid pressure increase. On the other hand, there can also be cases where the tectonics are inactive, as in ancient faults where no particular built-up stress exists. Injection into these areas can also lead to induced seismicity, but the magnitudes of the induced earthquakes will tend to be much smaller because of the lack of tectonic stress contribution.

The bottom line for induced seismicity related to sequestration is that it is a very real hazard. In well-chosen sites, the main impact is nuisance: earthquakes that might be felt but that usually do not do any measurable damage. In tectonically active areas, it is possible that injection could trigger a tectonic earthquake that could cause measurable damage.

Recently, some researchers have suggested that caprock integrity could be compromised by induced seismicity [10.39], but data on permeability of faults in shale caprock to support this assertion are currently not available.

Near-surface impacts due to leakage to the atmosphere

The ultimate failure of a geological carbon sequestration site would involve leakage of CO_2 to the atmosphere. Before entering and mixing in the atmosphere, the CO_2 would likely have an impact in the near-surface environment. **Figure 10.4.1** indicates that many of these near-surface HSE impacts can be severe because people and other animals and plants are present. **Figure 10.4.4** shows some of the features and environments in which impacts could occur.

Impacts to groundwater have already been discussed in this section. Above the water table is the unsaturated (so-called *vadose*) zone in which resides the roots of trees and plants and burrowing animals. Elevated CO_2 concentrations in the vadose zone can cause plant stress or even plant death. This occurred on a relatively large scale starting in the late 1980's at Horseshoe Lake in the Eastern Sierra mountains of California due to natural magmatic CO_2 [10.41].

Simulation studies have shown that it does not take a very large CO_2 leakage flux to create high concentrations in the vadose zone [10.42]. **Figure 10.4.5** shows simulation results [10.43] for a typical geological sequestration site containing 4×10^9 kg CO_2. Above this site, it is assumed that there is a cylindrical area (100 m) in which there is a leakage of CO_2. **Figure 10.4.6** shows simulation results for maximum leakage flux and

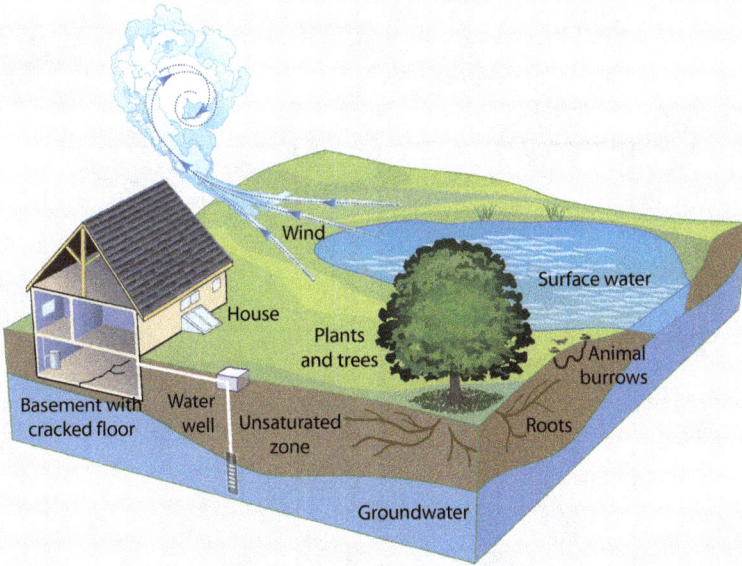

Figure 10.4.4 Sketch of near-surface features relevant to near-surface HSE impacts of CO_2 leakage

Figure by Walter Denn in Oldenburg and Unger [10.43].

corresponding CO_2 concentrations in the vadose zone for a variety of vadose zone properties. It was found that leakage rates on the order of several hundreds of tonnes (10^5–10^6 kg) of CO_2 per year over areas of radius 100 m would be required to exceed a leakage flux equal to the typical natural ecological CO_2 flux driven by photosynthesis. Another important property is the concentration of CO_2 in the shallow subsurface. If this concentration exceeds 30% by volume, trees will die. The right-hand side of **Figure 10.4.6** shows that concentrations in the shallow subsurface exceed 30% for these same leakage rates that produce low fluxes. The reason that low fluxes can create high concentrations in the vadose zone is that, unlike in the above-ground region where turbulence caused by wind and free-flowing air occurs, not many dispersive processes are active in the pore space of the vadose zone.

Once CO_2 is in the atmosphere, effects of density differences and turbulence will control mixing. As shown in **Figure 10.4.6**, CO_2 is a dense

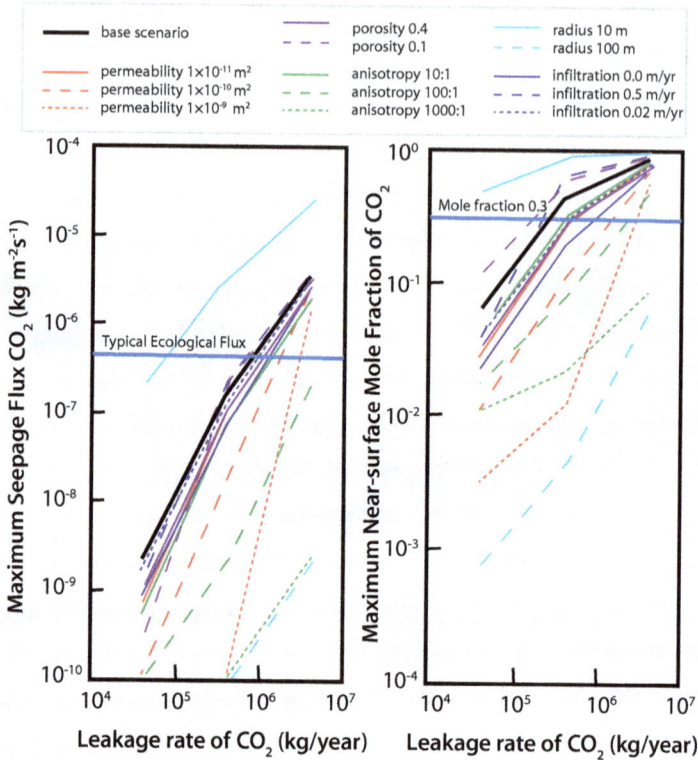

Figure 10.4.5 Simulation of CO_2 migration

Results of simulations of CO_2 migration in the vadose zone showing (a) maximum seepage flux including effects of variable permeability, infiltration of rainwater, porosity, and radius of the leakage area, with typical ecological flux shown by the horizontal blue line, and (b) maximum near-surface concentration of CO_2 in soil gas. The mole fraction equal to 0.3 is highlighted because at this concentration trees died at Horseshoe Lake, CA. *Figure redrawn from Oldenburg and Unger [10.43].*

gas at ambient conditions relative to air; as such it will tend to hug the ground and occupy low spaces. While normally associated with decreased mixing, dense gasses can also mix more effectively because they have their own inherent gravity driving force relative to passive gasses. The thought experiment shown in **Figure 10.4.7** elucidates this concept. Briefly, a density contrast between two gasses can, in many

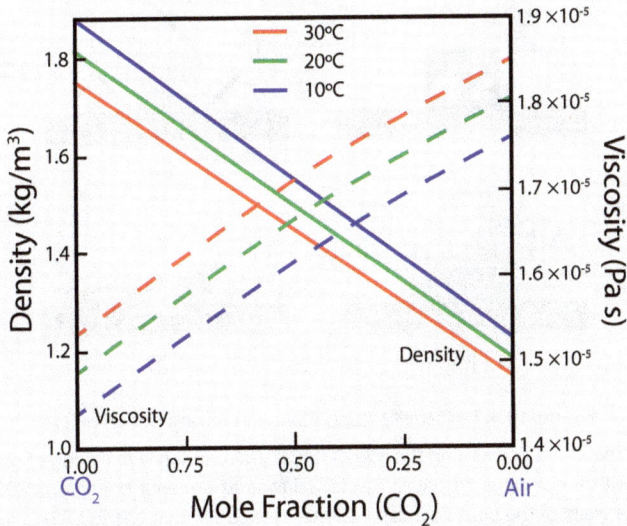

Figure 10.4.6 Density and viscosity of CO_2-air mixtures

Density and viscosity of CO_2-air mixtures at three different temperatures and at a pressure of 1 bar. *Figure adapted from Oldenburg and Unger [10.43]. Reprinted with permission from ASA, CSSA, SSSA.*

circumstances, provide an additional driving force for flow that enhances diffusion and turbulent mixing (dispersion). In other cases, the dense gas may flow downward into depressions in the ground surface or in a building (e.g., in basements) where winds are subdued and thereby resist mixing.

Research has been directed at the behavior of dense gasses in the context of liquefied natural gas (LNG) (e.g., Britter [10.44]). Although the larger density relative to air of LNG is caused by its very low temperature, this work is directly applicable to the leakage of CO_2, which is inherently more dense than air (molecular weight of 44 g/mol vs. 29 g/mol). Using a combination of theory and empiricism, Britter and McQuaid [10.45] developed some very useful expressions for predicting the dilution of a leaking dense gas in a wind field. With a wind speed of 2 m/s, CO_2 concentration would reduce by a factor of 50 after 100 m of travel above ground, and by a factor of 125 after 200 m of travel. Hence, CO_2

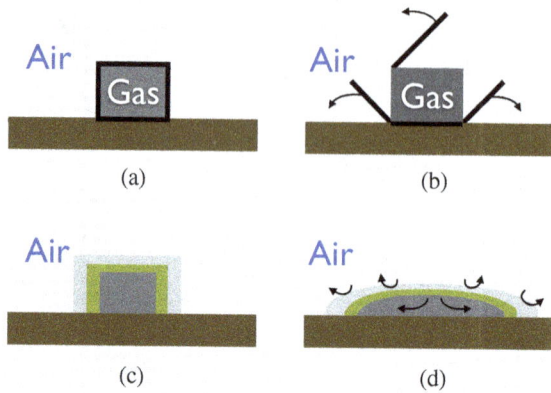

Figure 10.4.7   Density-dependence in mixing of two gasses

Thought experiment elucidating the role of density-dependence in mixing of two gasses. We start with a container of gas on a flat surface at ambient conditions (a). We then remove the walls of the container (b). With no wind, a passive gas will mix only by diffusion (c) whereas a dense gas will flow on its own and mix by dispersion (d).

dispersion by wind is quite effective, which leads to the conclusion that above-ground HSE impacts due to leaking CO_2 are primarily local and will not extend very far downstream due to turbulent mixing. This result is corroborated by more recent simulation studies of pipeline leakage involving very accurate pipeline leakage flux evolution and three-dimensional Navier-Stokes atmospheric dispersion processes [10.46].

From these studies, it appears that density effects in a CO_2 surface leakage scenario only need to be considered if the seepage area is large or the wind is very light. This finding is quantified in **Figure 10.4.8** through application of the Britter and McQuaid [10.45] density-dependence criterion for a variety of system properties. As shown in the figure, seepage fluxes of the order 10^4 times larger than those of the natural ecosystem exchange will be density-dependent, even for relatively large wind speeds, while lesser fluxes are often passive by this criterion.

In summary, there are recognized hazards of CO_2 injection, but they are generally local effects and they are considered manageable. The hazards of geological carbon sequestration should always be considered relative to other energy-related hazards (see **Question 10.4.1**) and most

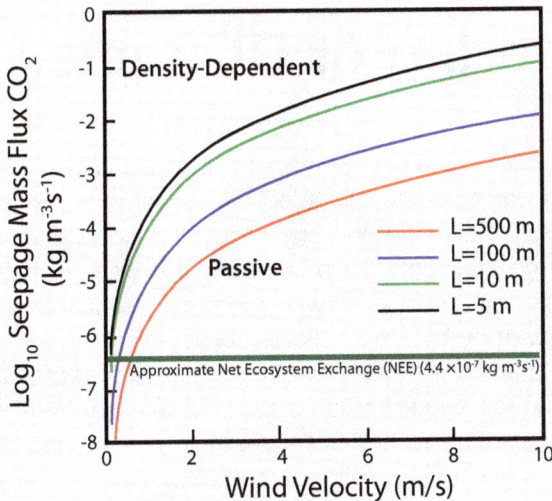

Figure 10.4.8 Active versus passive flows

Delineation of active (density-dependent) vs. passive (not density-dependent) flows as controlled by seepage flux and wind speed for various different scales of surface CO_2 seepage. The figure shows that large seepage fluxes (10^4 times that of natural plant CO_2 uptake) will produce density-dependent flows even for relatively high wind speed (e.g., 8 m/s) for seepage over 100 m length scales. *Figure adapted from Oldenburg and Unger* [10.43].

Question 10.4.1 Comparative environmental risks

Consider the HSE risks of current energy-related technologies as made obvious by recent failures such as the Deepwater Horizon drilling platform fire and Macondo Well blowout, the Big Branch coal mine fire, and the San Bruno, CA, natural gas pipeline explosion. How would potential failures of sequestration system components (pipelines, geological storage sites) compare with existing energy-related technology failures? Compare and contrast by considering effects such as flammability and explosion potential, proximity of population, etc.

importantly relative to doing nothing about greenhouse gas emissions, which entails very significant impacts to global environmental health.

Section 5

Monitoring of GCS sites

One way of reducing the possibility of unexpected HSE impacts is to keep a close watch on the CO_2 injection and storage processes so that irregularities can be found and any hazardous conditions mitigated. Of course other good reasons for monitoring also exist, including the need to quantify storage and trapping to meet the verification needs of cap and trade and/or carbon tax policies, to optimize storage efficiency, to quantify and map the various evolving storage processes, and to assure the public that the sequestration site is not leaking. In this section, we discuss the main approaches to sequestration monitoring of the subsurface and surface and we provide examples of their use.

Subsurface monitoring

Subsurface monitoring of fluids in the pore space of deep geological formations is very challenging due to the large thickness of overlying rock. The most direct method is to drill a well to the depth desired and sample the fluid directly. This approach is expensive and does not provide spatial coverage. Indirect geophysical methods provide good spatial coverage but are also expensive and often do not provide high spatial resolution (i.e., the distance between grid points for which we have data is large). **Figure 10.5.1** illustrates the tradeoff between spatial resolution and scale of monitoring. As shown, the highest resolution methods are limited in the scale of spatial coverage, while the methods that can cover the large areas needed for sequestration monitoring tend to yield low-resolution data.

In **Table 10.5.1** we list several different subsurface monitoring approaches [10.47]. These can be subdivided on the basis of whether they are intermittently or continuously applied. Because of the density and opacity of rock, most of the methods rely on indirect geophysical approaches whereby acoustic or electromagnetic energy is applied to the rock and the response is recorded and interpreted. The hydrocarbon exploration and extraction industries have motivated a great deal of experience and long history of development of these methods.

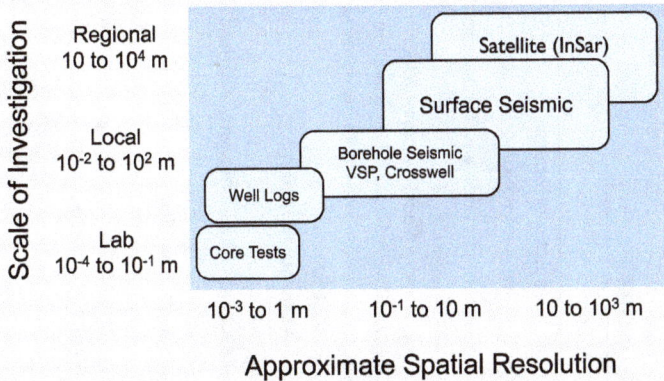

Figure 10.5.1   Scale of subsurface monitoring

Scale of investigation (or spatial coverage) vs. spatial resolution for monitoring of subsurface systems. *Figure adapted from Tom Daley (LBNL).*

Nevertheless, active research is being carried out by industry and academia to continue to improve deep subsurface monitoring. The needs of sequestration monitoring in particular are driving development of improved resolution to verify CO_2 storage. In this section, we focus on two of the more important methods (seismic monitoring and InSAR) and present examples of their use in geological carbon sequestration.

Seismic monitoring

Seismic monitoring is typically carried out by applying vibrational energy to the ground surface and recording and interpreting the reflections from deep structures at nearby instruments. **Figure 10.5.2** shows two different kinds of seismic sources, surface seismic sources, e.g., provided by a vibroseis truck, and well-based seismic sources, e.g., provided by a spinning eccentric mass. The resulting seismic signals can be recorded by seismometers either at the surface or in the well. The figure shows various acoustic signals propagating through the rock-fluid system and related reflections and refractions arising from contrasting seismic wave propagation properties of the rock-fluid system. The figure also illustrates the complexities involved in the collection and interpretation of data using this method.

Table 10.5.1 Examples of subsurface carbon sequestration monitoring approaches

Temporal Classification	Approach	Comment
Time-lapse snapshots	3D surface seismic	Expensive, fair spatial resolution, does not require wells.
	Vertical seismic profile (VSP)	Good vertical resolution. Limited to near-well area.
	Crosswell Tomography/Imaging	Good spatial resolution. Requires more than one well for seismic or electrical geophysical methods.
	Well logging (RST, Sonic, Resistivity, P/T)	RST = Reservoir Saturation Tool; Sonic = Local Seismic Velocity; Resistivity = Electrical Resistivity; P/T = Pressure and Temperature.
Continuous monitoring	U-Tube fluid sampling	Formally intermittent, but frequency can be large enough relative to fluid flow to be considered continuous.
	CASSM (Continuous Active-Source Seismic Monitoring)	Good for discerning temporal evolution of CO_2 plume.
	ERT (Electrical Resistance Tomography)	Good for discerning temporal evolution of fluid movement, phase saturation change.
	DTPS (Distributed Temperature Perturbation Sensing)	Good for estimating phase saturation around well.
	Deep Borehole Microseismic	Monitors fluid pressure migration due to injection through detection of induced microcracking and microseismic events.

Source: Daley et al. [10.47].

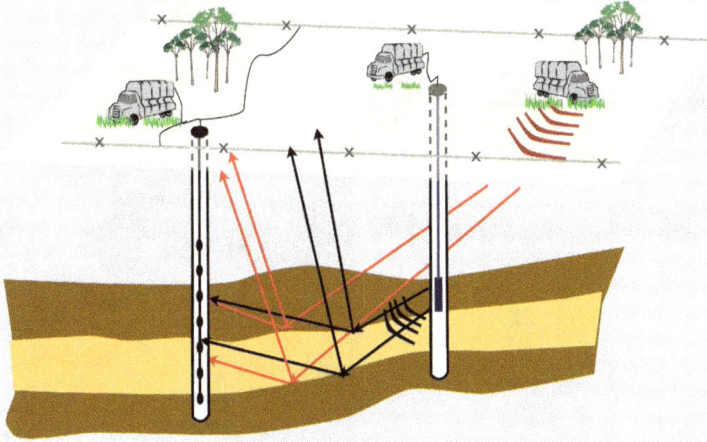

Figure 10.5.2 3D sketch of the ground surface

3D sketch of the ground surface, two wells, and a layered geological structure to illustrate the methods of surface and borehole seismic monitoring. Seismic waves can be induced on the surface (vibroseis truck, red waves and arrows) or in a well by a spinning eccentric mass (right-hand side well). The signals can be monitored either on the surface or in a well (left-hand side well).

Nonetheless, seismic monitoring is the main approach to large-scale delineation of CO_2 spreading in the deep subsurface and has been used successfully to monitor the Sleipner sequestration project that began injection of CO_2 in 1996. **Figure 10.5.3** shows a 3D sketch of the Sleipner system in which natural gas with about 9% CO_2 content is produced from a deep formation, processed at the off-shore platform using amine scrubbing to remove CO_2, and then sent by pipeline to market. The separated CO_2 is then injected into the shallower Utsira Formation for geological carbon sequestration.

Figure 10.5.4 shows a cross-section of seismic reflections in the Utsira Formation [10.48]. The darker bands are stronger reflections indicating large contrast in seismic wave speed due to CO_2 saturation. As shown, the area of the Utsira saturated with CO_2 increased over time as CO_2 was injected, migrated upward and spread out underneath the caprock. This is the best subsurface seismic record of a large-scale sequestration project. Note that while the data provide strong evidence that CO_2

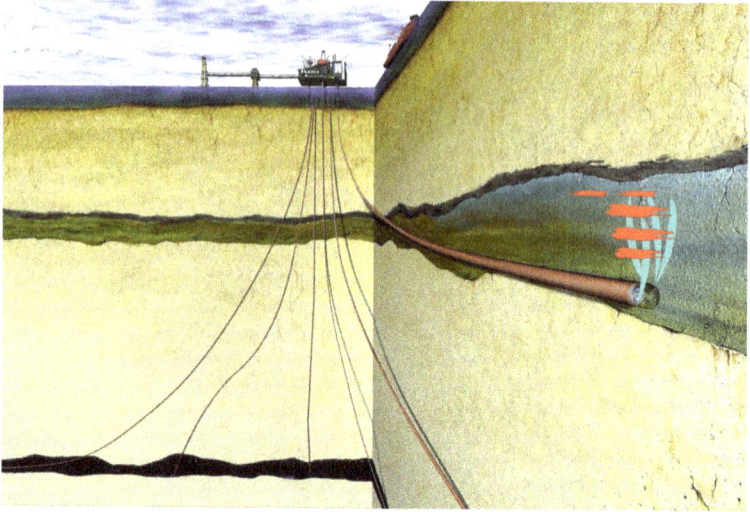

Figure 10.5.3 3D sketch of the injection of CO_2

Sleipner project: 3D sketch of the injection of CO_2 through a horizontal well into a formation above the one from which natural gas with 9% CO_2 is produced. The light blue pillars and red broad plumes represent the interpretation of how the buoyant CO_2 is filling the formation in response to repeating discontinuous low-permeability shale layers (broad plumes) within high-permeability sandstone (pillars). *Graphic by Statoil.*

is occupying the pore space, the resolution in terms of quantifying how much CO_2 is being stored, and where this storage is occurring, is not all that high.

An example of higher-resolution seismic monitoring in both space and time is provided by the Frio CO_2 injection test in the onshore Texas Gulf Coast region [10.49]. The Frio CO_2 injection test was carried out in two phases, in 2004 and 2006. Summarizing the test briefly, 1600 tonnes of CO_2 were injected over 10 days in 2004 into the Frio Formation at a depth of about 5,000 feet. In the second phase in 2006, approximately 300 tonnes of CO_2 were injected into a sandstone layer about 400 feet deeper than the larger injection of the first phase. Seismic sources and receivers were set up to provide temporal information on CO_2 migration [10.47]. **Figure 10.5.5** shows a sketch of the borehole-based seismic sources and receivers with seismic ray paths, along with a depiction of the sandstone layers (colored blue and green). **Figure 10.5.5** also shows the successful

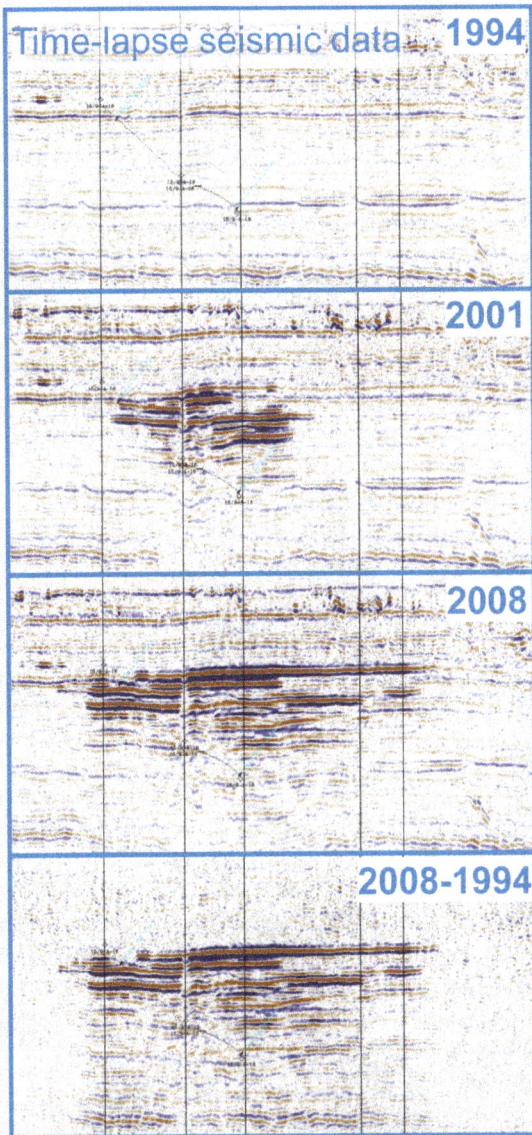

Figure 10.5.4 Time lapse of seismic reflections

Seismic reflections along a cross-section showing the change in acoustic properties of the Utsira Formation as CO_2 has been injected since 1996. The final frame shows the difference between the response in 2008 (after 12 years of injection) and the initial (pre-injection) system. *Graphic by Statoil. Reproduced with permission from Fjæran [10.48].*

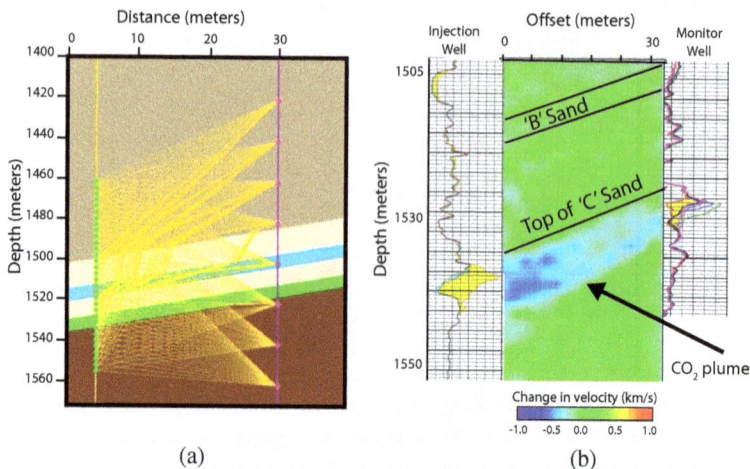

Figure 10.5.5 Seismic receiving and monitoring
(a) Sketch of seismic receivers, sources, and hypothetical seismic ray paths at the Frio CO_2 injection site, and (b) changes in seismic velocity as measured from seismic tomography, and confirmatory temporal RST well log data. (a): *Reproduced with permission from Tom Daley.* (b): *Reproduced from Daley et al.* [10.2], *with kind permission from Springer Science + Business Media.*

detection of high CO_2 saturation in the so-called C-sand as confirmed by logging with the Reservoir Saturation Tool (RST) in the injection and monitor wells.

The geophysical monitoring campaign at the Frio CO_2 injection test also included the use of a continuous seismic source approach that could monitor the movement of the plume between wells. This very successful demonstration is depicted in **Figure 10.5.6** by the yellow, green, red, and blue seismic ray paths emanating from a piezoelectric seismic source placed in the injection well and to hydrophone receivers in the observation well. In the right-hand frame of **Figure 10.5.6**, the delay times are shown, which are a measure of the changing seismic velocity related to how much CO_2 saturation is in the path of the seismic wave as a function of time for arrivals at the five receivers, each placed at different depths. As shown, the delay times nicely match the growth of the plume, with early delays occurring for the blue and red ray paths corresponding to a short plume at an early time, and subsequent later delays for the

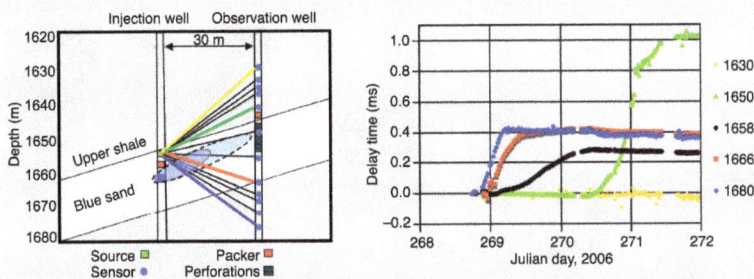

Figure 10.5.6 Seismic source approach

(Left) Sketch of ray paths from source in the injection well to receivers in the observation well. (Right) Temporal evolution of delay time for seismic energy arriving at various receivers in the observation well. The pattern of change in delay time matches well with a growing and spreading CO_2 plume. *Reproduced with permission from Daley et al.* [10.48].

other ray paths corresponding to the plume growing across the space between the injection and observation wells.

Remote sensing (InSAR)

Another large-scale monitoring approach that has proven itself very useful is InSAR (Interferometric Synthetic Aperture Radar). The concept behind this method is depicted in **Figure 10.5.7** which shows a satellite source and receiver and the ground surface. Briefly, the difference in phase between the transmitted and the received signal between subsequent reflections of the same point (ideally with correction using a static object on the ground surface known as a point scatterer) can be translated into a difference in distance which indicates deflection of the ground surface. This approach is used to monitor large-scale landslides and tectonic motions with very small surface movements.

The use of InSAR for geological carbon sequestration has been demonstrated as depicted in **Figure 10.5.8** for the In Salah CO_2 injection project [10.50]. Briefly, the In Salah project separates CO_2 from natural gas produced from several reservoirs in the area. The CO_2 is then injected into a low-permeability sandstone using long horizontal wells. The increased pressure in the injection formation propagates strain upward through the

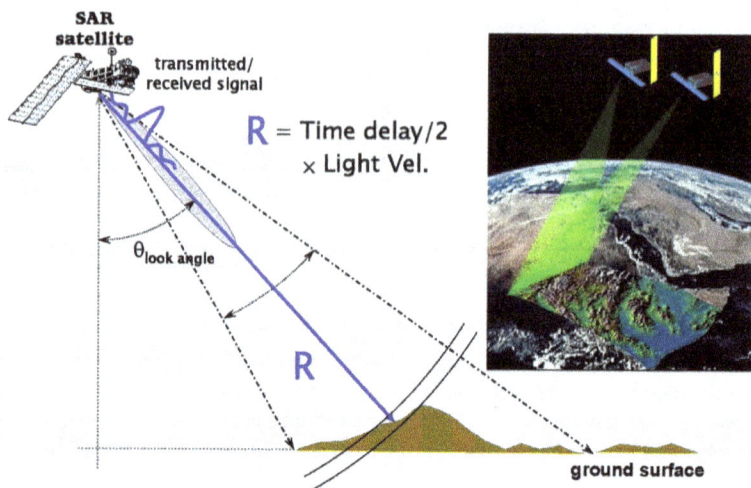

Figure 10.5.7 InSAR monitoring

Sketch of the concept and use of InSAR for monitoring small vertical motions over relatively large areas. *Reproduced with permission from the Geodesy and Seismology Group, University of Miami.*

overlying formations which causes surface uplift. These small surface uplifts can be detected using InSAR. The corresponding subsidence within the gas reservoir (west and south of wells KB-501 and KB-502, respectively) and in wadis undergoing erosion from streamflow (southwest part of the figure) shows up as positive range velocity (increasing distance to satellite). The uplift detected by InSAR reflects the pressurization-induced deformation caused by injection rather than the CO_2 phase saturation typically detected by seismic methods. InSAR may be complicated by vegetation and other land surface changes in more vegetated, farmed, or populated areas, giving rise to increased reliance on engineered point scatterers.

Surface gas flux (eddy-covariance and chamber methods)

Another class of monitoring methods at sequestration sites is surface monitoring to detect and locate CO_2 leakage into the atmosphere. A wide

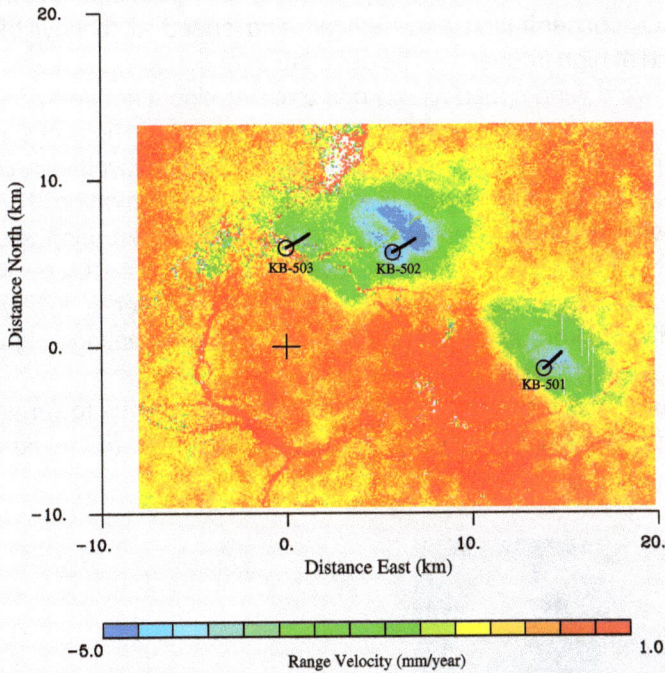

Figure 10.5.8 Range velocity for the In Salah site

Range velocity (mm/yr) as measured by InSAR at the In Salah CO_2 injection site (around the 3 injection wells indicated by circles). The negative range velocities depicted by the blue and green areas are caused by uplift of the surface due to pressure increases at depth from CO_2 injection. *Reproduced from Vasco et al.* [10.50], *with permission from John Wiley and Sons.*

variety of methods can potentially be used to monitor and detect leaking CO_2. The anomaly created by leaking CO_2 could be either a detectable leakage flux or an increase in CO_2 concentration, or both, e.g., in the shallow soil. The challenge of surface and near-surface monitoring is that the large background carbon cycle (e.g., plant photosynthesis and root respiration) signals can easily mask subtle sequestration leakage signals. Although there are many potential approaches to detecting and locating near-surface leakage (e.g., [10.51,10.52]), here we focus on two flux-measuring approaches: eddy-covariance monitoring and accumulation chamber monitoring.

Eddy-covariance monitoring is carried out by simultaneously measuring CO_2 concentration and vertical wind speed at (nearly) the same point and at high frequency. The covariation of concentration and upward or downward wind speed allows one to determine a net vertical CO_2 flux averaged over time and over an area related to the height of the tower. This approach can detect processes such as photosynthesis of plants and trees, decay of biological materials, and soil respiration. The detection of very small CO_2 leakage fluxes from a sequestration site would theoretically be possible, but the overprint of natural CO_2 cycling processes must be measured and time-averaged over a few years to develop a background curve in order to discern a leakage signal (e.g., Lewicki *et al.* [10.53]).

Figure 10.5.9 shows a photo of an eddy-covariance tower with various instruments including an infrared gas analyzer (IRGA) and sonic

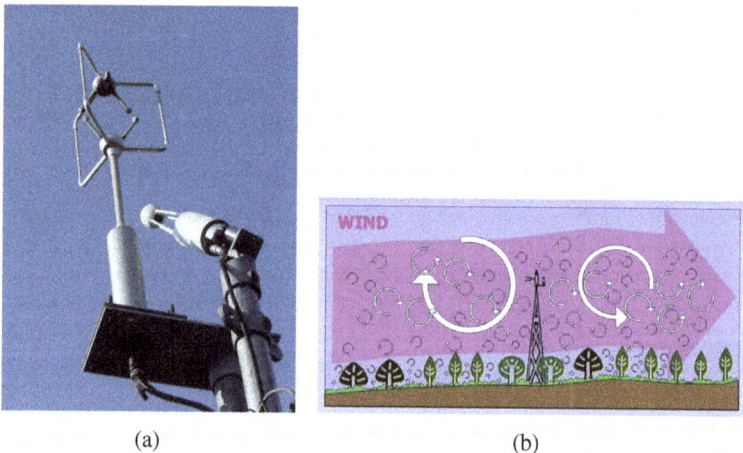

(a)   (b)

Figure 10.5.9   Eddy covariance method

(a) Photograph of an open path eddy covariance system, consisting of an ultrasonic anemometer to measure 3D air flow and infrared gas analyser (IRGA) to measure carbon dioxide. This system is situated at the top of a 20 m high tower above a forest. *Image by Veedar from Wiki Commons:* http://en.wikipedia.org/wiki/File:Eddy_Covariance_IRGA_Sonic.jpg

(b) A cartoon of the turbulent eddies that the eddy covariance method measures along with CO_2 concentration to estimate net vertical CO_2 flux. *Diagram created by George Burba:* http://en.wikipedia.org/wiki/File:EddyCovariance_diagram_1.jpg

anemometer. Eddy-covariance towers can range from 1 m in height to many hundreds of meters with higher towers providing greater spatial averaging. **Figure 10.5.9 (b)** depicts the turbulent eddies that the sonic anemometer measures at high frequency along with CO_2 concentration to estimate a net vertical CO_2 flux. The eddy-covariance method has proven capable of detecting a known leakage signal in the shallow-release test carried out in Montana [10.54].

The other proven flux-measuring approach uses an accumulation chamber. This simple device consists of a headspace sealed to the ground surface and connected by inlet and outlet tubes to a gas analyzer and a pressure transducer. The tubing carries gas from the chamber to the analyzer and back to the chamber while keeping the pressure constant in the chamber. The rate of change of CO_2 concentration in the chamber is a measure of the flux of CO_2 into the chamber. **Figure 10.5.10** shows an example of an accumulation chamber deployed on the soil surface. Accumulation chambers provide accurate measures of soil CO_2 flux, but many measurements are needed to provide spatial coverage (e.g., Lewicki *et al.* [10.53]).

In this chapter, we have taken a continuum scale view of geological carbon sequestration and reviewed the modeling approach to understanding sequestration processes, the methods of capacity assessment, potential hazards, and monitoring approaches. Overall, based on these

Figure 10.5.10   Accumulation chamber

Accumulation chamber (approximately 10 cm in diameter) deployed on the soil surface showing inlet and outlet tubing that connects to a gas analyzer. *Picture by USGS.*

studies using the best available information and experience in other fields, it appears that geological carbon sequestration is technically feasible. Pilot and industrial scale injection projects need to be carried out with careful monitoring to test the concepts further and thereby ensure that large-scale geological carbon sequestration is safe and effective.

Review

10.1. Reading self-test

1. Which statement is not correct for mixing in field-scale models?
   a. Most models assume that the system is well-mixed within each simulation block
   b. Capillarity fingering causes deviations from well-mixed behavior
   c. If a simulation grid block contains more than a reasonable threshold minimum amount of CO_2-rich phase, the entire aqueous phase in the grid block is assumed to be in equilibrium with the CO_2-rich phase
   d. Weathering reactions involving minerals, water, and CO_2 may be rate-limited by the slow diffusive mixing of these three components

2. Which statement is correct about grids in a field-scale simulation?
   a. Field-scale simulations are currently being replaced by analytical models
   b. A typical grid size in a field-scale simulation is 1 cm x 1 cm x 1 cm
   c. A single core sample represents a region of the formation large enough to capture the anisotropy of natural formations
   d. Field-scale carbon sequestration models treat each simulation grid block as a plug flow reactor
   e. Answers B and D
   f. None of the above

3. Which statement related to carbon sequestration induced seismicity is correct?
   a. The natural increase in tectonic stress is augmented by a CO_2 injection process which may trigger an earthquake

b. In well-chosen sequestration sites, it is advised to reinforce the structures of newly built houses to prevent minor damage from induced earthquakes

c. It is a well-known fact that caprock integrity will be compromised by induced seismicity

d. Induced seismicity occurs because of injection of gasses, not because of the extraction of gasses

4. Which statement related to the CO_2 leakage flux from a sequestration site and the typical natural ecological CO_2 flux driven by photosynthesis is not correct?

a. A typical ecological CO_2 flux is 4.4×10^{-7} kg s^{-1} m^{-2}

b. The ecological flux of CO_2 is equivalent to a leakage of ca 0.1% per year of a typical sequestration site containing 4×10^9 kg CO_2

c. Trees die if the CO_2 mole fraction above a leakage site exceeds 0.3

d. On the order of 10^5–10^6 kg of CO_2 per year, areas of radius 100 m would be required to exceed a leakage flux equal to the typical natural ecological CO_2 flux

e. The larger the radius over which leakage occurs, the higher the risk for high concentrations in the shallow subsurface

10.2. Carbon sequestration hazards

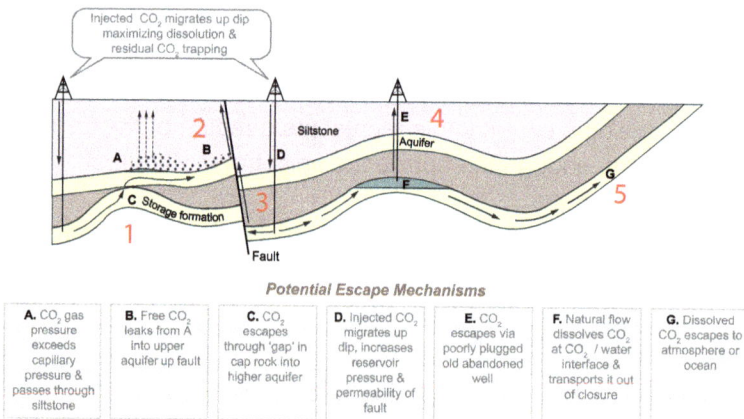

Potential Escape Mechanisms

| A. CO_2 gas pressure exceeds capillary pressure & passes through siltstone | B. Free CO_2 leaks from A into upper aquifer up fault | C. CO_2 escapes through 'gap' in cap rock into higher aquifer | D. Injected CO_2 migrates up dip, increases reservoir pressure & permeability of fault | E. CO_2 escapes via poorly plugged old abandoned well | F. Natural flow dissolves CO_2 at CO_2 / water interface & transports it out of closure | G. Dissolved CO_2 escapes to atmosphere or ocean |

Figure from [10.36].

1. Which remedial measure would be useful at the indicated location indicated by the numbers 1–5 on the previous page?
___ Remove CO_2
___ Lower injection pressure
___ Purify ground water
___ Re-plug well with cement
___ Intercept and reinject CO_2

2. Place the hazards at the correct depth and impact scale.

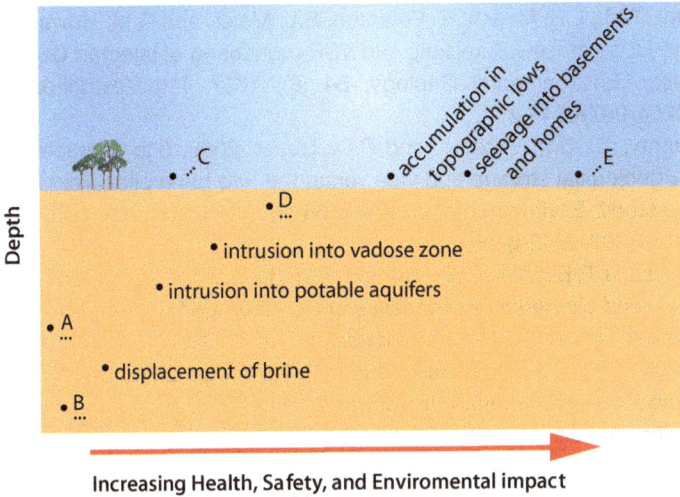

• accumulation in topographic lows
• seepage into basements and homes

• ...C
• ...D
• intrusion into vadose zone
• intrusion into potable aquifers
• ...A
• displacement of brine
• ...B
• ...E

Depth

Increasing Health, Safety, and Enviromental impact

___ Surface water
___ Induced seismicity
___ Root zone effect
___ Ground plume
___ Hydrocarbon reservoirs

Section 7

References

1. Underschultz, J., C. Boreham, T. Dance, *et al.*, 2011. CO_2 storage in a depleted gas field: An overview of the CO2CRC Otway Project and initial results. Int. J. Greenh. Gas Con., **5** (4), 922. http://dx.doi.org/10.1016/j.ijggc.2011.02.009

2. Daley, T.M., L.R. Myer, J.E. Peterson, E.L. Majer, and G.M. Hoversten, 2008. Time-lapse crosswell seismic and VSP monitoring of injected CO_2 in a brine aquifer. Environmental Geology, **54** (8), 1657. http://dx.doi.org/10.1007/s00254-007-0943-z

3. Doughty, C., B.M. Freifeld, and R.C. Trautz, 2008. Site characterization for CO_2 geological storage and vice versa: the Frio brine pilot, Texas, USA as a case study. Environmental Geology, **54** (8), 1635. http://dx.doi.org/10.1007/s00254-007-0942-0

4. Zhou, Q., J.T. Birkholzer, E. Mehnert, Y.-F. Lin, and K. Zhang, 2010. Modeling basin- and plume-scale processes of CO_2 storage for full-scale deployment. Groundwater, **48** (4), 494. http://dx.doi.org/10.1111/j.1745-6584.2009.00657.x

5. Kim, Y., J. Wan, T.J. Kneafsey, and T.K. Tokunaga, 2012. Dewetting of silica surfaces upon reactions with supercritical CO_2 and brine: Pore-scale studies in micromodels. Env. Sci. Technol., **46** (7), 4228. http://dx.doi.org/10.1021/es204096w

6. Sato, K., S. Mito, T. Horie, *et al.*, 2011. Monitoring and simulation studies for assessing macro- and meso-scale migration of CO_2 sequestered in an onshore aquifer: Experiences from the Nagaoka pilot site, Japan. Int. J. Greenh. Gas Con., **5** (1), 125. http://dx.doi.org/10.1016/j.ijggc.2010.03.003

7. Xu, T., J.A. Apps, K. Pruess, and H. Yamamoto, 2007. Numerical modeling of injection and mineral trapping of CO_2 with H_2S and SO_2 in a sandstone formation. Chemical Geology, **242** (3–4), 319. http://dx.doi.org/10.1016/j.chemgeo.2007.03.022

8. Armstrong, R. and J. Ajo-Franklin, 2011. Investigating biomineralization using synchrotron based X-ray computed microtomography. Geophysical Research Letters, **38** (8), L08406. http://dx.doi.org/10.1029/2011GL046916

9. Deng, H., B.R. Ellis, C.A. Peters, J.P. Fitts, D. Crandall, and G.S. Bromhal, 2013. Modifications of carbonate fracture hydrodynamic properties by CO_2-acidified brine flow. Energy & Fuels, **27** (8), 4221–4231. http://dx.doi.org/10.1021/ef302041s

10. Ellis, B.R., G.S. Bromhal, D.L. McIntyre, and C.A. Peters, 2011. Changes in caprock integrity due to vertical migration of CO_2-enriched brine. Energy Procedia, **4**, 5327. http://dx.doi.org/10.1016/j.egypro.2011.02.514

11. Elkhoury, J.E., P. Ameli, and R.L. Detwiler, 2013. Dissolution and deformation in fractured carbonates caused by flow of CO_2-rich brine under reservoir conditions. Int. J. Greenh. Gas Con., **16** (S1), S203. http://dx.doi.org/10.1016/j.ijggc.2013.02.023

12. Noiriel, C., B. Madé, and P. Gouze, 2007. Impact of coating development on the hydraulic and transport properties in argillaceous limestone fracture. Water Resources Research, **43** (9), W09406. http://dx.doi.org/10.1029/2006WR005379

13. Kharaka, Y.K., J.J. Thordsen, S.D. Hovorka, et al., 2009. Potential environmental issues of CO_2 storage in deep saline aquifers: Geochemical results from the Frio-I Brine Pilot test, Texas, USA. Applied Geochemistry, **24** (6), 1106. http://dx.doi.org/10.1016/j.apgeochem.2009.02.010

14. Espinoza, N., and J.C. Santamarina, 2010. Water-CO_2-mineral systems: Interfacial tension, contact angle, and diffusion — Implications to CO_2 geological storage. Water Resources Research, **46** (7), W07537. http://dx.doi.org/10.1029/2009WR008634

15. Shao, H., J.R. Ray, and Y.-S. Jun, 2011. Effects of organic ligands on supercritical CO_2-induced phlogopite dissolution and secondary mineral formation. Chemical Geology, **290** (3–4), 121. http://dx.doi.org/10.1016/j.chemgeo.2011.09.006

16. Janelle R. Thompson, Massachusetts Institute of Technology (private communication).

17. Morozova, D., M. Wandrey, M. Alawi, et al., 2010. Monitoring of the microbial community composition in saline aquifers during CO_2 storage by fluorescence in situ hybridization. Int. J. Greenh. Gas Con., **4** (6), 981. http://dx.doi.org/10.1016/j.ijggc.2009.11.014

18. Jenny A. Cappuccio, Lawrence Berkeley National Laboratory (private communication).

19. Eichhubl, P., N.C. Davatzes, and S.P. Becker, 2009. Structural and diagenetic control of fluid migration and cementation along the Moab fault, Utah. AAPG Bulletin, **93** (5), 653. http://dx.doi.org/10.1306/02180908080

20. Wenk, H.-R., M. Voltolini, M. Mazurek, L.R. Van Loon, and A. Vinsot, 2008. Preferred orientations and anisotropy in shales: Callovo-Oxfordian shale (France) and Opalinus Clay (Switzerland). Clays and Clay Minerals, **56** (3), 285–306. http://dx.doi.org/10.1346/ccmn.2008.0560301

21. Marschall, P., S. Horseman, and T. Gimmi, 2005. Characterisation of gas transport properties of the Opalinus Clay, a potential host rock formation for radioactive waste disposal. Oil & Gas Science and Technology — Rev. IFP, **60** (1), 121. http://dx.doi.org/10.2516/ogst:2005008

22. Keller, L.M., P. Schuetz, R. Erni, *et al.*, 2012. Characterization of multi-scale microstructural features in Opalinus Clays. Microporous and Mesoporous Materials, **170**, 83. http://dx.doi.org/10.1016/j.micromeso.2012.11.029

23. Keller, L.M., L. Holzer, R. Wepf, P. Gasser, B. Münch, and P. Marschall, 2011. On the application of focused ion beam nanotomography in characterizing the 3D pore space geometry of Opalinus clay. Physics and Chemistry of the Earth, **36** (17–18), 1539. http://dx.doi.org/10.1016/j.pce.2011.07.010

24. Bocquet, L., and E. Charlaix, 2010. Nanofluidics, from bulk to interfaces. Chemical Society Reviews, **39** (3), 1073. http://dx.doi.org/10.1039/b909366b

25. Botan, A., B. Rotenberg, V. Marry, P. Turq, and B. Noetinger, 2010. Carbon dioxide in montmorillonite clay hydrates: Thermodynamics, structure, and transport from molecular simulation. J. Phy. Chem. C, **114** (35), 14962. http://dx.doi.org/10.1021/jp1043305

26. Fernandez-Martinez, A., Y. Hu, B. Lee, Y.-S. Jun, and G.A. Waychunas, 2013. In situ determination of interfacial energies between heterogeneously nucleated $CaCO_3$ and quartz substrates: Thermodynamics of CO_2 mineral trapping. Environ. Sci. Technol., **47** (1), 102. http://dx.doi.org/10.1021/es3014826

27. Bildstein, O., C. Kervévan, V. Lagneau, *et al.*, 2010. Integrative modeling of caprock integrity in the context of CO_2 storage: Evolution of transport and geochemical properties and impact on performance and safety assessment. Oil & Gas Sci. Technol., — Rev IFP, **65** (3), 485. http://dx.doi.org/10.2516/ogst/2010006

28. Felmy, A.R., O. Qafoku, B.W. Arey, *et al.*, 2012. Reaction of water-saturated supercritical CO_2 with forsterite: Evidence for magnesite formation at low temperatures. Geochimica et Cosmochimica Acta, **91**, 271. http://dx.doi.org/10.1016/j.gca.2012.05.026

29. Bradshaw, J., S. Bachu, D. Bonijoly, *et al.*, 2007. CO_2 storage capacity estimation: Issues and development of standards. Int. J. Greenh. Gas Con., 1(1), 62. http://dx.doi.org/10.1016/S1750-5836(07)00027-8

30. Doughty, C., K. Pruess, S.M. Benson, S.D. Hovorka, P.R. Knox, and C.T. Green, 2001. "Capacity investigation of brine-bearing sands of the Frio formation for geologic sequestration of CO_2" in First National Conference on Carbon Sequestration. Morgantown: US Dept. of Energy, National Energy Technology Laboratory. pp. 14.

31. US Dept. of Energy, National Energy Technology Laboratory, 2008. Carbon Sequestration Atlas of the United States and Canada (4th ed.; Atlas IV) 10.32 from p. 61. http://www.netl.doe.gov/technologies/carbon_seq/refshelf/atlasIII/index.html, p. 142.

32. Causebrook, R., 2010. Overview of Capacity Estimation Methodologies for Saline Reservoirs, CCS Summer School of CAGS, Wuhan, PRC. http://www.cagsinfo.net/pdfs/summerschool/Lesson5/5-3Storage-capacity-assessment.PDF

33. Brennan, S.T., R.C. Burruss, M.D. Merrill, P.A. Freeman, and L.F. Ruppert, 2010. A probabilistic assessment methodology for the evaluation of geologic carbon dioxide storage. US Dept. of the Interior, US Geological Survey, Reston, Va. http://purl.fdlp.gov/GPO/gpo22655

34. Burruss, R.C., S.T. Brennan, P.A. Freeman, et al., 2009. Development of a Probabilistic Assessment Methodology for Evaluation of Carbon Dioxide Storage, USGS Open-File Report 2009-1035. http://pubs.usgs.gov/of/2009/1035/ofr2009-1035.pdf

35. Metz, B., O. Davidson, H. deConinck, M. Loos, and L. Meyer, 2005. IPCC Special Report on Carbon Dioxide Capture and Storage. http://www.ipcc.ch/pdf/special-reports/srccs/srccs_wholereport.pdf

36. Oldenburg, C.M., 2012. "Geologic carbon sequestration: sustainability and environmental risk" in Encyclopedia of Sustainability Science and Technology, edited by R.A. Meyers, pp. 4119–4133. Springer: New York.

37. Oldenburg, C.M., 2007. "Migration mechanisms and potential impacts of CO_2 leakage and seepage", in Carbon Capture and Sequestration: Integrating Technology, Monitoring, and Regulation, edited by E.J. Wilson and D. Gerald pp. 127–146. Iowa: Blackwell.

38. Apps, J.A., L. Zheng, Y. Zhang, T. Xu, and J.T. Birkholzer, 2010. Evaluation of potential changes in groundwater quality in response to CO_2 leakage from deep geologic storage. Transport Porous Med., **82** (1), 215. http://dx.doi.org/10.1007/S11242-009-9509-8

39. Zoback, M.D., and S.M. Gorelick, 2012. Earthquake triggering and large-scale geologic storage of carbon dioxide. P. Natl. Acad. Sci. USA, **109** (26), 10164. http://dx.doi.org/10.1073/Pnas.1202473109

40. Talwani, P., 1997. On the nature of reservoir-induced seismicity. Pure. Appl. Geophys., **150** (3–4), 473. http://dx.doi.org/10.1007/S000240050089

41. Gerlach, T.M., M.P. Doukas, K.A. McGee, and R. Kessler, 2001. Soil efflux and total emission rates of magmatic CO_2 at the Horseshoe Lake tree kill, Mammoth Mountain, California, 1995–1999. Chem. Geol., **177** (1–2), 101. http://dx.doi.org/10.1016/S0009-2541(00)00385-5

42. Oldenburg, C.M., and A.J.A. Unger, 2003. On leakage and seepage from geologic carbon sequestration sites. Vadose Zone J., **2** (3), 287. http://dx.doi.org/10.2136/vzj2003.2870

43. Oldenburg, C.M., and A.J.A. Unger, 2004. Coupled vadose zone and atmospheric surface-layer transport of carbon dioxide from geologic carbon sequestration sites. Vadose Zone J., **3** (3), 848. http://dx.doi.org/10.2136/vzj2004.0848

44. Britter, R.E., 1989. Atmospheric dispersion of dense gases. Annu. Rev. Fluid. Mech., **21**, 317. http://dx.doi.org/10.1146/Annurev.Fluid.21.1.317

45. Britter, R.E., and J.D. McQuaid, 1988. Workbook on the Dispersion of Dense Gases. England: Health & Safety Executive.

46. Mazzoldi, A., D. Picard, P.G. Sriram, and C.M. Oldenburg, 2013. Simulation-based estimates of safety distances for pipeline transportation of carbon dioxide. Greenhouse Gas Sci. Technol., **3** (1), 66. http://dx.doi.org/10.1002/ghg.1318

47. Daley, T.M., R.D. Solbau, J.B. Ajo-Franklin, and S.M. Benson, 2007. Continuous active-source seismic monitoring of CO_2 injection in a brine aquifer. Geophysics, **72** (5), A57. http://dx.doi.org/10.1190/1.2754716

48. Fjæran, T., 2012. Monitoring a CO_2 Storage. Seminar on Evaluation of CO_2 Storage Potential, ITB Bandung, 11–12 December 2012. http://www.ccop.or.th/eppm/projects/43/docs/TorFjaeran_Monitoring_CO2_Storage_ITB.pdf

49. Hovorka, S.D., C. Doughty, S.M. Benson, et al., 2006. Measuring permanence of CO_2 storage in saline formations: The Frio experiment. Environ. Geosci., **13** (2), 105–121.

50. Vasco, D.W., A. Rucci, A. Ferretti, et al., 2010. Satellite-based measurements of surface deformation reveal fluid flow associated with the geological storage of carbon dioxide. Geophysical Research Letters, **37** (3), L03303. http://dx.doi.org/10.1029/2009GL041544

51. Spangler, L.H., L.M. Dobeck, K.S. Repasky, et al., 2010. A shallow subsurface controlled release facility in Bozeman, Montana, USA, for testing near surface CO_2 detection techniques and transport models. Environ. Earth Sci., **60** (2), 227. http://dx.doi.org/10.1007/S12665-009-0400-2

52. Oldenburg, C.M., J.L. Lewicki, and R.P. Hepple, 2003. Near-Surface Monitoring Strategies for Geologic Carbon Dioxide Storage Verification. LBNL Technical Report. http://www.osti.gov/servlets/purl/840984-dTw752/native/

53. Lewicki, J.L., G.E. Hilley, and C.M. Oldenburg, 2005. An improved strategy to detect CO_2 leakage for verification of geologic carbon sequestration. Geophys. Res. Lett., **32** (19), L19403. http://dx.doi.org/10.1029/2005gl024281

54. Lewicki, J.L., G.E. Hilley, M.L. Fischer, et al., 2009. Eddy covariance observations of surface leakage during shallow subsurface CO_2 releases. J. Geophys. Res. Atmos., **114**, D12302. http://dx.doi.org/10.1029/2008jd011297

Chapter 11

Land Use and Geo-Engineering

Using Carbon Capture and Sequestration, we are able to reduce *greenhouse emissions*, but this technology will not reduce the *absolute CO_2 levels* in the atmosphere. For a more complete view on the control of climate change with carbon management, we introduce in this chapter techniques that actually reduce atmospheric CO_2 levels. We divide these techniques into two categories: land use and geo-engineering.

Section 1

Introduction

In Chapters 2 and 3, we discussed the lifetime of a CO_2 molecule in the atmosphere. We have seen that, although the exchange between CO_2 in the atmosphere and on the surface of the ocean is very rapid, the subsequent mixing of the CO_2-saturated surface layer with the deep parts of the oceans occurs on a time scale of over two hundred years. This multi-hundred year time scale assumes atmospheric CO_2 concentrations are sufficiently low to not exceed the buffering capacity of the ocean. If the buffering capacity is exceeded, it will take ten thousand to a hundred thousand years for the carbon-cycle reactions to have restored the atmospheric CO_2 levels to equilibrium values.

The extraordinary time scales associated with the carbon cycle make the ramifications of CO_2 emissions very different from those of anthropogenic emissions of other gasses. For example, flue gas components such as SO_x and NO_x, the principal components of acid rain, can be readily removed with available technologies. Historically, once flue gasses were cleaned of these components, acid rains diminished and within a few years the problems associated with acidification were mitigated. Similarly, the ban of chlorofluorocarbons (CFCs) has already led to some reduction of the polar ozone hole as the existing CFCs are slowly decomposing in the atmosphere. On the other hand, if we were to halt *all* CO_2 emissions today, it would take hundreds to thousands of years for CO_2 levels to return to pre-industrial values! In the previous chapters, we discussed techniques to reduce or at best stop CO_2 emissions. At present, these technologies have not been implemented on a scale broad enough to mitigate climate change. For this reason, we may need to employ technologies that *decrease* the existing CO_2 levels in the atmosphere. As we will see, these techniques involve large-scale human intervention in processes of the earth: **geo-engineering**.

No doubt some of our readers feel that we have already geo-engineered our planet enough. Indeed, we are in the middle of two ongoing global geo-engineering projects. The first experiment, using fossil fuels for energy, burns the carbon that nature has sequestered in plants over millions of years in a period of a few hundred years. The other geo-engineering experiment changes land use so as to accommodate a world of 9 billion humans or more. Changes in land use are responsible for 15% of the total greenhouse emissions, while the burning of fossil fuels (along with the making of cement) accounts for the remaining 90%. These "geo-engineering projects" supply the world population with food and energy, with the consequence of emitting large amounts of CO_2 to the atmosphere.

Modern geo-engineering aims to "reverse engineer" the negative effects of the increasing CO_2 levels. The first thing that comes to mind is stopping, or even reversing, the negative impact of changes in land use. To impact climate change this would need to be done on a very large scale — one that would encompass most of the land that was changed to provide food for an ever-increasing population in the first place. Other geo-engineering technologies that are currently being considered include the creation of large-scale devices to reflect a significant fraction of the sunlight reaching the earth and enhancing the uptake of CO_2 by the ocean. At present none of these schemes is employed. In fact, most ideas are in a very early stage of development and many are controversial. A recent report of the Royal Society gives a good overview of the status of the various projects [11.1]. The two key recommendations of this report are:

- Parties should make increased efforts toward mitigating and adapting to climate change, and in particular to agreeing to global emissions reductions of at least 50% of 1990 levels by 2050 and more thereafter. Nothing now known about geo-engineering options gives any reason to diminish these efforts.
- Further research and development of geo-engineering options should be undertaken to investigate whether low-risk methods can be made available *if it becomes necessary to reduce the rate of warming* in this century. This should include appropriate observations, the development and use of climate models, and carefully planned and executed experiments.

Given the substantive infrastructure already in existence today that creates CO_2 emissions (i.e., the fossil fuel industry and industrial scale farming), geo-engineering will never be a sensible alternative to preventing emissions. Geo-engineering is, however, the only "plan B" we have — and the indications for "plan A" (cessation or mitigation of all CO_2 emissions) are not all that positive. The idea of doing some research into a plan B is important. There is more than a little irony in the fact that all research funding aimed at mitigating CO_2 emissions, including finding renewable energy sources, is a negligible fraction of the costs of current energy production. As the authors of this text do research on energy, you may have correctly sensed some frustration in these words. Completely unrelated to the present topics, but to put energy research in perspective, consider the US Department of Energy's Solar Energy Hub. It is one of the largest concerted DOE renewable energy projects, and it seeks to develop devices that convert sunlight into fuel, one of the holy grails in energy research. The scientific challenges that need to be solved are enormous. The total budget for five years of research is about 125 million dollars, about the same amount that Hollywood typically invests in the making of a single movie, such as *Angels and Demons*. (In *Angels and Demons* the source of energy was, admittedly, much more spectacular!)

Section 2

Land use

Introduction

In our introductory section we noted that of the 10 Gt of annual global anthropogenic carbon emissions, 15%, or 1.5 Gt, results from land use change. In the 19th century land use changes in Europe and North America contributed significantly to CO_2 emissions, only to be swamped by fossil fuel use in the 20th century (see **Figure 11.2.1**). Before 1850, wood was the most common energy source; turning forest into agricultural land served two purposes: energy production and increased cropland for farming. Agricultural techniques improved enormously, allowing Europeans and North Americans to make more intelligent use of the land and thereby increase its productivity. Indeed, we see that in these continents changes

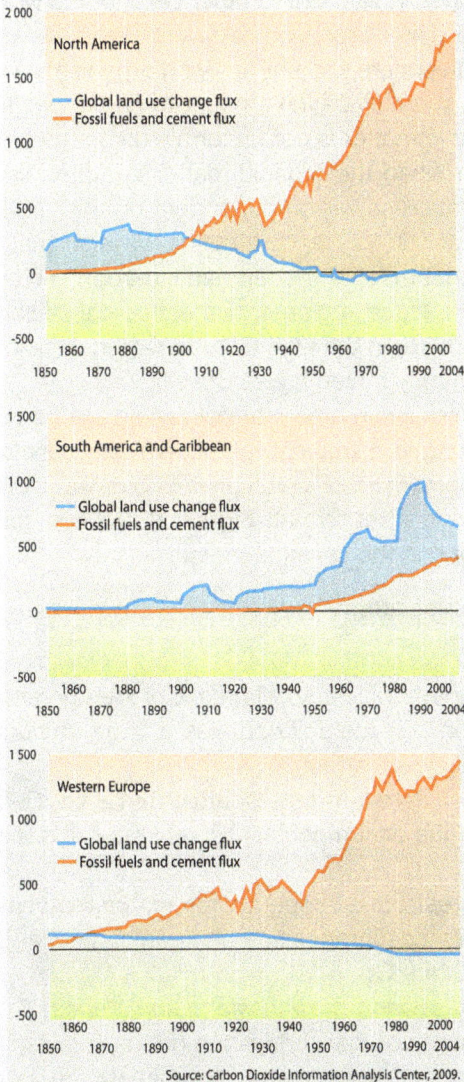

Figure 11.2.1 Historical CO_2 emissions by region

CO_2 emissions units are in millions of metric tonnes (Mt). *Figures by Riccardo Pravettoni in Trumper et al.* [11.2], *reproduced with permission from GRID-Arendal.*

in land use have resulted in net carbon sinks. Unfortunately, the positive effects in these countries are more than compensated by emissions associated with changes in land use in other parts of the world.

In Chapter 3 (The Carbon Cycle), we stated that 2,100 Gt of carbon is stored in the biosphere and soil (in living organisms and organic matter in soil), which is about twice the amount of carbon we find in the atmosphere. This organic matter is not uniformly distributed over the planet; in each part of the world the organic matter is stored in a different form. Changes in the land use will be dependent on this [11.2].

Figure 11.2.2 shows the distribution of terrestrial carbon over the planet. We see that most of the terrestrial carbon is stored in the tropics and in the high altitude regions. In the tropical regions the carbon is mainly biomass, while in the high altitude regions it is in the form of **permafrost** (permanently frozen layers of soil).

Managing land use in a permafrost region will be very different from managing land use in a tropical forest. We can consider the terrestrial world as being comprised of seven distinct biomes, each of which stores carbon in a very different format. **Figure 11.2.3** summarizes the relative amounts of carbon in the seven biomes.

The different biomes are:

- *Tundras*: Tundras contain 155.4 Gt carbon. Only slow growing, sturdy plants can survive this harsh climate. Cold temperatures prevent plant decomposition and the soil contains a large amount of frozen dead plant materials (permafrost).
- *Boreal forests*: Boreal forests contain 384.2 Gt. Because of the low temperatures the decomposition is very slow and most of the carbon is in the soil.
- *Temperate forests*: In a temperate forest, the temperatures are higher, resulting in rapid decomposition of organic material. The total amount of carbon is 314.9 Gt.
- *Temperate grasslands, savannas and shrublands*: The typical vegetation in these parts of the world is too dry for forests, yet these regions get more rain than forests. Grazing animals are an essential part of grassland ecosystems. Often these lands can be successfully converted to crop production with little impact on the carbon balance. These areas contain less biomass per unit surface than forests. The total amount of carbon in this biome is 183.7 Gt.

Global map of terrestrial carbon stocks

Figure 11.2.2 Distribution of terrestrial carbon

Map courtesy of UNEP-WCMC, based on biomass carbon data from Ruesch and Gibbs [11.3], and soil carbon data generated from the Harmonized World Soil Database (version 1.1) [11.4, 11.5] and FAO/IIASA/ISRIC/ISS-CAS/JRC [11.6].

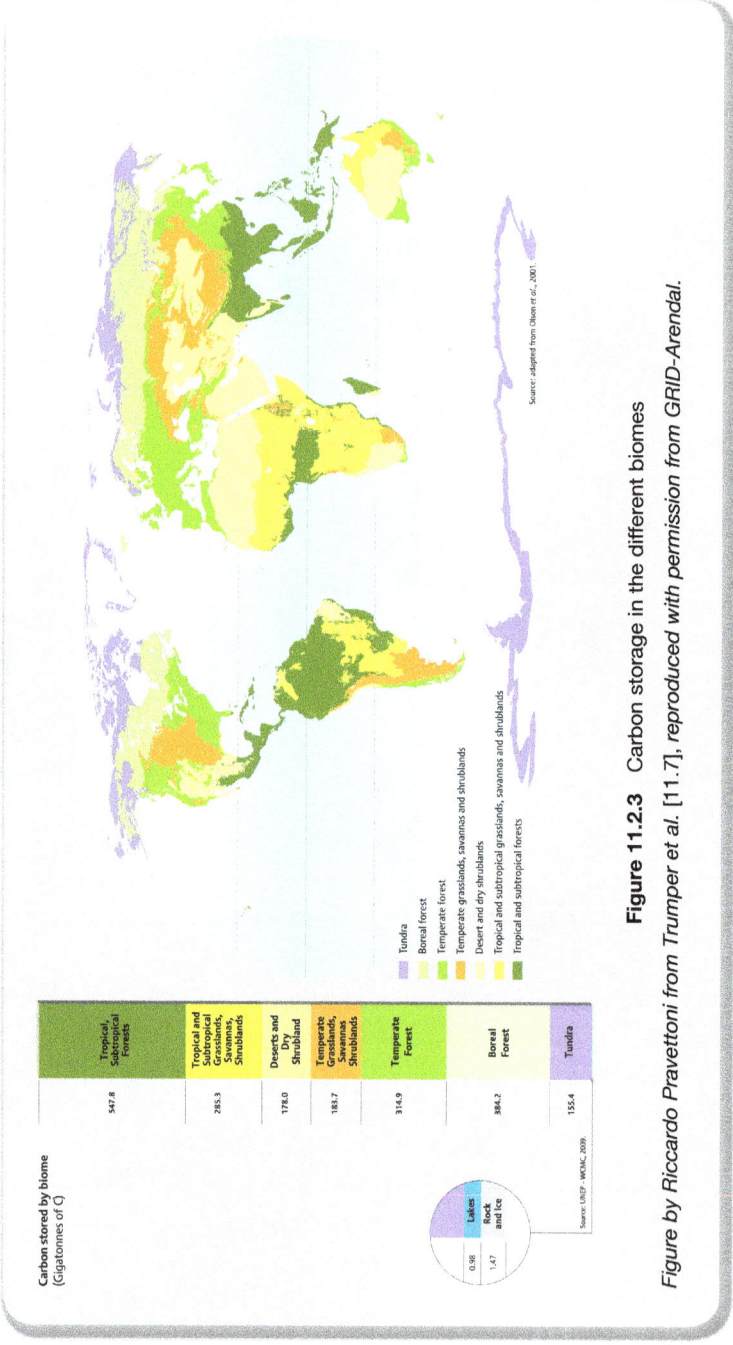

Figure 11.2.3 Carbon storage in the different biomes

Figure by Riccardo Pravettoni from Trumper et al. [11.7], reproduced with permission from GRID-Arendal.

Carbon stored by biome
(Gigatonnes of C)

Biome	Value
Tropical, Subtropical Forests	547.8
Tropical and Subtropical Grasslands, Savannas, Shrublands	285.3
Deserts and Dry Shrubland	178.0
Temperate Grasslands, Savannas, Shrublands	181.7
Temperate Forest	314.9
Boreal Forest	384.2
Tundra	155.4
Lakes	0.98
Rock and Ice	1.47

Source: UNEP -WCMC, 2009.

Tundra
Boreal forest
Temperate forest
Temperate grasslands, savannas and shrublands
Desert and dry shrublands
Tropical and subtropical grasslands, savannas and shrublands
Tropical and subtropical forests

Source: adapted from Olson et al., 2001

- *Deserts and dry shrublands*: Due to low or very seasonal precipitation, the amount of biomass per unit area in these regions is very small. The total amount of carbon in these desert areas is 178 Gt.
- *Tropical and subtropical grasslands, savannas and shrublands*: Savannas are the largest component of the earth's vegetation. The total amount of carbon in this biome is 285.3 Gt.
- *Tropical and subtropical forests*: This biome contains 547.8 Gt of carbon and thus contains the largest amount of carbon. The high temperatures enable rapid growth. Most of the carbon is in biomass.

Tundras

The tundras are mainly found in the arctic (see **Figure 11.2.3**). The very low temperatures cause the plants to grow very slowly during a short period of the year. The cold temperatures also make the decomposition process even slower, resulting in a large amount of carbon that has accumulated over the years. The total carbon pool in the tundra is twice as much as in the atmosphere.

The tundras have been affected very little by land use and have little potential to gain more carbon. The permafrost associated with tundras, however, is predicted to be a major contributor to climate change. Increasing global temperatures can cause the permafrost to decompose, yielding large additional emissions of CO_2 and CH_4. This positive feedback loop could give rise to significantly higher future temperatures than are currently predicted by the climate models.

Forests

The boreal, temperate, and tropical forests are important sinks for carbon. The boreal forest has the second highest reservoir of carbon, mainly because modest temperatures prevent decomposition. This low decomposition rate and the continuous growth make these forests a net, albeit small, carbon sink.

In the temperate forest, the decomposition rate is much larger, which also yields very fertile soil. This fertile soil is attractive for conversion into cropland. In many regions, this conversion process has stopped; in a few cases, it has even been reverted. Returning croplands back to temperate forests will return these biomes to carbon sinks.

Tropical forests are the most important carbon reservoirs and are active carbon sinks owing to the rapid growth of biomass in the warm and humid climates. These tropical forests have suffered from deforestation, causing significant carbon emissions. A loss of 6.5–14.8 million ha per year (the current rate of deforestation) gives an emission of 0.8–2.2 Gt C per year. Halting or even reversing the deforestation will continue to be an important topic of discussion.

Peatlands

Peatland soils contain a large amount of carbon (estimated to be 550 Gt, scattered over the planet). Peatlands occur in water-rich regions where standing water prevents the decomposition of organic material. If converted to croplands, the peatlands are drained and the decomposition process begins, with large amounts of carbon emissions as a result. Peatlands are not a biome, but because of their specific properties, it is important to discuss them separately. Consider, for example, the use of palm-oil for biofuels in Malaysia and Indonesia. The total carbon emissions from deforestation and drainage of the peatland to make it suitable for palm oil plantations is much more than what can be gained by using the biofuel. In this case, conventional fossil fuels are actually a better alternative!

Summary

In **Table 11.2.1**, a summary is given of the properties of the various biomes. The message of this section is that if we can stop deforestation, we reduce global emissions. For example, halting deforestation in the tropical forests will reduce emissions by at least 10%. Whether we are able to halt or reverse the deforestation will depend on whether we are able to grow enough food. In Europe and North America, research on crop efficiencies and agricultural techniques has resulted in increased yields by an order of magnitude. It is encouraging to see that the reversal of deforestation has already taken place in the temperate forests of these countries. If similar efficiency gains can be obtained in the tropical regions, then reverse geo-engineering could significantly impact carbon emissions.

Table 11.2.1 Carbon in the natural biomes

	Vegetation growth	Vegetation decomposition	Carbon source or sink	Current carbon storage (t/ha)	Where carbon is mostly stored	Main threats for potential carbon emission
Tundra	slow	slow	sink	≈ 258	permafrost	rising temperatures
Boreal forest	slow	slow	sink	soil: 116–343 vegetation: 61–93	soil	fires, logging, mining
Temperate forest	fast	fast	sink	156–320	biomass above and below ground	historical losses high but largely ceased
Temperate grassland	intermediate	slow	likely sink	soil: 113 vegetation: 8	soil	historical losses high but largely ceased
Desert and dry shrubland	slow	slow	sink (but uncertain)	desert soil: 14–102 dryland soil < 226 vegetation: 2–30	soil	land degradation
Savanna and tropical grassland	fast	fast	sink	soil: < 174 vegetation: < 88	soil	fires with subsequent conversion to pasture or grazing land
Tropical forest	fast	fast	sink	soil: 94–191 vegetation: 170–250	above ground vegetation	deforestation and forest degradation
Peatland	slow	slow	sink	1,450	soil	drainage, conversion, fire

Based on data from Trumper et al. [11.2].

Section 3

Geo-engineering: carbon dioxide removal

To geo-engineer the CO_2 currently in the atmosphere, two types of technologies are possible. We can either enhance the uptake of CO_2 by driving natural sequestration processes, or capture CO_2 directly from the air with subsequent sequestration. It is important to note that for either of these two strategies to have any impact, the scale needs to be comparable to that of present-day annual CO_2 emissions. Of the current proposed geo-engineering ideas, we will discuss the three we feel are most worthy of consideration: enhancing the weathering process, enhancing the uptake by the oceans, and direct capture from air.

Enhanced biomass

Changing the land use, which we discussed in the previous section, is in essence a form of geo-engineering in which we restore nature's capacity to uptake larger amounts of CO_2. In this section, we discuss some alternative biological routes to sequester CO_2 that have often been investigated in this context.

One may argue that use of bioenergy and biofuels would be an obvious way to *restore* nature's uptake of CO_2. However, in this chapter we focus on technologies that allow us to *reduce* CO_2 levels. The conversion of biomass to fuels is a very important development in renewable fuels; each year one harvests exactly the same amount of carbon as is emitted. The utilization of biofuels therefore contributes to a CO_2-neutral economy, but does not reduce CO_2 levels. A simple way to achieve a net reduction in CO_2 levels is, for example, to use biomass to make hydrogen and subsequently react the hydrogen with CO_2.

Storage of CO_2 as biomass is not equivalent to geological storage. If done correctly, the ultimate fate of geologically stored CO_2 is mineralization, i.e., formation of limestone as in the Cliffs of Dover. Biomass, however, will eventually decompose and become CO_2. For example, if a tree dies the time scale to decompose and emit CO_2 can be tens of years in the colder climates, or as short as one year in the rainforest of Costa Rica. The essence of biomass carbon storage is that if a tree dies, a new tree will grow, and this keeps the CO_2 level constant. If we want to reduce CO_2 levels by growing more trees, the effect is permanent only if one ensures this forest will flourish for many years (see also **Question 11.3.1**).

> ### Question 11.3.1 CO_2 in greenhouses
>
> CO_2 is often used in greenhouses to enhance biomass production. A farmer has the idea to use the CO_2 from a local natural gas fired power plant because the CO_2 will increase the farmer's biomass by 20%. To finance the operation the farmer has the idea of including an annual carbon credit for the CO_2 he has sequestered. How would you advise him to do this calculation? Hint: find a country in which carbon credits are issued and ascertain how the credit regulations define the time scale for carbon sequestration.

One way to engineer this process is to prevent the decomposition of biomass. For example, if a tree dies and we then dump it in the deep ocean, or some other place where we can prevent the natural decomposition of the tree, we can permanently sequester the carbon. This sounds like a simple scheme but, as you may have guessed by now, nothing is simple if we need to carry this out on a scale that impacts the climate. On such a large scale, one has to carefully analyze the ancillary impacts. For example, the trees have to be transported and buried; such processes cost energy and one would like to ensure that the net effect of the entire process is reduction in the CO_2 levels. The other important factor is that if trees decompose it is not only CO_2 that is released but also minerals and nutrients. If these now are buried in the ocean instead of on land it will create large perturbations of the ecosystems.

An alternative to burying the biomass is to convert it to *biochar* (see **Box 11.3.1**). As the carbon atoms in biochar are bound together much more strongly than in plant matter, biochar is much more resistant to decomposition by micro-organisms. The fact that biochar has been found in archeological sites indicates it can be stable for hundreds to thousands of years. Biochar mixed with soils can improve agricultural productivity. Converting biomass to biochar can be an alternative to the burning of biomass in combination with CCS (see **Figure 11.3.1**) [11.8].

Box 11.3.1 Biochar

Biochar is obtained by pyrolysis of biomass. Pyrolysis is heating in a low- (or no-) oxygen environment. The lack of oxygen prevents the combustion of the biomass. Pyrolysis is the process that occurs when we roast food, giving it the brown color. If we roast at too high a temperature, the food becomes black and we have produced a form of biochar. In fact, pyrolysis has been used for centuries to produce charcoal.

To convert biomass, we can use different forms of pyrolysis. Depending on the temperature and the length of the pyrolysis process we can produce a mixture of bio-oils, biochar, and syngas. High temperature pyrolysis is also known as gasification. Temperatures of 400–500°C produce more char, while temperatures above 700°C favor the yield of liquid and gas fuel components. The figures show some examples of biochar.

Left: This char has a carbon content of 87% and has been shown to detoxify the more serious aflotoxins plaguing US agricultural lands.

Right: Biochar made from rotting palm trees, from weeds which are otherwise burned in the open air, and from cassava stems, are gathered and burned by farmers to avoid uncontrolled growing in the farm and soil nutrient depletion.

Images courtesy of Christophe Steiner: http://www.biochar.org

From a technological point of view, no fundamental difficulties prevent the operation of a large scale biochar program in which we permanently sequester the carbon that would otherwise be emitted to the atmosphere. At present, however, our knowledge of the potential negative effects is incomplete and much more research is needed.

Enhanced weathering

In nature's carbon cycle (Chapter 3), the temperature of the earth is regulated by CO_2 levels. If the temperature is high the weathering of rocks

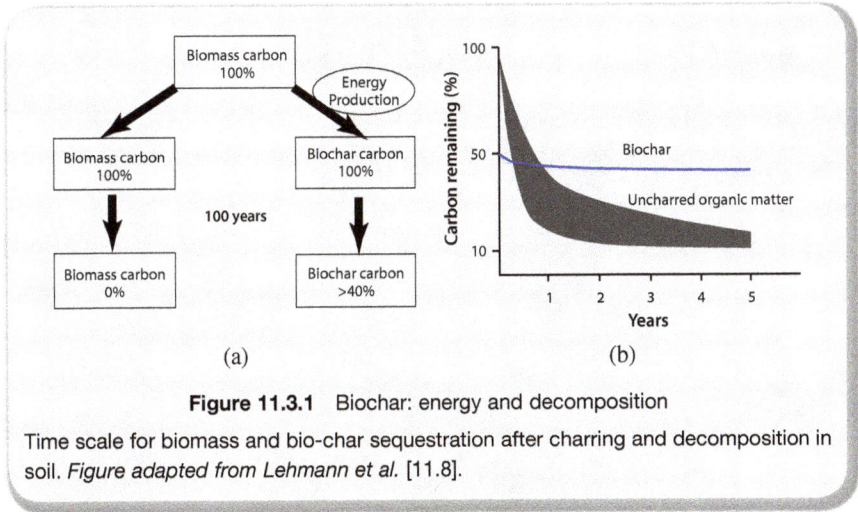

Figure 11.3.1 Biochar: energy and decomposition

Time scale for biomass and bio-char sequestration after charring and decomposition in soil. *Figure adapted from Lehmann et al.* [11.8].

occurs faster, causing the CO_2 levels to decrease. The time scale of this process is several orders of magnitude slower than our rate of emission. If we could accelerate the natural weathering process, we would be able to reduce CO_2 levels.

The idea is to add abundant minerals (e.g., olivine, **Box 11.3.2**) to soil used for agriculture [11.9]. The scale of the operation of course would have to be enormous. One would need to mine, grind, transport, and spread these rocks over fields. The volume of olivine needed would be on the order of 7 km^3 per year, which is about twice the amount of coal we mine. At present, little is known on the potential impacts of these weathering reactions on the soil. Alternative proposals involve conducting the weathering reaction in a chemical engineering plant and then releasing the resulting bicarbonate solutions into the sea.

An advantage of these methods is that all chemicals are already present in large quantities in the soils and the oceans. Of course, one has to mitigate the effects of the large concentrations. For every CO_2 molecule sequestered, one needs a mineral molecule. As a consequence, the amount of material we need will be enormous, most likely exceeding in mass the amount of CO_2 we need to sequester. Mining such enormous quantities of material would have significant impacts on the environment, would be expensive, and would create ancillary energy (and thus carbon) costs.

Box 11.3.2 Olivine weathering

Olivine is a mineral (see figure) with the formula $(Mg,Fe)_2SiO_4$. It is a common mineral in the earth's subsurface but weathers quickly on the surface. The weathering reaction is:

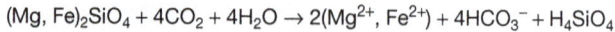

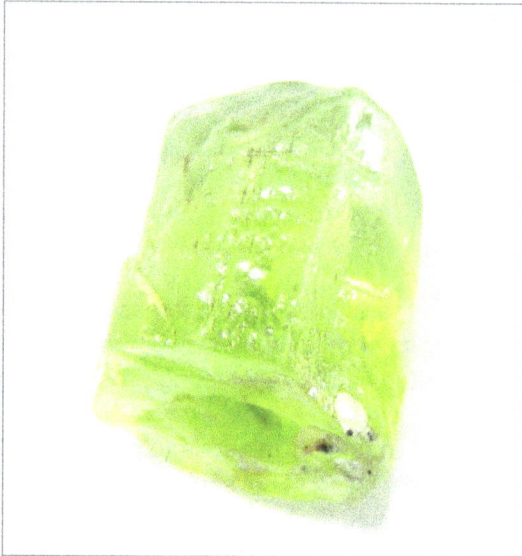

$$(Mg, Fe)_2SiO_4 + 4CO_2 + 4H_2O \rightarrow 2(Mg^{2+}, Fe^{2+}) + 4HCO_3^- + H_4SiO_4$$

Image by Azuncha (2006), WikiCommons: http://en.wikipedia.org/wiki/File:Peridot2.jpg

Direct air capture

In Chapter 4, we computed the minimal energy required to separate CO_2 from flue gas. These calculations showed that separation from a stream with a lower CO_2 concentration requires more energy per CO_2 molecule. For example, compared to capturing a CO_2 molecule from a coal-fired power plant, the minimum energy to capture the same molecule from air will be a factor of five higher.

The idea in **Direct Air Capture** is similar to that in capturing CO_2 from flue gasses; we flow air over a material that selectively adsorbs CO_2. Because of the low concentration of CO_2 in air, we need a material that can adsorb at very low partial pressures. At present, two types of processes are proposed:

- Adsorption into highly alkaline solutions (see **Box 11.3.3**) [11.10]
- Adsorption onto solids [11.11,11.12]

The design of flue gas capture processes requires capture of 90% of the inlet CO_2; Direct Air Capture processes, however, are not subject to

---

Box 11.3.3 Carbon capture with sodium hydroxide

In Chapter 5, we saw that adding a base to water will significantly enhance the CO_2 solubility in the mixture because of the reaction of CO_2 with the hydroxide:

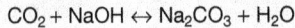

$$CO_2 + NaOH \leftrightarrow Na_2CO_3 + H_2O$$

This is a very exothermic reaction that is efficient in capturing very small concentrations of CO_2. The idea is to use this process to capture CO_2 directly from the air. Similar to capturing CO_2 from flue gasses using absorption, we need to regenerate the solvent and compress the CO_2 for, say, geological sequestration.

The regeneration of sodium carbonate into sodium hydroxide is called causticization and is a well-known industrial process. The process involves three steps. First, Na_2CO_3 is reacted with lime, $Ca(OH)_2$, to form NaOH and lime mud ($CaCO_3$):

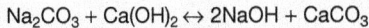

$$Na_2CO_3 + Ca(OH)_2 \leftrightarrow 2NaOH + CaCO_3$$

$$\Delta H_{100°C} = -5.3 \text{ kJ/mol } CO_2$$

The calcium carbonate precipitates in this reaction and is subsequently calcined (heated with air) to recover the CO_2:

$$CaCO_3 \leftrightarrow CaO(s) + CO_2$$

$$\Delta H_{900°C} = 179 \text{ kJ/mol } CO_2$$

The CaO is converted back to $Ca(OH)_2$ through quicklime hydration:

$$CaO(s) + H_2O \leftrightarrow Ca(OH)_2,$$

$$\Delta H_{100°C} = -65 \text{ kJ/mol } CO_2$$

The enthalpy of absorption of CO_2 from air into sodium hydroxide solution for a 1 M solution at standard conditions is −109.4 kJ/mol CO_2. Hence, the minimum energy to recover the CO_2 is also 109.4 kJ/mol CO_2. We see that the calcination step already takes 179 kJ/mol CO_2, which indicates that the required energy for conventional causticization is far beyond the thermodynamic minimum. Therefore, this process consumes a large amount of energy.

that same restriction. Nevertheless, compared to capture from flue gasses, the costs of Direct Air Capture are significantly higher [11.13].

Ocean fertilization

In Chapter 3, we saw that the oceans are responsible for a significant uptake of CO_2. In most parts of the oceans the CO_2 uptake is limited by nutrients [11.14]. If we were to artificially supply these nutrients, we would be able to enhance the CO_2 uptake. The limiting nutrients that have been studied include N, P, and Fe. To estimate the effects of these nutrients on CO_2 uptake, we consider their relative amounts used by algae in building their organic tissue. These are expressed by the characteristic Redfield ratios of the nutrient elements, C:N:P:Fe. For algae, the ratios are typically 106:16:1:0.001. This implies that for every additional P atom, one will able to sequester 106 carbon atoms. The Redfield ratio shows that Fe has the largest impact, and therefore most research has focused on the effects of increasing the Fe content of the ocean. Of course, a key concern is that adding the amounts needed to impact the CO_2 levels of the atmosphere could produce large, potentially undesirable effects on ecosystems. These possible effects are poorly understood.

Ocean up- or downwelling

The slow mixing of the surface layer of the ocean and the deep ocean is responsible for the very long time (>200 years, see Chapter 3) required before the CO_2 levels in the atmosphere will decrease. If we could enhance the mixing time by up- or downwelling water from the deep ocean reservoir to the surface, we would be able to shorten this time. The amount of water that would lead to sequestration of ~0.02 Gt C/yr is 1 million m^3/s. This flow rate exceeds the known volume of all the major rivers of the world combined!

Section 4

Geo-engineering: solar radiation management

In the effort to reduce global warming, an alternative to the geo-engineering techniques that reduce the CO_2 levels in the atmosphere, which we focused on in the previous section, is to reduce the amount of sunlight that is reaching the earth. We have seen in Chapter 2 that a doubling of the CO_2 concentration would correspond to a radiative forcing of 4 W/m^2. The temperature increase can therefore be compensated if we are able to reduce the solar radiation by an equivalent amount. The methods that have been proposed include:

Reflecting part of the sunlight back into space: to achieve a reduction of 4 W/m^2 we need to reduce the incoming sunlight by 1.8%. The idea is to bring giant reflectors into orbit around the earth that would reflect about 2% of the incoming sunlight. The size of these reflectors would need to be on the order of a million square kilometers [11.15]. Clearly, building such a reflector is not a short-term solution. However, if such a reflector could be built, it is one of the few options that would give an instantaneous reduction of the global temperature.

Increase the concentration of aerosols in the atmosphere: in Chapter 2 we saw that the effect of large eruptions of volcanos is a reduction of global temperatures through increasing levels of sulphate aerosols in the stratosphere, which increases the stratosphere's natural albedo effects. The idea is to mimic this cooling effect from volcanos by introducing hydrogen sulphide (H_2S) or sulphur dioxide (SO_2) into the stratosphere as gas [11.16, 11.17], where it is converted into sulphate particles with characteristic sizes on the order of several tenths of a micron.

Albedo effect at the surface of the earth: in Chapter 2, we also saw that the earth reflects part of the radiation from the sun. This reflection depends on the nature of the surface (see **Table 11.4.1**). To compensate for the 4 W/m^2, we would need to increase the average reflection of the earth's surface from ~107 to ~111 W/m^2. This would imply an increase in

Table 11.4.1 Albedo coefficients of surfaces on the earth

Surface	Albedo coefficient
Ocean	0.1
Land Surface	0.2–0.3
Ice and snow	0.6–0.8
Average	0.15

The albedo coefficient is defined as the ratio of reflected radiation from the surface to incident radiation upon it. *Data from* [11.16].

the mean surface albedo of the planet from 0.15 to about 0.17. This does not look like a large change, but because most of the surface of the earth is ocean, for which it is impractical to change its reflectivity, the net change on "usable" surface would therefore be much larger. The required increase over the land area would be on average 8%. In practice, this is further limited in that only on a small percentage of the total land surface can one effectively modify its characteristics (see **Question 11.4.1**).

We emphasize that changing the amount of sunlight reaching the earth's surface does not resolve the underlying cause of global warming; it only mitigates the negative effects of increasing global temperatures. This also implies that as soon as we stop applying any solar radiation management techniques, the temperature would quite rapidly increase if we do not control CO_2 emissions.

Question 11.4.1 10% of the surface

Estimate what the albedo coefficient needs to be to compensate for the 4 W/m^2 of radiative forces, if 10% of the total land surface can be modified.

Section 5

CO_2 utilization

In this book, we have focused on carbon capture and geological seques-tration. In the sequestration sections, the CO_2 was viewed as a waste that must be "put somewhere." At present, there is a broad concern that any process should focus on recycling rather than storing it as waste. This has motivated many suggestions about a more "positive" use of CO_2 compared to storage in geological formations.

In Chapter 1, we already saw that a chemical conversion of CO_2 into our universally useful product Dreamium™ would require so much resource material that we would saturate any market or deplete any supply.

Because petroleum is the present feedstock for production of many chemicals, one can envision replacing petroleum with CO_2 as a feedstock for carbon-containing chemicals. We have to realize, however, that only 7% of petroleum-derived carbon is used by the chemical industry; the other 93% is used as fuel. Of course, any reduction in use of fossil fuels is important and we most likely will need to use every available technol-ogy to reduce CO_2 emissions, but the scope of using CO_2 as a source of recyclable carbon for the chemical industry is too limited to offer a true solution.

In our discussion on the use of CO_2, we have not mentioned the idea of recycling CO_2 back into a fuel. Interesting research is presently being conducted on the catalytic conversion of CO_2 into CO, with particular emphasis on the use of renewable energy sources for this conversion. What are the consequences of such a scheme? If we have the technol-ogy to convert CO_2 to CO using renewable energy sources, why not use that renewable energy in the first place and avoid using fossil fuels to generate electricity? In generating electricity, we do not address trans-portation fuels. In an energy economy in which renewable energy is abundant, however, we can imagine upgrading CO_2 from Direct Air Capture into a transportation fuel.

Section 6

Outlook

Many of the ideas discussed in this chapter may at first glance look more like science fiction than serious science. We have used seven chapters in this text to explain the challenges in carbon capture and sequestration. The message of the present chapter is that these challenges in CCS are modest compared to what must be done to reduce absolute CO_2 levels in the atmosphere. Indeed, given what we know now, any CO_2 molecule we emit today will undoubtedly be much more expensive to capture later.

Our geo-engineering experiment of emitting into the atmosphere in less than two centuries the amount of CO_2 that nature has sequestered over a period of a million years creates a very large burden on our future. Consider the words from the 2011 World Energy Outlook: *"Delaying action is a false economy. For every $1 of investment in the power sector avoided before 2020, an additional $4.3 would need to be spent after 2020 to compensate for the higher emissions."*

Do we want to risk paying such a price in the future and delaying action now while some doubts about the exact extent and consequences of climate change continue to pervade the discussions?

Well, what will you do if you feel some pain on your chest and are short of breath? I guess you will visit your doctor and if you do not like the message of changing your lifestyle, you will ask for a second opinion, and a third, and another one, and yet another one, and another ... and another... until you have visited the doctor who finally tells you that the experts are creating a conspiracy, all your symptoms are fluctuations and there is nothing wrong with you! And you had almost risked wasting precious resources in adopting a healthy lifestyle...

Section 7
References

1. Shepherd, J. *et al.*, 2009. Geo-engineering the Climate: Science, Governance and Uncertainty. London: Royal Society.
2. Trumper, K., M. Bertzky, B. Dickson, G. van der Heijden, M. Jenkins, and P. Manning, 2009. The Natural Fix? The Role of Ecosystems in Climate Mitigation. A UNEP Rapid Response Assessment. http://www.grida.no/files/publications/natural-fix/BioseqRRA_scr.pdf (Figures 11.2.1 and 11.2.3 in this book are by Riccardo Pravettoni.)
3. Ruesch A., and H.K. Gibbs, 2008. New IPCC Tier-1 Global Biomass Carbon Map for the Year 2000. Available online from the Carbon Dioxide Information Analysis Center, http://cdiac.ornl.gov. Oak Ridge National Laboratory, Oak Ridge, Tennessee.
4. Scharlemann J.P.W., R. Hiederer, and V. Kapos, in prep. Global Map of Terrestrial Soil Organic Carbon Stocks. A 1-km dataset derived from the Harmonized World Soil Database. UNEP-WCMC & EU-JRC: Cambridge UK.
5. Kapos V., C. Ravilious, A. Campbell, *et al.*, 2008. Carbon and Biodiversity: a Demonstration Atlas. UNEP-WCMC: Cambridge, UK.
6. FAO/IIASA/ISRIC/ISS-CAS/JRC, 2009. Harmonized World Soil Database (version 1.1). FAO, Rome, Italy and IIASA, Laxenburg, Austria.
7. Pravettoni, R., United Nations Environment Programme, GRID-Arendal, 2009. Carbon stored by biome. http://www.grida.no/graphicslib/detail/carbon-stored-by-biome_9082
8. Lehmann, J., J. Gaunt, and M. Rondon, 2006. Bio-char sequestration in terrestrial ecosystems — a review. Mitigation and Adaptation Strategies for Global Change, **11** (2), 395. http://dx.doi.org/10.1007/s11027-005-9006-5
9. Schuiling, R.D. and P. Krijgsman, 2006. Enhanced weathering: An effective and cheap tool to sequester CO_2. Climatic Change, **74** (1–3), 349. http://dx.doi.org/10.1007/S10584-005-3485-Y
10. Stolaroff, J.K., D.W. Keith, and G.V. Lowry, 2008. Carbon dioxide capture from atmospheric air using sodium hydroxide spray. Environ. Sci. Technol., **42** (8), 2728. http://dx.doi.org/10.1021/Es702607w
11. Lackner, KS., 2009. Capture of carbon dioxide from ambient air. Eur. Phys. J-Spec. Top., **176**, 93. http://dx.doi.org/10.1140/Epjst/E2009-01150-3
12. Lackner, K.S., 2003. Climate change: A guide to CO_2 sequestration. Science, **300** (5626), 1677. http://dx.doi.org/10.1126/science.1079033

13. Socolow, R., M. Desmond, R. Aines, *et al.*, 2011. Direct Air Capture of CO_2 with Chemicals: A Technology Assessment for the APS Panel on Public Affairs. US: American Physical Society.

14. Aumont, O. and L. Bopp, 2006. Globalizing results from ocean in situ iron fertilization studies. Global Biogeochem. Cycles, **20** (2), GB2017. http://dx.doi.org/10.1029/2005gb002591

15. Hoffert, M.I., K. Caldeira, G. Benford, *et al.*, 2002. Advanced technology paths to global climate stability: energy for a greenhouse planet. Science, **298** (5595), 981. http://dx.doi.org/10.1126/science.1072357

16. Crutzen, P.J., 2006. Albedo enhancement by stratospheric sulfur injections: A contribution to resolve a policy dilemma? Climatic Change, **77** (3–4), 211. http://dx.doi.org/10.1007/S10584-006-9101-Y

17. Wigley, T.M.L., 2006. A combined mitigation/geoengineering approach to climate stabilization. Science, **314** (5798), 452. http://dx.doi.org/10.1126/Science.1131728

Chapter 12

List of Symbols

Section 1

List of symbols

A	area
A	Helmholtz free energy
C_p	heat capacity at constant pressure
c_i	concentration component i [mol/m^3]
D_i	diffusion coefficient component i
E	energy
g	gravitational constant
H	enthalpy
h	molar enthalpy
Δh_i	the heat of adsorption of component i
I	intensity of light
J	energy of a radiating body
j_i	flux per unit area component i [mol/s m^2]
k_B	Boltzmann's constant
k_i	mass transfer coefficient for component i
k	hopping rate
k	permeability of a rock
K_i	Henry constant for component i
K_i	equilibrium constant chemical reaction
$1/K_i$	solubility
L	thickness
m	mass
N_A	Avogadro's number
n_i	flux [m^3/s]
P	plate number
P	permeability
P'	permeance
p	pressure
p_i	partial pressure of component i
Q	heat
q_i	amount adsorbed per unit volume of component i
r	reaction rate

R	gas constant ($N_A k_B$)
R	radius of curvature
S	entropy
S_i	phase separation of component i
s	molar entropy/entropy per mole
t	time
T	temperature
U	internal energy
u	molar internal energy
u	velocity
V	volume
W	work
w	work per mole
v	molar volume
x_i	mole fraction
x_i	mole fraction in the liquid phase
y_i	mole fraction in the gas phase

Greek symbols

α	ideal separation factor
β	reciprocal temperature $1/k_B T$
Γ	thermodynamic factor
γ	surface tension
ε	void fraction
η	Carnot efficiency
θ	fractional occupancy
θ	stage cut
θ	contact angle
κ	solubility
Λ	Thermal de Broglie wavelength
λ	heat transfer coefficient
μ_i	chemical potential component i
μ	viscosity
ρ	density (number of molecules per unit volume)
σ	Stefan-Boltzmann constant
σ	adsorption constant

σ   loading, number of adsorbed molecules of component i

Φ, ϕ flux [mol/s]

Sub- and superscripts

abs   absorption
ads   adsorption
des   desorption
ex   excess
i   component
IG   ideal gas
mix   mixture
par   parasitic

Chapter 13

Credits

THANKS!

Section 1

Cover figure captions

- The cover illustrates the global aspect of carbon capture and sequestration. The jar contains a molecular view of the metal organic framework (MOF) MG-MOF-74, which is studied by many groups for carbon capture. *Cover designed by Wayne Keefe (wkeefe@mac.com), based on the research by A. Dzubak, L.-C. Lin, J. Kim, J.A. Swisher, R. Poloni, S.N. Maximoff, B. Smit, and L. Gagliardi, 2012. Ab-initio carbon capture in open-site metal organic frameworks. Nat. Chem., 4, 810–816.*

- The half title page gives an illustration of the Metal Organic Framework Mg-MOF-74 in which CO_2 molecules are adsorbed. *Image prepared by Li-Chiang Lin and Roberta Poloni.*

- The preface shows the Campanile Bell Tower of UC Berkeley. *Photo by Alan Nyiri, courtesy of the Atkinson Photographic Archive.* http://gallery.berkeley.edu/viewphoto.php?&albumId=199002&imageId=6126278&page=2&imagepos=53

- Chapter 1: **Energy and Electricity**: Coal power plant in Datteln (Germany) at the Dortmund-Ems-Kanal. *Image by Arnold Paul (Wikimedia Commons, 2006).* http://commons.wikimedia.org/wiki/File:Coal_power_plant_Datteln_2.jpg

- Chapter 2: **The Atmosphere and Climate Modeling**: Climate modeling simulation. *Image courtesy of NASA.* http://www.isgtw.org/sites/default/files/img_2011/climate-modelling.jpg

- Chapter 3: **The Carbon Cycle**: This picture is a very simplified representation of the contemporary global carbon cycle, showing components of the Global Carbon Cycle (alternate), for more details about this natural flux between the terrestrial biosphere and the atmosphere and between the marine biosphere and the atmosphere, see: Office of Biological and Environmental Research of the US Department of Energy Office of Science, science.energy.gov/ber/ *Image by US DOE, 2008. Carbon Cycling and Biosequestration, prepared by the Biological and Environmental Research Information System, Oak Ridge National*

Laboratory, genomicscience.energy.gov/ and genomics.energy.gov/:
This picture was featured on the cover of the report from the March
2008 Workshop, DOE/SC-108, US DOE, Office of Science.

- Chapter 4: **Introduction to Carbon Capture**: Schwarze Pumpe plant Brandenburg, Germany. *Reproduced with permission, © Bureau de Recherches Géologiques et Minières — Vattenfall.* http://www.brgm.eu/content/geological-storage-co$_2$-safety-is-priority
- Chapter 5: **Absorption**: Bottoms, R.R. (Girdler Corp.), "Separating acid gases," U.S. Patent 1783901, 1930.
- Chapter 6: **Adsorption**: This picture gives an illustration of the zeolite FAU; the red and white sticks show the Si, O framework and the surface gives the energy of a CO_2 molecule at a particular position. *Prepared by Dr. Richard Martin.*
- Chapter 7: **Membranes**: The image gives a cross-section of a scanning electron microscopy — micrographs of a functionalized polyaniline-based composite membranes. *Prepared by Natalia V. Blinova and Frantisek Svec.*
- Chapter 8: **Introduction to Geological Sequestration**: *Image from Curtis M. Oldenburg.*
- Chapter 9: **Fluids and Rocks**: Scanning electron microscopy (SEM) images of a sandstone sample from the Cranfield CO_2 sequestration pilot site in Mississippi (6 by 2.4 mm area) with pore space and mineralogy identified by electron microscopy and X-ray spectroscopy techniques. *Images by Landrot, G., J.B. Ajo-Franklin, L. Yang, S. Cabrini, and C.I. Steefel, 2012. Measurement of accessible reactive surface area in a sandstone, with application to CO_2 mineralization. Chemical Geology, 113, 318–319. http://dx.doi.org/10.1016/j.chemgeo. 2012.05.010, reproduced with permission from Elsevier.*
- Chapter 10: **Large-Scale Geological Carbon Sequestration**: *Figure courtesy of Earth Science Division, LBNL.*
- Chapter 11: **Land Use and Geo-Engineering**: *Earth image courtesy of NASA.*

Glossary

Absorber

The part of an adsorption process equipment in which the gas mixture that needs to be separated is brought into contact with the solvent.

See also stripper.

Absorption

Removing CO_2 (or any other component) from a mixture using the differences in the solubilities of the components of the mixture in a solvent. In the book, absorption always refers to the use of liquid solvents to separate components from a gas mixture.

See also scrubbing.

Adsorption

Removing CO_2 (or any other component) from a mixture using the differences in adsorption of the different components in a solid. In the book, absorption always refers to a solid.

Albedo effect

Reflection of the sunlight by the surface of the earth.

Aquifer

A formation of permeable rocks saturated with water, with a degree of permeability that allows water withdrawal through wells.

Barrer

Unit of permeance of a material. The definition of 1 Barrer:

$$1\,\text{Barrer} = \frac{10^{-10}\,(\text{cm}^3\text{gas})\,(\text{STP})\,(\text{cm thickness})}{(\text{cm}^2\text{membrane area})\,\sec\,(\text{cm Hg pressure})}$$

Here "cm^3 gas (STP)" represents the quantity of gas that would take up one cubic centimeter at standard temperature and pressure (STP) as calculated via the ideal gas law, i.e., the molar volume. The "cm thickness" represents the thickness of the material whose permeability is being measured, and "cm^2 membrane area" is the surface area of that material. The conversion to SI units is 1 Barrer = 3.348×10^{-19} kmol m/ (m^{2}s Pa).

Beer's law

Law describing how the intensity (I) of light decreases if it passes through a medium (gas):

$$\frac{I}{I_0} = e^{-\rho\sigma l},$$

where ρ is the density of the gas, σ the absorption coefficient of the substance, and l the distance the light travels through the material (i.e., the path length).

Bicarbonate

CO_2 dissolved in water gives bicarbonate (HCO_3^-):

$$CO_2\,(aq) + H_2O \rightleftharpoons HCO_3^- + H^+$$

See also carbonate.

Biological pump

In ocean surface layers, organisms use sunlight to convert CO_2 into biomass. If these surface organisms die and sink to the deep part of the ocean before they decompose, there is a net flux of organic carbon from the atmosphere toward the deep ocean. This flux is referred to as the "biological pump."

Breakthrough curves

These curves give the point at which a solid absorbent material cannot adsorb CO_2 anymore. Upon adsorption, a CO_2 front will form and slowly move through the adsorption column until it has reached the end of the column. At this point, the CO_2 breaks through the adsorber and the adsorber needs to be regenerated.

Buoyancy

The difference in density that gives a resulting upward or downward force.

Calcite

One of the most abundant forms of carbon ($CaCO_3$).

Capillary pressure

Pressure difference between the gas and liquid phases in a pore caused by the capillary forces.

Capillary trapping

In capillary trapping or residual trapping, part of the CO_2 becomes immobilized as small bubbles at the trailing edge of the mobile CO_2 plume, while the plume itself continues to migrate to the highest point in the formation.
See also residual trapping.

Caprock

A fine-textured rock that has a very limited permeability to CO_2 and is used as a seal to prevent CO_2 from reaching the surface.

Carbonate

CO_2 dissolved in water gives bicarbonate, which can lose another proton to give carbonate (CO_3^{2-}):

$$CO_2(aq) + H_2O \rightleftarrows HCO_3^- + H^+$$

$$HCO_3^- + H^+ \rightleftarrows CO_3^{2-} + 2H^+$$

See also bicarbonate.

CCS

Carbon Capture and Sequestration/Storage. In Europe, one mainly uses the term storage.

Chemical looping

The idea of chemical looping is to separate the combustion process into two separate reactors. Oxygen is taken from the air by the first reactor and then transported to the other reactor, where combustion takes place and CO_2 is produced. In this scheme, there is no mixing of N_2 with flue gas, therefore the separation only involves CO_2 and H_2O.

Clastic sedimentary rocks

Clastic sedimentary rocks are formed from eroded fragments of older rocks. These rock fragments are transported by water or wind, then deposited (in the process of sedimentation). The grains of these sediments are subsequently consolidated into a rock (in the process of lithification).

Collective diffusion coefficient

See Maxwell-Stefan diffusion coefficient. *See also* Darken-corrected diffusion coefficient.

Concurrent flow

Flow pattern in which the two flows run parallel to each other and in the same direction.

See also countercurrent flow.

Countercurrent flow

Flow pattern in which the two flows run parallel to each other but in opposite directions.

See also concurrent flow.

Darcy's law

Gives the relation between the flux of matter through a porous media and the pressure:

$$j = -\frac{k}{\mu}\nabla p,$$

where ∇p is the pressure gradient, μ the viscosity of the fluid, and k is the permeability of the rock.

Darken-corrected diffusion coefficient

See Maxwell-Stefan diffusion coefficient. *See also* collective diffusion coefficient.

Direct air capture

Process by which CO_2 is removed directly from the air.

Diurnal cycle

Daily cycle. In this book, the diurnal cycle refers to the daily photosynthesis cycle of plants.

Dreamium

Hypothetical molecule consisting of CO_2 chemically bound to another molecule ZZ. The nice, but very unrealistic, thing about dreamium is that one can use any molecule ZZ to investigate the large-scale production of a hypothetical product made from CO_2.

Emission scenarios

Prediction of future carbon emissions. In this text, we often refer to the IPCC scenarios A1, A2, B1, B2.

Enhanced oil recovery

Injection of CO_2 into oil fields with declining primary productivity. The recovery of oil is enhanced because the CO_2 decreases the viscosity.

Entropy of mixing

Change in entropy when we mix pure components.

Facilitated transport

A diffusion mechanism in which a chemical reaction enhances the permeability of one of the components without decreasing its selectivity.

Faint young sun paradox

The observation that the sun was much weaker when the earth was formed than it is now. Such a weak sun would result in a temperature of the earth that would be too low for liquid water to be on earth. The paradox is that water was observed at these times.

Feed

A gas stream that enters a specific process unit such as a membrane unit or an adsorber.

Fick-diffusion coefficient

The diffusion coefficient that is associated with transport of mass caused by a difference in concentration. This is the diffusion coefficient we use in most practical applications.

See Fick's law. See also Fick-diffusion coefficient.

Fick's law

States that the flux of material is proportional to the gradient of the concentration. The proportionality constant is the (transport or Fick) diffusion coefficient.

$$j_{CO_2} = -D_{CO_2} \frac{dc_{CO_2}}{dz}$$

See also Fick-diffusion coefficient.

Fixed bed

Adsorption process using a solid adsorption material that is fixed in the adsorber.

Flue gas

In this text we refer to flue gas as the combustion exhaust gas produced at power plants. The composition of flue gas depends on what is being burned, but it will usually consist of mostly nitrogen, carbon dioxide, and water vapor, as well as oxygen. It may further contain small percentages of carbon monoxide, nitrogen oxides, and sulfur oxides, or particulate matter.

Fluidized bed

Adsorption process using a solid adsorption material that is moving in the adsorber. If a gas is flowing through the solid at certain conditions, this mixture can behave as a fluid.

Fracking

The propagation of fractures in a rock layer by a pressurized fluid. Induced hydraulic fracturing or hydrofracturing is a technique used to release petroleum, natural gas (including shale gas, tight gas, and coal seam gas), or other substances for extraction.

See also hydraulic fracturing.

Fracture permeability

Fractures can impart very high permeability to otherwise low-permeability rocks. In fact, even a single fracture with a small aperture can impart a fairly large overall permeability to a rock. It has been shown that fractures provide permeability k (in m^2) according to the so-called Cubic Law, which follows from the flow of a fluid between two parallel plates:

$$k = \frac{Nb^3}{12},$$

where N is the number of fractures per meter and b is the fracture aperture.

Geo-engineering

Large-scale human intervention in processes of the earth.

Greenhouse gas

A greenhouse gas (GHG) is a gas in an atmosphere that absorbs and emits radiation within the thermal infrared range. The primary greenhouse gasses in the earth's atmosphere are water vapor, carbon dioxide, methane, nitrous oxide, and ozone.

Heat of absorption

The heat that is released if we absorb a gas in a liquid absorbent.

See *also* heat of desorption.

Heat of adsorption

The heat that is released if we adsorb a gas in a solid adsorbent.
See also heat of desorption.

Heat of desorption

The negative of the heat of adsorption or absorption. The heat of adsorption is the heat that is released if we adsorb a gas in a solid, or the heat of absorption in the case of absorption in a liquid.
See also heat of adsorption.

Henry coefficient

See Henry constant.

Henry constant

Henry constant or coefficient gives the solubility or adsorption of a gas in a liquid or solid. The definition of the Henry coefficient differs from one application to another, as do the units.

Hockey stick curve

The famous curve that shows the average global temperature over the years. The flat part indicates a constant temperature and in recent years an upward trend is observed.

Hydraulic fracturing ("fracking")

The ability to create high-permeability vertical fractures. It is the basis for the relatively new approach to extracting hydrocarbons.
See also fracking.

Hydrostatic pressure

The equilibrium pressure due to gravity alone; most fluids in sedimentary basins are at this pressure.

Hysteresis

The observation that the state of a system not only depends on its current condition but also on its history. Here, we look into the hysteresis in the permeability and saturation of CO_2 in the rocks.

Ideal separation factor

The maximum separation that can be achieved. For a binary mixture, this factor is defined as:

$$\alpha_{CO_2,N_2} = \frac{x_{CO_2,P} / x_{N_2,P}}{x_{CO_2,R} / x_{N_2,R}}$$

IGCC

See integrated gasification combined cycle.

Imbibition

The displacement of one fluid by another; here we mainly use it to displace CO_2 by water.

Inorganic carbon cycle

Part of the carbon cycle that does not involve photosynthesis but weathering and volcanos.

Integrated gasification combined cycle

The integrated gasification combined cycle (IGCC) process relies on the conversion of coal into syngas. Syngas is a mixture of CO and H_2, and is formed by partially oxidizing coal in a coal gasifier. A water gas shift reactor, which converts water and carbon monoxide into hydrogen and carbon dioxide, is used to increase the amount of hydrogen in the mixture. Before the fuel goes into the burner, the CO_2 is separated from the H_2, and the H_2 is subsequently burned to produce the heat for generating the steam. In the IGCC process, the CO_2 separation involves a mixture of hydrogen, carbon monoxide, and carbon dioxide at high pressure.

Ionic liquid

Liquid consisting of positively and negatively charged molecular ions. Because of the molecular structure they melt at much lower temperatures compared to salts.

IPCC

Intergovernmental Panel on Climate Change.

Irreducible saturation

See residual saturation.

Langmuir isotherm

Adsorption isotherm that follows Langmuir's equation:

$$\theta(p) = \frac{\sigma(p)}{\sigma_{max}} = \frac{bp}{1+bp},$$

where p is the partial pressure, θ is the fractional occupancy, σ is the loading (in mol/kg adsorbent) and σ_{max} is the saturation loading.

Life-cycle analysis

A life-cycle analysis of a process is aimed to obtain a comprehensive estimate of the overall costs or environmental impact of a process, starting from building the factory and ending with the disposal of all chemicals.

Limestone

Limestone is a sedimentary rock composed largely of the minerals calcite and aragonite, which are different crystal forms of calcium carbonate ($CaCO_3$). Limestone makes up about 10% of the total volume of all sedimentary rocks.

Maxwell-Stefan diffusion coefficient

The Maxwell-Stefan (or Darken-corrected, or collective) diffusion coefficient is the diffusion coefficient that relates transport of mass to a gradient in the chemical potential. This is a more fundamental way of describing the diffusion coefficient and typically follows from a molecular simulation. If we know how the concentration is related to the chemical potential, we can easily convert this diffusion coefficient into the Fick diffusion coefficient.

See also collective diffusion coefficient, Darken-corrected diffusion coefficient, and Fick-diffusion coefficient.

McCabe-Thiele method

Graphical method developed by McCabe and Thiele to estimate the number of (hypothetical) plates that are needed to achieve a separation process.

MEA

See monoethanol amine.

Metal Organic Frameworks

Metal Organic Frameworks (MOFs) are metal/organic hybrid solids built from organic linkers and nodes made out of inorganic metals (or metal-containing clusters).

Mineral trapping

Dissolved CO_2 will react with minerals such as feldspars to liberate cations (e.g., Mg^{2+}, Fe^{2+}, Ca^{2+}), and these cations then become available to react with CO_3^{2-} in the aqueous phase to form carbonate minerals. As carbonates are the thermodynamically most stable form of carbon, *mineral trapping* is the final fate of injected CO_2.

Minimum work

Thermodynamic concept that describes the minimum amount of work that needs to be performed to operate a process without violating the first and second laws of thermodynamics.

MOF

See Metal Organic Frameworks (MOFs).

Monoethanol amine

Monoethanol amine (MEA), or C_2H_7NO, is the reference fluid for CO_2 scrubbing.

Mtoe

A Mtoe stands for 10^6 tonnes of equivalent oil, i.e., the produced energy is equivalent to the energy released by burning 10^6 tonnes of oil (in SI units 1,000 Mtoe = 42 EJ; E = exa = 10^{18}).

Navier-Stokes equations

The fundamental equations that describe the flow of a fluid based on conservation of mass and momentum.

Nuclear Magnetic Resonance (NMR)

A spectroscopic technique that uses the unique magnetic properties of some atoms.

Oxy-combustion

Process in which coals are burned with nearly pure oxygen. The resulting flue gas contains only water and CO_2 which can be easily separated by condensation.

Packed column

Type of absorber in which a packing ensures optimal contact between the gas and the liquid.

See also plate tower.

Palaeocene-Eocene Thermal Maximum

The Paleocene and the Eocene eras are separated by a peak in the ocean temperatures. This peak is known as the Palaeocene-Eocene Thermal Maximum (PETM).

Parasitic energy

A Carbon Capture and Sequestration process costing energy that cannot be used to, for example, produce electricity. This loss of electricity is often expressed in the form of parasitic energy.

Permafrost

Permanently frozen layers of soil.

Permeability

A material property that indicates the ease with which molecules are transported through a material. The permeability depends on the solubility and diffusion coefficient.

Permeance

The permeance P' is the permeability P per unit thickness L of a material and gives the ease with which a gas can flow through a material:

$$j = \frac{P'}{L}(p_R - p_p) = p(p_R - p_p),$$

where j is the flux and $p_R - p_P$ the pressure difference of the gas between the retentate and permeate.

Permeate

The part of a gas mixture that passes through a membrane.

PETM

See Palaeocene-Eocene Thermal Maximum.

Phase saturation

The fraction of the pure volume that is occupied by phase i is called the phase saturation (e.g., brine saturation or CO_2 saturation):

$$S_i = \frac{V_i}{V_p} = \frac{V_i}{\phi V_T},$$

where V_i is the total volume of phase i and V_p is the pore volume. In the second equation, ϕ is the porosity and V_T is the total volume of the rock.

Plate tower

Type of absorber in which plates ensure optimal contact between the gas and the liquid.
    See also packed column.

Porosity

The volume fraction of a rock that is occupied by pore space (ϕ).

Post-combustion carbon capture

Process in which the CO_2 is removed after the burning of the fossil fuels.

Pre-combustion carbon capture

Process in which gasses are separated before the combustion processes, which makes the separation of CO_2 after the combustion easier.

Pressure swing adsorption

Adsorption or absorption process where one uses a pressure swing: adsorption at high pressure and desorption at low pressure to achieve a separation.

See also temperature swing adsorption.

Radiative forcing

A measure of how the energy balance of the earth-atmosphere system is influenced when factors that affect climate are altered, given as the measured rate of energy change per unit area (in W/m^2) of the globe.

Relative permeability

The relative permeability K_{ij} of phase i in the presence of phase j, is defined as the effective permeability divided by the single phase permeability.

Residual phase trapping

See capillary trapping.

Residual saturation

The lowest point of saturation: because of the capillary forces not all gasses can be removed, giving a residual or irreversible saturation.

See also irreducible saturation.

Residual trapping

See capillary trapping.

Retentate

The part of a gas mixture that does not pass through a membrane.

Robeson plot

In a Robeson plot, the selectivity of a mixture is plotted as a function of the permeation of the component with the largest permeability. The aim is to have a membrane with both a high selectivity and permeability.

Scrubbing

Separating a gas mixture by liquid absorption.
    See also absorption.

Self-diffusion coefficient

This coefficient characterizes the diffusion of a single molecule in a fluid of identical molecules. This type of diffusion at the molecular level can be measured by labeling some of the molecules (e.g., by using Nuclear Magnetic Resonance spectroscopy or NMR).

Sensible heat

The heat required to increase the temperature of the solvent to the desorption conditions.

Solubility trapping

Once trapped as small bubbles, the brine-CO_2 surface area increases significantly and this allows CO_2 to dissolve in the brine. The process of CO_2 dissolving in brine is called solubility trapping.

Spray column

Type of absorber in which the spraying of droplets ensure optimal contact between the gas and the liquid.

Stage cut

The stage cut θ is defined as the fraction of the feed that flows through a membrane:

$$\theta = \frac{j_P}{j_F}$$

Note that $\theta \leq 1$.

Stefan-Boltzmann's law

A law describing the relation between the temperature of an object (T) and the energy (J) this body emits through blackbody radiation:

$$J = \sigma T^4,$$

where σ is the Stefan-Boltzmann constant.

Stratigraphic trapping

See structural trapping.

Stripper

Equipment of an absorption process in which the absorbed gas is removed/stripped from the solvent.
   See also absorber.

Structural trapping

Structural trapping of CO_2 occurs when the caprock prevents the CO_2 from moving upward.
   See also stratigraphic trapping.

Syngas

A mixture of CO and H_2 formed, for example, by partially oxidizing coal in a coal gasifier.

Temperature swing adsorption

Adsorption or absorption process where one uses a temperature swing: adsorption at low temperature and desorption at high temperature to achieve a separation.

See also pressure swing adsorption.

Van't Hoff equation

Thermodynamic relation describing the temperature dependence of the Henry coefficient, H:

$$\frac{d \ln H}{dT} = \frac{\Delta h_{ads}}{RT^2},$$

where Δh_{ads} is the heat of adsorption.

Weathering reactions

Reaction of the weak acid formed by CO_2 and water with rocks:

$$CO_2 + CaAl_2Si_2O_8 + 2H_2O \rightleftarrows CaCO_3 + Al_2Si_2O_5(OH)_4$$

Wetting

Wetting is the ability of a fluid to wet a substrate. If a fluid wets a substrate it will spread as a thin film, while a non-wetting fluid will form a droplet. Fluids can also partially wet a substrate, and this wetting behavior is quantified by the contact angle.

Working capacity

The amount of material that is removed in one cycle for an adsorption or absorption process, usually the difference between the amount adsorbed at adsorption conditions minus the amount that stays in the material at desorption conditions.

Young-Laplace equation

Relation between the pressure difference across the inside and the outside of a curved interface. For a cylindrical pore and a fully wetting fluid (zero contact angle):

$$p_{gas} - p_{liquid} = \frac{2}{R}\gamma_{LG},$$

where R is the radius of the pore and γ_{LG} the liquid-gas surface tension.

Zeolites

Zeolites are nanoporous materials. The basic building blocks of zeolites are corner-sharing TO_4 groups, where the T-atom is usually Si, Al, or in some cases also P. At present, over 200 different zeolite structures are known, each of which has a similar chemical composition, but a very different pore topology.

See also Zeolitic Imidazolate Frameworks (ZIFs).

Zeolitic Imidazolate Frameworks

Zeolitic Imidazolate Frameworks (ZIFs) are a special kind of Metal Organic Framework in which the size of the linker and angle between linkers are tuned in such a way that they mimic the Si–O–Si bond in zeolites and hence have the same pore topologies as zeolites.

ZIF

See Zeolitic Imidazolate Frameworks. *See also* zeolites.

Answers

Chapter 1

1.1. Reading self-test

1. d
2. e
3. d
4. b
5. b
6. b
7. b
8. b
9. c

1.2. CO_2 emissions

1. b
2. a
3. d

1.3. Global energy consumption

1. b
2. A — Oil; B — Coal; C — Natural gas; D — Nuclear energy; E — Hydroelectricity; and F — Renewables

1.4. Capturing CO_2

1. b
2. A — Power from coal; B — Iron and steel; C — Refineries; D — Cement; E — Power from oil; and F — Power from gas

Chapter 2

2.1. Reading self-test

1. a
2. b
3. d
4. b
5. b
6. e
7. a
8. c
9. c
10. b

Chapter 3

3.1. Reading self-test

1. d
2. c
3. c
4. d
5. c
6. d
7. a
8. c
9. b
10. b

Chapter 4

4.1. Reading self-test

1. b
2. a
3. d
4. c
5. d
6. d
7. b
8. b

4.2. Parasitic energy

1. A — Separating CO_2 from the air; B — Burning natural gas;
    C — Burning coal; D — IGCC
2. c

Chapter 5

5.1. Reading self-test

1. b
2. a
3. e
4. c
5. a
6. b
7. d
8. d
9. c
10. f

5.2. Absorption

1. A — Nitrogen; B — Flue gas; C — Absorber;
    D — Heat exchanger; E — Stripper;
    F — Condenser

Chapter 6

6.1. Reading self-test

1. b
2. f
3. d
4. a
5. d
6. b
7. e
8. f
9. b

10. b

Chapter 7

7.1. Reading self-test

1. e
2. c
3. A — Feed; B — Retentate; C — Permeate
4. A — 114.7 MW; B — 46.0 MW; C — 46.6 MW
5. c
6. d
7. a
8. c
9. a
10. a

Chapter 8

8.1. Reading self-test

1. d
2. d
3. c
4. b
5. d
6. c

Chapter 9

9.1. Reading self-test

1. d
2. b
3. c
4. c
5. a

9.2. Mineral dissolution

1. b and d

9.3. Capillary breakthrough pressure

1. c, d and e

9.4. Residual CO_2 saturation

1. a and d

Chapter 10

10.1. Reading self-test

1. c
2. f
3. a
4. e

10.2. Carbon sequestration hazards

1. $\underline{1}$ Remove CO_2
   $\underline{3}$ Lower injection pressure
   $\underline{2}$ Purify ground water
   $\underline{4}$ Re-plug well with cement
   $\underline{5}$ Intercept & reinject CO_2
2. A — Hydrocarbon reservoirs; B — Induced seismicity; C — Surface water; D — Root zone effect; E — Ground plume

Index

supercritical CO_2 4
surface tension 401–403, 406
syngas 145

temperature swing operation 238
thermodynamic coefficient 312–314
thermohaline circulation 115, 116
transport diffusion coefficient 320,
  321, 340
transition state theory 337, 338
tundra 524, 527, 529

upwelling 398, 399

Van't Hoff equation 242

water-energy nexus 166
weathering reactions 119, 121,
  533
wedges 10–13
wetting 401, 422, 423, 473, 481
working capacity 208, 214, 216

Young-Laplace equation 403, 406,
  408, 409

zeolites 303, 311, 320, 344, 346,
  476
Zeolitic Imidazolate Frameworks
  (ZIFs) 255, 257, 258, 264

www.ingramcontent.com/pod-product-compliance
Lightning Source LLC
Chambersburg PA
CBHW070712220326
41598CB00026B/3697